Studies in Algebra and Number Theory

ADVANCES IN MATHEMATICS
SUPPLEMENTARY STUDIES, VOLUME 6

EDITED BY

Gian-Carlo Rota

Department of Mathematics
Massachusetts Institute of Technology
Cambridge, Massachusetts

With the Editorial Board
of *Advances in Mathematics*

ACADEMIC PRESS New York San Francisco London 1979
A Subsidiary of Harcourt Brace Jovanovich, Publishers

ACADEMIC PRESS, INC.
111 Fifth Avenue, New York, New York 10003

United Kingdom Edition published by
ACADEMIC PRESS, INC. (LONDON) LTD.
24/28 Oval Road, London NW1 7DX

Library of Congress Cataloging in Publication Data

Main entry under title:

Studies in algebra and number theory.

(Advances in mathematics : Supplementary studies ; v. 6)
Includes bibliographies.
1. Algebra--Addresses, essays, lectures. 2. Numbers, Theory of --Addresses, essays, lectures. I. Rota, Gian-Carlo, (Date) II. Series.
QA155.2.S78 512 79-4638
ISBN 0-12-599153-3

PRINTED IN THE UNITED STATES OF AMERICA

79 80 81 82 9 8 7 6 5 4 3 2 1

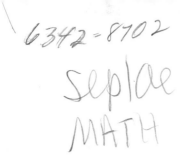

Studies in Algebra and Number Theory

ADVANCES IN MATHEMATICS
SUPPLEMENTARY STUDIES, VOLUME 6

ADVANCES IN
Mathematics
SUPPLEMENTARY STUDIES

EDITED BY Gian-Carlo Rota

Contents

Explicit Class Field Theory in Global Function Fields

David R. Hayes

Some Diophantine Equations Related to the Quadratic Form $ax^2 + by^2$

Edward A. Bender and Norman P. Herzberg

The Left Regular Representation of a p-Adic Algebraic Group Is Type 1

Elliot C. Gootman and Robert R. Kallman

Lattices in Semisimple Groups over Local Fields

Gopal Prasad

Commutative *R*-Subalgebras of *R*-Infinite *R*-Algebras and the Schmidt Problem for *R*-Algebras

Thomas J. Laffey

List of Contributors

Numbers in parentheses indicate the pages on which the authors' contributions begin.

EDWARD A. BENDER (219), University of California at San Diego, La Jolla, California 92093

ELLIOT C. GOOTMAN (273), Department of Mathematics, University of Georgia, Athens, Georgia 30602

DAVID R. HAYES (173), Departments of Mathematics and Statistics, University of Massachusetts, Amherst, Massachusetts 01002

NORMAN P. HERZBERG (219), Institute for Defense Analyses, Princeton, New Jersey 08540

ROBERT R. KALLMAN (273), Department of Mathematics, University of Florida, Gainesville, Florida 32611

THOMAS J. LAFFEY (357), Department of Mathematics, University College, Belfield, Dublin, Ireland

RONALD L. LIPSMAN (143), Department of Mathematics, University of Maryland, College Park, Maryland 20742

R. J. PLYMEN (159), Department of Mathematics, University of Manchester, Manchester, England

GOPAL PRASAD (285), Tata Institute of Fundamental Research, Bombay, India

GARTH WARNER (1), Department of Mathematics, University of Washington, Seattle, Washington 98195

Preface

The supplementary volumes of the journal *Advances in Mathematics* are issued from time to time to facilitate publication of papers already accepted for publication in the journal. The volumes will deal in general—but not always—with papers on related subjects, such as algebra, topology, foundations, etc., and are available individually and independently of the journal.

STUDIES IN ALGEBRA AND NUMBER THEORY
ADVANCES IN MATHEMATICS SUPPLEMENTARY STUDIES, VOL. 6

Selberg's Trace Formula for Nonuniform Lattices: The R-Rank One Case†

GARTH WARNER

Department of Mathematics, University of Washington, Seattle, Washington

Contents

1. INTRODUCTION

Let G be a noncompact connected simple Lie group of split-rank 1; let Γ be a discrete subgroup of G such that the volume of G/Γ is finite but such that G/Γ is not compact. For example, the pair (G, Γ) where $G = \mathbf{SL}(2, \mathbf{R})$, $\Gamma = \mathbf{SL}(2, \mathbf{Z})$ satisfies these hypotheses. Call $L_{G/\Gamma}$ the left regular representation of G on the Hilbert space $L^2(G/\Gamma)$—then a central problem in the theory of automorphic forms relative to the pair (G, Γ) is the decomposition of $L_{G/\Gamma}$ into irreducible unitary representations. As a first step one proves, using the theory of Eisenstein series, that $L^2(G/\Gamma)$ admits an orthogonal decomposition

$$L^2(G/\Gamma) = L_{\mathrm{d}}^2(G/\Gamma) \oplus L_{\mathrm{c}}^2(G/\Gamma),$$

$L_{\mathrm{d}}^2(G/\Gamma)$ (respectively $L_{\mathrm{c}}^2(G/\Gamma)$) being an $L_{G/\Gamma}$-invariant subspace of $L^2(G/\Gamma)$ in which $L_{G/\Gamma}$ decomposes discretely (respectively continuously). Call $L_{G/\Gamma}^{\mathrm{d}}$ (respectively $L_{G/\Gamma}^{\mathrm{c}}$) the restriction of $L_{G/\Gamma}$ to $L_{\mathrm{d}}^2(G/\Gamma)$ (respectively $L_{\mathrm{c}}^2(G/\Gamma)$). It turns out that $L_{G/\Gamma}^{\mathrm{c}}$ can be written as a direct integral over the "principal

† Supported in part by NSF Grant MPS-75-08549.

ISBN 0-12-599153-3

series" representations of G, so one's understanding of $L^{\mathrm{c}}_{G/\Gamma}$ is essentially complete. As for $L^{\mathrm{d}}_{G/\Gamma}$, if $\hat{G}$ is the set of unitary equivalence classes of irreducible unitary representations of G and if m_U is the multiplicity with which a given U in $\hat{G}$ occurs in $L^{\mathrm{d}}_{G/\Gamma}$ (necessarily finite), then

$$L^{\mathrm{d}}_{G/\Gamma} = \sum_{U \in \hat{G}} \oplus\, m_U U.$$

To describe $L^{\mathrm{d}}_{G/\Gamma}$, therefore, one must determine those U for which $m_U > 0$ together with an explicit formula for m_U. Apart from a few numerical examples, nothing is known about this important problem. One method of attack is to develop a "trace formula" of "Selberg type"; it is to this question that the present article is addressed. The basic idea behind what is going on here is not difficult to describe; on the other hand, the actual execution of the method and justification of the details is rather lengthy. Suppose that α is a smooth integrable function on G—then

$$L_{G/\Gamma}(\alpha) = \int_G \alpha(x) L_{G/\Gamma}(x)\, d_G(x)$$

is an integral operator on $L^2(G/\Gamma)$ which, however, need not be of the trace class. Write $L^{\mathrm{d}}_{G/\Gamma}(\alpha)$ (respectively $L^{\mathrm{c}}_{G/\Gamma}(\alpha)$) for the restriction of $L_{G/\Gamma}(\alpha)$ to $L^2_{\mathrm{d}}(G/\Gamma)$ (respectively $L^2_{\mathrm{c}}(G/\Gamma)$)—then

$$L_{G/\Gamma}(\alpha) = L^{\mathrm{d}}_{G/\Gamma}(\alpha) + L^{\mathrm{c}}_{G/\Gamma}(\alpha).$$

If α is sufficiently regular (say in an appropriate Schwartz space in the sense of Harish-Chandra), then both $L^{\mathrm{d}}_{G/\Gamma}(\alpha)$ and $L^{\mathrm{c}}_{G/\Gamma}(\alpha)$ are integral operators; moreover, it can be shown that $L^{\mathrm{d}}_{G/\Gamma}(\alpha)$ is of the trace class, the trace being computable by integrating its kernel over the diagonal. We shall refer to this result as the first stage of the Selberg trace formula. The next step in the analysis is the computation of the integral giving the trace of $L^{\mathrm{d}}_{G/\Gamma}(\alpha)$. A key role here is played by the Poisson summation formula. The net result is that the trace of $L^{\mathrm{d}}_{G/\Gamma}(\alpha)$ can be expressed in terms of certain distributions on G, e.g., orbital integrals (or perturbations thereof). This is the Selberg trace formula in its second stage. The third and final stage of the Selberg trace formula consists in the explicit determination of the Fourier transforms, in the sense of Harish-Chandra, of the aforementioned distributions. This is the most difficult step in the analysis. For general G, we shall only be able to give a complete answer in the special case when α is bi-invariant under a maximal compact subgroup; this will suffice, though, for certain important applications which will be considered elsewhere. It should be stressed that the only obstacle to having a satisfactory theory in general is the computation of the Fourier transform of a single, albeit complicated, noncentral distribution. Once this has been done, a number of important consequences will follow, e.g., explicit formulas for the multiplicities of the integrable discrete series in $L^2_{\mathrm{d}}(G/\Gamma)$.

The thrust of the present paper, then, is to give a complete and detailed proof of the Selberg trace formula for nonuniform lattices Γ in a simple split-rank 1 group G, at least through the second stage. The investigation depends heavily on the Garland–Raghunathan reduction theory and the theory of Eisenstein series, both of which are reviewed in Section 2. In Section 3 we discuss the spectral decomposition of $L^2(G/\Gamma)$, establishing in particular the orthogonal decomposition

$$L^2(G/\Gamma) = L^2_{\mathrm{d}}(G/\Gamma) \oplus L^2_{\mathrm{c}}(G/\Gamma)$$

mentioned above. Sections 4–6 are devoted to the proof of the Selberg trace formula in its second stage (subject to a certain technical restriction on Γ). The analysis in Section 6 is carried out under the assumption that α has compact support. For the applications, it is necessary to know that the Selberg trace formula is valid for suitable classes of noncompactly supported functions, e.g., the K-finite matrix coefficients of the integrable discrete series. Such an extension is made in Section 8. Section 7, which is of a preliminary nature, serves to establish the convergence of certain Epstein-like zeta functions. In Section 9, the Selberg trace formula in its third stage is developed for "class one" functions. We terminate in Section 10 with a list of open problems and indicate a number of avenues for further research.

At this point it is perhaps appropriate to make some remarks of a historical nature. The whole subject originated with Selberg's [27a] famous paper (although Delsarte had apparently anticipated some of the ideas years before this). Selberg only made explicit statements about $\mathbf{SO}(2)\backslash\mathbf{SL}(2,\mathbf{R})/\Gamma$ and did not give any proofs there. Selberg did give, however, complete proofs for this case in an unpublished manuscript which has had a fairly wide circulation. Independently, proofs have been provided in this special situation by a number of people, including (at least) Faddev, Kalinin and Venkov, Kubota, Lax and Phillips, Langlands, and the author. No progress of any real significance was made for some fifteen years until the work of Langlands appeared (cf. Jacquet and Langlands [16]). Langlands deals with $\mathbf{GL}(2)$ in the adele picture and gives a comprehensive outline of how the trace formula should go in that setting; complete details were later supplied by Duflo and Labesse [6]. Langlands' methods differ somewhat in detail from those of Selberg (although not, of course, in spirit) and are more susceptible to generalization. They were in fact developed by Arthur [1a] in the adele picture for semisimple algebraic groups of rank 1 over a number field. Much of our treatment is directly motivated by the work of Arthur and Langlands. Finally, we should mention that Venkov [32] has recently studied the case $\mathbf{SO}(n)\backslash\mathbf{SO}(n,1)/\Gamma$.

2. Eisenstein Series

Let G be a noncompact connected semisimple Lie group with finite center; let K be a maximal compact subgroup of G. We shall assume that $\operatorname{rank}(G/K) = 1$. In addition it will be supposed that G is simple and is embedded in the simply connected complex analytic group corresponding to the complexification of the Lie algebra $\mathfrak{g}$ of G.

Let Γ be a discrete subgroup of G such that the volume of G/Γ is finite but such that G/Γ is not compact. Under these circumstances, the reduction theory of Garland and Raghunathan [11, pp. 304–306] is applicable and may be described as follows. Relative to some Iwasawa decomposition $G = K \cdot A \cdot N$ of G, there is a parabolic subgroup P of G with Langlands decomposition $P = M \cdot A \cdot N$ (M the centralizer of A in K) having certain properties which we shall now enumerate. Let λ be the unique simple root of the pair (G, A) implicit in the choice of N; let $\xi_\lambda : A \to \mathbf{R}^+$ be the associated quasi-character of A. Given $t > 0$, put

$$A[t] = \{a \in A : \xi_\lambda(a) \leqslant t\}, \qquad A(t) = \{a \in A : \xi_\lambda(a) < t\}.$$

For any compact neighborhood ω of 1 in N, the set $\mathfrak{S}_{t,\omega} = K \cdot A[t] \cdot \omega$ is called a *Siegel domain* in G (relative to P) while the set $\mathfrak{C}_t = K \cdot A(t) \cdot N$ is called a *cylindrical domain* in G (relative to P). Let r be the number of Γ-inequivalent cusps—then one can choose elements $k_1 = 1, \ldots, k_r$ in K such that the conjugates $P_i = {}^{k_i}P$ form a complete set of representatives for the Γ-cuspidal parabolic subgroups of $G \bmod \Gamma$. Each P_i admits a Langlands decomposition $P_i = M_i \cdot A_i \cdot N_i$ where $M_i = {}^{k_i}M$, $A_i = {}^{k_i}A$, $N_i = {}^{k_i}N$ ($i = 1, \ldots, r$). Let $\kappa_i = k_i^{-1}$, $\mathfrak{s} = \{\kappa_i\}$—then one can find a Siegel domain $\mathfrak{S}_{t_0,\omega_0}$ such that the set $\mathfrak{S} = \mathfrak{S}_{t_0,\omega_0} \cdot \mathfrak{s}$ has the following properties:

(i) $\mathfrak{S} \cdot \Gamma = G$;
(ii) $\{\gamma \in \Gamma : \mathfrak{S}\gamma \cap \mathfrak{S} \neq \varnothing\}$ is finite.

We remark that once t_0 and ω_0 have been shown to exist, they can then be replaced by any $t > t_0$, $\omega \supset \omega_0$ without affecting either statement (i) or statement (ii). For $1 \leqslant i \leqslant r$,

$$N_i/N_i \cap \Gamma$$

is compact. Because $\mathfrak{s}$ is finite, it can be assumed that ω_0 is chosen in such a way that $k_i\omega_0 k_i^{-1} = \kappa_i^{-1}\omega_0\kappa_i$ contains a fundamental domain for the group $N_i \cap \Gamma$ acting to the right on N_i. Using this hypothesis on ω_0, one can then produce a ${}_0t < t_0$ such that:

(iii) $K \cdot A[{}_0t] \cdot \omega_0 \cdot \kappa_i \cap K \cdot A[{}_0t] \cdot \omega_0 \cdot \kappa_j\gamma = \varnothing$ for $\kappa_i, \kappa_j \in \mathfrak{s}$ $(i \neq j)$ and for $\gamma \in \Gamma$;

(iv) $K \cdot A[{}_0t] \cdot \omega_0 \cdot \kappa_i \cap K \cdot A[{}_0t] \cdot \omega_0 \cdot \kappa_i\gamma \neq \varnothing$ for $\kappa_i \in \mathfrak{s}$ and for $\gamma \in \Gamma$, only if $\gamma \in M_i \cdot N_i$.

Speaking roughly, these two properties say that it is possible to separate the cusps of Γ. A simple argument (cf. Raghunathan [23a, p. 289]) then gives that the sets

$$K \cdot A[{}_0t] \cdot N\kappa_i \qquad (1 \leqslant i \leqslant r)$$

are mutually disjoint and that, moreover, for any i

$$K \cdot A[{}_0t] \cdot N\kappa_i \cap K \cdot A[{}_0t] \cdot N\kappa_i\gamma \neq \varnothing$$

if and only if $\gamma \in M_i \cdot N_i$. Consequently (cf. Raghunathan [23a, p. 290]) there exists a compact subset Ω_{0t} of G such that the complement of the Γ-saturation $\Omega_{0t} \cdot \Gamma$ of Ω_{0t} in G decomposes into a finite number of mutually disjoint Γ-saturated open sets $\mathfrak{O}_1, \ldots, \mathfrak{O}_r$ where each $\mathfrak{O}_i$ contains $\mathfrak{C}_{0t}\kappa_i$ as an open and closed subset and is in fact the Γ-saturation of $\mathfrak{C}_{0t}\kappa_i$. In other words

$$G = \Omega_{0t} \cdot \Gamma \cup \bigcup_{i=1}^{r} \mathfrak{C}_{0t}\kappa_i \cdot \Gamma \qquad \text{(disjoint union)},$$

so

$$G/\Gamma = \pi(\Omega_{0t}) \cup \bigcup_{i=1}^{r} \pi(\mathfrak{C}_{0t}\kappa_i) \qquad \text{(disjoint union)},$$

where $\pi: G \to G/\Gamma$ is the natural projection. As a final point in this circle of ideas, we mention that the set of conjugacy classes of maximal unipotent subgroups of Γ is finite and, in fact, that every maximal unipotent subgroup of Γ is conjugate to a unique $N_i \cap \Gamma$; for details, see Raghunathan [23b, p. 202].

We turn now to the theory of Eisenstein series on G/Γ. Complete proofs can be found in Harish-Chandra [13a] or Langlands [19a]. Harish-Chandra [13a] explicitly treats only the case when Γ is arithmetic; this is done because Borel's reduction theory is then applicable. But, using the Garland–Raghunathan reduction theory, one can carry over the theory virtually word for word to the general case. Alternatively, it is easy to verify that the Garland–Raghunathan reduction theory implies that the axioms assumed by Langlands [19a] are in force in the present case so that one can quote Langlands [19a] directly.

Keeping to the above notations, identify M with $M \cdot N/N$. It is known that $\Gamma \cap P \subset M \cdot N$ (cf. Garland and Raghunathan [11, p. 295]). Put $\Gamma_M = \Gamma \cap M \cdot N/\Gamma \cap N$—then Γ_M is a discrete subgroup of M, hence is finite, M being compact. Let M^* be the normalizer of A in K—then $W(A) = M^*/M$ is the Weyl group of the pair (G, A). Since $\operatorname{rank}(G/K) = 1$, $W(A)$ is of order 2. Let $\hat{M}$ be the set of unitary equivalence classes of irreducible unitary representations of M. For each $\sigma \in \hat{M}$, let ξ_σ denote the character of σ, $d(\sigma)$ the degree of σ, and $\chi_\sigma = d(\sigma)\xi_\sigma$. The group $W(A)$ operates to the left on $\hat{M}$ in the obvious way. Write

$$L^2(M/\Gamma_M) = \sum_{\sigma \in \hat{M}} \oplus\, n(\sigma, \Gamma_M)\sigma,$$

the $n(\sigma, \Gamma_M)$ being certain nonnegative integers. Given an orbit ϑ in $W(A)\backslash\hat{M}$, pick $\sigma \in \vartheta$ and set

$$S_\vartheta = n(\sigma, \Gamma_M)\sigma + n(w\sigma, \Gamma_M)w\sigma,$$

w the nontrivial element in $W(A)$. Then

$$L^2(M/\Gamma_M) = \sum_{\vartheta \in W(A)\backslash\hat{M}} \oplus\, S_\vartheta.$$

Let $\hat{K}$ be the set of unitary equivalence classes of irreducible unitary representations of K. For each $\delta \in \hat{K}$, let ξ_δ denote the character of δ, $d(\delta)$ the degree of δ, and $\chi_\delta = d(\delta)\xi_\delta$. Fix a class $\delta \in \hat{K}$. Denote by $L^2(K; \delta)$ the subspace of $L^2(K)$ consisting of those functions which transform under the left regular representation according to δ. Given $\vartheta \in W(A)\backslash\hat{M}$ such that S_ϑ is nontrivial, let $\mathscr{E}(\vartheta, \delta)$ be the set of all continuous functions $\Phi: G \to \mathbf{C}$ such that:

(i) Φ is right invariant under $(\Gamma \cap P) \cdot A \cdot N$;
(ii) for every $x \in G$, the function

$$m \mapsto \Phi(xm)$$

belongs to S_ϑ;
(iii) for every $x \in G$, the function

$$k \mapsto \Phi(kx)$$

belongs to $L^2(K; \delta)$.

It is known that $\mathscr{E}(\vartheta, \delta)$ is a finite-dimensional Hilbert space of analytic functions with inner product

$$(\Phi, \Psi) = \int_K \int_{M/\Gamma_M} \Phi(km)\overline{\Psi(km)}\, d_K(k)\, d_M(m).$$

This is proved formally in Langlands [19a, p. 50] and is actually a consequence of some observations made in the next section.

Let $\mathfrak{a}$ be the Lie algebra of A—then we shall agree to equip the dual of $\mathfrak{a}$ with the usual Euclidean structure derived from the Killing form. Since $\lambda/|\lambda|$ is a unit vector for this structure, a complex number s becomes a linear function on $\mathfrak{a}$ through the identification $s \leftrightarrow s(\lambda/|\lambda|)$. In particular ρ, the sum of the positive roots of the pair (G, A) (counted with multiplicity) divided by 2, is identified with its length $|\rho|$. For any $x \in G$, we denote by $H(x)$ that element of $\mathfrak{a}$ such that $x \in K\exp(H(x)) \cdot N$. This said, let $\Phi \in \mathscr{E}(\vartheta, \delta)$—then, attached to Φ is the *Eisenstein series*

$$E(P:\Phi:s:x) = \sum_{\gamma \in \Gamma/\Gamma \cap P} e^{(s-|\rho|)(H(x\gamma))}\Phi(x\gamma).$$

$E(P:\Phi:s:x)$ is a C^∞ function on $\{s: \mathrm{Re}(s) < -|\rho|\} \times G$ which is holomorphic in s and right invariant under Γ. For every $x \in G$, the function

$$k \mapsto E(P:\Phi:s:kx) \qquad (\mathrm{Re}(s) < -|\rho|)$$

belongs to $L^2(K;\delta)$. If $\mathfrak{Z}$ is the center of the universal enveloping algebra $\mathfrak{G}$ of $\mathfrak{g}_c$ ($\mathfrak{g}_c$ the complexification of the Lie algebra $\mathfrak{g}$ of G), then $E(P:\Phi:s:x)$ is $\mathfrak{Z}$-finite. All these assertions are detailed in Harish-Chandra [13a, pp. 26–31].

Modulo obvious notational changes, the definitions and results indicated above carry over to each of the P_i ($i = 1, \ldots, r$). Fix i and j—then $P_i = M_i \cdot A_i \cdot N_i$, $P_j = M_j \cdot A_j \cdot N_j$ and the orbit spaces

$$W(A_i)\backslash\hat{M}_i, \qquad W(A_j)\backslash\hat{M}_j$$

are in a canonical one-to-one correspondence. Corresponding orbits are said to be *associate*. Introduce the set $W(A_i, A_j)$ of all bijections $w: A_i \to A_j$ such that $wa_i = xa_ix^{-1}$ ($a_i \in A_i$) for some $x \in G$. Now fix $\delta \in \hat{K}$ and associate orbits $\vartheta_i \in W(A_i)\backslash\hat{M}_i$, $\vartheta_j \in W(A_j)\backslash\hat{M}_j$. Let $\Phi_i \in \mathscr{E}(\vartheta_i, \delta)$—then the integral

$$\int_{N_j/N_j \cap \Gamma} E(P_i:\Phi_i:s:xn_j)\, d_{N_j}(n_j),$$

known as the *constant term* of the Eisenstein series $E(P_i:\Phi_i:s:x)$ along P_j and denoted by $E_{P_j}(P_i:\Phi_i:s:x)$, is computable and in fact

$$E_{P_j}(P_i:\Phi_i:s:x) = \sum_{w \in W(A_i, A_j)} e^{(ws-|\rho|)(H_j(x))} \cdot (c_{P_j|P_i}(w:s)\Phi_i)(x).$$

Here

$$c_{P_j|P_i}(w:s): \mathscr{E}(\vartheta_i, \delta) \to \mathscr{E}(\vartheta_j, \delta)$$

is a certain linear transformation which, as a function of s, is defined and holomorphic in the region $\mathrm{Re}(s) < -|\rho|$ (cf. Harish-Chandra [13a, p. 44]). Let $\vartheta = \vartheta_1, \vartheta_2, \ldots, \vartheta_r$ be a complete collection of associate orbits. Put

$$\mathscr{E}(\vartheta, \delta) = \sum_{i=1}^{r} \oplus\, \mathscr{E}(\vartheta_i, \delta),$$

where the column vector

$$\mathbf{\Phi} = \begin{pmatrix} \Phi_1 \\ \vdots \\ \Phi_r \end{pmatrix} \in \mathscr{E}(\vartheta, \delta)$$

has norm $\|\mathbf{\Phi}\|^2 = \sum_{i=1}^{r} \|\Phi_i\|^2$. For any complex number s such that $\mathrm{Re}(s) < -|\rho|$ we want to define a linear transformation

$$\mathbf{c}_{\vartheta,\delta}(s): \mathscr{E}(\vartheta, \delta) \to \mathscr{E}(\vartheta, \delta).$$

This is done as follows. Let

$$\mathbf{\Phi} = \begin{pmatrix} \Phi_1 \\ \vdots \\ \Phi_r \end{pmatrix} \in \mathscr{E}(\vartheta, \delta).$$

Then it is enough to define $\mathbf{c}_{\vartheta,\delta}(s)\Phi_i$ ($i = 1, \ldots, r$), which in turn is completely prescribed when $(\mathbf{c}_{\vartheta,\delta}(s)\Phi_i)_j$ is defined ($j = 1, \ldots, r$). Put

$$(\mathbf{c}_{\vartheta,\delta}(s)\Phi_i)_j = c_{P_j|P_i}(k_j w k_i^{-1} : k_i s)\Phi_i \in \mathscr{E}(\vartheta_j, \delta),$$

w the nontrivial element in $W(A)$. Interchanging i and j, we thus have

$$\begin{aligned}(\mathbf{c}_{\vartheta,\delta}(s)\mathbf{\Phi})_i &= \left(\sum_j \mathbf{c}_{\vartheta,\delta}(s)\Phi_j\right)_i \\ &= \sum_j c_{P_i|P_j}(k_i w k_j^{-1} : k_j s)\Phi_j,\end{aligned}$$

which can be interpreted as saying that the image of $\mathbf{\Phi}$ under $\mathbf{c}_{\vartheta,\delta}(s)$ is simply obtained by formal matrix multiplication. Fundamental to the theory is the fact that $\mathbf{c}_{\vartheta,\delta}$ can be meromorphically continued to the whole s-plane. Its poles in the half-plane $\mathrm{Re}(s) < 0$ lie in the set $\{s \in \mathbf{R} : -|\rho| \leqslant s < 0\}$; there are but finitely many of them there and they are all simple. Along the imaginary axis, $\mathbf{c}_{\vartheta,\delta}$ is holomorphic. Another basic point is that $\mathbf{c}_{\vartheta,\delta}$ satisfies a functional equation, viz.

$$\mathbf{c}_{\vartheta,\delta}(s)\mathbf{c}_{\vartheta,\delta}(-s) = I,$$

I the identity operator. Because the adjoint $\mathbf{c}_{\vartheta,\delta}(s)^*$ of $\mathbf{c}_{\vartheta,\delta}(s)$ is $\mathbf{c}_{\vartheta,\delta}(\bar{s})$, it follows that if s is pure imaginary, then

$$\mathbf{c}_{\vartheta,\delta}(s)\mathbf{c}_{\vartheta,\delta}(-s) = \mathbf{c}_{\vartheta,\delta}(s)\mathbf{c}_{\vartheta,\delta}(\bar{s}) = \mathbf{c}_{\vartheta,\delta}(s)\mathbf{c}_{\vartheta,\delta}(s)^* = I,$$

so $\mathbf{c}_{\vartheta,\delta}$ is unitary along the imaginary axis. Given

$$\mathbf{\Phi} = \begin{pmatrix} \Phi_1 \\ \vdots \\ \Phi_r \end{pmatrix} \in \mathscr{E}(\vartheta, \delta),$$

put

$$\mathbf{E}(\mathbf{\Phi} : s : x) = \sum_{i=1}^{r} E(P_i : \Phi_i : s : x).$$

One can prove that $\mathbf{E}(\boldsymbol{\Phi}:s:x)$, as a function of s, can be meromorphically continued to the whole s-plane; moreover, its poles are poles of $\mathbf{c}_{\vartheta,\delta}$. Finally,

$$\mathbf{E}(\boldsymbol{\Phi}:s:x) = \mathbf{E}(\mathbf{c}_{\vartheta,\delta}(s)\boldsymbol{\Phi}:-s:x).$$

The proofs of the above results are set down in Harish-Chandra [13a, pp. 89–105].

Eisenstein series, while right invariant under Γ, do not lie in $L^2(G/\Gamma)$. Their significance in the spectral decomposition of $L^2(G/\Gamma)$ will be explained in the next section. To produce functions in $L^2(G/\Gamma)$, it is convenient to introduce the notion of theta series. This will now be done.

Fix an orbit $\vartheta \in W(A)\backslash\hat{M}$ such that S_ϑ is nontrivial. Fix a class $\delta \in \hat{K}$. Let $\phi: G \to \mathbf{C}$ be a differentiable function such that:

(i) ϕ is right invariant under $(\Gamma \cap P) \cdot N$;
(ii) for every $x \in G$, the function

$$m \mapsto \phi(xm)$$

belongs to S_ϑ;
(iii) for every $x \in G$, the function

$$k \mapsto \phi(kx)$$

belongs to $L^2(K;\delta)$.

Then one can associate with ϕ a differentiable function

$$\check{\phi}: A \to \mathscr{E}(\vartheta,\delta),$$

that is, a differentiable function

$$\check{\phi}: A \times G \to \mathbf{C}$$

with the property that for each $a \in A$, the function $x \mapsto \check{\phi}(a:x)$ belongs to $\mathscr{E}(\vartheta,\delta)$. Explicitly, if k_x is the K-component of x in the Iwasawa decomposition $G = K \cdot A \cdot N$, then $\check{\phi}(a:x) = \phi(k_x a)$. The set of all ϕ for which $\check{\phi}$, as a function from A to $\mathscr{E}(\vartheta,\delta)$, is of compact support is denoted by $\check{\mathscr{V}}(\vartheta,\delta)$. It can be shown without difficulty that the correspondence $\phi \leftrightarrow \check{\phi}$ serves to identify $\mathscr{V}(\vartheta,\delta)$ with $C_c^\infty(A) \otimes \mathscr{E}(\vartheta,\delta)$. Let $\widehat{C_c^\infty(A)}$ be the set of Fourier–Laplace transforms of $C_c^\infty(A)$. Suppose that $\phi \in \mathscr{V}(\vartheta,\delta)$—then there is associated with ϕ in a canonical way an element $\hat{\phi} \in \widehat{C_c^\infty(A)} \otimes \mathscr{E}(\vartheta,\delta)$, called the *Fourier transform* of ϕ, such that

$$\phi(x) = \frac{1}{2\pi} \int_{\mathrm{Re}(s)=s_0} \hat{\phi}(s:x) e^{(s-|\rho|)(H(x))} \, |ds|,$$

s_0 any complex number. Explicitly:

$$\hat{\phi}(s:x) = \int_A \check{\phi}(a:x)e^{-(s-|\rho|)(H(a))}\, d_A(a),$$

d_A being appropriate Haar measure on A. In passing, note that $\hat{\phi}(s:x)$, as a function of s, is rapidly decreasing along vertical lines. Consider now the series

$$\Theta_\phi(x) = \sum_{\gamma \in \Gamma/\Gamma \cap P} \phi(x\gamma) \qquad (x \in G).$$

One can prove that it is absolutely convergent and that its sum lies in $L^2(G/\Gamma)$ (cf. Harish-Chandra [13a, p. 32]). Θ_ϕ is known as the *theta series* attached to ϕ. If $\operatorname{Re}(s) < -|\rho|$, then

$$\Theta_\phi(x) = \frac{1}{2\pi}\int_{\operatorname{Re}(s)=s_0} E(P:\hat{\phi}(s:?):s:x)\,|ds|.$$

In addition, for arbitrary $\phi_1, \phi_2 \in \mathscr{V}(\vartheta, \delta)$, the *incomplete scalar product formula* obtains (cf. Harish-Chandra [13a, p. 46]):

$$(\Theta_{\phi_1}, \Theta_{\phi_2}) = \frac{1}{2\pi}\int_{\operatorname{Re}(s)=s_0} \{(\hat{\phi}_1(s:?), \hat{\phi}_2(-\bar{s}:?))$$
$$+ (c_{P|P}(w:s)\hat{\phi}_1(s:?), \hat{\phi}_2(\bar{s}:?))\}\,|ds| \qquad (\operatorname{Re}(s) < -|\rho|).$$

The inner product on the left of the equal sign is calculated in $L^2(G/\Gamma)$, whereas the inner products on the right under the integral sign are taken in $\mathscr{E}(\vartheta, \delta)$. It should also be pointed out that the formula presupposes certain normalizations of the relevant invariant measures which will be made explicit later. Now fix a real number $R > |\rho|$. Let T_R be the set of all complex numbers s such that $|\operatorname{Re}(s)| < R$. Call $\mathscr{H}_R(\vartheta, \delta)$ the set of all bounded holomorphic functions Φ on T_R with values in $\mathscr{E}(\vartheta, \delta)$ such that $p\Phi$ is bounded for every polynomial function p on $\mathbf{C}$—then it is clear that

$$\mathscr{H}_R(\vartheta, \delta) \supset \widehat{C_c^\infty(A)} \otimes \mathscr{E}(\vartheta, \delta).$$

Given $\Phi \in \mathscr{H}_R(\vartheta, \delta)$, put

$$\phi(x) = \frac{1}{2\pi}\int_{\operatorname{Re}(s)=s_0} \Phi(s:x)e^{(s-|\rho|)(H(x))}\,|ds| \qquad (|\operatorname{Re}(s)| < R).$$

Then, utilizing a limit process, it can be shown that it is possible to associate with ϕ an element $\Theta_\phi \in L^2(G/\Gamma)$. Moreover, for arbitrary Φ_1, Φ_2 in $\mathscr{H}_R(\vartheta, \delta)$ the incomplete scalar product formula for $(\Theta_{\phi_1}, \Theta_{\phi_2})$ is still valid (the $\hat{\phi}_i$ being replaced by Φ_i $(i = 1, 2)$). The approximation arguments leading to these conclusions are spelled out in Harish-Chandra [13a, pp. 76–79].

Just as for Eisenstein series, the preceding definitions and results admit obvious formulations in terms of each of the P_i $(i = 1, \ldots, r)$. Fix $\delta \in \hat{K}$ and associate orbits $\vartheta = \vartheta_1, \vartheta_2, \ldots, \vartheta_r$. Employing evident notations, let $\mathscr{V}(\vartheta, \delta)$ stand for the direct sum of the $\mathscr{V}(\vartheta_i, \delta)$. An element $\boldsymbol{\phi} \in \mathscr{V}(\vartheta, \delta)$ has components $\phi_1, \ldots, \phi_r$ and attached to each ϕ_i is its Fourier transform $\hat{\phi}_i$ where

$$\hat{\phi}_i \in \widehat{C_c^\infty(A_i)} \otimes \mathscr{E}(\vartheta_i, \delta)$$

and is given by

$$\hat{\phi}_i(s:x) = \int_{A_i} \check{\phi}_i(a_i:x) e^{-(s-|\rho|)(H_i(a_i))} \, d_{A_i}(a_i).$$

It will be convenient to refer to the column vector

$$\hat{\boldsymbol{\phi}} = \begin{pmatrix} \hat{\phi}_1 \\ \vdots \\ \hat{\phi}_r \end{pmatrix}$$

as the *Fourier transform* of $\boldsymbol{\phi}$. Put

$$\boldsymbol{\Theta}_{\boldsymbol{\phi}} = \sum_{i=1}^{r} \Theta_{\phi_i}.$$

For arbitrary $\boldsymbol{\phi}_1, \boldsymbol{\phi}_2 \in \mathscr{V}(\vartheta, \delta)$, one then has the *complete scalar product formula* (cf. Harish-Chandra [13a, p. 86]):

$$(\boldsymbol{\Theta}_{\boldsymbol{\phi}_1}, \boldsymbol{\Theta}_{\boldsymbol{\phi}_2}) = \frac{1}{2\pi} \int_{\mathrm{Re}(s)=s_0} \{(\hat{\boldsymbol{\phi}}_1(s:?), \hat{\boldsymbol{\phi}}_2(-\bar{s}:?)) \\ + (\mathbf{c}_{\vartheta,\delta}(s)\hat{\boldsymbol{\phi}}_1(s:?), \hat{\boldsymbol{\phi}}_2(\bar{s}:?))\} \, |ds| \qquad (\mathrm{Re}(s) < -|\rho|).$$

The inner product on the left of the equal sign is calculated in $L^2(G/\Gamma)$, whereas the inner products on the right under the integral sign are taken in $\mathscr{E}(\vartheta, \delta)$. When written out in full, the complete scalar product formula thus asserts that

$$\int_{G/\Gamma} \boldsymbol{\Theta}_{\boldsymbol{\phi}_1}(x) \overline{\boldsymbol{\Theta}_{\boldsymbol{\phi}_2}(x)} \, d_G(x)$$

$$= \frac{1}{2\pi} \int_{\mathrm{Re}(s)=s_0} \left\{ \sum_{i=1}^{r} \int_K \int_{M_i/\Gamma_{M_i}} \hat{\phi}_{1i}(s:km_i) \overline{\hat{\phi}_{2i}(-\bar{s}:km_i)} \, d_K(k) \, d_{M_i}(m_i) \right.$$

$$+ \sum_{i=1}^{r} \int_K \int_{M_i/\Gamma_{M_i}} \left[\sum_{j=1}^{r} (c_{P_i|P_j}(k_i w k_j^{-1}:k_j s)\hat{\phi}_{1j})(s:km_i) \right]$$

$$\left. \times \overline{\hat{\phi}_{2i}(\bar{s}:km_i)} \, d_K(k) \, d_{M_i}(m_i) \right\} |ds| \qquad (\mathrm{Re}(s) < -|\rho|).$$

The normalizations of the various invariant measures which serve to ensure equality will be given later. For $R > |\rho|$, denote by $\mathscr{H}_R(\vartheta, \delta)$ the direct sum of the $\mathscr{H}_R(\vartheta_i, \delta)$. We shall view the elements $\mathbf{\Phi}$ of $\mathscr{H}_R(\vartheta, \delta)$ as bounded holomorphic functions on T_R with values in $\mathscr{E}(\vartheta, \delta)$ such that $p\mathbf{\Phi}$ is bounded for every polynomial function p on $\mathbf{C}$. A function $\mathbf{\Phi} \in \mathscr{H}_R(\vartheta, \delta)$ has components $\Phi_1, \ldots, \Phi_r$. To Φ_i there is attached a function ϕ_i and thence an element Θ_{ϕ_i} in $L^2(G/\Gamma)$. Let $\boldsymbol{\phi}$ denote the vector whose ith component is ϕ_i. Put

$$\mathbf{\Theta}_{\boldsymbol{\phi}} = \sum_{i=1}^{r} \Theta_{\phi_i}.$$

Then: For arbitrary $\mathbf{\Phi}_1, \mathbf{\Phi}_2$ in $\mathscr{H}_R(\vartheta, \delta)$, the complete scalar product formula for $(\mathbf{\Theta}_{\boldsymbol{\phi}_1}, \mathbf{\Theta}_{\boldsymbol{\phi}_2})$ is still valid (the $\hat{\boldsymbol{\phi}}_i$ being replaced by $\mathbf{\Phi}_i$ $(i = 1, 2)$).

We recall that $\mathbf{c}_{\vartheta,\delta}$ has but finitely many poles in the half-plane $\operatorname{Re}(s) < 0$; they are all simple and lie in the set $\{s \in \mathbf{R} : -|\rho| \leqslant s < 0\}$, say at $s_1, \ldots, s_{n(\vartheta,\delta)}$. Let $\gamma_{\vartheta,\delta}(s_i)$ be the residue of $-2\pi \cdot \mathbf{c}_{\vartheta,\delta}$ at $s = s_i$ $(1 \leqslant i \leqslant n(\vartheta, \delta))$. Returning to the complete scalar product formula, let us shift the line of integration from $\operatorname{Re}(s) = s_0 < -|\rho|$ to the imaginary axis. To justify this, it is necessary to keep in mind that the norm of $\mathbf{c}_{\vartheta,\delta}$ is bounded at infinity in the strip $\{s : s_0 \leqslant \operatorname{Re}(s) \leqslant 0\}$ (cf. Harish-Chandra [13a, p. 107]). This being the case, Cauchy's theorem then gives

$$(\mathbf{\Theta}_{\boldsymbol{\phi}_1}, \mathbf{\Theta}_{\boldsymbol{\phi}_2}) = \frac{1}{2\pi} \sum_{i=1}^{n(\vartheta,\delta)} (\gamma_{\vartheta,\delta}(s_i)\mathbf{\Phi}_1(s_i : ?), \mathbf{\Phi}_2(s_i : ?))$$
$$+ \frac{1}{2\pi} \int_{\operatorname{Re}(s)=0} \{(\mathbf{\Phi}_1(s : ?), \mathbf{\Phi}_2(s : ?)) + (\mathbf{c}_{\vartheta,\delta}(s)\mathbf{\Phi}_1(s : ?), \mathbf{\Phi}_2(\bar{s} : ?))\} \, |ds|.$$

Here, for definiteness, $\mathbf{\Phi}_1$ and $\mathbf{\Phi}_2$ are holomorphic functions on the complex plane with values in $\mathscr{E}(\vartheta, \delta)$ such that $p\mathbf{\Phi}_i$ $(i = 1, 2)$ is bounded on any T_R for every polynomial p on $\mathbf{C}$. Because $\mathbf{c}_{\vartheta,\delta}$ is a unitary operator along the imaginary axis, the last integral can be written

$$\frac{1}{\pi} \int_{-\infty}^{\infty} (\tfrac{1}{2}[\mathbf{\Phi}_1(\sqrt{-1}\eta : ?) + \mathbf{c}_{\vartheta,\delta}(-\sqrt{-1}\eta)\mathbf{\Phi}_1(-\sqrt{-1}\eta : ?)],$$
$$\tfrac{1}{2}[\mathbf{\Phi}_2(\sqrt{-1}\eta : ?) + \mathbf{c}_{\vartheta,\delta}(-\sqrt{-1}\eta)\mathbf{\Phi}_2(-\sqrt{-1}\eta : ?)]) \, d\eta \qquad (s = \xi + \sqrt{-1}\eta).$$

LEMMA 2.1. *The operator $\gamma_{\vartheta,\delta}(s_i)$ $(1 \leqslant i \leqslant n(\vartheta, \delta))$ is positive semidefinite.*

Proof. Since the adjoint $\mathbf{c}_{\vartheta,\delta}(s)^*$ of $\mathbf{c}_{\vartheta,\delta}(s)$ is $\mathbf{c}_{\vartheta,\delta}(\bar{s})$, $\mathbf{c}_{\vartheta,\delta}$ is self-adjoint along the real axis. Therefore $\gamma_{\vartheta,\delta}(s_i)$ $(1 \leqslant i \leqslant n(\vartheta, \delta))$ is a self-adjoint operator.

Now fix i—then, to finish the proof, we must show that $(\gamma_{\vartheta,\delta}(s_i)\mathbf{\Phi}, \mathbf{\Phi}) \geqslant 0$ for all $\mathbf{\Phi} \in \mathscr{E}(\vartheta, \delta)$. Given $\mathbf{\Phi} \in \mathscr{E}(\vartheta, \delta)$, define holomorphic functions $\mathbf{\Phi}_n : \mathbf{C} \to \mathscr{E}(\vartheta, \delta)$ $(n = 1, 2, \ldots)$ by the rule

$$\mathbf{\Phi}_n(s : ?) = \frac{\prod_{j \neq i} (s - s_j)}{\prod_{j \neq i} (s_i - s_j)} \cdot e^{n(s^2 - s_i^2)} \cdot \mathbf{\Phi}(?).$$

Since $p\mathbf{\Phi}_n$ is bounded on any T_R for every polynomial p on $\mathbf{C}$, $\mathbf{\Phi}_n$ may be inserted into the formula above. The left-hand side is, of course, a nonnegative quantity depending on n. On the right-hand side, we get $(2\pi)^{-1}(\gamma_{\vartheta,\delta}(s_i)\mathbf{\Phi}, \mathbf{\Phi})$ plus the L^2-norm squared of

$$\tfrac{1}{2}[\mathbf{\Phi}_n(\sqrt{-1}\eta : ?) + \mathbf{c}_{\vartheta,\delta}(-\sqrt{-1}\eta)\mathbf{\Phi}_n(-\sqrt{-1}\eta : ?)]$$

taken with respect to the measure $d\eta/\pi$. Thanks to the unitarity of $\mathbf{c}_{\vartheta,\delta}$ along the imaginary axis, the latter term is dominated by the L^2-norm squared of $\mathbf{\Phi}_n(\sqrt{-1}\eta : ?)$, a quantity which goes to zero as $n \to \infty$. It therefore follows that

$$(\gamma_{\vartheta,\delta}(s_i)\mathbf{\Phi}, \mathbf{\Phi}) \geqslant 0,$$

as we wished to prove. ■

The fact that the operators $\gamma_{\vartheta,\delta}(s_i)$ $(1 \leqslant i \leqslant n(\vartheta, \delta))$ are positive semidefinite has an interesting consequence which was first noticed by Arthur [1a, p. 351] in the adelic setting. Call an element $\sigma \in \hat{M}$ *unramified* if $w\sigma \neq \sigma$ and *ramified* if $w\sigma = \sigma$. Here, as always, w is the nontrivial element of $W(A)$. Let σ be an unramified class in $\hat{M}$—then the Hilbert space $\mathscr{E}(\vartheta, \delta)$ admits an evident orthogonal direct sum decomposition

$$\mathscr{E}(\vartheta, \delta) = \mathscr{E}_\sigma(\vartheta, \delta) \oplus \mathscr{E}_{w\sigma}(\vartheta, \delta),$$

and it is known from the general theory that $\mathbf{c}_{\vartheta,\delta}(s)$ maps $\mathscr{E}_\sigma(\vartheta, \delta)$ to $\mathscr{E}_{w\sigma}(\vartheta, \delta)$ and vice versa (cf. Harish-Chandra [13a, p. 44]). Choose an orthonormal basis $\{e_\alpha\}$ for $\mathscr{E}(\vartheta, \delta)$. It will be assumed that a given e_α lies in either $\mathscr{E}_\sigma(\vartheta, \delta)$ or $\mathscr{E}_{w\sigma}(\vartheta, \delta)$. The trace of $\gamma_{\vartheta,\delta}(s_i)$ is then

$$\sum_\alpha (\gamma_{\vartheta,\delta}(s_i)e_\alpha, e_\alpha).$$

Because of the choice of the e_α and the orthogonality of $\mathscr{E}_\sigma(\vartheta, \delta)$ and $\mathscr{E}_{w\sigma}(\vartheta, \delta)$, this expression must be zero. On the other hand, the preceding lemma says that $\gamma_{\vartheta,\delta}(s_i)$ is positive semidefinite. It therefore follows that $\gamma_{\vartheta,\delta}(s_i) = 0$. The upshot thus is that in the unramified case, $\mathbf{c}_{\vartheta,\delta}$ is actually holomorphic throughout the region to the left of the imaginary axis. One should observe, however, that this remark is not as attractive as first appears since it can very well happen that every element in $\hat{M}$ is ramified. Indeed, under the present hypotheses, there are two possibilities:

(i) G has a compact Cartan subgroup;
(ii) G does not have a compact Cartan subgroup.

Thanks to the classification, in the first case it is always true that every element of $\hat{M}$ is ramified. On the other hand, in the second case, there are always elements of $\hat{M}$ which are unramified.

We shall conclude this section with a review of some basic definitions. Given $x \in G$, let a_x be the A-component of x in the Iwasawa decomposition $G = K \cdot A \cdot N$. A continuous function $f: G/\Gamma \to \mathbf{C}$ is said to be *slowly increasing* if there is a constant $C > 0$ and an integer $n \geqslant 0$ such that

$$|f(x\kappa_i)| \leqslant C(\xi_\lambda(a_x))^{-n}$$

for all $x \in \mathfrak{S}_{t_0,\omega_0}$, $\kappa_i \in \mathfrak{s}$. On the other hand, f is said to be *rapidly decreasing* if for every integer n there is a constant $C_n > 0$ such that

$$|f(x\kappa_i)| \leqslant C_n(\xi_\lambda(a_x))^n$$

for all $x \in \mathfrak{S}_{t_0,\omega_0}$, $\kappa_i \in \mathfrak{s}$. A differentiable function $f: G/\Gamma \to \mathbf{C}$ is said to be an *automorphic form* if:

(i) the functions $x \mapsto f(kx)$ $(k \in K)$ lie in a finite-dimensional vector space, that is, f is left K-finite;
(ii) the functions $x \mapsto Zf(x)$ $(Z \in \mathfrak{Z})$ form a finite-dimensional vector space, that is, f is $\mathfrak{Z}$-finite;
(iii) f is slowly increasing.

An automorphic form is necessarily an analytic function. Examples of automorphic forms include the Eisenstein series. The *constant term* of a continuous function $f: G/\Gamma \to \mathbf{C}$ along the Γ-cuspidal parabolic subgroup P_i is the continuous function $f_{P_i}: G \to \mathbf{C}$ defined by the rule

$$f_{P_i}(x) = \int_{N_i/N_i \cap \Gamma} f(xn_i)\, d_{N_i}(n_i) \qquad (i = 1, \ldots, r).$$

For the proof of the following important fact, the reader is referred to Harish-Chandra [13a, p. 11].

LEMMA 2.2. *Suppose that f is an automorphic form—then, for each i between 1 and r, the function $f - f_{P_i}$ is rapidly decreasing on $\mathfrak{S}_{t_0,\omega_0}\kappa_i$, that is, for every integer n there is a constant $C_n > 0$ such that*

$$|f(x\kappa_i) - f_{P_i}(x\kappa_i)| \leqslant C_n(\xi_\lambda(a_x))^n$$

for all $x \in \mathfrak{S}_{t_0,\omega_0}$.

3. Spectral Decomposition of $L^2(G/\Gamma)$

Agreeing to retain the definitions and notations set down in the preceding section, we shall consider here the decomposition of the left regular representation $L_{G/\Gamma}$ of G on the Hilbert space $L^2(G/\Gamma)$. It will be shown that $L^2(G/\Gamma)$ admits both a "continuous spectrum" and a "discrete spectrum." The continuous spectrum can be completely described using Eisenstein series and admits a direct integral decomposition in terms of the unitary representations of G induced from the finite-dimensional irreducible unitary representations of the Γ-cuspidal parabolic subgroups P_i ($i = 1, \ldots, r$), that is, in terms of "principal series" representations of G. On the other hand, the discrete spectrum is hardly understood at all, although the Selberg trace formula, which is the main tool of investigation, does provide some information.

A function $f \in L^2(G/\Gamma)$ is said to be a *cusp form* if each of the functions f_{P_i} defined by the rule

$$f_{P_i}(x) = \int_{N_i/N_i \cap \Gamma} f(xn_i)\, d_{N_i}(n_i) \qquad (i = 1, \ldots, r)$$

vanishes for almost all $x \in G$. Call $L_0^2(G/\Gamma)$ the set of all cusp forms in $L^2(G/\Gamma)$—then $L_0^2(G/\Gamma)$ is a closed subspace of $L^2(G/\Gamma)$ which is invariant under the representation $L_{G/\Gamma}$. Write $L_{G/\Gamma}^0$ for the restriction of $L_{G/\Gamma}$ to $L_0^2(G/\Gamma)$—then it can be shown that $L_{G/\Gamma}^0$ decomposes discretely into an orthogonal direct sum of irreducible unitary representations of G, the multiplicity of any particular irreducible unitary representation in this decomposition being finite. In other words, if $\hat{G}$ is the set of unitary equivalence classes of irreducible unitary representations of G and if m_U is the multiplicity with which a given U in $\hat{G}$ occurs in $L_{G/\Gamma}^0$, then

$$L_{G/\Gamma}^0 = \sum_{U \in \hat{G}} \oplus\, m_U U.$$

A central problem in the theory is the determination of those U for which $m_U > 0$ together with an explicit formula for m_U.

Let us review the method used to establish the discrete decomposability of $L_{G/\Gamma}^0$. The key is the following result due to Langlands [19a, p. 41].

Lemma 3.1. *Let $\alpha \in C_c^1(G)$—then the operator $L_{G/\Gamma}^0(\alpha)$ on $L_0^2(G/\Gamma)$ given by the prescription*

$$L_{G/\Gamma}^0(\alpha)f = \int_G \alpha(x) L_{G/\Gamma}^0(x) f\, d_G(x) = \alpha * f \qquad (f \in L_0^2(G/\Gamma))$$

is of the Hilbert–Schmidt class.

Actually Langlands only explicitly proves that $L^0_{G/\Gamma}(\alpha)$ $(\alpha \in C^1_c(G))$ is a compact operator. But the improved assertion is an immediate consequence of what is said there. In fact Langlands [19a, p. 40] proves that there exists a constant $C > 0$, depending only on α, and positive continuous functions ϕ_i $(i = 1, \ldots, r)$, square integrable over $\mathfrak{S}_{t_0,\omega_0}$, such that

$$|L^0_{G/\Gamma}(\alpha)f(x\kappa_i)| \leqslant C\phi_i(x)\|f\|_2 \qquad (f \in L^2_0(G/\Gamma))$$

for all $x \in \mathfrak{S}_{t_0,\omega_0}$, $\kappa_i \in \mathfrak{s}$. This being the case, extend $L^0_{G/\Gamma}(\alpha)$ to a bounded operator T_α on $L^2(G/\Gamma)$ by rendering it trivial on the orthogonal complement of $L^2_0(G/\Gamma)$. For each $x \in G$, the assignment $f \mapsto T_\alpha f(x)$ is a continuous linear functional on $L^2(G/\Gamma)$ so there is an element $K^0_\alpha(x, ?)$ in $L^2(G/\Gamma)$ such that

$$T_\alpha f(x) = \int_{G/\Gamma} K^0_\alpha(x, y) f(y)\, d_G(y).$$

To conclude that $L^0_{G/\Gamma}(\alpha)$ is of the Hilbert–Schmidt class, it would be enough to verify that

$$\int_{G/\Gamma}\int_{G/\Gamma} |K^0_\alpha(x, y)|^2\, d_G(x)\, d_G(y) < +\infty.$$

There is, however, a difficulty here since one does not know a priori that K^0_α is measurable on $G/\Gamma \times G/\Gamma$. To get around this point, observe that the assignment $x \mapsto K^0_\alpha(x, ?)$ is weakly continuous, hence weakly measurable. Owing then to the existence of the ϕ_i and the fact that square integrability on G/Γ amounts to square integrability on each of the $\mathfrak{S}_{t_0,\omega_0}\kappa_i$ $(i = 1, \ldots, r)$, one can now appeal to a measure-theoretic generality (cf. Lang [18, pp. 232–234]) which says there exists a kernel $T_\alpha \in L^2(G/\Gamma \times G/\Gamma)$ such that for almost all x

$$T_\alpha f(x) = \int_{G/\Gamma} T_\alpha(x, y) f(y)\, d_G(y).$$

Therefore $L^0_{G/\Gamma}(\alpha)$ is indeed Hilbert–Schmidt.

Because $L^0_{G/\Gamma}$ represents the elements of $C^1_c(G)$ by compact (and even Hilbert–Schmidt) operators, the discrete decomposability of $L^0_{G/\Gamma}$ now follows from a simple generality (cf. Warner [34a, p. 248]).

There is no known method of constructing all the elements of $L^2_0(G/\Gamma)$. Generalizing the classical notion of Poincaré series, one can at least say the following.

LEMMA 3.2. *Let f be a differentiable function in $L^1(G)$. Suppose that f is K-finite and $\mathfrak{Z}$-finite. Put*

$$p_f(x) = \sum_{\gamma \in \Gamma} f(x\gamma) \qquad (x \in G).$$

Then the series defining p_f is absolutely and uniformly convergent on compact subsets of G/Γ. In addition there exists a constant $C > 0$ such that

$$\sum_{\gamma \in \Gamma} |f(x\gamma)| \leqslant C\|f\|_1 \qquad (x \in G),$$

so p_f is a bounded function on G/Γ. If, moreover, f is in $L^2(G)$, then p_f is a cusp form on G/Γ.

Proof. The assertions regarding the convergence and boundedness of p_f are due to Harish-Chandra. Proofs may be found in Baily and Borel [2, pp. 491–492]. Because p_f is bounded on G/Γ, one has, of course, $p_f \in L^2(G/\Gamma)$. Assume now that $f \in L^2(G)$ (in addition to the other hypotheses)—then f is necessarily a cusp form on G (cf. Harish-Chandra [13b, p. 538]). This means, in particular, that

$$\int_{N_i} f(xn_i)\, d_{N_i}(n_i) = 0 \qquad (i = 1, \ldots, r)$$

for all $x \in G$. To prove that $p_f \in L_0^2(G/\Gamma)$, we must show that

$$\int_{N_i/N_i \cap \Gamma} p_f(xn_i)\, d_{N_i}(n_i) = 0 \qquad (i = 1, \ldots, r)$$

for all $x \in G$. It is permissible to write

$$\begin{aligned}\int_{N_i/N_i \cap \Gamma} p_f(xn_i)\, d_{N_i}(n_i) &= \int_{N_i/N_i \cap \Gamma} \left(\sum_{\gamma \in \Gamma} f(xn_i\gamma)\right) d_{N_i}(n_i) \\ &= \int_{N_i/N_i \cap \Gamma} \left(\sum_{\gamma \in \Gamma \cap N_i \backslash \Gamma} \sum_{\delta \in \Gamma \cap N_i} f(xn_i\,\delta\gamma)\right) d_{N_i}(n_i) \\ &= \sum_{\gamma \in \Gamma \cap N_i \backslash \Gamma} \left(\int_{N_i} f(xn_i\gamma)\, d_{N_i}(n_i)\right).\end{aligned}$$

In the last summation, γ runs through a set of representatives for $\Gamma \cap N_i \backslash \Gamma$. Since the space of cusp forms on G is stable under both left and right translations, we have

$$\int_{N_i} f(xn_i\gamma)\, d_{N_i}(n_i) = 0$$

for all $x \in G$. Therefore p_f is a cusp form on G/Γ. ■

We shall refer to p_f as the *Poincaré series* attached to f. In passing, one should note that the problem of determining those f for which p_f is nonzero appears to be very difficult. Natural candidates would include the K-finite matrix coefficients of the integrable discrete series for G (which, of course, exist if and only if $\operatorname{rank}(G) = \operatorname{rank}(K)$).

By $L^2(\vartheta, \delta)$, we shall understand the closed subspace of $L^2(G/\Gamma)$ spanned by the functions $\boldsymbol{\Theta}_\phi(\boldsymbol{\phi} \in \mathscr{V}(\vartheta, \delta))$. If $\boldsymbol{\Phi} \in \mathscr{H}_R(\vartheta, \delta)$, then $\boldsymbol{\Theta}_\phi \in L^2(\vartheta, \delta)$. It is known that

$$L^2(\vartheta', \delta') \perp L^2(\vartheta'', \delta'')$$

unless $\vartheta' = \vartheta''$, $\delta' = \delta''$ (cf. Harish-Chandra [13a, p. 84]). Call ${}_0L^2(G/\Gamma)$ the orthogonal complement to $L_0^2(G/\Gamma)$ in $L^2(G/\Gamma)$—then it can be shown that (cf. Harish-Chandra [13a, p. 89])

$${}_0L^2(G/\Gamma) = \sum_{\vartheta,\delta} \oplus\, L^2(\vartheta, \delta).$$

Moreover, if ϑ is fixed, then

$$\sum_{\delta} \oplus\, L^2(\vartheta, \delta)$$

is $L_{G/\Gamma}$-invariant.

Our objective now will be to obtain an orthogonal decomposition of $L^2(\vartheta, \delta)$ by utilizing the complete scalar product formula. These results are buried in Langlands' [19a] infamous "Chapter 7" (cf., too, Langlands [19b, pp. 247–252]). The discussion here parallels the adelic case (cf. Arthur [1a, pp. 341–345]).

Consider the vector space of all functions f_{d} from $[-|\rho|, 0[$ to $\mathscr{E}(\vartheta, \delta)$. Since the operator $\boldsymbol{\gamma}_{\vartheta,\delta}(s_i)$ $(1 \leqslant i \leqslant n(\vartheta, \delta))$ is positive semidefinite (cf. Lemma 2.1), one can equip this space with a positive semidefinite inner product according to the rule

$$(f_{\mathrm{d}}, f_{\mathrm{d}}) = \frac{1}{2\pi} \sum_{i=1}^{n(\vartheta,\delta)} (\boldsymbol{\gamma}_{\vartheta,\delta}(s_i) f_{\mathrm{d}}(s_i), f_{\mathrm{d}}(s_i)).$$

Factoring out by the null vectors gives a finite-dimensional Hilbert space which we shall denote by $\mathscr{L}_{\mathrm{d}}^2(\vartheta, \delta)$. On the other hand, call $\mathscr{L}_{\mathrm{c}}^2(\vartheta, \delta)$ the space of square integrable functions f_{c} from $\sqrt{-1}\,\mathbf{R}$ to $\mathscr{E}(\vartheta, \delta)$ such that

$$f_{\mathrm{c}}(-\sqrt{-1}\,\eta) = \mathbf{c}_{\vartheta,\delta}(\sqrt{-1}\,\eta) f_{\mathrm{c}}(\sqrt{-1}\,\eta)$$

equipped with the inner product

$$(f_{\mathrm{c}}, f_{\mathrm{c}}) = \frac{1}{\pi} \int_{-\infty}^{\infty} (f_{\mathrm{c}}(\sqrt{-1}\,\eta), f_{\mathrm{c}}(\sqrt{-1}\,\eta))\, d\eta.$$

Write

$$\mathscr{L}^2(\vartheta, \delta) = \mathscr{L}_{\mathrm{d}}^2(\vartheta, \delta) \oplus \mathscr{L}_{\mathrm{c}}^2(\vartheta, \delta).$$

Given $\mathbf{\Phi} \in \widehat{C_c^\infty(A)} \otimes \mathscr{E}(\vartheta, \delta)$, let

$$\mathbf{\Phi}_d = \mathbf{\Phi}|[-|\rho|, 0[\in \mathscr{L}_d^2(\vartheta, \delta),$$

$$\mathbf{\Phi}_c = \tfrac{1}{2}[\mathbf{\Phi}(\sqrt{-1}\eta : ?) + \mathbf{c}_{\vartheta,\delta}(-\sqrt{-1}\eta)\mathbf{\Phi}(-\sqrt{-1}\eta : ?)] \in \mathscr{L}_c^2(\vartheta, \delta).$$

Then, on the basis of what has been said above, it is clear that the correspondence

$$(\mathbf{\Phi}_d, \mathbf{\Phi}_c) \leftrightarrow \mathbf{\Theta}_\phi$$

sets up an isometry between dense subspaces of $\mathscr{L}^2(\vartheta, \delta)$ and $L^2(\vartheta, \delta)$, so there is an isometric isomorphism

$$I_{\vartheta,\delta} : \mathscr{L}^2(\vartheta, \delta) \to L^2(\vartheta, \delta).$$

Put

$$L_d^2(\vartheta, \delta) = I_{\vartheta,\delta}(\mathscr{L}_d^2(\vartheta, \delta)), \qquad L_c^2(\vartheta, \delta) = I_{\vartheta,\delta}(\mathscr{L}_c^2(\vartheta, \delta)).$$

Then we have

$$L^2(\vartheta, \delta) = L_d^2(\vartheta, \delta) \oplus L_c^2(\vartheta, \delta).$$

The reader should view the subscripts "d" and "c" as standing for "discrete" and "continuous," respectively. Justification for this notation will appear presently.

Suppose given $f_d \in \mathscr{L}_d^2(\vartheta, \delta)$, $f_c \in \mathscr{L}_c^2(\vartheta, \delta)$—then it turns out that it is possible to express $I_{\vartheta,\delta}(f_d)$ and $I_{\vartheta,\delta}(f_c)$ in terms of Eisenstein series and their residues, a preliminary discussion being needed for the latter.

Let $\mathbf{\Phi} \in \mathscr{E}(\vartheta, \delta)$. Write

$$\mathbf{E}_i(\mathbf{\Phi} : x) = -2\pi \cdot \operatorname*{Res}_{s=s_i} \mathbf{E}(\mathbf{\Phi} : s : x) \qquad (i = 1, \ldots, n(\vartheta, \delta)).$$

LEMMA 3.3. *$\mathbf{E}_i(\mathbf{\Phi} : x)$ $(i = 1, \ldots, n(\vartheta, \delta))$ is a square integrable automorphic form which is orthogonal to $L_0^2(G/\Gamma)$.*

Before giving the proof, let us recall an estimate which will be used constantly in what follows. Since $\mathbf{E}(\mathbf{\Phi} : s : x)$ is an automorphic form, there is a constant $C > 0$ and an integer $n \geqslant 0$ such that

$$|\mathbf{E}(\mathbf{\Phi} : s : x\kappa_i)| \leqslant C\|\mathbf{\Phi}\|(\xi_\lambda(a_x))^{-n}$$

for all $x \in \mathfrak{S}_{t_0,\omega_0}$, $\kappa_i \in \mathfrak{s}$. A priori, C and n depend on $\mathbf{\Phi}$ and s. Let ω be a compact subset of the region where $\mathbf{c}_{\vartheta,\delta}$ is holomorphic—then it can be shown that C and n may always be chosen in such a way that the estimate is uniform in $\mathbf{\Phi}$ and uniform in s ($s \in \omega$) (cf. Harish-Chandra [13a, p. 105]).

Proof of Lemma 3.3. Fix an index i_0 between 1 and $n(\vartheta, \delta)$. Since $\mathbf{\Phi} \dot{\in} \mathscr{E}(\vartheta, \delta)$, we have

$$\mathbf{E}(\mathbf{\Phi}:s:x) = \int_K \chi_\delta(k)\mathbf{E}(\mathbf{\Phi}:s:kx)\,d_K(k),$$

a relation which is clearly preserved upon taking residues. There is an ideal $\mathfrak{I}$ in $\mathfrak{Z}$ of finite codimension such that

$$Z\mathbf{E}(\mathbf{\Phi}:s:?) = 0$$

for all $Z \in \mathfrak{I}$ and this too is unchanged upon taking residues. To complete the verification that $\mathbf{E}_{i_0}(\mathbf{\Phi}:x)$ is an automorphic form, we must check the growth condition. Relative to an appropriately small contour $\mathscr{C}_{i_0}$ around s_{i_0}, use Cauchy's theorem to write

$$\mathbf{E}_{i_0}(\mathbf{\Phi}:x) = \sqrt{-1}\oint_{\mathscr{C}_{i_0}} \frac{\mathbf{E}(\mathbf{\Phi}:s:x)}{s - s_{i_0}}\,ds.$$

In view of what has been said above, there is a constant $C > 0$ and an integer $n \geqslant 0$ such that

$$|\mathbf{E}(\mathbf{\Phi}:s:x\kappa_i)| \leqslant C\|\mathbf{\Phi}\|(\xi_\lambda(a_x))^{-n}$$

for all $x \in \mathfrak{S}_{t_0,\omega_0}$, $\kappa_i \in \mathfrak{s}$, $\mathbf{\Phi} \in \mathscr{E}(\vartheta, \delta)$, $s \in \mathscr{C}_{i_0}$. It is thus immediate that $\mathbf{E}_{i_0}(\mathbf{\Phi}:x)$ is slowly increasing. Therefore $\mathbf{E}_{i_0}(\mathbf{\Phi}:x)$ is an automorphic form. Let us now turn to the square integrability assertion. Since $\mathbf{E}_{i_0}(\mathbf{\Phi}:x)$ is right invariant under Γ, we need only check square integrability on each of the $\mathfrak{S}_{t_0,\omega_0}\kappa_i$. Let $\mathbf{E}_{i_0,P_i}(\mathbf{\Phi}:x)$ be the constant term of $\mathbf{E}_{i_0}(\mathbf{\Phi}:x)$ along P_i—then, since $\mathbf{E}_{i_0}(\mathbf{\Phi}:x)$ is an automorphic form,

$$\mathbf{E}_{i_0}(\mathbf{\Phi}:x\kappa_i) - \mathbf{E}_{i_0,P_i}(\mathbf{\Phi}:x\kappa_i) \qquad (x \in \mathfrak{S}_{t_0,\omega_0})$$

is rapidly decreasing on $\mathfrak{S}_{t_0,\omega_0}\kappa_i$ (cf. Lemma 2.2), hence is square integrable there. In this connection, let us keep in mind that the Haar measure of $\mathfrak{S}_{t_0,\omega_0}$ is finite, as results immediately from the integral formula

$$d_G(x) = a^{2\rho}d_K(k)\,d_A(a)\,d_N(n).$$

So, to finish the proof of the square integrability of $\mathbf{E}_{i_0}(\mathbf{\Phi}:x)$, it will be enough to show that $\mathbf{E}_{i_0,P_i}(\mathbf{\Phi}:x)$ is square integrable over $\mathfrak{S}_{t_0,\omega_0}\kappa_i$. Using the definitions, one calculates without difficulty that

$$\mathbf{E}_{P_i}(\mathbf{\Phi}:s:x) = e^{(s-|\rho|)(H_i(x))}\cdot\Phi_i(x) + e^{(-s-|\rho|)(H_i(x))}\cdot(\mathbf{c}_{\vartheta,\delta}(s)\mathbf{\Phi})_i(x).$$

Taking the residue at $s = s_{i_0}$ then gives

$$\mathbf{E}_{i_0,P_i}(\mathbf{\Phi}:x) = e^{(-s_{i_0}-|\rho|)(H_i(x))}\cdot(\boldsymbol{\gamma}_{\vartheta,\delta}(s_{i_0})\mathbf{\Phi})_i(x).$$

Consider

$$\int_{\mathfrak{S}_{t_0,\omega_0}} |\mathbf{E}_{i_0,P_i}(\mathbf{\Phi}:x\kappa_i)|^2\, d_G(x).$$

If $x = kan$ ($k \in K$, $a \in A$, $n \in N$), then $x\kappa_i = kk_i^{-1} \cdot k_iak_i^{-1} \cdot k_ink_i^{-1} \in K \cdot A_i \cdot N_i$. Since $(\gamma_{\vartheta,\delta}(s_{i_0})\mathbf{\Phi})_i$ is right invariant under $A_i \cdot N_i$, the square integrability of our function is seen to be equivalent to the finiteness of

$$\int_{-\infty}^{\log t_0/|\lambda|} e^{(-2s_{i_0}-2|\rho|)t} \cdot e^{2|\rho|t}\, dt,$$

which is certainly true, s_{i_0} being negative. Therefore $\mathbf{E}_{i_0}(\mathbf{\Phi}:x)$ belongs to $L^2(G/\Gamma)$. The final claim is that $\mathbf{E}_{i_0}(\mathbf{\Phi}:x)$ is orthogonal to $L_0^2(G/\Gamma)$. For each s which is not a pole of $\mathbf{c}_{\vartheta,\delta}$, $\mathbf{E}(\mathbf{\Phi}:s:x)$ is an automorphic form. Consequently, for every automorphic cusp form f, the function

$$\mathbf{E}(\mathbf{\Phi}:s:x)\overline{f(x)}$$

is integrable over G/Γ (cf. Harish-Chandra [13a, p. 15]). If $\operatorname{Re}(s) < -|\rho|$, then

$$\begin{aligned}
&\int_{G/\Gamma} \mathbf{E}(\mathbf{\Phi}:s:x)\overline{f(x)}\, d_G(x)\\
&= \sum_{i=1}^{r} \int_{G/\Gamma} \Bigg(\sum_{\gamma \in \Gamma/\Gamma \cap P_i} e^{(s-|\rho|)(H_i(x\gamma))} \cdot \Phi_i(x\gamma) \Bigg) \overline{f(x)}\, d_G(x)\\
&= \sum_{i=1}^{r} \int_{G/\Gamma \cap P_i} (e^{(s-|\rho|)(H_i(x))} \cdot \Phi_i(x))\overline{f(x)}\, d_G(x)\\
&= \sum_{i=1}^{r} \int_K \int_{M_i/\Gamma_{M_i}} \int_{A_i} \int_{N_i/N_i \cap \Gamma} (e^{(s+|\rho|)(H_i(a_i))} \cdot \Phi_i(km_i))\\
&\quad \times \overline{f(km_ia_in_i)}\, d_K(k)\, d_{M_i}(m_i)\, d_{A_i}(a_i)\, d_{N_i}(n_i)\\
&= 0,
\end{aligned}$$

f being a cusp form. Owing to the estimate on $\mathbf{E}(\mathbf{\Phi}:s:x)$ quoted earlier and the fact that f is a cusp form, hence rapidly decreasing, one sees that the integral

$$\int_{G/\Gamma} \mathbf{E}(\mathbf{\Phi}:s:x)\overline{f(x)}\, d_G(x)$$

is uniformly convergent on compacta in s, the poles of $\mathbf{c}_{\vartheta,\delta}$ being, of course, excluded. This means, in particular, that the residue at $s = s_{i_0}$ can be computed by taking the appropriate limit under the integral sign. Analytic continuation then implies that

$$\int_{G/\Gamma} \mathbf{E}_{i_0}(\mathbf{\Phi}:x)\overline{f(x)}\, d_G(x) = 0.$$

Since $L_0^2(G/\Gamma)$ admits an orthonormal basis composed of automorphic forms (cf. Harish-Chandra [13a, p. 15]), it therefore follows that $\mathbf{E}_{i_0}(\mathbf{\Phi}:x)$ is orthogonal to $L_0^2(G/\Gamma)$. ■

Let us return to the problem of computing the images of $\mathscr{L}_{\mathrm{d}}^2(\vartheta, \delta)$ and $\mathscr{L}_{\mathrm{c}}^2(\vartheta, \delta)$ in $L^2(\vartheta, \delta)$ under $I_{\vartheta,\delta}$.

LEMMA 3.4. *Suppose given:*

(i) *a function* $f_{\mathrm{d}} \in \mathscr{L}_{\mathrm{d}}^2(\vartheta, \delta)$;
(ii) *a compactly supported function* $f_{\mathrm{c}} \in \mathscr{L}_{\mathrm{c}}^2(\vartheta, \delta)$.

Then, for almost all $x \in G$,

$$(I_{\vartheta,\delta} f_{\mathrm{d}})(x) + (I_{\vartheta,\delta} f_{\mathrm{c}})(x)$$
$$= \frac{1}{2\pi} \sum_{i=1}^{n(\vartheta,\delta)} \mathbf{E}_i(f_{\mathrm{d}}(s_i):x) + \frac{1}{2\pi} \int_{\mathrm{Re}(s)=0} \mathbf{E}(f_{\mathrm{c}}(s):s:x)\,|ds|.$$

One should note that the integral

$$\int_{\mathrm{Re}(s)=0} \mathbf{E}(f_{\mathrm{c}}(s):s:x)\,|ds|$$

exists. In fact, on the basis of an estimate quoted earlier, there is a constant $C > 0$ and an integer $n \geqslant 0$ such that

$$|\mathbf{E}(f_{\mathrm{c}}(s):s:x\kappa_i)| \leqslant C\|f_{\mathrm{c}}(s)\|(\xi_\lambda(a_x))^{-n}$$

for all $x \in \mathfrak{S}_{t_0,\omega_0}$, $\kappa_i \in \mathfrak{s}$, $s \in \operatorname{spt}(f_{\mathrm{c}})$, whence

$$\left|\int_{\mathrm{Re}(s)=0} \mathbf{E}(f_{\mathrm{c}}(s):s:x\kappa_i)\,|ds|\right| \leqslant C(\xi_\lambda(a_x))^{-n} \int_{\operatorname{spt}(f)} \|f_{\mathrm{c}}(s)\|\,|ds|$$
$$\leqslant C \cdot \sqrt{\operatorname{meas}(\operatorname{spt}(f_{\mathrm{c}}))} \cdot \|f_{\mathrm{c}}\|_2 \cdot (\xi_\lambda(a_x))^{-n},$$

a finite quantity.

It will be convenient to preface the proof of Lemma 3.4 with a preliminary estimate.

SUBLEMMA. *Fix an index i between* 1 *and r. Let* ω *be a compact subset of the region where* $\mathbf{c}_{\vartheta,\delta}$ *is holomorphic. Then: For every integer n there is a constant* $C_n > 0$ *such that*

$$|\mathbf{E}(\mathbf{\Phi}:s:x\kappa_i) - \mathbf{E}_{P_i}(\mathbf{\Phi}:s:x\kappa_i)| \leqslant C_n\|\mathbf{\Phi}\|(\xi_\lambda(a_x))^{-n}$$

for all $x \in \mathfrak{S}_{t_0,\omega_0}$, $\mathbf{\Phi} \in \mathscr{E}(\vartheta, \delta)$, $s \in \omega$.

[The reader will observe that while Lemma 2.2 guarantees that

$$\mathbf{E}(\mathbf{\Phi}:s:x\kappa_i) - \mathbf{E}_{P_i}(\mathbf{\Phi}:s:x\kappa_i)$$

is rapidly decreasing on $\mathfrak{S}_{t_0,\omega_0}\kappa_i$, the estimate there does not take into account the first two variables.]

Proof. Let $I_c^\infty(G)$ be the compactly supported C^∞ functions α on G which are K-central, i.e., $\alpha(kxk^{-1}) = \alpha(x)$ for every $k \in K$. At those s for which $\mathbf{E}(\mathbf{\Phi}:s:x)$ is holomorphic, there is a canonically defined representation $\mathbf{U}^{\vartheta,s}$ of $I_c^\infty(G)$ on $\mathscr{E}(\vartheta, \delta)$, explicated later in this section, such that

$$\alpha * \mathbf{E}(\mathbf{\Phi}:s:?)(x) = \mathbf{E}(\mathbf{U}^{\vartheta,s}(\alpha)\mathbf{\Phi}:s:x).$$

This said, for each s an easy variant of an argument of Langlands [19a, p. 47] ensures the existence of an α_s such that $\mathbf{U}^{\vartheta,s}(\alpha_s) = 1$. Fix a point $s_0 \in \omega$. Write α_0 in place of α_{s_0} and let $\mathscr{N}(s_0)$ be a relatively compact neighborhood of s_0 such that $\det(\mathbf{U}^{\vartheta,s}(\alpha_0)) \neq 0$ $(s \in \mathscr{N}(s_0))$ and on which $\mathbf{E}(\mathbf{\Phi}:s:x)$ is holomorphic. We remark, in passing, that the assignment $s \mapsto \mathbf{U}^{\vartheta,s}(\alpha_0)$ is continuous. For s varying in $\mathscr{N}(s_0)$, one then has the following relations:

$$\mathbf{E}(\mathbf{\Phi}:s:x) = \alpha_0 * \mathbf{E}(\mathbf{U}^{\vartheta,s}(\alpha_0)^{-1}\mathbf{\Phi}:s:?)(x),$$
$$\mathbf{E}_{P_i}(\mathbf{\Phi}:s:x) = \alpha_0 * \mathbf{E}_{P_i}(\mathbf{U}^{\vartheta,s}(\alpha_0)^{-1}\mathbf{\Phi}:s:?)(x).$$

Because $\mathbf{U}^{\vartheta,s}$ is a representation of $I_c^\infty(G)$, these formulas continue to hold when α_0 is replaced by a positive integral power α_0^ν. Select a constant $C > 0$ and an integer $n \geqslant 0$ such that

$$|\mathbf{E}(\mathbf{\Phi}:s:x\kappa_i)| \leqslant C\|\mathbf{\Phi}\|(\xi_\lambda(a_x))^{-n}$$

for all $x \in \mathfrak{S}_{t_0,\omega_0}$, $\mathbf{\Phi} \in \mathscr{E}(\vartheta, \delta)$, $s \in \mathscr{N}(s_0)$. Let ν be an arbitrary positive integer—then we have

$$\begin{aligned}&|\mathbf{E}(\mathbf{\Phi}:s:x\kappa_i) - \mathbf{E}_{P_i}(\mathbf{\Phi}:s:x\kappa_i)|\\ &\quad = |\alpha_0^\nu * \mathbf{E}(\mathbf{U}^{\vartheta,s}(\alpha_0^\nu)^{-1}\mathbf{\Phi}:s:?)(x\kappa_i) - \alpha_0^\nu * \mathbf{E}_{P_i}(\mathbf{U}^{\vartheta,s}(\alpha_0^\nu)^{-1}\mathbf{\Phi}:s:?)(x\kappa_i)|\end{aligned}$$

for all $x \in \mathfrak{S}_{t_0,\omega_0}$, $\mathbf{\Phi} \in \mathscr{E}(\vartheta, \delta)$, $s \in \mathscr{N}(s_0)$. According to a principle spelled out in Langlands [19a, p. 45], there is a constant $M > 0$ which allows one to estimate the second term uniformly by

$$M\|\mathbf{U}^{\vartheta,s}(\alpha_0^\nu)^{-1}\mathbf{\Phi}\|(\xi_\lambda(a_x))^{-n+\nu},$$

which itself can be estimated uniformly by

$$N\|\mathbf{\Phi}\|(\xi_\lambda(a_x))^{-n+\nu}$$

for some other constant $N > 0$. Changing the notation, we have thus proved that for every integer n there is a constant $C_n > 0$ such that

$$|\mathbf{E}(\mathbf{\Phi}:s:x\kappa_i) - \mathbf{E}_{P_i}(\mathbf{\Phi}:s:x\kappa_i)| \leqslant C_n\|\mathbf{\Phi}\|(\xi_\lambda(a_x))^{-n}$$

for all $x \in \mathfrak{S}_{t_0,\omega_0}$, $\mathbf{\Phi} \in \mathscr{E}(\vartheta, \delta)$, $s \in \mathscr{N}(s_0)$. Since s_0 is an arbitrary point in ω and ω is compact, we are done. ■

Proof of Lemma 3.4. Let f denote the function on the right-hand side of the purported equality and put

$$g = f - I_{\vartheta,\delta} f_{\mathrm{d}} - I_{\vartheta,\delta} f_{\mathrm{c}}.$$

The assertion then is $g(x) = 0$ for almost all x. To prove this, it will clearly be enough to show that g is an element of $L^2(G/\Gamma)$ which is orthogonal to $L_0^2(G/\Gamma)$ and has the property that the constant terms g_{P_i} are zero almost everywhere. In view of the definitions and the preceding lemma, the square integrability of g will follow when it is shown that

$$\int_{\mathrm{Re}(s)=0} \mathbf{E}(f_{\mathrm{c}}(s):s:x)\,|ds|$$

is square integrable over each $\mathfrak{S}_{t_0,\omega_0}\kappa_i$. The constant term along P_i of

$$\int_{\mathrm{Re}(s)=0} \mathbf{E}(f_{\mathrm{c}}(s):s:x)\,|ds|$$

is

$$\int_{\mathrm{Re}(s)=0} \mathbf{E}_{P_i}(f_{\mathrm{c}}(s):s:x)\,|ds|$$

and we claim that

$$\int_{\mathrm{Re}(s)=0} (\mathbf{E}(f_{\mathrm{c}}(s):s:x\kappa_i) - \mathbf{E}_{P_i}(f_{\mathrm{c}}(s):s:x\kappa_i))\,|ds| \qquad (x \in \mathfrak{S}_{t_0,\omega_0})$$

is rapidly decreasing on $\mathfrak{S}_{t_0,\omega_0}$, hence is square integrable there. Owing to the sublemma, for every integer n there is a constant $C_n > 0$ such that

$$|\mathbf{E}(f_{\mathrm{c}}(s):s:x\kappa_i) - \mathbf{E}_{P_i}(f_{\mathrm{c}}(s):s:x\kappa_i)| \leqslant C_n \|f_{\mathrm{c}}(s)\| (\xi_\lambda(a_x))^n$$

for all $x \in \mathfrak{S}_{t_0,\omega_0}$, $s \in \mathrm{spt}(f_{\mathrm{c}})$. The claim thus follows upon making the obvious estimates. To complete our verification, we need only establish the square integrability of

$$\int_{\mathrm{Re}(s)=0} \mathbf{E}_{P_i}(f_{\mathrm{c}}(s):s:x)\,|ds|$$

over $\mathfrak{S}_{t_0,\omega_0}\cdot \kappa_i$. Write

$$\begin{aligned}\int_{\mathrm{Re}(s)=0} &\mathbf{E}_{P_i}(f_{\mathrm{c}}(s):s:x)\,|ds| \\ &= \int_{\mathrm{Re}(s)=0} (e^{(s-|\rho|)(H_i(x))} \cdot f_{\mathrm{c}}(s)_i(x) \\ &\qquad + e^{(-s-|\rho|)(H_i(x))} \cdot (\mathbf{c}_{\vartheta,\delta}(s) f_{\mathrm{c}}(s))_i(x))\,|ds|.\end{aligned}$$

Since

$$\mathbf{c}_{\vartheta,\delta}(s) f_{\mathrm{c}}(s) = f_{\mathrm{c}}(-s) \qquad (\mathrm{Re}(s) = 0),$$

the last term above becomes

$$2\int_{\mathrm{Re}(s)=0}(e^{(s-|\rho|)(H_i(x))}\cdot f_c(s)_i(x))\,|ds|.$$

We need only study this integral when $x\in\mathfrak{S}_{t_0,\omega_0}\kappa_i$. Let $\Phi_1,\ldots,\Phi_p$ be an orthonormal basis for $\mathscr{E}(\vartheta_i,\delta)$—then there are compactly supported square integrable functions $f_1,\ldots,f_p$ on $\sqrt{-1}\,\mathbf{R}$ such that

$$f_c(\sqrt{-1}\,\eta)_i=\sum_{j=1}^{p}f_j(\sqrt{-1}\,\eta)\Phi_j.$$

Owing to the right invariance of the Φ_j under $A_i\cdot N_i$, it suffices to show that

$$e^{-|\rho|t}\sum_{j=1}^{p}\Phi_j(k)\int_{-\infty}^{\infty}e^{\sqrt{-1}\,\eta t}f_j(\sqrt{-1}\,\eta)\,d\eta$$

is square integrable over $K\times[-\infty,\log t_0/|\lambda|]$. The Jacobian $e^{2|\rho|t}$ will cancel out the factor $e^{-2|\rho|t}$. Because the Φ_j are continuous functions on K and the

$$\int_{-\infty}^{\infty}e^{\sqrt{-1}\,\eta t}f_j(\sqrt{-1}\,\eta)\,d\eta$$

are Fourier transforms of compactly supported L^2-functions on $\sqrt{-1}\,\mathbf{R}$, the square integrability contention is now clear. Therefore g is indeed an element of $L^2(G/\Gamma)$. The next assertion is that g is orthogonal to $L_0^2(G/\Gamma)$. The only term for which there can be any question is again

$$\int_{\mathrm{Re}(s)=0}\mathbf{E}(f_c(s):s:x)\,|ds|$$

which, however, is easy to deal with (cf. the proof of Lemma 3.3). The remaining contention is that the constant terms g_{P_i} are zero almost everywhere. To establish this, let $\{\mathbf{\Phi}^n\}$ be a sequence of functions in $\widehat{C_c^\infty(A)}\otimes\mathscr{E}(\vartheta,\delta)$ such that

$$\lim_{n\to\infty}(\mathbf{\Phi}_d^n,\mathbf{\Phi}_c^n)=(f_d,f_c)$$

in $\mathscr{L}^2(\vartheta,\delta)$. The constant term along P_i of the theta series $\mathbf{\Theta}_{\Phi^n}$ is given by

$$\int_{N_i/N_i\cap\Gamma}\mathbf{\Theta}_{\Phi^n}(xn_i)\,d_{N_i}(n_i)=\frac{1}{2\pi}\int_{\mathrm{Re}(s)=s_0}\mathbf{E}_{P_i}(\mathbf{\Phi}^n(s:?):s:x)\,|ds|\qquad(s_0<-|\rho|),$$

which, upon shifting the line of integration to the imaginary axis, becomes

$$\frac{1}{2\pi}\sum_{j=1}^{n(\vartheta,\delta)}e^{(-s_j-|\rho|)(H_i(x))}(\gamma_{\vartheta,\delta}(s_j)\mathbf{\Phi}_d^n(s_j:?))_i(x)$$

$$+\frac{1}{\pi}\int_{\mathrm{Re}(s)=0}e^{(s-|\rho|)(H_i(x))}\cdot\mathbf{\Phi}_c^n(s:x)_i\,|ds|.$$

Comparing this expression with the formula for $f_{P_i}(x)$ (which is easy to write down on the basis of earlier work), a series of simple arguments allows one to arrange things in such a way as to ensure that

$$\lim_{n\to\infty} (\boldsymbol{\Theta}_{\phi^n})_{P_i}(x) = f_{P_i}(x)$$

for almost all $x \in G$. The details are straightforward and will be left to the reader. On the other hand,

$$\lim_{n\to\infty} \boldsymbol{\Theta}_{\phi^n} = I_{\vartheta,\delta} f_{\mathrm{d}} + I_{\vartheta,\delta} f_{\mathrm{c}}$$

in $L^2(G/\Gamma)$, so some subsequence of $\{\boldsymbol{\Theta}_{\phi^n}\}$ must converge to $I_{\vartheta,\delta} f_{\mathrm{d}} + I_{\vartheta,\delta} f_{\mathrm{c}}$ almost everywhere. As passage to a subsequence will not affect the other conditions, we shall not change the notation. It is then immediate that

$$\lim_{n\to\infty} (\boldsymbol{\Theta}_{\phi^n})_{P_i}(x) = (I_{\vartheta,\delta} f_{\mathrm{d}} + I_{\vartheta,\delta} f_{\mathrm{c}})_{P_i}(x)$$

for almost all $x \in G$. Therefore $g_{P_i} = 0$ [a.e.]. ∎

One obvious consequence of the above result is the fact that the space spanned by the $\mathbf{E}_i(\boldsymbol{\Phi} : x)$ ($\boldsymbol{\Phi} \in \mathscr{E}(\vartheta, \delta)$), i running between 1 and $n(\vartheta, \delta)$, is precisely $L_{\mathrm{d}}^2(\vartheta, \delta)$.

Put now

$$\mathscr{L}_{\mathrm{d}}^2 = \sum_{\vartheta,\delta} \oplus \mathscr{L}_{\mathrm{d}}^2(\vartheta, \delta), \qquad \mathscr{L}_{\mathrm{c}}^2 = \sum_{\vartheta,\delta} \oplus \mathscr{L}_{\mathrm{c}}^2(\vartheta, \delta),$$

$${}_0L_{\mathrm{d}}^2(G/\Gamma) = \sum_{\vartheta,\delta} \oplus L_{\mathrm{d}}^2(\vartheta, \delta), \qquad {}_0L_{\mathrm{c}}^2(G/\Gamma) = \sum_{\vartheta,\delta} \oplus L_{\mathrm{c}}^2(\vartheta, \delta).$$

Write

$$I_{\mathrm{d}} = \sum_{\vartheta,\delta} \oplus I_{\vartheta,\delta}^{\mathrm{d}}, \qquad I_{\mathrm{c}} = \sum_{\vartheta,\delta} \oplus I_{\vartheta,\delta}^{\mathrm{c}},$$

where

$$I_{\vartheta,\delta}^{\mathrm{d}} = I_{\vartheta,\delta} | \mathscr{L}_{\mathrm{d}}^2(\vartheta, \delta), \qquad I_{\vartheta,\delta}^{\mathrm{c}} = I_{\vartheta,\delta} | \mathscr{L}_{\mathrm{c}}^2(\vartheta, \delta).$$

Then

$$I_{\mathrm{d}} : \mathscr{L}_{\mathrm{d}}^2 \to L^2(G/\Gamma), \qquad I_{\mathrm{c}} : \mathscr{L}_{\mathrm{c}}^2 \to L^2(G/\Gamma).$$

Let I_{d}^* (respectively I_{c}^*) be the adjoint of I_{d} (respectively I_{c})—then $I_{\mathrm{d}} I_{\mathrm{d}}^*$ (respectively $I_{\mathrm{c}} I_{\mathrm{c}}^*$) is the orthogonal projection of $L^2(G/\Gamma)$ onto ${}_0L_{\mathrm{d}}^2(G/\Gamma)$ (respectively ${}_0L_{\mathrm{c}}^2(G/\Gamma)$). The space ${}_0L^2(G/\Gamma)$, which we recall is the orthogonal complement in $L^2(G/\Gamma)$ of $L_0^2(G/\Gamma)$, equals

$${}_0L_{\mathrm{d}}^2(G/\Gamma) \oplus {}_0L_{\mathrm{c}}^2(G/\Gamma).$$

It will turn out that ${}_0L^2_{\mathrm{d}}(G/\Gamma)$ and ${}_0L^2_{\mathrm{c}}(G/\Gamma)$ are $L_{G/\Gamma}$-invariant. This being so, let us agree to write $L^2_{\mathrm{d}}(G/\Gamma)$ for $L^2_0(G/\Gamma) \oplus {}_0L^2_{\mathrm{d}}(G/\Gamma)$, $L^2_{\mathrm{c}}(G/\Gamma)$ for ${}_0L^2_{\mathrm{c}}(G/\Gamma)$—then we have the orthogonal decomposition

$$L^2(G/\Gamma) = L^2_{\mathrm{d}}(G/\Gamma) \oplus L^2_{\mathrm{c}}(G/\Gamma).$$

The subscripts "d" and "c" are meant to suggest "discrete" and "continuous," respectively. It has in fact been pointed out above that $L^2_0(G/\Gamma)$ is discretely decomposable with finite multiplicities; the same is true of ${}_0L^2_{\mathrm{d}}(G/\Gamma)$, as will be shown below. On the other hand, $L^2_{\mathrm{c}}(G/\Gamma)$ can be written as a direct integral of a family of unitary representations of G varying continuously with the parameter. We shall now take up the details lying behind these assertions.

Given an orbit ϑ and a complex variable s, one has a natural representation (ϑ, s) of P on S_ϑ:

M operates by the left regular representation L_{M/Γ_M};
A operates via multiplication by a^{-s} ($\equiv e^{-s(\log a)}$);
N operates trivially.

Call $U^{\vartheta,s}$ the associated induced representation of G: G thus operates by left translation on the Hilbert space of those Borel functions $f: G \to S_\vartheta$ which satisfy

$$f(xman) = a^{s-\rho} L_{M/\Gamma_M}(m^{-1}) f(x)$$

and have the property that

$$\|f\|^2 = \int_K \left(\int_{M/\Gamma_M} f(k)(m)\overline{f(k)(m)}\, d_M(m) \right) d_K(k) < +\infty.$$

$U^{\vartheta,s}$ is unitary when s is pure imaginary. Let $\mathscr{E}^{\vartheta,s}$ be the representation space of $U^{\vartheta,s}$. Write $\mathscr{E}^{\vartheta,s}(\delta)$ for the δth-isotypic component of $\mathscr{E}^{\vartheta,s}$—then we claim that

$$\mathscr{E}^{\vartheta,s}(\delta) \simeq \mathscr{E}(\vartheta, \delta).$$

To see this, let $f \in \mathscr{E}^{\vartheta,s}(\delta)$. Let $\Phi^s_f: G \to \mathbf{C}$ be defined by the rule

$$\Phi^s_f(x) = a_x^{\rho - s} \cdot f(x)(1).$$

Keeping in mind the identifications

$$M/\Gamma_M \sim M \cdot N/(\Gamma \cap P) \cdot N \sim P/(\Gamma \cap P) \cdot A \cdot N,$$

it is then a strictly routine matter to verify that the correspondence $f \mapsto \Phi^s_f$ implements a linear isomorphism between $\mathscr{E}^{\vartheta,s}(\delta)$ and $\mathscr{E}(\vartheta, \delta)$. Write

$$\mathscr{E}(\vartheta) = \sum_\delta \oplus\, \mathscr{E}(\vartheta, \delta).$$

Then $\mathscr{E}(\vartheta)$ can be identified with the Hilbert space of those Borel functions $\Phi: G/(\Gamma \cap P) \cdot A \cdot N \to \mathbf{C}$ which have the property that for every $x \in G$, the function

$$m \mapsto \Phi(xm)$$

belongs to S_ϑ, and is such that

$$\|\Phi\|^2 = \int_K \int_{M/\Gamma_M} \Phi(km)\overline{\Phi(km)}\, d_K(k)\, d_M(m) < +\infty.$$

It is easy to transport the action of $U^{\vartheta,s}$ on $\mathscr{E}^{\vartheta,s}$ to $\mathscr{E}(\vartheta)$. Explicitly, for $\Phi \in \mathscr{E}(\vartheta)$,

$$U^{\vartheta,s}(x)\Phi$$

is the function

$$y \mapsto a_{x^{-1}y}^{s-\rho} \cdot a_y^{-(s-\rho)} \cdot \Phi(x^{-1}y) \qquad (x, y \in G).$$

For later use, we remark in passing that the Hilbert space adjoint of $U^{\vartheta,s}(x)$ is $U^{\vartheta,-\bar{s}}(x^{-1})$ (cf. Warner [34a, p. 446]). Another point worth noticing is this. Suppose that s is pure imaginary (so $U^{\vartheta,s}$ is unitary)—then we have

$$U^{\vartheta,s} = \begin{cases} n(\sigma,\Gamma_M)U^{\sigma,s} \oplus n(w\sigma,\Gamma_M)U^{w\sigma,s} & \text{if} \quad \sigma \text{ is unramified,} \\ n(\sigma,\Gamma_M)U^{\sigma,s} & \text{if} \quad \sigma \text{ is ramified,} \end{cases}$$

where $\vartheta = \{\sigma, w\sigma\}$ and $U^{\sigma,s}$, $U^{w\sigma,s}$ are "principal series" representations of G (cf. Warner [34a, pp. 448–449]). Let $[\delta:\sigma]$ (respectively $[\delta:w\sigma]$) be the number of times that σ (respectively $w\sigma$) occurs in $\delta|M$—then it is well known (cf. Warner [34a, p. 450]) that the restriction of $U^{\sigma,s}$ (respectively $U^{w\sigma,s}$) to K contains δ precisely $[\delta:\sigma]$ (respectively $[\delta:w\sigma]$) times. Consequently

$$\dim(\mathscr{E}(\vartheta,\delta))$$
$$= d(\delta) \times \begin{cases} n(\sigma,\Gamma_M)[\delta:\sigma] + n(w\sigma,\Gamma_M)[\delta:w\sigma] & \text{if} \quad \sigma \text{ is unramified,} \\ n(\sigma,\Gamma_M)[\delta:\sigma] & \text{if} \quad \sigma \text{ is ramified.} \end{cases}$$

Let now $\vartheta = \vartheta_1, \vartheta_2, \ldots, \vartheta_r$ be a complete collection of associate orbits. Put

$$\mathscr{E}(\vartheta) = \sum_{i=1}^{r} \oplus\, \mathscr{E}(\vartheta_i), \qquad \mathscr{E} = \sum_{\vartheta} \oplus\, \mathscr{E}(\vartheta).$$

Then, for each value of the complex variable s, there is a canonically defined representation $\mathbf{U}^s$ of G on the Hilbert space $\mathscr{E}$: Each of the spaces $\mathscr{E}(\vartheta)$ is $\mathbf{U}^s$-stable and the corresponding representation of G, $\mathbf{U}^{\vartheta,s}$, is decomposed by the $U^{\vartheta_i,s}$, i.e.,

$$\mathbf{U}^{\vartheta,s} = \sum_{i=1}^{r} \oplus U^{\vartheta_i,s}.$$

One has

$$\mathbf{U}^s(x)^* = \mathbf{U}^{-\bar{s}}(x^{-1}) \qquad (x \in G),$$

the star denoting Hilbert space adjoint. $\mathbf{U}^s$ is therefore unitary when s is pure imaginary.

Given a finite subset F of $\hat{K}$, write

$$\mathscr{E}(\vartheta, F) = \sum_{\delta \in F} \oplus \mathscr{E}(\vartheta, \delta).$$

If $\mathbf{\Phi} = \sum_{\delta \in F} \mathbf{\Phi}_\delta$ $(\mathbf{\Phi}_\delta \in \mathscr{E}(\vartheta, \delta))$ is an element of $\mathscr{E}(\vartheta, F)$, then we put by definition

$$\mathbf{E}(\mathbf{\Phi}:s:x) = \sum_{\delta \in F} \mathbf{E}(\mathbf{\Phi}_\delta:s:x).$$

LEMMA 3.5. *Suppose that α is a left K-finite function in $C_c^\infty(G)$. Fix a pair (ϑ, δ) and an element $\mathbf{\Phi} \in \mathscr{E}(\vartheta, \delta)$. Then*

$$L_{G/\Gamma}(\alpha)\mathbf{E}(\mathbf{\Phi}:s:?)(x) = \mathbf{E}(\mathbf{U}^{\vartheta,s}(\alpha)\mathbf{\Phi}:s:x)$$

as meromorphic functions of s.

Proof. Since α is a left K-finite function, there exists a finite subset F of $\hat{K}$ such that

$$\alpha = \bar{\chi}_F * \alpha,$$

where $\bar{\chi}_F = \sum_{\Delta \in F} \bar{\chi}_\Delta$. The operator

$$\mathbf{U}^{\vartheta,s}(\bar{\chi}_F) = \int_K \bar{\chi}_F(k)\mathbf{U}^{\vartheta,s}(k)\, d_K(k)$$

projects $\mathscr{E}(\vartheta)$ onto $\mathscr{E}(\vartheta, F)$, so

$$\mathbf{U}^{\vartheta,s}(\alpha)\mathbf{\Phi} = \mathbf{U}^{\vartheta,s}(\bar{\chi}_F * \alpha)\mathbf{\Phi}$$

is a vector in $\mathscr{E}(\vartheta, F)$. The right-hand side of the claimed equality thus is meaningful. On the other hand, the left-hand side admits the obvious formal interpretation. This said, write

$$\mathbf{U}^{\vartheta,s}(\bar{\chi}_F * \alpha)\mathbf{\Phi} = \sum_{\Delta \in F} \mathbf{\Phi}_\Delta,$$

where

$$\mathbf{\Phi}_\Delta = \int_G (\bar{\chi}_\Delta * \alpha)(x)\mathbf{U}^{\vartheta,s}(x)\mathbf{\Phi}\, d_G(x).$$

Assume now that $\operatorname{Re}(s) < -|\rho|$. By definition, we have

$$\mathbf{E}(\mathbf{U}^{\vartheta,s}(\alpha)\mathbf{\Phi}:s:x) = \sum_{\Delta \in F} \mathbf{E}(\mathbf{\Phi}_\Delta:s:x).$$

In turn

$$\mathbf{E}(\mathbf{\Phi}_\Delta:s:x) = \sum_{i=1}^{r} E(P_i:\Phi_\Delta^i:s:x),$$

Φ_Δ^i the ith component of $\mathbf{\Phi}_\Delta$ in $\mathscr{E}(\vartheta_i, \Delta)$. Explicating $E(P_i:\Phi_\Delta^i:s:x)$ gives

$$\begin{aligned} E(P_i:\Phi_\Delta^i:s:x) &= \sum_{\gamma \in \Gamma/\Gamma \cap P_i} e^{(s-|\rho|)(H_i(x\gamma))} \cdot \Phi_\Delta^i(x\gamma) \\ &= \sum_{\gamma \in \Gamma/\Gamma \cap P_i} e^{(s-|\rho|)(H_i(x\gamma))} \cdot \int_G (\bar{\chi}_\Delta * \alpha)(y) U^{\vartheta_i,s}(y)\Phi_i(x\gamma)\, d_G(y), \end{aligned}$$

Φ_i the ith component of $\mathbf{\Phi}$ in $\mathscr{E}(\vartheta_i, \delta)$. Using the definition of $U^{\vartheta_i,s}$, we find that

$$\begin{aligned} &\int_G (\bar{\chi}_\Delta * \alpha)(y) U^{\vartheta_i,s}(y)\Phi_i(x\gamma)\, d_G(y) \\ &\quad = e^{-(s-|\rho|)(H_i(x\gamma))} \cdot \int_G (\bar{\chi}_\Delta * \alpha)(y) e^{(s-|\rho|)(H_i(y^{-1}x\gamma))}\Phi_i(y^{-1}x\gamma)\, d_G(y). \end{aligned}$$

Thus

$$E(P_i:\Phi_\Delta^i:s:x) = \int_G (\bar{\chi}_\Delta * \alpha)(y) E(P_i:\Phi_i:s:y^{-1}x)\, d_G(y)$$

and so

$$\mathbf{E}(\mathbf{\Phi}_\Delta:s:x) = \int_G (\bar{\chi}_\Delta * \alpha)(y) \mathbf{E}(\mathbf{\Phi}:s:y^{-1}x)\, d_G(y).$$

Finally then

$$\begin{aligned} \mathbf{E}(\mathbf{U}^{\vartheta,s}(\alpha)\mathbf{\Phi}:s:x) &= \sum_{\Delta \in F} \mathbf{E}(\mathbf{\Phi}_\Delta:s:x) \\ &= \int_G (\bar{\chi}_F * \alpha)(y) \mathbf{E}(\mathbf{\Phi}:s:y^{-1}x)\, d_G(y) \\ &= L_{G/\Gamma}(\alpha)\mathbf{E}(\mathbf{\Phi}:s:?)(x). \end{aligned}$$

This equality has been established under the assumption that $\operatorname{Re}(s) < -|\rho|$. The extension to general s is handled by analytic continuation. ■

Given an orbit ϑ, let us agree to write $\mathscr{E}_K(\vartheta)$ for the algebraic direct sum of the $\mathscr{E}(\vartheta, \delta)$—then we can define an endomorphism

$$\mathbf{c}_\vartheta(s): \mathscr{E}_K(\vartheta) \to \mathscr{E}_K(\vartheta)$$

by the requirement

$$\mathbf{c}_\vartheta(s) \,|\, \mathscr{E}(\vartheta, \delta) = \mathbf{c}_{\vartheta,\delta}(s).$$

LEMMA 3.6. *Suppose that α is a left K-finite function in $C_c^\infty(G)$—then, for any orbit ϑ,*

$$\mathbf{c}_\vartheta(s)\mathbf{U}^{\vartheta,s}(\alpha) = \mathbf{U}^{\vartheta,-s}(\alpha)\mathbf{c}_\vartheta(s)$$

as meromorphic functions of s.

Proof. Observe first that the assumption "α left K-finite" ensures that both $\mathbf{U}^{\vartheta,s}(\alpha)$ and $\mathbf{U}^{\vartheta,-s}(\alpha)$ are in fact endomorphisms of $\mathscr{E}_K(\vartheta)$. This being so, fix δ and $\mathbf{\Phi} \in \mathscr{E}(\vartheta, \delta)$—then, using the functional equation for Eisenstein series, we derive that

$$\begin{aligned}\mathbf{E}(\mathbf{c}_\vartheta(s)\mathbf{U}^{\vartheta,s}(\alpha)\mathbf{\Phi}: -s:x) &= \mathbf{E}(\mathbf{U}^{\vartheta,s}(\alpha)\mathbf{\Phi}:s:x)\\ &= L_{G/\Gamma}(\alpha)\mathbf{E}(\mathbf{\Phi}:s:?)(x)\\ &= L_{G/\Gamma}(\alpha)\mathbf{E}(\mathbf{c}_\vartheta(s)\mathbf{\Phi}:-s:?)(x)\\ &= \mathbf{E}(\mathbf{U}^{\vartheta,-s}(\alpha)\mathbf{c}_\vartheta(s)\mathbf{\Phi}:-s:x).\end{aligned}$$

Since $\mathbf{E}(?:s:x)$ is not identically zero in s and x, it therefore follows that

$$\mathbf{c}_\vartheta(s)\mathbf{U}^{\vartheta,s}(\alpha)\mathbf{\Phi} = \mathbf{U}^{\vartheta,-s}(\alpha)\mathbf{c}_\vartheta(s)\mathbf{\Phi}$$

for all $\mathbf{\Phi} \in \mathscr{E}(\vartheta, \delta)$, thereby proving the lemma. ■

Suppose now that s is pure imaginary—then $\mathbf{U}^{\vartheta,s}$ and $\mathbf{U}^{\vartheta,-s}$ are unitary representations which, moreover, are equivalent. Because each

$$\mathbf{c}_{\vartheta,\delta}(s): \mathscr{E}(\vartheta, \delta) \to \mathscr{E}(\vartheta, \delta)$$

is unitary, one may actually view $\mathbf{c}_\vartheta(s)$ as operating on the orthogonal direct sum of the $\mathscr{E}(\vartheta, \delta)$. The preceding lemma may then be interpreted as saying that $\mathbf{c}_\vartheta(s)$ is a unitary intertwining operator for the equivalent representations $\mathbf{U}^{\vartheta,s}$ and $\mathbf{U}^{\vartheta,-s}$. Agreeing to write $\mathscr{E}^?(\vartheta)$ in place of $\mathscr{E}(\vartheta)$ to emphasize that $\mathscr{E}^?(\vartheta)$ is the representation space for $\mathbf{U}^{\vartheta,?}$, there is thus a commutative diagram

$$\begin{array}{ccc}\mathscr{E}^s(\vartheta) & \xrightarrow{\mathbf{c}_\vartheta(s)} & \mathscr{E}^{-s}(\vartheta)\\ \Big\downarrow{\scriptstyle \mathbf{U}^{\vartheta,s}} & & \Big\downarrow{\scriptstyle \mathbf{U}^{\vartheta,-s}}\\ \mathscr{E}^s(\vartheta) & \xrightarrow[\mathbf{c}_\vartheta(s)]{} & \mathscr{E}^{-s}(\vartheta)\end{array}$$

Fix an orbit ϑ, and consider

$$\sum_\delta \oplus L_d^2(\vartheta, \delta), \qquad \sum_\delta \oplus L_c^2(\vartheta, \delta).$$

Since the left K-finite functions are dense in $C_c^\infty(G)$, Lemma 3.5 implies that each of these subspaces is $L_{G/\Gamma}$-invariant. Therefore ${}_0L_d^2(G/\Gamma)$ and $L_c^2(G/\Gamma)$ are $L_{G/\Gamma}$-invariant. We claim that ${}_0L_d^2(G/\Gamma)$ is discretely decomposable with finite multiplicities. To see this, it suffices to show that the corresponding representation of G on ${}_0L_d^2(G/\Gamma)$ is K-finite (cf. Warner [34a, pp. 334–335]). Fix δ—then it is a question of verifying that the dimension of

$$\sum_{\vartheta} \oplus L_d^2(\vartheta, \delta)$$

is finite. We have seen above that $L_d^2(\vartheta, \delta)$ is spanned by the

$$\mathbf{E}_i(\mathbf{\Phi} : x)(\mathbf{\Phi} \in \mathscr{E}(\vartheta, \delta)),$$

i running between 1 and $n(\vartheta, \delta)$. Since $n(\vartheta, \delta) = 0$ unless $\vartheta = \{\sigma\}$ with σ ramified (cf. the discussion following Lemma 2.1), it follows that

$$\dim(L_d^2(\vartheta, \delta)) \leqslant r \cdot d(\delta) \cdot (d(\sigma) \cdot [\delta : \sigma] \cdot n(\{\sigma\}, \delta)).$$

Write

$$\delta | M = \sum_{i=1}^{m(\delta)} [\delta : \sigma_i] \sigma_i.$$

Then $d(\sigma_i)$, $[\delta : \sigma_i] \leqslant d(\delta)$, so the dimension of the δth-isotypic component of ${}_0L_d^2(G/\Gamma)$ cannot exceed

$$r \cdot d(\delta)^3 \cdot \sum n(\{\sigma_i\}, \delta),$$

the sum being taken over the ramified σ_i in $\delta | M$. Therefore ${}_0L_d^2(G/\Gamma)$ is discretely decomposable with finite multiplicities. Let us turn now to the decomposition of $L_c^2(G/\Gamma)$. Call $L_{G/\Gamma}^c$ the restriction of $L_{G/\Gamma}$ to $L_c^2(G/\Gamma)$—then we claim that

$$L_c^2(G/\Gamma) \simeq \sum_{\vartheta} \oplus \int_{\mathrm{Re}(s) \geqslant 0} \oplus \mathscr{E}^s(\vartheta) |ds|, \qquad L_{G/\Gamma}^c \simeq \sum_{\vartheta} \oplus \int_{\mathrm{Re}(s) \geqslant 0} \oplus \mathbf{U}^{\vartheta, s} |ds|.$$

In essence, the claim is merely a summary of results already obtained. In fact, fix a ϑ and consider the direct integral

$$\int_{\mathrm{Re}(s) \geqslant 0} \oplus \mathscr{E}^s(\vartheta) |ds|.$$

The δth-isotypic component is

$$\int_{\mathrm{Re}(s) \geqslant 0} \oplus \mathscr{E}^s(\vartheta, \delta) |ds|.$$

There is an obvious isometric isomorphism

$$\int_{\mathrm{Re}(s) \geqslant 0} \oplus \mathscr{E}^s(\vartheta, \delta) |ds| \to \mathscr{L}_c^2(\vartheta, \delta),$$

namely the rule which assigns to each

$$f \in \int_{\mathrm{Re}(s) \geqslant 0} \oplus \mathscr{E}^s(\vartheta, \delta)\,|ds|$$

the function

$$f_c \in \mathscr{L}_c^2(\vartheta, \delta)$$

defined by

$$f_c(\sqrt{-1}\,\eta) = \begin{cases} (\pi/2)^{1/2} \cdot f(\sqrt{-1}\,\eta) & \text{if} \quad \eta \geqslant 0, \\ (\pi/2)^{1/2} \cdot \mathbf{c}_{\vartheta,\delta}(-\sqrt{-1}\,\eta) f(-\sqrt{-1}\,\eta) & \text{if} \quad \eta < 0. \end{cases}$$

Therefore

$$\int_{\mathrm{Re}(s) \geqslant 0} \oplus \mathscr{E}^s(\vartheta)\,|ds|$$

is isometrically isomorphic with

$$\sum_{\delta} \oplus \mathscr{L}_c^2(\vartheta, \delta).$$

Let α be a left K-finite function in $C_c^\infty(G)$—then the equivariance of our mappings is expressed by the commutativity of the diagram

$$\begin{array}{ccc}
\int_{\mathrm{Re}(s) \geqslant 0} \oplus \mathscr{E}^s(\vartheta)\,|ds| & \xrightarrow{\int_{\mathrm{Re}(s) \geqslant 0} \mathbf{U}^{\vartheta,s}(\alpha)|ds|} & \int_{\mathrm{Re}(s) \geqslant 0} \oplus \mathscr{E}^s(\vartheta)\,|ds| \\
\downarrow & & \downarrow \\
\sum_{\delta} \oplus \mathscr{L}_c^2(\vartheta, \delta) & & \sum_{\delta} \oplus \mathscr{L}_c^2(\vartheta, \delta) \\
\downarrow & & \downarrow \\
\sum_{\delta} \oplus L_c^2(\vartheta, \delta) & \xrightarrow[L_{G/\Gamma}(\alpha)]{} & \sum_{\delta} \oplus L_c^2(\vartheta, \delta)
\end{array}$$

The claim thus follows upon summing over ϑ.

4. Removal of the Continuous Spectrum

In order to orient ourselves, let us begin with a few words of motivation. We have seen that the theory of Eisenstein series leads to the orthogonal decomposition

$$L^2(G/\Gamma) = L_d^2(G/\Gamma) \oplus L_c^2(G/\Gamma),$$

where the spectrum in $L^2_{\mathrm{d}}(G/\Gamma)$ is discrete and the spectrum in $L^2_{\mathrm{c}}(G/\Gamma)$ is continuous. Since $L^2_{\mathrm{c}}(G/\Gamma)$ admits an explicit direct integral decomposition in terms of multiples of the principal series, there is little more that can be said in this direction. The chief remaining problem in the theory is thus the study of $L^2_{\mathrm{d}}(G/\Gamma)$, where, unfortunately, little progress has been made. Suppose that α is a sufficiently regular integrable function on G (in a sense to be made precise below)—then $L_{G/\Gamma}(\alpha)$ is an integral operator on $L^2(G/\Gamma)$ with kernel

$$K_\alpha(x, y) = \sum_{\gamma \in \Gamma} \alpha(x\gamma y^{-1}).$$

Because of the presence of the continuous spectrum, $L_{G/\Gamma}(\alpha)$ is not of the trace class. The idea then is to remove the contribution of the continuous spectrum to K_α so as to come up with a trace class operator. What actually happens is this. Write

$$L_{G/\Gamma}(\alpha) = L^{\mathrm{d}}_{G/\Gamma}(\alpha) + L^{\mathrm{c}}_{G/\Gamma}(\alpha),$$

where $L^{\mathrm{d}}_{G/\Gamma}(\alpha)$ (respectively $L^{\mathrm{c}}_{G/\Gamma}(\alpha)$) is the restriction of $L_{G/\Gamma}(\alpha)$ to $L^2_{\mathrm{d}}(G/\Gamma)$ (respectively $L^2_{\mathrm{c}}(G/\Gamma)$). Then

$$L^{\mathrm{d}}_{G/\Gamma}(\alpha) = L_{G/\Gamma}(\alpha) - L^{\mathrm{c}}_{G/\Gamma}(\alpha).$$

It turns out that $L^{\mathrm{c}}_{G/\Gamma}(\alpha)$ is also an integral operator with a kernel which is directly expressible in terms of Eisenstein series. When this is removed from $L_{G/\Gamma}(\alpha)$, what is left, i.e., $L^{\mathrm{d}}_{G/\Gamma}(\alpha)$, is an integral operator which, moreover, is of the trace class. In this way, Selberg's trace formula becomes available for the study of $L^2_{\mathrm{d}}(G/\Gamma)$.

For our purposes, it will not suffice to work just with $C_c^\infty(G)$. Instead, we shall need a larger class, namely Harish-Chandra's space of integrable rapidly decreasing functions, denoted by $\mathscr{C}^1(G)$. Given $x \in G$, let

$$\sigma(x) = \text{distance between } K \text{ and } xK \text{ in } G/K,$$

$$\Xi(x) = \int_K a_{xk}^{-\rho}\, dk.$$

For any $\alpha \in C^\infty(G)$, write

$${}_{D_1}|\alpha|_{r,D_2} = \sup_{x \in G} (1 + \sigma(x))^r\, \Xi^{-2}(x) |\alpha(D_1; x; D_2)| \quad (D_1, D_2 \in \mathfrak{G}; r \in \mathbf{R}).$$

The symbol $\alpha(D_1; x; D_2)$ $(x \in G;\ D_1, D_2 \in \mathfrak{G})$ is standard notation in the subject (cf. Warner [34b, p. 104]). Call $\mathscr{C}^1(G)$ the set of all α in $C^\infty(G)$ such that ${}_{D_1}|\alpha|_{r,D_2} < +\infty$ for all $r \in \mathbf{R}$ and pairs $(D_1, D_2) \in \mathfrak{G} \times \mathfrak{G}$. Topologize $\mathscr{C}^1(G)$ by means of the seminorms ${}_{D_1}|?|_{r,D_2}$—then $\mathscr{C}^1(G)$ becomes a locally convex, complete, Hausdorff, topological vector space. The inclusion $C_c^\infty(G) \hookrightarrow \mathscr{C}^1(G)$ is continuous with dense range. There are also continuous inclusions $\mathscr{C}^1(G) \hookrightarrow L^i(G)$ $(i = 1, 2)$.

Let ϕ be a continuous function on G—then ϕ is said to be of *regular growth* if there exist a nonnegative integrable function ϕ_0 on G and a compact symmetric neighborhood $\mathscr{N}$ of the identity in G such that for all $x \in G$ we have

$$|\phi(x)| \leqslant \int_{\mathscr{N}} \phi_0(yx)\, d_G(y).$$

Call $\chi_{\mathscr{N}}$ the characteristic function of $\mathscr{N}$—then the condition of regular growth can be expressed by the relation

$$|\phi| \leqslant \chi_{\mathscr{N}} * \phi_0.$$

Any function of regular growth is clearly integrable. If ϕ is a continuous compactly supported function, then ϕ is of regular growth. In fact let $\mathscr{N}$ be any compact symmetric neighborhood of the identity in G—then we can take for ϕ_0 the function defined by the rule

$$\phi_0(x) = \frac{1}{\operatorname{meas}(\mathscr{N})} \cdot \max_{y \in \mathscr{N}} |\phi(y^{-1}x)|.$$

Another point is this. Suppose it is known that $|\phi| \leqslant \phi_0'$ where ϕ_0' itself is of regular growth, i.e., $\phi_0' \leqslant \chi_{\mathscr{N}} * {}^{\backprime}\phi_0$—then $\phi * \phi$ is of regular growth. The following simple proof was shown to us by R. Gangolli:

$$\begin{aligned} |\phi * \phi| &\leqslant |\phi| * |\phi| \\ &\leqslant \phi_0' * \phi_0' \\ &\leqslant (\chi_{\mathscr{N}} * {}^{\backprime}\phi_0) * \phi_0' = \chi_{\mathscr{N}} * ({}^{\backprime}\phi_0 * \phi_0'). \end{aligned}$$

LEMMA 4.1. *Let $\alpha \in \mathscr{C}^1(G)$—then α is of regular growth.*

Proof. This result is known (cf. Gangolli–Warner [10, pp. 335–337]) but we shall give here a simpler proof based on a suggestion of Harish-Chandra. Let $\mathscr{N}$ be any compact symmetric neighborhood of the identity in G. Fix a real number $r > 0$ such that (cf. Warner [34b, p. 156])

$$\int_G \vdash\!\circ\!\dashv^2(y)(1 + \sigma(y))^{-r}\, d_G(y) < \infty.$$

Fix a real number $M > 0$ such that (cf. Warner [34b, p. 153])

$$\vdash\!\circ\!\dashv^2(x) \leqslant M \vdash\!\circ\!\dashv^2(yx) \qquad (x \in G,\ y \in \mathscr{N}).$$

Fix a real number $N > 0$ such that (cf. Warner [34b, p. 67])

$$(1 + \sigma(x))^{-r} \leqslant N(1 + \sigma(yx))^{-r} \qquad (x \in G,\ y \in \mathcal{N}).$$

Then it is clear that

$$\vdash\!\circ\!\dashv^2(x)(1 + \sigma(x))^{-r} \leqslant \frac{M \cdot N}{\operatorname{meas}(\mathcal{N})} \cdot \int_{\mathcal{N}} \vdash\!\circ\!\dashv^2(yx)(1 + \sigma(yx))^{-r}\, d_G(y).$$

Since there exists a constant $C > 0$ such that

$$|\alpha(x)| \leqslant C \cdot \vdash\!\circ\!\dashv^2(x)(1 + \sigma(x))^{-r} \qquad (x \in G),$$

it follows that α is of regular growth. ■

Remark. It is important to note that the above proof shows that any element of $\mathscr{C}^1(G)$ admits a majorant which itself is of regular growth. So, in view of the observation made directly before Lemma 4.1, given $\alpha \in \mathscr{C}^1(G)$, the convolution $\alpha * \alpha$, which a priori need not lie in $\mathscr{C}^1(G)$, is of regular growth. [In reality, one can prove that $\mathscr{C}^1(G)$ is closed under convolution; see the Appendix to this section.]

LEMMA 4.2. *Let α be of regular growth—then there is a constant $C > 0$ and an integer $n \geqslant 0$ such that*

$$\sum_{\gamma \in \Gamma} |\alpha(x\kappa_i\gamma y)| \leqslant C(\xi_\lambda(a_x))^{-n}$$

for all $x \in \mathfrak{S}_{t_0,\omega_0}$, $\kappa_i \in \mathfrak{s}$, $y \in G$.

Proof. By definition, there exist a nonnegative integrable function α_0 on G and a compact symmetric neighborhood $\mathcal{N}$ of the identity in G such that for all $x \in G$

$$|\alpha(x)| \leqslant \int_{\mathcal{N}} \alpha_0(yx)\, d_G(y).$$

This being so, fix $x \in \mathfrak{S}_{t_0,\omega_0}$, $\kappa_i \in \mathfrak{s}$, $y \in G$—then

$$|\alpha(x\kappa_i\gamma y)| \leqslant \int_{\mathcal{N}} \alpha_0(zx\kappa_i\gamma y)\, d_G(z).$$

Now obviously

$$\mathcal{N}x\kappa_i\gamma_1 y \cap \mathcal{N}x\kappa_i\gamma_2 y \neq \varnothing \Leftrightarrow \gamma_2\gamma_1^{-1} \in \kappa_i^{-1}x^{-1}\mathcal{N}^{-1}\mathcal{N}x\kappa_i.$$

Because Γ is discrete, the intersection

$$\Gamma \cap \kappa_i^{-1}x^{-1}\mathcal{N}^{-1}\mathcal{N}x\kappa_i$$

is finite and in fact (cf. Harish-Chandra [13a, p. 9]) its cardinality is bounded by

$$M(\xi_\lambda(a_x))^{-n}$$

for some $M > 0$ and some integer $n \geqslant 0$. It therefore follows that

$$\sum_{\gamma \in \Gamma} |\alpha(x\kappa_i\gamma y)| \leqslant M(\xi_\lambda(a_x))^{-n} \cdot \int_G \alpha_0(z)\, d_G(z),$$

which completes the proof of the lemma. ∎

Let $\alpha \in \mathscr{C}^1(G)$. Write

$$K_\alpha(x, y) = \sum_{\gamma \in \Gamma} \alpha(x\gamma y^{-1}).$$

Then according to what has been said above, the series defining K_α is absolutely convergent. In fact an easy variant of the preceding argument implies that the convergence is actually uniform on compacta, so K_α is continuous and even smooth. For any $f \in L^2(G/\Gamma)$, we have

$$(L_{G/\Gamma}(\alpha)f)(x) = \int_{G/\Gamma} K_\alpha(x, y) f(y)\, d_G(y)$$

which means that $L_{G/\Gamma}(\alpha)$ is an integral operator on $L^2(G/\Gamma)$ with kernel K_α. K_α may be viewed as a function on $G/\Gamma \times G/\Gamma$ The estimate provided by Lemma 4.2 then says that K_α is slowly increasing. This turns out to be a key point, as will become apparent during the proof of the following theorem.

THEOREM 4.3. *Let* $\alpha \in \mathscr{C}^1(G)$*—then the operator* $L^0_{G/\Gamma}(\alpha)$ *is of the trace class.*

[The reader should compare this assertion with Lemma 3.1.]

We shall preface the proof with an important technical result which was kindly communicated to us by G. B. Folland.

THEOREM 4.4. *Let G be a Lie group. Let Δ be an elliptic operator on G. Given an integer* $p \geqslant 1$, *there exist an integer* $N \geqslant 1$ *and* $\mu \in C_c^p(G)$, $\nu \in C_c^\infty(G)$ *such that*

$$\Delta^N \cdot \mu = \delta + \nu$$

in the sense of the theory of distributions on G.

[Here δ is the Dirac measure at the identity on G.]

Proof. Let k be the order of Δ and n the dimension of G. Choose $N > (n + p)/k$. The assertion being local, we shall work in a coordinate patch $\mathscr{N} \simeq \mathbf{R}^n$ centered at the identity of G with coordinates $x_1, \ldots, x_n$. Given $f \in C_c^\infty(\mathscr{N})$, write $\hat{f}$ for its Fourier transform:

$$\hat{f}(\xi) = (2\pi)^{-n/2} \int e^{-\sqrt{-1}x \cdot \xi} f(x)\, dx,$$

dx being Lebesgue measure. Let P be the transpose of Δ^N on $\mathcal{N}$ with respect to Lebesgue measure, i.e.,

$$\int Pf(x) \cdot g(x)\,dx = \int f(x) \cdot \Delta^N g(x)\,dx$$

for $f, g \in C_c^\infty(\mathcal{N})$. Since P is elliptic, there is a pseudodifferential operator Q of order Nk such that $QP - I$ is a smoothing operator. In precise terms,

$$Qf(x) = (2\pi)^{-n/2} \int e^{\sqrt{-1}x \cdot \xi} \sigma(x, \xi) \hat{f}(\xi)\,d\xi,$$

where $\sigma \in C^\infty(\mathbf{R}^n \times \mathbf{R}^n)$ with $|\sigma(x, \xi)| = 0(\|\xi\|^{-Nk})$ as $\|\xi\| \to \infty$, and

$$QPf(x) = f(x) + (2\pi)^{-n/2} \int e^{\sqrt{-1}x \cdot \xi} \tau(x, \xi) \hat{f}(\xi)\,d\xi,$$

where $\tau \in C^\infty(\mathbf{R}^n \times \mathbf{R}^n)$ with $|\tau(x, \xi)| = 0(\|\xi\|^{-M})$ for all $M > 0$ as $\|\xi\| \to \infty$. Set

$$\begin{aligned} u(x) &= (2\pi)^{-n} \int e^{-\sqrt{-1}x \cdot \xi} \sigma(0, \xi)\,d\xi \\ &= (2\pi)^{-n/2} \hat{\sigma}(0, x). \end{aligned}$$

Owing to the estimate $|\sigma(0, \xi)| \leqslant C\|\xi\|^{-Nk}$ (some $C > 0$) and the fact that $Nk > n + p$, the integral defining u is absolutely convergent, as are the integrals

$$(\partial/\partial x)^q u(x) = (2\pi)^{-n} \int e^{-\sqrt{-1}x \cdot \xi} (-\sqrt{-1}\,\xi)^q \sigma(0, \xi)\,d\xi$$

for all $q = (q_1, \ldots, q_n)$ with $\sum q_i \leqslant p$. Therefore, $u \in C^p(\mathcal{N})$. Similar reasoning implies that the function v defined by the rule

$$v(x) = (2\pi)^{-n} \int e^{-\sqrt{-1}x \cdot \xi} \tau(0, \xi)\,d\xi = (2\pi)^{-n/2} \hat{\tau}(0, x)$$

is in $C^\infty(\mathcal{N})$. If now $f \in C_c^\infty(\mathcal{N})$, then

$$(2\pi)^{-n/2} \int \sigma(0, \xi) \widehat{Pf}(\xi)\,d\xi = QPf(0) = f(0) + (2\pi)^{-n/2} \int \tau(0, \xi) \hat{f}(\xi)\,d\xi.$$

Because Fourier transformation is a unitary operation, we thus have

$$\int u(x) Pf(x)\,dx = f(0) + \int v(x) f(x)\,dx,$$

so, f being arbitrary,

$$\Delta^N \cdot u = \delta + v$$

on $\mathcal{N}$. To complete the proof, choose $\phi \in C_c^\infty(\mathcal{N})$ with $\phi = 1$ on a neighborhood of 0, and put $\mu = \phi u$—then $\mu \in C_c^p(\mathcal{N})$ and $v = \Delta^N \cdot (\phi u) - \delta \in C_c^\infty(\mathcal{N})$ (since $[\Delta^N, \phi]$ vanishes near 0 while u is smooth away from 0). μ and v extend trivially to all of G, and we are done. ∎

Let $D \in \mathfrak{G}$—then D may be viewed in the usual way as either a right or left invariant differential operator on G, the two possibilities being distinguished by writing either D_{R} or D_{L}. It is well known that one can find in $\mathfrak{G}$ an element Δ which is elliptic and, moreover, K-central (cf. Warner [34a, p. 269]). This said, fix an integer $p \geqslant 1$—then, in the notations of the preceding theorem, we have

$$\Delta_{\mathrm{R}}^{N} \cdot \mu = \delta + \nu.$$

Since Δ is K-central, one can assume in addition that both μ and ν are K-central. Let now α be a C^{∞} function on G—then

$$\alpha = \alpha * \Delta_{\mathrm{R}}^{N} \cdot \mu - \alpha * \nu,$$

the star denoting convolution of distributions. The operation $\Delta_{\mathrm{R}}^{N} \cdot \mu$ can be regarded as convolution of distributions (cf. Warner [34a, p. 491]):

$$\Delta_{\mathrm{R}}^{N} \cdot \mu = \Delta_{\mathrm{R}}^{N} * \mu.$$

It then follows immediately that

$$\alpha = (\alpha * \Delta_{\mathrm{R}}^{N}) * \mu - \alpha * \nu = (\Delta_{\mathrm{L}}^{N}\alpha) * \mu - \alpha * \nu.$$

In other words:

LEMMA 4.5. *Let* $\alpha \in C^{\infty}(G)$*—then there exist* $\beta \in C^{\infty}(G)$, $\mu \in C_{\mathrm{c}}^{p}(G)$, $\nu \in C_{\mathrm{c}}^{\infty}(G)$ *such that*

$$\alpha = \beta * \mu + \alpha * \nu.$$

Proof of Theorem 4.3. Let $\alpha \in \mathscr{C}^{1}(G)$—then, by the above, we can write $\alpha = \beta * \mu + \alpha * \nu$, where, of course $\beta \in \mathscr{C}^{1}(G)$ (β is a certain derivative of α). We have

$$L_{G/\Gamma}^{0}(\alpha) = L_{G/\Gamma}^{0}(\beta) \cdot L_{G/\Gamma}^{0}(\mu) + L_{G/\Gamma}^{0}(\alpha) \cdot L_{G/\Gamma}^{0}(\nu).$$

Thanks to Lemma 3.1, both $L_{G/\Gamma}^{0}(\mu)$ and $L_{G/\Gamma}^{0}(\nu)$ are of the Hilbert–Schmidt class. So, to conclude that $L_{G/\Gamma}^{0}(\alpha)$ is of the trace class, we need only establish that $L_{G/\Gamma}^{0}$ represents the elements of $\mathscr{C}^{1}(G)$ by Hilbert–Schmidt operators. However, because of the estimate provided by Lemma 4.2, this can be done by following an argument of Harish-Chandra [13a, pp. 13–14] (who considers only the case $C_{\mathrm{c}}^{\infty}(G)$) line by line. ■

Remark. The assignment

$$\alpha \mapsto \operatorname{tr}(L_{G/\Gamma}^{0}(\alpha)) \qquad (\alpha \in \mathscr{C}^{1}(G))$$

is continuous in the topology of $\mathscr{C}^{1}(G)$. The proof depends on an unpublished observation of Harish-Chandra which goes as follows. In the notations of Warner [34a, p. 269], take for Δ the operator $1 - \omega_{-} - \omega_{+}$. Given $U \in \hat{G}$, write $[U:\delta]$ for the multiplicity of δ in $U|K$. Suppose that $[U:\delta] \geqslant 1$ so that the δth-isotypic component of U is nontrivial—then, as is well known, $U(\Delta)$ operates as a scalar ($\geqslant 1$) there, call it $\Delta_{U,\delta}$. On the other hand, if $[U:\delta] = 0$, then we set $\Delta_{U,\delta} = 1$. This said, write

$$L_{G/\Gamma}^0 = \sum_{U \in \hat{G}} \oplus\, m_U U.$$

Then, in the notations of Theorem 4.4, we claim that

$$\sum_{U \in \hat{G}} m_U \sum_{\delta \in \hat{K}} d(\delta)[U:\delta]\, \Delta_{U,\delta}^{-2N} < +\infty.$$

To verify this, pick an orthonormal basis $\{e_{U,i}\}$ for the representation space of U by picking an orthonormal basis $\{e_{U,i} : i \in I_U(\delta)\}$ in each δth-isotypic component. Since $L_{G/\Gamma}^0(\mu)$ is Hilbert–Schmidt, we have then

$$\sum_{U \in \hat{G}} m_U \sum_{\delta \in \hat{K}} \sum_{i \in I_U(\delta)} (U(\mu)e_{U,i}, U(\mu)e_{U,i}) < +\infty.$$

It will be convenient to assume that

$$\Delta_{\mathrm{L}}^N \cdot \mu = \delta + \nu.$$

Then

$$\begin{aligned}
&\sum_{U \in \hat{G}} m_U \sum_{\delta \in \hat{K}} \sum_{i \in I_U(\delta)} (U(\mu)e_{U,i}, U(\mu)e_{U,i}) \\
&\quad = \sum_{U \in \hat{G}} m_U \sum_{\delta \in \hat{K}} \sum_{i \in I_U(\delta)} \Delta_{U,\delta}^{-2N}(U(\mu * \Delta_{\mathrm{L}}^N)e_{U,i}, U(\mu * \Delta_{\mathrm{L}}^N)e_{U,i}) \\
&\quad = \sum_{U \in \hat{G}} m_U \sum_{\delta \in \hat{K}} \sum_{i \in I_U(\delta)} \Delta_{U,\delta}^{-2N}((U(\delta) + U(\nu))e_{U,i}, (U(\delta) + U(\nu))e_{U,i}) \\
&\quad = \sum_{U \in \hat{G}} m_U \sum_{\delta \in \hat{K}} d(\delta)[U:\delta]\, \Delta_{U,\delta}^{-2N} + C_\nu,
\end{aligned}$$

C_ν a finite constant, $L_{G/\Gamma}^0(\nu)$ being of the trace class. Hence the claim. Let now $\alpha \in \mathscr{C}^1(G)$—then

$$\begin{aligned}
&\operatorname{tr}(L_{G/\Gamma}^0(\alpha)) \\
&\quad = \sum_{U \in \hat{G}} m_U \sum_{\delta \in \hat{K}} \sum_{i \in I_U(\delta)} (U(\alpha)e_{U,i}, e_{U,i}) \\
&\quad = \sum_{U \in \hat{G}} m_U \sum_{\delta \in \hat{K}} \sum_{i \in I_U(\delta)} \Delta_{U,\delta}^{-2N} \cdot \int_G (\Delta_{\mathrm{L}}^{2N}\alpha)(x)(U(x)e_{U,i}, e_{U,i})\, d_G(x).
\end{aligned}$$

The integration by parts implicit in the passage from the second to the third equality can be justified in the well-known way (cf. Eguchi [8, p. 171]). Owing to the claim above, it is clear that there exists a positive constant C with the property that

$$|\operatorname{tr}(L_{G/\Gamma}^0(\alpha))| \leqslant C \|\Delta_{\mathrm{L}}^{2N}\alpha\|_1$$

for all $\alpha \in \mathscr{C}^1(G)$. The assignment

$$\alpha \mapsto \operatorname{tr}(L^0_{G/\Gamma}(\alpha)) \qquad (\alpha \in \mathscr{C}^1(G))$$

is therefore continuous in the topology of $\mathscr{C}^1(G)$.

THEOREM 4.6. *Let α be a left K-finite function in $\mathscr{C}^1(G)$—then the operator $L^{\mathrm{d}}_{G/\Gamma}(\alpha)$ is of the trace class.*

Proof. One has

$$L^{\mathrm{d}}_{G/\Gamma}(\alpha) = L_{G/\Gamma}(\alpha)|L^2_{\mathrm{d}}(G/\Gamma).$$

According to Theorem 4.3, $L_{G/\Gamma}(\alpha)$ is of the trace class when restricted to the space of cusp forms. On the other hand, in view of the discussion at the end of Section 3 and the assumption of left K-finiteness on α, $L_{G/\Gamma}(\alpha)$ is of finite rank when restricted to the space spanned by the residues of Eisenstein series. Therefore $L^{\mathrm{d}}_{G/\Gamma}(\alpha)$ is of the trace class. ∎

Fix now a K-finite function α in $\mathscr{C}^1(G)$. The operator $L^{\mathrm{d}}_{G/\Gamma}(\alpha)$ is of the trace class and

$$L^{\mathrm{d}}_{G/\Gamma}(\alpha) = L_{G/\Gamma}(\alpha) - L^{\mathrm{c}}_{G/\Gamma}(\alpha),$$

where $L^{\mathrm{c}}_{G/\Gamma}(\alpha) = L_{G/\Gamma}(\alpha) I_{\mathrm{c}} I^*_{\mathrm{c}}$, $I_{\mathrm{c}} I^*_{\mathrm{c}}$ being the orthogonal projection of $L^2(G/\Gamma)$ onto $L^2_{\mathrm{c}}(G/\Gamma)$ (cf. Section 3). $L_{G/\Gamma}(\alpha)$ is an integral operator with kernel K_α. The same is true of $L^{\mathrm{c}}_{G/\Gamma}(\alpha)$ whose kernel K^{c}_α can be described as follows. Let $\{e_n : n \in I\}$ be an orthonormal basis for $\mathscr{E}$ where, without loss of generality, each $e_n \in \mathscr{E}(\vartheta, \delta)$ for some orbit ϑ and some K-type δ. Put

$$\mathbf{U}^s_{mn}(\alpha) = (\mathbf{U}^s(\alpha)e_n, e_m) \qquad (m, n \in I).$$

Set

$$K^{\mathrm{c}}_\alpha(x, y : s) = \frac{1}{4\pi} \sum_{m,\, n \in I} \mathbf{U}^s_{mn}(\alpha) \cdot \mathbf{E}(e_m : s : x)\overline{\mathbf{E}(e_n : s : y)},$$

$$K^{\mathrm{c}}_\alpha(x, y) = \int_{\mathrm{Re}(s)=0} K^{\mathrm{c}}_\alpha(x, y : s)\,|ds|.$$

Due to our assumption on α, the sum on the right-hand side of the first equality is finite. In these notations, the remainder of this section will be devoted to proving the following theorem.

THEOREM 4.7. *Let α be a K-finite function in $\mathscr{C}^1(G)$—then $L^{\mathrm{c}}_{G/\Gamma}(\alpha)$ is an integral operator with kernel K^{c}_α. In addition, the kernel*

$$K^{\mathrm{d}}_\alpha = K_\alpha - K^{\mathrm{c}}_\alpha$$

is integrable over the diagonal and its integral is the trace of $L^{\mathrm{d}}_{G/\Gamma}(\alpha)$, *i.e.*,

$$\operatorname{tr}(L^{\mathrm{d}}_{G/\Gamma}(\alpha)) = \int_{G/\Gamma} K^{\mathrm{d}}_{\alpha}(x, x)\, d_G(x).$$

The operator $L^{\mathrm{c}}_{G/\Gamma}(\alpha)$ is best discussed by first introducing certain $L^{\mathrm{c}}_{G/\Gamma}$-invariant subspaces of $L^2_{\mathrm{c}}(G/\Gamma)$ and deriving a formula for the kernel of $L^{\mathrm{c}}_{G/\Gamma}(\alpha)$ on them. The complete description of $L^{\mathrm{c}}_{G/\Gamma}(\alpha)$ will then be obtained by a limiting process from these partial descriptions. Given a finite collection $\mathscr{F} = \{\vartheta\}$ of orbits, write

$$\mathscr{L}^2_{\mathrm{c}}(\mathscr{F}) = \sum_{\vartheta \in \mathscr{F}} \oplus \sum_{\delta \in \hat{K}} \oplus \mathscr{L}^2_{\mathrm{c}}(\vartheta, \delta).$$

For any $T > 0$, let $\mathscr{L}^2_{\mathrm{c}}(\mathscr{F}, T)$ be the subspace of $\mathscr{L}^2_{\mathrm{c}}(\mathscr{F})$ consisting of those elements whose projection onto any one of the summands $\mathscr{L}^2_{\mathrm{c}}(\vartheta, \delta)$ is a function with support in the interval $[-\sqrt{-1}\,T, \sqrt{-1}\,T]$. Call $L^2_{\mathrm{c}}(G/\Gamma; \mathscr{F}, T)$ the image of $\mathscr{L}^2_{\mathrm{c}}(\mathscr{F}, T)$ under the map $I_{\mathrm{c};\mathscr{F},T}(I_{\mathrm{c};\mathscr{F},T} = I_{\mathrm{c}} \,|\, \mathscr{L}^2_{\mathrm{c}}(\mathscr{F}, T))$—then it is clear that $L^2_{\mathrm{c}}(G/\Gamma; \mathscr{F}, T)$ is a $L_{G/\Gamma}$-invariant subspace of $L^2_{\mathrm{c}}(G/\Gamma)$. The restriction $L^{\mathrm{c}}_{G/\Gamma}(\alpha; \mathscr{F}, T)$ of $L^{\mathrm{c}}_{G/\Gamma}(\alpha)$ to $L^2_{\mathrm{c}}(G/\Gamma; \mathscr{F}, T)$ is given by $L_{G/\Gamma}(\alpha) I_{\mathrm{c};\mathscr{F},T} I^*_{\mathrm{c};\mathscr{F},T}$, $I_{\mathrm{c};\mathscr{F},T} I^*_{\mathrm{c};\mathscr{F},T}$ being the orthogonal projection of $L^2(G/\Gamma)$ onto $L^2_{\mathrm{c}}(G/\Gamma; \mathscr{F}, T)$.

SUBLEMMA. *Let* $\mathscr{F} = \{\vartheta\}$ *be a finite collection of orbits. Let* $I(\vartheta)$ *be the indices* n *such that* $e_n \in \mathscr{E}(\vartheta, \delta)$ *for some* δ. *Fix* $T > 0$—*then* $L^{\mathrm{c}}_{G/\Gamma}(\alpha; \mathscr{F}, T)$ *is an integral operator on* $C^\infty_{\mathrm{c}}(G/\Gamma)$ *with kernel*

$$K^{\mathrm{c}}_{\alpha}(x, y : \mathscr{F}, T) = \sum_{\vartheta \in \mathscr{F}} \int_{-\sqrt{-1}\,T}^{\sqrt{-1}\,T} K^{\mathrm{c}}_{\alpha}(x, y : s : \vartheta)\, |ds|,$$

where

$$K^{\mathrm{c}}_{\alpha}(x, y : s : \vartheta) = \frac{1}{4\pi} \sum_{m,\, n \in I(\vartheta)} \mathbf{U}^s_{mn}(\alpha) \cdot \mathbf{E}(e_m : s : x) \overline{\mathbf{E}(e_n : s : y)}.$$

Proof. Let $f \in C^\infty_{\mathrm{c}}(G/\Gamma)$—then we have

$$L^{\mathrm{c}}_{G/\Gamma}(\alpha; \mathscr{F}, T) f = L_{G/\Gamma}(\alpha) I_{\mathrm{c};\mathscr{F},T} I^*_{\mathrm{c};\mathscr{F},T} f.$$

By definition, $I^*_{\mathrm{c};\mathscr{F},T} f \in \mathscr{L}^2_{\mathrm{c}}(\mathscr{F}, T)$. Let $(I^*_{\mathrm{c};\mathscr{F},T} f)_{\vartheta,\delta}$ be the component of $I^*_{\mathrm{c};\mathscr{F},T} f$ in $\mathscr{L}^2_{\mathrm{c}}(\vartheta, \delta)$—then

$$I^*_{\mathrm{c};\mathscr{F},T} f = \sum_{\vartheta,\delta} (I^*_{\mathrm{c};\mathscr{F},T} f)_{\vartheta,\delta}.$$

Thus

$$I_{\mathrm{c};\mathscr{F},T} I^*_{\mathrm{c};\mathscr{F},T} f = \sum_{\vartheta,\delta} f_{\vartheta,\delta},$$

where $f_{\vartheta,\delta} = I^{c}_{\vartheta,\delta}((I^{*}_{c;\mathscr{F},T}f)_{\vartheta,\delta})$ and so

$$L_{G/\Gamma}(\alpha)I_{c;\mathscr{F},T}I^{*}_{c;\mathscr{F},T}f = \sum_{\vartheta,\delta} L_{G/\Gamma}(\alpha)f_{\vartheta,\delta},$$

the sum on the right being finite in view of our standing supposition on α. Owing to Lemma 3.4,

$$f_{\vartheta,\delta} = \frac{1}{2\pi}\int_{-\sqrt{-1}T}^{\sqrt{-1}T} \mathbf{E}((I^{*}_{c;\mathscr{F},T}f)_{\vartheta,\delta}(s):s:?)\,|ds|.$$

Lemma 3.5 then implies that

$$L_{G/\Gamma}(\alpha)f_{\vartheta,\delta} = \frac{1}{2\pi}\int_{-\sqrt{-1}T}^{\sqrt{-1}T} \mathbf{E}(\mathbf{U}^{\vartheta,s}(\alpha)\cdot(I^{*}_{c;\mathscr{F},T}f)_{\vartheta,\delta}(s):s:?)\,|ds|.$$

Write

$$(I^{*}_{c;\mathscr{F},T}f)_{\vartheta,\delta}(s) = \sum_{n\in I(\vartheta,\delta)} ((I^{*}_{c;\mathscr{F},T}f)_{\vartheta,\delta}(s), e_n)e_n,$$

$I(\vartheta,\delta)$ the subset of $I(\vartheta)$ consisting of those indices n such that $e_n \in \mathscr{E}(\vartheta,\delta)$. Then

$$\begin{aligned}
&\mathbf{U}^{\vartheta,s}(\alpha)\cdot(I^{*}_{c;\mathscr{F},T}f)_{\vartheta,\delta}(s)\\
&\quad= \sum_{n\in I(\vartheta,\delta)} ((I^{*}_{c;\mathscr{F},T}f)_{\vartheta,\delta}(s), e_n)\cdot\mathbf{U}^{\vartheta,s}(\alpha)e_n\\
&\quad= \sum_{m\in I(\vartheta)}\sum_{n\in I(\vartheta,\delta)} ((I^{*}_{c;\mathscr{F},T}f)_{\vartheta,\delta}(s), e_n)\cdot(\mathbf{U}^{\vartheta,s}(\alpha)e_n, e_m)e_m\\
&\quad= \sum_{m\in I(\vartheta)}\sum_{n\in I(\vartheta,\delta)} ((I^{*}_{c;\mathscr{F},T}f)_{\vartheta,\delta}(s), e_n)\cdot\mathbf{U}^{s}_{mn}(\alpha)e_m.
\end{aligned}$$

Using the definitions and Lemma 3.4, we calculate without difficulty that

$$((I^{*}_{c;\mathscr{F},T}f)_{\vartheta,\delta}(s), e_n) = \frac{1}{2}\int_{G/\Gamma} f(y)\overline{\mathbf{E}(e_n:s:y)}\,d_G(y) \qquad (n\in I(\vartheta,\delta)).$$

Therefore

$$\begin{aligned}
&L_{G/\Gamma}(\alpha)f_{\vartheta,\delta}\\
&\quad= \int_{G/\Gamma}\left(\int_{-\sqrt{-1}T}^{\sqrt{-1}T}\frac{1}{4\pi}\sum_{m\in I(\vartheta)}\sum_{n\in I(\vartheta,\delta)} \mathbf{U}^{s}_{mn}(\alpha)\cdot\mathbf{E}(e_m:s:?)\overline{\mathbf{E}(e_n:s:y)}\right)f(y)\,d_G(y).
\end{aligned}$$

The conclusion of the sublemma now follows by addition. ■

LEMMA 4.8. *There is a constant $C > 0$ and an integer $n \geqslant 0$ such that*

$$\sum_{\vartheta}\int_{\mathrm{Re}(s)=0} |K^{c}_{\alpha}(x\kappa_i, y\kappa_i:s:\vartheta)|\,|ds| \leqslant C(\xi_\lambda(a_x))^{-n}(\xi_\lambda(a_y))^{-n}$$

for all $x, y \in \mathfrak{S}_{t_0,\omega_0}$, $\kappa_i \in \mathfrak{s}$.

Proof. Since α is K-finite, there is a finite subset F of $\hat{K}$ such that $\alpha = \bar{\chi}_F * \alpha * \bar{\chi}_F (\bar{\chi}_F = \sum_{\delta \in F} \bar{\chi}_\delta)$. We have seen above that $\alpha = \beta * \mu + \alpha * \nu$ where $\beta \in \mathscr{C}^1(G)$, $\mu \in C_c^p(G)$, $\nu \in C_c^\infty(G)$. Because μ, ν are K-central, $\bar{\chi}_F * \mu = \mu * \bar{\chi}_F$, $\bar{\chi}_F * \nu = \nu * \bar{\chi}_F$. There is therefore no loss of generality in assuming that β, μ, and ν are themselves K-finite: $\beta = \bar{\chi}_F * \beta * \bar{\chi}_F$, $\mu = \bar{\chi}_F * \mu * \bar{\chi}_F$, $\nu = \bar{\chi}_F * \nu * \bar{\chi}_F$. This said, fix an orbit ϑ—then there is a finite-dimensional subspace $\mathscr{E}_K^\alpha(\vartheta)$ of $\mathscr{E}_K(\vartheta)$ which contains the ranges and the orthogonal complements of the kernels of $\mathbf{U}^{\vartheta,s}(\alpha)$, $\mathbf{U}^{\vartheta,s}(\beta)$, $\mathbf{U}^{\vartheta,s}(\mu)$, $\mathbf{U}^{\vartheta,s}(\nu)$. Given s and x, there exists a vector $\mathbf{E}^\alpha(s:x) \in \mathscr{E}_K^\alpha(\vartheta)$ with the property that

$$(\mathbf{\Phi}, \mathbf{E}^\alpha(s:x)) = \mathbf{E}(\mathbf{\Phi}:s:x)$$

for all $\mathbf{\Phi} \in \mathscr{E}_K^\alpha(\vartheta)$. In what follows, we shall pretend that the same is true of arbitrary elements in $\mathscr{E}_K(\vartheta)$ but, as will be apparent, this convention is merely a notational convenience. Let us now estimate $|K_\alpha^c(x\kappa_i, y\kappa_i:s:\vartheta)|$:

$$\begin{aligned}
&|K_\alpha^c(x\kappa_i, y\kappa_i:s:\vartheta)| \\
&= \frac{1}{4\pi}\left|\sum_{m,\, n \in I(\vartheta)} \mathbf{U}_{mn}^s(\alpha) \cdot \mathbf{E}(e_m:s:x\kappa_i)\overline{\mathbf{E}(e_n:s:y\kappa_i)}\right| \\
&= \frac{1}{4\pi}\left|\sum_{m,\, n \in I(\vartheta)} (\mathbf{U}^{\vartheta,s}(\alpha)e_n, e_m) \cdot \mathbf{E}(e_m:s:x\kappa_i)\overline{\mathbf{E}(e_n:s:y\kappa_i)}\right| \\
&= \frac{1}{4\pi}\left|\sum_{n \in I(\vartheta)} \mathbf{E}\left(\sum_{m \in I(\vartheta)} (\mathbf{U}^{\vartheta,s}(\alpha)e_n, e_m)e_m:s:x\kappa_i\right)\overline{\mathbf{E}(e_n:s:y\kappa_i)}\right| \\
&= \frac{1}{4\pi}\left|\sum_{n \in I(\vartheta)} \mathbf{E}(\mathbf{U}^{\vartheta,s}(\alpha)e_n:s:x\kappa_i)\overline{\mathbf{E}(e_n:s:y\kappa_i)}\right| \\
&= \frac{1}{4\pi}\left|\sum_{n \in I(\vartheta)} (\mathbf{U}^{\vartheta,s}(\alpha)e_n, \mathbf{E}^\alpha(s:x\kappa_i))(\mathbf{E}^\alpha(s:y\kappa_i), e_n)\right| \\
&= \frac{1}{4\pi}\left|\sum_{n \in I(\vartheta)} (e_n, \mathbf{U}^{\vartheta,s}(\alpha)^*\mathbf{E}^\alpha(s:x\kappa_i))(\mathbf{E}^\alpha(s:y\kappa_i), e_n)\right| \\
&= \frac{1}{4\pi}|(\mathbf{E}^\alpha(s:y\kappa_i), \mathbf{U}^{\vartheta,s}(\alpha)^*\mathbf{E}^\alpha(s:x\kappa_i))| \\
&= \frac{1}{4\pi}|(\mathbf{E}^\alpha(s:y\kappa_i), \mathbf{U}^{\vartheta,s}(\mu)^*\mathbf{U}^{\vartheta,s}(\beta)^*\mathbf{E}^\alpha(s:x\kappa_i)) \\
&\qquad + (\mathbf{E}^\alpha(s:y\kappa_i), \mathbf{U}^{\vartheta,s}(\nu)^*\mathbf{U}^{\vartheta,s}(\alpha)^*\mathbf{E}^\alpha(s:x\kappa_i))| \\
&\leqslant \frac{1}{4\pi}\{|(\mathbf{U}^{\vartheta,s}(\mu)\mathbf{E}^\alpha(s:y\kappa_i), \mathbf{U}^{\vartheta,s}(\beta)^*\mathbf{E}^\alpha(s:x\kappa_i))| \\
&\qquad + |(\mathbf{U}^{\vartheta,s}(\nu)\mathbf{E}^\alpha(s:y\kappa_i), \mathbf{U}^{\vartheta,s}(\alpha)^*\mathbf{E}^\alpha(s:x\kappa_i))|\}.
\end{aligned}$$

Since s is pure imaginary, one has $\mathbf{U}^{\vartheta,s}(\alpha)^* = \mathbf{U}^{\vartheta,s}(\alpha^*), \ldots, \mathbf{U}^{\vartheta,s}(\nu)^* = \mathbf{U}^{\vartheta,s}(\nu^*)$ where, of course, $\alpha^*(?) = \overline{\alpha(?^{-1})}, \ldots, \nu^*(?) = \overline{\nu(?^{-1})}$. Using the Schwarz inequality, we thus see that $|K_\alpha^c(x\kappa_i, y\kappa_i : s : \vartheta)|$ is bounded by the product of

$$\frac{1}{2\sqrt{\pi}}(\mathbf{U}^{\vartheta,s}(\mu^* * \mu)\mathbf{E}^\alpha(s : y\kappa_i), \mathbf{E}^\alpha(s : y\kappa_i))^{1/2}$$

and

$$\frac{1}{2\sqrt{\pi}}(\mathbf{U}^{\vartheta,s}(\beta * \beta^*)\mathbf{E}^\alpha(s : x\kappa_i), \mathbf{E}^\alpha(s : x\kappa_i))^{1/2}$$

plus the product of

$$\frac{1}{2\sqrt{\pi}}(\mathbf{U}^{\vartheta,s}(\nu^* * \nu)\mathbf{E}^\alpha(s : y\kappa_i), \mathbf{E}^\alpha(s : y\kappa_i))^{1/2}$$

and

$$\frac{1}{2\sqrt{\pi}}(\mathbf{U}^{\vartheta,s}(\alpha * \alpha^*)\mathbf{E}^\alpha(s : x\kappa_i), \mathbf{E}^\alpha(s : x\kappa_i))^{1/2}.$$

Let $\alpha_{\mathrm{H}} = \alpha * \alpha^*$, $\beta_{\mathrm{H}} = \beta * \beta^*$, $\mu_{\mathrm{H}} = \mu^* * \mu$, $\nu_{\mathrm{H}} = \nu^* * \nu$—then $|K_\alpha^c(x\kappa_i, y\kappa_i : s : \vartheta)|$ is majorized by

$$K_{\alpha_{\mathrm{H}}}^c(x\kappa_i, x\kappa_i : s : \vartheta)^{1/2} K_{\nu_{\mathrm{H}}}^c(y\kappa_i, y\kappa_i : s : \vartheta)^{1/2}$$
$$+ K_{\beta_{\mathrm{H}}}^c(x\kappa_i, x\kappa_i : s : \vartheta)^{1/2} K_{\mu_{\mathrm{H}}}^c(y\kappa_i, y\kappa_i : s : \vartheta)^{1/2}.$$

The functions α_{H}, β_{H}, μ_{H}, ν_{H} are continuous K-finite positive Hermitian elements in $L^1(G)$ and they all are of regular growth. It is clear from the proof of the sublemma that the assertion made there holds true for each of them. Applying the Schwarz inequality in the evident manner, we thus see that it will be enough to establish our estimate for a kernel of the form

$$K_\phi^c(x\kappa_i, x\kappa_i : s : \vartheta),$$

where $\phi = \psi * \psi^*$, ψ a continuous K-finite element in $L^1(G)$ of regular growth. According to the sublemma,

$$\sum_{\vartheta \in \mathscr{F}} \int_{-\sqrt{-1}T}^{\sqrt{-1}T} K_\phi^c(x\kappa_i, x\kappa_i : s : \vartheta)\,|ds|$$

is the value on the diagonal of the kernel of $L_{G/\Gamma}^c(\phi; \mathscr{F}, T)$ (considered as an operator on $C_c^\infty(G/\Gamma)$). Now $L_{G/\Gamma}^c(\phi; \mathscr{F}, T)$ is the restriction of the positive semidefinite operator $L_{G/\Gamma}(\phi)$ to the invariant subspace $L_c^2(G/\Gamma; \mathscr{F}, T)$ so its kernel (considered as an operator on $C_c^\infty(G/\Gamma)$), which is continuous, can be bounded on the diagonal by the kernel of $L_{G/\Gamma}(\phi)$:

$$\sum_{\vartheta \in \mathscr{F}} \int_{-\sqrt{-1}T}^{\sqrt{-1}T} K_{\phi}^{c}(x\kappa_i, x\kappa_i : s : \vartheta) \, |ds| \leqslant \sum_{\gamma \in \Gamma} \phi(x\kappa_i \gamma \kappa_i^{-1} x^{-1}),$$

the summation on the right being majorized in turn by (cf. Lemma 4.2)

$$C(\xi_\lambda(a_x))^{-n}.$$

As this estimate is independent of $\mathscr{F}$ and T, the proof is complete. ■

We can now prove part of Theorem 4.7. Since $L_{G/\Gamma}^{d}(\alpha)$ is of the trace class, there exists a function T_α^{d} in $L^2(G/\Gamma \times G/\Gamma)$ such that

$$(L_{G/\Gamma}^{d}(\alpha)f, g) = \int_{G/\Gamma} \int_{G/\Gamma} T_\alpha^{d}(x, y) f(y) \overline{g(x)} \, d_G(y) \, d_G(x)$$

for all $f, g \in C_c^\infty(G/\Gamma)$. On the other hand, we have

$$(L_{G/\Gamma}(\alpha)f, g) = \int_{G/\Gamma} \int_{G/\Gamma} K_\alpha(x, y) f(y) \overline{g(x)} \, d_G(y) \, d_G(x)$$

for all $f, g \in C_c^\infty(G/\Gamma)$. It is clear from the definitions that K_α^{c} is measurable. Moreover, thanks to the estimate provided by Lemma 4.8, we have

$$(L_{G/\Gamma}^{c}(\alpha)f, g) = \int_{G/\Gamma} \int_{G/\Gamma} K_\alpha^{c}(x, y) f(y) \overline{g(x)} \, d_G(y) \, d_G(x)$$

for all $f, g \in C_c^\infty(G/\Gamma)$. Because $C_c^\infty(G/\Gamma) \otimes C_c^\infty(G/\Gamma)$ is dense in $L^2(G/\Gamma \times G/\Gamma)$, it then follows that

$$T_\alpha^{d} = K_\alpha - K_\alpha^{c}$$

almost everywhere in $G/\Gamma \times G/\Gamma$. In other words: K_α^{d} is square integrable over $G/\Gamma \times G/\Gamma$ and represents $L_{G/\Gamma}^{d}(\alpha)$. These considerations also imply that $L_{G/\Gamma}^{c}(\alpha)$ is an integral operator with kernel K_α^{c}.

Remark. The reader will note that one cannot conclude at once from Lemma 4.8 that $L_{G/\Gamma}^{c}(\alpha)$ is an integral operator with kernel K_α^{c}. The point is that the estimate only guarantees convergence of the relevant integral for a suitably restricted class of functions, e.g., $C_c^\infty(G/\Gamma)$, or, more generally, the rapidly decreasing functions. Abstractly, if $T: L^2(X, \mu) \to L^2(X, \mu)$ is a bounded linear transformation which is represented by a kernel on, e.g., $C_c^\infty(X)$, then it is false in general that T is represented by the given kernel on all of $L^2(X, \mu)$. The following examples were suggested by G. B. Folland. (1) Take $X = \mathbf{R}$, $d\mu(x) = dx$ and consider the kernel associated with the Fourier transform: $K(x, y) = e^{\sqrt{-1}xy}$; (2) Take $X =]0, 1[$, $d\mu(x) = dx/\sqrt{x}$, and consider the kernel associated with the "restricted" Mellin transform: $K(x, y) = x^{\sqrt{-1}y}$.

It remains to prove that K_α^{d} is integrable over the diagonal with

$$\operatorname{tr}(L_{G/\Gamma}^{\mathrm{d}}(\alpha)) = \int_{G/\Gamma} K_\alpha^{\mathrm{d}}(x, x)\, d_G(x).$$

LEMMA 4.9. *The kernel K_α^{d} is continuous in each variable separately.*

Proof. Since $K_\alpha(x, y)$ is continuous in (x, y), we need only consider $K_\alpha^{\mathrm{c}}(x, y)$. In the notations introduced during the proof of Lemma 4.8, for any $\mathscr{F}$ and T, the integral $I_\alpha(x, y : \mathscr{F}, T)$ defined by

$$\left[\sum_{\mathcal{O} \in \mathscr{F}} \int_{\substack{s \in \sqrt{-1}\mathbf{R} \\ |s| \geq T}} + \sum_{\mathcal{O} \in \mathscr{C}\mathscr{F}} \int_{-\sqrt{-1}T}^{\sqrt{-1}T} + \sum_{\mathcal{O} \in \mathscr{C}\mathscr{F}} \int_{\substack{s \in \sqrt{-1}\mathbf{R} \\ |s| \geq T}}\right] |K_\alpha^{\mathrm{c}}(x, y : s : \mathcal{O})|\, |ds|$$

is bounded by

$$I_{\alpha_{\mathrm{H}}}(x, x : \mathscr{F}, T)^{1/2} I_{\nu_{\mathrm{H}}}(y, y : \mathscr{F}, T)^{1/2} + I_{\beta_{\mathrm{H}}}(x, x : \mathscr{F}, T)^{1/2} I_{\mu_{\mathrm{H}}}(y, y : \mathscr{F}, T)^{1/2}.$$

Keeping in mind that the estimate derived in Lemma 4.8 is applicable to α_{H}, ν_{H}, β_{H}, and μ_{H}, it follows that for fixed x (or y) the integral $K_\alpha^{\mathrm{c}}(x, y)$ converges uniformly for y (or x) varying in compacta of G/Γ. Therefore the kernel $K_\alpha^{\mathrm{c}}(x, y)$ is continuous in each variable separately. ∎

Thanks to this lemma, the proof of Theorem 4.7 can now be completed by appealing to the following important criterion, which was kindly communicated to us by M. S. Osborne.

THEOREM 4.10. *Let X be a locally compact, second countable Hausdorff space, μ a regular Borel measure on X. Let E be a finite-dimensional Hermitian vector bundle over X. Let $K \in L^2(E \otimes E^*)$; let $T_K : \Gamma L^2(E) \to \Gamma L^2(E)$ be the associated integral operator. Suppose that K is separately continuous; suppose that T_K is of the trace class. Then: The function* tr_K, $x \mapsto \operatorname{tr}(K(x, x))$, *is in $L^1(X, \mu)$ and*

$$\operatorname{tr}(T_K) = \int_X \operatorname{tr}_K(x)\, d\mu(x).$$

Proof. Since T_K is trace class, hence compact, a well known generality (cf. Ringrose [25, p. 71]) guarantees the existence of positive real numbers λ_n and orthonormal sequences $\{\varphi_n\}$, $\{\psi_n\}$ in $\Gamma L^2(E)$ such that

$$K = \sum_n \lambda_n(\varphi_n \otimes \psi_n^*),$$

where $\sum_n \lambda_n < +\infty$. It follows from this that

$$\operatorname{tr}(T_K) = \sum_n \lambda_n \cdot \int_X (\varphi_n(x), \psi_n(x))\, d\mu(x).$$

Because $\sum_n \lambda_n < +\infty$, monotone convergence implies that

$$\int_X \sum_n \lambda_n \|\varphi_n(x)\|^2 \, d\mu(x) = \sum_n \lambda_n = \int_X \sum_n \lambda_n \|\psi_n(x)\|^2 \, d\mu(x).$$

Thus

$$\sum_n \lambda_n \|\varphi_n(x)\|^2, \qquad \sum_n \lambda_n \|\psi_n(x)\|^2$$

are finite almost everywhere and so

$$\sum_n \lambda_n \|\varphi_n(x)\|^2, \qquad \sum_n \lambda_n \|\psi_n(x)\|^2$$

converge almost everywhere. As X is σ-compact, we can write

$$X = \bigcup_{k=1}^{\infty} C_k,$$

with C_k compact, $C_k \subset C_{k+1}$. Owing to the theorems of Lusin and Egoroff, there are compact sets $C_k' \subset C_k$, $C_k' \subset C_{k+1}'$ such that

(i) φ_n and ψ_n are continuous on C_k';
(ii) $\sum_n \lambda_n \|\varphi_n\|^2$, $\sum_n \lambda_n \|\psi_n\|^2$ are uniformly convergent on C_k';
(iii) $\mu(C_k - C_k') < 1/k$.

For all $x, y \in C_k'$, we have

$$\left[\sum_{n=s}^{t} \lambda_n \|\varphi_n(x)\| \cdot \|\psi_n(y)\|\right]^2 \leqslant \left[\sum_{n=s}^{t} \lambda_n \|\varphi_n(x)\|^2\right] \cdot \left[\sum_{n=s}^{t} \lambda_n \|\psi_n(y)\|^2\right],$$

so, by (ii),

$$\tilde{K}(x, y) = \sum_n \lambda_n (\varphi_n(x) \otimes \psi_n^*(y))$$

converges absolutely uniformly on $C_k' \times C_k'$, thus is continuous there. It therefore follows that $K = \tilde{K}$ almost everywhere on $C_k' \times C_k'$. This implies that

$$\int_{C_k'} \int_{C_k'} \|K(x, y) - \tilde{K}(x, y)\| \, d\mu(x) \, d\mu(y) = 0.$$

Let

$$S_k = \{y \in C_k' \colon \int_{C_k'} \|K(x, y) - \tilde{K}(x, y)\| \, d\mu(x) \neq 0\}.$$

Obviously, $\mu(S_k) = 0$. If $y \notin S_k$, then

$$\int_{C_k'} \|K(x, y) - \tilde{K}(x, y)\| \, d\mu(x) = 0,$$

whence $K(x, y) = \tilde{K}(x, y)$ almost everywhere on C'_k. Keeping in mind the separate continuity of K, we deduce that $K(x, y) = \tilde{K}(x, y)$ for all x in the support of $\mu | C'_k$ and all y not in S_k. Put

$$\tilde{X} = \bigcup_{k=1}^{\infty} \tilde{X}_k, \qquad \tilde{X}_k = \operatorname{spt}(\mu | C'_k) - \bigcup_{l=1}^{\infty} S_l.$$

Then: $K = \tilde{K}$ on $\tilde{X} \times \tilde{X}$. We claim now that $\mu(X - \tilde{X}) = 0$. To see this, note first that

$$\begin{aligned}\mu(\tilde{X}_k) &= \mu(\operatorname{spt}(\mu | C'_k)) \\ &= \mu(C'_k) > \mu(C_k) - 1/k.\end{aligned}$$

Therefore

$$\begin{aligned}\mu(X - \tilde{X}) &= \lim_{k\to\infty} \mu[(X - \tilde{X}) \cap C_k] \\ &= \lim_{k\to\infty} \mu(C_k - \tilde{X}) \\ &\leqslant \lim_{k\to\infty} \mu(C_k - \tilde{X}_k) \leqslant \lim_{k\to\infty} 1/k = 0,\end{aligned}$$

as claimed. We have

$$K(\tilde{x}, \tilde{x}) = \sum_n \lambda_n(\varphi_n(\tilde{x}) \otimes \psi_n^*(\tilde{x}))$$

for all $\tilde{x} \in \tilde{X}$. Monotone convergence then gives

$$\begin{aligned}\int_X |\operatorname{tr}_K(x)|\, d\mu(x) &\leqslant \sum_n \lambda_n \cdot \int_X |(\varphi_n(x), \psi_n(x))|\, d\mu(x) \\ &\leqslant \sum_n \lambda_n < +\infty,\end{aligned}$$

so, by dominated convergence,

$$\begin{aligned}\operatorname{tr}(T_K) &= \sum_n \lambda_n \cdot \int_X (\varphi_n(x), \psi_n(x))\, d\mu(x) \\ &= \int_X \operatorname{tr}_K(x)\, d\mu(x).\end{aligned}$$

Hence the theorem. ■

Appendix

Our objective here will be to establish the following result.

LEMMA. *The space $\mathscr{C}^1(G)$ is closed under convolution.*

We remark that Trombi and Varadarajan [30a, p. 246] have proved that $\mathscr{C}^1(K\backslash G/K)$ is closed under convolution by utilizing Fourier transforms. Such a proof cannot be generalized. Instead we shall use the Sobelev lemma (cf. Trombi and Varadarajan [30b, p. 298]):

SUBLEMMA. *Suppose that α is a C^∞ function on G with the property that*

$$\int_G |\alpha(D_1; x; D_2)|\, d_G(x) < +\infty$$

for all $D_1, D_2 \in \mathfrak{G}$—then

$$\sup_{x \in G} \Xi^{-2}(x) \cdot |\alpha(x)| < +\infty.$$

In the definition of the space $\mathscr{C}^1(G)$, $1 + \sigma$ (which is not C^∞) may be replaced by any other function having the same order of growth. There exists, in fact, an analytic K-bi-invariant function ϕ on G such that:

(i) $\phi(x) = \phi(x^{-1})$ for all $x \in G$;

(ii) $M(1 + \sigma(x)) \leqslant \phi(x) \leqslant N(1 + \sigma(x))$ for all $x \in G$, M and N being suitable positive constants;

(iii) $\sup_{x \in G} |\phi(D_1; x; D_2)| < +\infty$ provided at least one of the D_i is in $\mathfrak{G}\mathfrak{g}$.

One may, e.g., take $\phi = 1 - \log(\Xi)$ (cf. Varadarajan [31, p. 69]). By way of notation, put $\phi_r = \Xi^2\phi^{-r}(r \in \mathbf{R})$—then $\phi_r \in L^1(G)$ for all sufficiently large r (cf. Warner [34b, p. 156]).

Passing now to the proof of the lemma, fix $\alpha, \beta \in \mathscr{C}^1(G)$. Since invariant derivatives can be moved through the convolution sign (cf. Warner [34b, p. 162]), it will be enough to prove the following: For any $r_0 > 0$, $|\alpha * \beta|_{r_0} < +\infty$. If $r > 0$, then we have

$$\frac{(1+\sigma)^{r_0}(x)}{\Xi^2(x)} \cdot \int_G |\alpha(xy)| \cdot |\beta(y^{-1})|\, d_G(y)$$

$$\leqslant \frac{(1+\sigma)^{r_0}(x)}{\Xi^2(x)} \cdot \int_G \frac{(1+\sigma)^r(xy)}{\Xi^2(xy)} \cdot |\alpha(xy)| \cdot \frac{(1+\sigma)^r(y^{-1})}{\Xi^2(y^{-1})} \cdot |\beta(y^{-1})|$$

$$\times \frac{\Xi^2(xy) \cdot \Xi^2(y^{-1})}{(1+\sigma)^r(xy) \cdot (1+\sigma)^r(y^{-1})}\, d_G(y)$$

$$\leqslant |\alpha|_r \cdot |\beta|_r \cdot \frac{(1+\sigma)^{r_0}(x)}{\Xi^2(x)} \cdot \int_G \frac{\Xi^2(xy) \cdot \Xi^2(y^{-1})}{(1+\sigma)^r(xy) \cdot (1+\sigma)^r(y^{-1})}\, d_G(y).$$

These estimates make it clear that in order to complete the proof of our lemma we need only show that for any $r_0 > 0$, there is an $r > 0$ such that

$$\sup_{x \in G} \Xi^{-2}(x) \cdot (\phi^{r_0}(x) \cdot (\phi_r * \phi_r)(x)) < +\infty.$$

This being the case, fix $r_0 > 0$—then, in view of the sublemma, we are reduced to establishing the existence of an $r > 0$ with the property that $\phi^{r_0} \cdot (\phi_r * \phi_r)$, and all its invariant derivatives, are integrable on G. We have

$$\begin{aligned}
&\int_G \int_G \phi^{r_0}(x)\phi_r(xy)\phi_r(y^{-1})\, d_G(y)\, d_G(x) \\
&\quad = \int_G \int_G \phi^{r_0}(x)\phi_r(xy^{-1})\phi_r(y)\, d_G(y)\, d_G(x) \\
&\quad = \int_G \left(\int_G \phi^{r_0}(x)\phi_r(xy^{-1})\, d_G(x) \right) \phi_r(y)\, d_G(y) \\
&\quad = \int_G \left(\int_G \phi^{r_0}(xy)\phi_r(x)\, d_G(x) \right) \phi_r(y)\, d_G(y) \\
&\quad \leqslant N^{r_0} \int_G \left(\int_G (1 + \sigma(xy))^{r_0}\phi_r(x)\, d_G(x) \right) \phi_r(y)\, d_G(y) \\
&\quad \leqslant N^{r_0} \int_G \left(\int_G (1 + \sigma(x))^{r_0}\phi_r(x)\, d_G(x) \right) (1 + \sigma(y))^{r_0}\phi_r(y)\, d_G(y) \\
&\quad \leqslant \frac{N^{r_0}}{M^{2r_0}} \cdot \|\phi^{r_0} \cdot \phi_r\|_1^2 < +\infty
\end{aligned}$$

provided r is chosen sufficiently large. Let, then, $r^0 > 0$ be such that

$$\phi_{r^0} \in L^1(G), \qquad \phi^{r_0} \cdot (\phi_{r^0} * \phi_{r^0}) \in L^1(G).$$

We claim that all the invariant derivatives of $\phi^{r_0} \cdot (\phi_{r^0} * \phi_{r^0})$ are also in $L^1(G)$. To see this, we first observe that given D_1, D_2, there exist $C_{D_1, D_2} > 0$ such that

$$|\phi_{r^0}(D_1 ; x; D_2)| \leqslant C_{D_1, D_2} \phi_{r^0}(x)$$

for all $x \in G$. Indeed, it is known that an estimate of this kind obtains for $\vdash\!\circ\!\dashv$ (cf. Warner [34b, p. 361]), so our contention follows from the fact that the invariant derivatives of ϕ are bounded. The claim then follows readily, thereby finishing the proof of the lemma.

5. Classification of the Elements of Γ

An element $\gamma \in \Gamma$ is said to be *central* if γ lies in the center Z_Γ of Γ. The proof of the following fact appears in Garland and Raghunathan [11, p. 318].

LEMMA 5.1. *The center of Γ is a subgroup of the center of G, hence is finite.*

A noncentral element $\gamma \in \Gamma$ is said to be *elliptic* if γ is G-conjugate to an element of K. Since Γ operates in the obvious way on the symmetric space $K\backslash G$, the elliptic elements in Γ can be characterized as those noncentral elements which admit a fixed point in $K\backslash G$ when viewed as a transformation of $K\backslash G$. Any elliptic element in Γ is clearly of finite order, and therefore lies in a finite subgroup of Γ. Elliptic elements are necessarily semisimple.

LEMMA 5.2. *There are but finitely many Γ-conjugacy classes of elliptic elements in Γ.*

Proof. This is essentially known (cf. Borel [3a, p. 14]) but we shall give the proof for sake of completeness. Let $\pi: G \to K\backslash G$ denote the natural projection—then $\pi(\mathfrak{S}) = \pi(1) \cdot \mathfrak{S}$ and:

(i) $\pi(\mathfrak{S}) \cdot \Gamma = K\backslash G$;
(ii) $\{\gamma \in \Gamma : \pi(\mathfrak{S})\gamma \cap \pi(\mathfrak{S}) \neq \emptyset\}$ is finite.

Since the subgroup of Γ which operates trivially on $K\backslash G$ is finite, for the purposes of the present proof it can be supposed that Γ operates effectively on $K\backslash G$. Let now F be any finite subgroup of Γ—then F belongs to a maximal compact subgroup of G, thus has a fixed point $p \in K\backslash G$. Select a $\gamma \in \Gamma$ such that $p\gamma \in \pi(\mathfrak{S})$—then $\gamma^{-1}F\gamma$ leaves a point of $\pi(\mathfrak{S})$ fixed, and so is contained in the finite set determined by (ii) above. Hence the lemma. ∎

A noncentral, nonelliptic, semisimple element of Γ is said to be *hyperbolic*. A noncentral element of Γ which is neither semisimple nor unipotent will be called *loxodromic*.

Given $\gamma \in \Gamma$, let us agree to write $\{\gamma\}_\Gamma$ (respectively $\{\gamma\}_G$) for its Γ-conjugacy class (respectively G-conjugacy class)—then the above definitions classify the $\{\gamma\}_\Gamma$:

$$\begin{cases} \text{semisimple:} & \begin{cases} \text{central} \\ \text{elliptic} \\ \text{hyperbolic} \end{cases} \\ \text{unipotent} & \\ \text{loxodromic} & \end{cases}$$

The following lemma was kindly communicated to us by M. S. Raghunathan.

LEMMA 5.3. *Suppose that $\gamma \in \Gamma$ has the property that $\{\gamma\}_\Gamma \cap P_i = \emptyset$ for all i between 1 and r—then γ is, of necessity, semisimple.*

Proof. Suppose that γ is not semisimple—then $\{\gamma\}_G$ is not closed in G (cf. Warner [34a, p. 121]). Let $\gamma = \gamma_s\gamma_u$ be the Jordan decomposition of γ (so γ_s is semisimple, γ_u is unipotent, and $\gamma_s\gamma_u = \gamma_u\gamma_s$)—then $\gamma_u \neq 1$. Let $\{x_n\}$ be a sequence of elements in G such that $\lim_{n\to\infty} x_n\gamma x_n^{-1} = \gamma_s$ (cf. Warner [34a, p. 121]). If $\{x_n\}$ is relatively compact mod Γ, then, by passing to a subsequence if necessary, we can find $\theta_n \in \Gamma$ such that $x_n = y_n\theta_n$, $\lim_{n\to\infty} y_n = y$. Thus

$$\lim_{n\to\infty} x_n\gamma x_n^{-1} = \lim_{n\to\infty} y_n(\theta_n\gamma\theta_n^{-1})y_n^{-1},$$

and so, as $\{y_n\}$ is a convergent sequence, the same must be true of $\{\theta_n\gamma\theta_n^{-1}\}$. Therefore $\theta_n\gamma\theta_n^{-1} = \theta \in \Gamma$ for all sufficiently large n. This implies that $y\theta y^{-1} = \gamma_s$. Because θ is a conjugate of γ, it follows that γ must be semisimple, i.e., that $\gamma_u = 1$, a contradiction. Assume now that $\{x_n\}$ has no convergent subsequence mod Γ. According to a result of Kazdan and Margolis (cf. Raghunathan [23b, p. 180]), there is then a sequence of unipotents $\theta_n \in \Gamma - \{1\}$ such that $x_n\theta_n x_n^{-1}$ converges to 1. Consider the elements $\theta_n, \gamma\theta_n\gamma^{-1}, \gamma^2\theta_n\gamma^{-2}, \ldots, \gamma^d\theta_n\gamma^{-d}$, $d = \dim(G)$. The inverse images of these elements in $\mathfrak{g}$ under the exponential map span a linear subspace E, say, which is stable under γ. On the other hand, for large n, x_n conjugates these elements into any given neighborhood of the identity of G. It therefore follows that the Lie algebra spanned by E is the Lie algebra of a connected unipotent subgroup U, say, of G (cf. Raghunathan [23b, p. 182]). Since γ normalizes U, it normalizes $U \cap \Gamma$. The latter is a nontrivial unipotent subgroup of Γ, hence is contained in a unique maximal unipotent subgroup of Γ, the Zariski closure of which is the unipotent radical of a Γ-cuspidal parabolic subgroup of G containing γ, a contradiction once again. ∎

We remark that the converse to Lemma 5.3 is not true in general since it can very well happen that a given $\Gamma \cap P_i$ contains noncentral semisimple elements (cf. Cohn [5]).

Given $\gamma \in \Gamma$, let us agree to write Γ_γ (respectively G_γ) for its Γ-centralizer (respectively G-centralizer).

LEMMA 5.4. *Suppose that $\gamma \in \Gamma$ has the property that $\{\gamma\}_\Gamma \cap P_i = \emptyset$ for all i between 1 and r—then Γ_γ is a lattice in G_γ.*

We shall defer the proof momentarily and pass first to a technical result. For any i between 1 and r, put

$$a_i(t) = \exp(tH_i) \qquad (t \in \mathbf{R}),$$

where H_i is the element in the Lie algebra of A_i such that $\lambda(H_i) = |\lambda|$.

SUBLEMMA. *Let C be a compact subset of G—then there is a number $\epsilon_C > 0$, independent of i, with the property that if*

$$a_i(t)\gamma a_i(-t) \in C$$

for some $\gamma \in \Gamma$ and some $t < \log \epsilon_C$, then necessarily $\gamma \in \Gamma \cap P_i$.

Proof. To simplify, we might just as well suppose that $i = 1$ and then drop it from the notation. Given $0 < \alpha < \beta$, let

$$A[\alpha, \beta] = \{a \in A : \alpha \leqslant \xi_\lambda(a) \leqslant \beta\}.$$

Then there exist $0 < \alpha < \beta$ and a compact subset ω of N such that

$$C \subset K \cdot A[\alpha, \beta] \cdot \omega.$$

There is clearly no loss of generality in supposing that t_0 and ω_0 have been so chosen that

$$\left.\begin{matrix}\beta\\ 1\end{matrix}\right\} < t_0, \qquad \omega \subset \omega_0.$$

If $a(t)\gamma a(-t) \in C$, then

$$a(t)\gamma \in K \cdot A[\alpha, t_0]a(t) \cdot a(-t)\omega_0 a(t) \subset K \cdot A[\alpha, t_0]a(t) \cdot N.$$

Take now ϵ_C to be any positive number such that

$$\log \epsilon_C < |\lambda|^{-1} \cdot \log({}_0t/t_0).$$

Then

$$t < \log \epsilon_C \Rightarrow A[\alpha, t_0]a(t) \subset A({}_0t),$$

i.e.,

$$t < \log \epsilon_C \Rightarrow a(t) \in K \cdot A({}_0t) \cdot N\gamma^{-1}.$$

On the other hand,

$$t < \log \epsilon_C \Rightarrow a(t) \in K \cdot A({}_0t) \cdot N.$$

Therefore

$$K \cdot A({}_0t) \cdot N \cap K \cdot A({}_0t) \cdot N\gamma^{-1} \neq \varnothing.$$

Thus $\gamma^{-1} \in \Gamma \cap P$ and so $\gamma \in \Gamma \cap P$, as desired. ■

Proof of Lemma 5.4. Let Γ_S be the subset of Γ comprised of those elements $\gamma \in \Gamma$ with the property stated in our lemma—then it is clear that Γ_S is a union of Γ-conjugacy classes. This being so, let $\alpha \in C_c^\infty(G)$—then it is clear that the assignment

$$x \mapsto \sum_{\gamma \in \Gamma_S} |\alpha(x\gamma x^{-1})|$$

is a well-defined function on G/Γ and we claim that

$$\int_{G/\Gamma} \left(\sum_{\gamma \in \Gamma_S} |\alpha(x\gamma x^{-1})| \right) d_G(x) < +\infty.$$

Of course we need only check integrability on the $\mathfrak{S}_{t_0,\omega_0}\kappa_i$. To simplify, take $i = 1$ and then drop it from the notation. Consider

$$\int_{\mathfrak{S}_{t_0,\omega_0}} \left(\sum_{\gamma \in \Gamma_S} |\alpha(x\gamma x^{-1})| \right) d_G(x).$$

Since $\mathfrak{S}_{t_0,\omega_0} = K \cdot A[t_0] \cdot \omega_0$, the integrand is nonzero only if there exist $k \in K$, $a \in A[t_0]$, $n \in \omega_0$, and $\gamma \in \Gamma_S$ such that

$$kan\gamma n^{-1}a^{-1}k^{-1} \in \operatorname{spt}(\alpha),$$

i.e., only if

$$an\gamma n^{-1}a^{-1} \in K \cdot \operatorname{spt}(\alpha) \cdot K,$$

i.e., only if

$$a\gamma a^{-1} \in \Omega_0 \cdot K \cdot \operatorname{spt}(\alpha) \cdot K \cdot \Omega_0^{-1},$$

where $\Omega_0^{-1} = \{ana^{-1} : a \in A[t_0],\ n \in \omega_0\}$, a relatively compact subset of N (cf. Borel and Harish-Chandra [4, p. 501]). In view of the sublemma, and the fact that $\gamma \notin \Gamma \cap P$, it follows immediately that the integrand is compactly supported on $\mathfrak{S}_{t_0,\omega_0}$, thus is integrable there. Hence the claim. Let now

$$(S) \sum_{\{\gamma\}_\Gamma}$$

stand for a sum over the Γ-conjugacy classes of the elements of Γ_S. Then we have

$$\int_{G/\Gamma} \left(\sum_{\gamma \in \Gamma_S} |\alpha(x\gamma x^{-1})| \right) d_G(x) = (S) \sum_{\{\gamma\}_\Gamma} \int_{G/\Gamma_\gamma} |\alpha(x\gamma x^{-1})| \, d_G(x).$$

Fix $\gamma \in \Gamma_S$. Since γ is semisimple (cf. Lemma 5.3), the homogeneous space G/G_γ carries a G-invariant measure d_{G/G_γ}, say (cf. Warner [34a, p. 120]). Let $\alpha_\gamma \in C_c^\infty(G)$ be such that

$$\int_{G/\Gamma_\gamma} |\alpha_\gamma(x\gamma x^{-1})| \, d_G(x) > 0.$$

Then

$$\int_{G/\Gamma_\gamma} |\alpha_\gamma(x\gamma x^{-1})|\, d_G(x) = \mathrm{vol}(G_\gamma/\Gamma_\gamma) \cdot \int_{G/G_\gamma} |\alpha_\gamma(x\gamma x^{-1})|\, d_{G/G_\gamma}(x).$$

This shows that $\mathrm{vol}(G_\gamma/\Gamma_\gamma)$ is finite, i.e., that Γ_γ is a lattice in G_γ, as desired. ■

6. The Selberg Trace Formula

Let α be a K-finite function in $\mathscr{C}^1(G)$—then, as we have seen in Section 4, $L_{G/\Gamma}^{\mathrm{d}}(\alpha)$ is a trace class operator and, moreover,

$$\mathrm{tr}(L_{G/\Gamma}^{\mathrm{d}}(\alpha)) = \int_{G/\Gamma} K_\alpha^{\mathrm{d}}(x, x)\, d_G(x).$$

The objective of the present section will be to calculate the integral on the right. The upshot of the computation is that $\mathrm{tr}(L_{G/\Gamma}^{\mathrm{d}}(\alpha))$ can be written as a certain sum (in general infinite), the terms of which are distributions on G, e.g., orbital integrals. In essence, this is the Selberg trace formula, at least in its second stage (cf. the Introduction).

Recall that the center Z_Γ of Γ is actually a central subgroup of G (cf. Lemma 5.1). Consequently: $Z_\Gamma \subset \Gamma \cap M_i$ $(i = 1, \ldots, r)$. We shall now make an assumption on Γ which will be in force throughout the remainder of this article.

Assumption. For each i between 1 and r,

$$\Gamma \cap P_i = Z_\Gamma \cdot \Gamma \cap N_i.$$

This assumption does in fact restrict the generality somewhat although, under very general circumstances, it is true that a given Γ contains a normal subgroup of finite index with the stated property (cf. Garland and Raghunathan [11, p. 310]). We make the assumption in order to keep certain combinatorial problems to a minimum. One should perhaps note that our assumption does not preclude the existence of elliptic elements in Γ, as the example $G = \mathbf{SL}(2, \mathbf{R})$ already shows.

For the time being it will be supposed that α is a K-finite element in $C_c^\infty(G)$; the passage to $\mathscr{C}^1(G)$ will be carried out later.

Let Γ_S be the subset of Γ comprised of those elements γ with the property that $\{\gamma\}_\Gamma \cap P_i = \varnothing$ for every i (cf. Lemma 5.3); let Γ_P be the subset of Γ comprised of those noncentral elements γ with the property that $\{\gamma\}_\Gamma \cap P_i \neq \varnothing$ for at least one i. In view of what has been said in Section 5, the elements of Γ_S are semisimple whereas the elements of Γ_P are parabolic, i.e., either unipotent or loxodromic. We can then write

$$\begin{aligned} K_\alpha(x,x) &= \sum_{\gamma\in\Gamma} \alpha(x\gamma x^{-1}) \\ &= \sum_{z\in Z_\Gamma} \alpha(z) + \sum_{\gamma\in\Gamma_S} \alpha(x\gamma x^{-1}) + \sum_{\gamma\in\Gamma_P} \alpha(x\gamma x^{-1}). \end{aligned}$$

The first and second terms of the sum will be allowed to stand. As for the third, we have the following lemma.

LEMMA 6.1. *In the above notations*

$$\sum_{\gamma\in\Gamma_P} \alpha(x\gamma x^{-1}) = \sum_{i=1}^{r} \sum_{\delta\in\Gamma/\Gamma\cap P_i} \sum_{z\in Z_\Gamma} \sum_{\substack{\eta\in\Gamma\cap N_i \\ \eta\neq 1}} \alpha(x\,\delta z\eta\,\delta^{-1}x^{-1}).$$

We shall preface the proof with a preliminary result.

SUBLEMMA. *Let N', N'' be maximal unipotent subgroups of G—then either $N' = N''$ or $N' \cap N'' = \{1\}$. Let z', $z'' \in Z_G$, the center of G—then*

$$z'N' \cap z''N'' \neq \varnothing \Rightarrow z' = z''.$$

Proof. The first assertion is well known and is a simple consequence of the Bruhat lemma (cf. Raghunathan [23b, p. 194]). Turning to the second, suppose there exist $n' \in N'$, $n'' \in N''$ with the property that $z'n' = z''n''$—then $zn' = n''$ where $z = z'(z'')^{-1} \in Z_G$. Now z is semisimple and commutes with the unipotent element n', thus, n'' being unipotent, the theory of the Jordan decomposition implies that $z = 1$, i.e., $z' = z''$. ■

Proof of Lemma 6.1. Let $\gamma\in\Gamma_P$—then, by definition, $\gamma = \tilde{\gamma}z\eta\tilde{\gamma}^{-1}$ for some $\tilde{\gamma}\in\Gamma$, $z\in Z_\Gamma$, $\eta\in\Gamma\cap N_i$. Write $\tilde{\gamma} = \delta p$ where $\delta\in\Gamma/\Gamma\cap P_i$, $p\in\Gamma\cap P_i$—then $\gamma = \delta zp\eta p^{-1}\delta^{-1}$. Since $p\eta p^{-1}\in\Gamma\cap N_i$, it follows that every term on the left-hand side is picked up at least once by the terms on the right-hand side. Every term on the right-hand side is obviously in Γ_P. Therefore, to finish the proof, we have only to show that a given term in Γ_P admits a unique representation of the form $\delta z\eta\,\delta^{-1}$. Suppose then that

$$\delta'z'\eta'\delta'^{-1} = \delta''z''\eta''\delta''^{-1},$$

where $\delta'\in\Gamma/\Gamma\cap P_{i'}$, $\delta''\in\Gamma/\Gamma\cap P_{i''}$, $z', z''\in Z_\Gamma$, $\eta'\in\Gamma\cap N_{i'}$, $\eta''\in\Gamma\cap N_{i''}$ $(1\leqslant i', i''\leqslant r)$. Because $\delta'N_{i'}\delta'^{-1}$, $\delta''N_{i''}\delta''^{-1}$ are maximal unipotent subgroups of G, it follows from the sublemma that $z' = z''$. Neither η' nor η'' is equal to 1. Thus, by the sublemma once more, $\delta'N_{i'}\delta'^{-1} = \delta''N_{i''}\delta''^{-1}$ and so $\delta'(\Gamma\cap N_{i'})\delta'^{-1} = \delta''(\Gamma\cap N_{i''})\delta''^{-1}$. This, of course, implies that $i' = i'' = i$, say. But then $\delta''^{-1}\delta'$ normalizes N_i, hence lies in $\Gamma\cap P_i$, which can be the case only if $\delta' = \delta''$, so trivially $\eta' = \eta''$. ■

Fix a number $\epsilon: 0 < \epsilon < {}_0t$. We shall view ϵ as a parameter which will be allowed to shrink whenever convenient. Call $\chi_{\epsilon,i}$ the characteristic function of the set $\mathfrak{C}_\epsilon \kappa_i$ $(1 \leqslant i \leqslant r)$. Write

$$\sum_{\gamma \in \Gamma_P} \alpha(x\gamma x^{-1}) = I_\alpha(x:\epsilon) + J_\alpha(x:\epsilon),$$

where

$$I_\alpha(x:\epsilon) = \sum_{i=1}^{r} \sum_{\delta \in \Gamma/\Gamma \cap P_i} \sum_{z \in Z_\Gamma} \sum_{\substack{\eta \in \Gamma \cap N_i \\ \eta \neq 1}} \alpha(x\delta z\eta\delta^{-1}x^{-1}) \cdot \chi_{\epsilon,i}(x\delta),$$

$$J_\alpha(x:\epsilon) = \sum_{i=1}^{r} \sum_{\delta \in \Gamma/\Gamma \cap P_i} \sum_{z \in Z_\Gamma} \sum_{\substack{\eta \in \Gamma \cap N_i \\ \eta \neq 1}} \alpha(x\delta z\eta\delta^{-1}x^{-1}) \cdot (1 - \chi_{\epsilon,i}(x\delta)).$$

Returning to K_α, we thus have that

$$K_\alpha(x, x) = \sum_{z \in Z_\Gamma} \alpha(z) + \sum_{\gamma \in \Gamma_S} \alpha(x\gamma x^{-1}) + I_\alpha(x:\epsilon) + J_\alpha(x:\epsilon).$$

Our next task will be to break up K_α^{c}. Since $K_\alpha^{\mathrm{d}} = K_\alpha - K_\alpha^{\mathrm{c}}$, we shall then have a decomposition of K_α^{d} suitable for the calculation of

$$\int_{G/\Gamma} K_\alpha^{\mathrm{d}}(x, x)\, d_G(x).$$

Let $\mathbf{\Phi} \in \mathscr{E}(\vartheta, \delta)$—then attached to $\mathbf{\Phi}$ is the Eisenstein series $\mathbf{E}(\mathbf{\Phi}:s:x)$ with constant term

$$\mathbf{E}_{P_i}(\mathbf{\Phi}:s:x) = e^{(s - |\rho|)(H_i(x))} \cdot \Phi_i(x) + e^{(-s - |\rho|)(H_i(x))} \cdot (\mathbf{c}_{\vartheta,\delta}(s)\mathbf{\Phi})_i(x).$$

For any $x \in G$, the series

$$\sum_{\delta \in \Gamma/\Gamma \cap P_i} \mathbf{E}_{P_i}(\mathbf{\Phi}:s:x\delta) \cdot \chi_{\epsilon,i}(x\delta)$$

is a finite sum. This follows from the fact that the set of all γ in Γ for which $\mathfrak{S}\gamma \cap \mathfrak{C}_\epsilon \kappa_i$ is nonempty is finite modulo $\Gamma \cap P_i$ (cf. Harish-Chandra [13a, p. 95]). Suppose in particular that $x \in \mathfrak{C}_\epsilon \kappa_i$—then in this case our series simply becomes the constant term $\mathbf{E}_{P_i}(\mathbf{\Phi}:s:x)$ once again. Write

$$\mathbf{E}_\epsilon'(\mathbf{\Phi}:s:x) = \sum_{i=1}^{r} \sum_{\delta \in \Gamma/\Gamma \cap P_i} \mathbf{E}_{P_i}(\mathbf{\Phi}:s:x\delta) \cdot \chi_{\epsilon,i}(x\delta),$$

$$\mathbf{E}_\epsilon''(\mathbf{\Phi}:s:x) = \mathbf{E}(\mathbf{\Phi}:s:x) - \mathbf{E}_\epsilon'(\mathbf{\Phi}:s:x).$$

Speaking intuitively, $\mathbf{E}_\epsilon'(\mathbf{\Phi}:s:x)$ (respectively $\mathbf{E}_\epsilon''(\mathbf{\Phi}:s:x)$) isolates the "bad" (respectively "good") part of $\mathbf{E}(\mathbf{\Phi}:s:x)$ at infinity.

Form now

$$K_\alpha^{\prime,c}(x:s:\epsilon) = \frac{1}{4\pi} \sum_{m,\, n \in I} \mathbf{U}_{mn}^s(\alpha) \cdot \mathbf{E}_\epsilon'(e_m : s : x)\overline{\mathbf{E}_\epsilon'(e_n : s : x)},$$

$$K_\alpha^{\prime,c}(x:\epsilon) = \int_{\mathrm{Re}(s)=0} K_\alpha^{\prime,c}(x:s:\epsilon)\,|ds|.$$

Then one has the following important estimate (cf. Lemma 4.8).

LEMMA 6.2. *There is a constant $C > 0$ and an integer $n \geqslant 0$ such that*

$$\sum_{\vartheta} \int_{\mathrm{Re}(s)=0} \frac{1}{4\pi} \left| \sum_{m,\, n \in I(\vartheta)} \mathbf{U}_{mn}^s(\alpha) \cdot \mathbf{E}_\epsilon'(e_m : s : x\kappa_i)\overline{\mathbf{E}_\epsilon'(e_n : s : x\kappa_i)} \right| |ds| \leqslant C(\xi_\lambda(a_x))^{-n}$$

for all $x \in \mathfrak{S}_{t_0,\omega_0}$, $\kappa_i \in \mathfrak{s}$.

Proof. Let $\delta_k \in \Gamma/\Gamma \cap P_k$, $\delta_l \in \Gamma/\Gamma \cap P_l$ $(k \neq l)$—then, since $\mathfrak{C}_\epsilon \kappa_k \cdot \Gamma \cap \mathfrak{C}_\epsilon \kappa_l \cdot \Gamma = \varnothing$, it is clear that

$$\chi_{\epsilon,k}(x\kappa_i\,\delta_k) \cdot \chi_{\epsilon,l}(x\kappa_i\,\delta_l) = 0.$$

Let $\delta_{j_1}, \delta_{j_2} \in \Gamma/\Gamma \cap P_j$ $(j_1 \neq j_2)$—then, since $\mathfrak{C}_\epsilon \kappa_j \cap \mathfrak{C}_\epsilon \kappa_j \gamma \neq \varnothing$ if and only if $\gamma \in \Gamma \cap P_j$, it is clear that

$$\chi_{\epsilon,j}(x\kappa_i\,\delta_{j_1}) \cdot \chi_{\epsilon,j}(x\kappa_i\,\delta_{j_2}) = 0.$$

If ϵ is chosen sufficiently small, as we suppose, then $\mathfrak{S}_{t_0,\omega_0}\kappa_k \cap \mathfrak{C}_\epsilon \kappa_l \cdot \Gamma = \varnothing$ provided $k \neq l$ (cf. Raghunathan [23b, p. 202]). The expression to be estimated is thus majorized by

$$\sum_{\delta \in \Gamma/\Gamma \cap P_i} \sum_{\vartheta} \int_{\mathrm{Re}(s)=0} \frac{1}{4\pi} \left| \sum_{m,\, n \in I(\vartheta)} \mathbf{U}_{mn}^s(\alpha) \cdot \mathbf{E}_{P_i}(e_m : s : x\kappa_i\,\delta)\overline{\mathbf{E}_{P_i}(e_n : s : x\kappa_i\,\delta)} \right| |ds|$$
$$\times\ \chi_{\epsilon,i}(x\kappa_i\,\delta).$$

At the cost of decreasing ϵ still further, one can force δ into $\Gamma \cap P_i$ (cf. Raghunathan [23b, p. 202]). Agreeing to make this adjustment and bearing in mind that $\mathbf{E}_{P_i}$ is obtained from $\mathbf{E}$ by integrating over a compact set, we find that the contention of the present lemma then follows from Lemma 4.8. ■

Let us set

$$K_\alpha^{\prime\prime,c}(x:\epsilon) = K_\alpha^{c}(x,x) - K_\alpha^{\prime,c}(x:\epsilon).$$

Then we can decompose $K_\alpha^{d}(x,x)$ according to the following scheme:

$$\sum_{z \in Z_\Gamma} \alpha(z), \qquad \sum_{\gamma \in \Gamma_S} \alpha(x\gamma x^{-1}),$$

$$I_\alpha(x:\epsilon) - K_\alpha^{\prime,c}(x:\epsilon), \qquad J_\alpha(x:\epsilon), \qquad -K_\alpha^{\prime\prime,c}(x:\epsilon).$$

We shall refer to these expressions as the *central*, *semisimple*, and *first*, *second*, and *third* parabolic terms, respectively. In order to calculate

$$\int_{G/\Gamma} K_\alpha^{\mathrm{d}}(x,x)\,d_G(x),$$

it is clearly enough to calculate the integral over G/Γ of each of the five terms separately. Unfortunately, things are not quite as simple as this. The problem lies with the first and third parabolic terms in that it seems difficult to establish directly that they are integrable over G/Γ. The situation can, however, be saved. We shall introduce below the notion of "weak integrability." Any integrable function is weakly integrable. During the ensuing discussion, it will be shown that the central, semisimple, and second parabolic terms are integrable, hence weakly integrable. In addition, the first parabolic term will turn out to be weakly integrable. Therefore, since K_α^{d} is integrable, it follows that the third parabolic term is weakly integrable. We shall, as we go along, calculate the integrals of the integrable terms and the weak integrals of the weakly integrable terms. This, of course, will suffice for the explicit evaluation of

$$\int_{G/\Gamma} K_\alpha^{\mathrm{d}}(x,x)\,d_G(x).$$

At this point we had best commit ourselves on a choice of the relevant invariant measures. It will be assumed that the Haar measure on G has been pre-assigned. Assuming that Γ has been given counting measure, the total volume $\mathrm{vol}(G/\Gamma)$ of G/Γ is then fixed by requiring that the integral formula

$$\int_G = \int_{G/\Gamma} \sum_{\Gamma}$$

be valid. The Haar measure on K will be so normalized that the total volume of K is one. Let i be an index between 1 and r. The Haar measure on M_i will be so normalized that the total volume of M_i is one. If it is stipulated that Z_Γ is equipped with counting measure, then an invariant measure on M_i/Z_Γ is determined by demanding that the integral formula

$$\int_{M_i} = \int_{M_i/Z_\Gamma} \sum_{Z_\Gamma}$$

hold true; this determination, however, is not the one that we shall actually use (see below). Assign to A_i the Haar measure obtained by exponentiating normalized Lebesgue measure (relative to the Euclidean structure associated with the Killing form). Fix once and for all a subgroup Ξ_i of finite index $\imath(i)$ in $N_i \cap \Gamma$ such that $L_i = \log(\Xi_i)$ is a lattice in $\mathfrak{n}_i$, the Lie algebra of N_i; that it is possible to do this follows from a generality due to Moore [21, p. 155]. Assuming that L_i has been given counting measure and that $\mathfrak{n}_i/L_i$ has been given total volume $\iota(i)$, the Haar measure on $\mathfrak{n}_i$ is then fixed by requiring that the integral formula

$$\int_{\mathfrak{n}_i} = \int_{\mathfrak{n}_i/L_i} \sum_{L_i}$$

be valid. Exponentiate this particular choice of Haar measure on $\mathfrak{n}_i$ to a Haar measure on N_i—then it is still the case that

$$\int_{N_i} = \int_{N_i/\Xi_i} \sum_{\Xi_i}$$

if, as we suppose, the total volume of N_i/Ξ_i is $\iota(i)$. Moreover, one has

$$\int_{N_i} = \int_{N_i/N_i \cap \Gamma} \sum_{N_i \cap \Gamma},$$

provided, of course, the total volume of $N_i/N_i \cap \Gamma$ is one. With these agreements, there is determined a positive constant $\mathbf{c}_G^i$ such that

$$\int_G f(x)\, d_G(x) = \mathbf{c}_G^i \int_K \int_{M_i} \int_{A_i} \int_{N_i} f(km_i a_i n_i) e^{2|\rho|(H_i(a_i))}\, d_K(k)\, d_{M_i}(m_i)\, d_{A_i}(a_i)\, d_{N_i}(n_i)$$

for all $f \in C_c(G)$. In the preceding and the following, we shall assume that M_i/Z_Γ has been given that invariant measure in which the total volume of M_i/Z_Γ is $\mathbf{c}_G^i/[Z_\Gamma]$; of course this normalization agrees with the one determined by the integral formula

$$\int_{M_i} = \int_{M_i/Z_\Gamma} \sum_{Z_\Gamma}$$

only when $\mathbf{c}_G^i = 1$ but this will not matter. If it is stipulated that $\Gamma \cap P_i$ is equipped with counting measure, then the invariant measure on $G/\Gamma \cap P_i$ is determined by demanding that the integral formula

$$\int_G = \int_{G/\Gamma \cap P_i} \sum_{\Gamma \cap P_i}$$

hold true. One then has that

$$\int_{G/\Gamma \cap P_i} f(x)\, d_G(x)$$
$$= \int_K \int_{M_i/Z_\Gamma} \int_{A_i} \int_{N_i/N_i \cap \Gamma} f(km_i a_i n_i) e^{2|\rho|(H_i(a_i))}\, d_K(k)\, d_{M_i}(m_i)\, d_{A_i}(a_i)\, d_{N_i}(n_i)$$

for all $f \in C_c(G/\Gamma \cap P_i)$.

Before proceeding to the actual calculations, we need to dispense with one more preliminary, namely the notion of "weak integrability." Let

$0 < t \leqslant {}_0t$—then G/Γ admits the partition

$$G/\Gamma = \pi(\Omega_t) \cup \bigcup_{i=1}^{r} \pi(\mathfrak{C}_t\kappa_i) \qquad \text{(disjoint union)},$$

where $\pi: G \to G/\Gamma$ is the natural projection (cf. Section 2). Call $\pi_i: G \to G/\Gamma \cap P_i$ the natural projection, $\pi_{P_i}: G/\Gamma \cap P_i \to G/\Gamma$ the canonical map. Clearly

$$\pi_{P_i}: \pi_i(\mathfrak{C}_t\kappa_i) \to \pi(\mathfrak{C}_t\kappa_i)$$

is a homeomorphism. Suppose now that f is an integrable function on G/Γ—then

$$\begin{aligned}\int_{G/\Gamma} f(x)\,d_G(x) &= \int_{\pi(\Omega_t)} f(x)\,d_G(x) + \sum_{i=1}^{r} \int_{\pi(\mathfrak{C}_t\kappa_i)} f(x)\,d_G(x)\\ &= \int_{\pi(\Omega_t)} f(x)\,d_G(x) + \sum_{i=1}^{r} \int_{\pi_i(\mathfrak{C}_t\kappa_i)} f \circ \pi_{P_i}(x)\,d_G(x).\end{aligned}$$

Let $\tilde{\chi}_{t,i}$ be the characteristic function of $\pi_i(\mathfrak{C}_t\kappa_i)$—then

$$\begin{aligned}\int_{\pi_i(\mathfrak{C}_t\kappa_i)} f \circ \pi_{P_i}(x)\,d_G(x) &= \int_{G/\Gamma \cap P_i} \tilde{\chi}_{t,i}(x) f \circ \pi_{P_i}(x)\,d_G(x)\\ &= \int_K \int_{M_i/Z_\Gamma} \int_{A_i(t)} \int_{N_i/N_i \cap \Gamma} f \circ \pi_{P_i}(k m_i a_i n_i)\\ &\quad \times e^{2|\rho|(H_i(a_i))}\,d_K(k)\,d_{M_i}(m_i)\,d_{A_i}(a_i)\,d_{N_i}(n_i).\end{aligned}$$

Here, of course,

$$A_i(t) = \{a_i \in A_i : \xi_\lambda(a_i) < t\}.$$

Because f is integrable, all the integrals above are absolutely convergent. In addition,

$$\int_{G/\Gamma} f(x)\,d_G(x) = \lim_{t \downarrow 0} \int_{\pi(\Omega_t)} f(x)\,d_G(x).$$

Suppose now that f is a locally integrable function on G/Γ—then we shall say that f is *weakly integrable* on G/Γ if for some $0 < t \leqslant {}_0t$ the integral

$$\int_{A_i(t)} \left| \int_K \int_{M_i/Z_\Gamma} \int_{N_i/N_i \cap \Gamma} f \circ \pi_{P_i}(k m_i a_i n_i)\,d_K(k)\,d_{M_i}(m_i)\,d_{N_i}(n_i) \right| e^{2|\rho|(H_i(a_i))}\,d_{A_i}(a_i)$$

is finite for all i. The issue of weak integrability is thus one of iterated integrals. If f is weakly integrable, then the quantity

$$\int_{\pi(\Omega_{0^t})} f(x)\,d_G(x) + \sum_{i=1}^{r} \int_{A_i(_0t)} \left(\int_K \int_{M_i/Z_\Gamma} \int_{N_i/N_i\cap\Gamma} f \circ \pi_{P_i}(km_ia_in_i)\,d_K(k)\,d_{M_i}(m_i)\,d_{N_i}(n_i) \right) \times e^{2|\rho|(H_i(a_i))}\,d_{A_i}(a_i)$$

is finite. We shall call it the *weak integral* of f and denote it by the symbol

$$(w)\int_{G/\Gamma} f(x)\,d_G(x).$$

Obviously

$$(w)\int_{G/\Gamma} f(x)\,d_G(x) = \int_{\pi(\Omega_t)} f(x)\,d_G(x) + \sum_{i=1}^{r} \int_{A_i(t)} \left(\int_K \int_{M_i/Z_\Gamma} \int_{N_i/N_i\cap\Gamma} f \circ \pi_{P_i}(km_ia_in_i)\,d_K(k)\,d_{M_i}(m_i)\,d_{N_i}(n_i) \right) \times e^{2|\rho|(H_i(a_i))}\,d_{A_i}(a_i)$$

for all $0 < t \leqslant {}_0t$. Therefore

$$(w)\int_{G/\Gamma} f(x)\,d_G(x) = \lim_{t\downarrow 0} \int_{\pi(\Omega_t)} f(x)\,d_G(x).$$

Having taken care of the preliminaries, let us pass to the calculation of

$$\int_{G/\Gamma} K_\alpha^{\mathrm{d}}(x, x)\,d_G(x).$$

The central term offers no difficulty:

$$\int_{G/\Gamma} \left(\sum_{z\in Z_\Gamma} \alpha(z) \right) d_G(x) = \mathrm{vol}(G/\Gamma) \left(\sum_{z\in Z_\Gamma} \alpha(z) \right).$$

The calculation of the semisimple term is also immediate (cf. the proof of Lemma 5.4):

$$\int_{G/\Gamma} \left(\sum_{\gamma\in\Gamma_S} \alpha(x\gamma x^{-1}) \right) d_G(x) = (S) \sum_{\{\gamma\}_\Gamma} \mathrm{vol}(G_\gamma/\Gamma_\gamma) \cdot \int_{G/G_\gamma} \alpha(x\gamma x^{-1})\,d_{G/G_\gamma}(x).$$

We shall now investigate the first parabolic term. Our objective will be to show that

$$(w)\int_{G/\Gamma} (I_\alpha(x:\epsilon) - K_\alpha^{\prime,c}(x:\epsilon))\,d_G(x)$$

exists and that, moreover,

$$\lim_{\epsilon \downarrow 0} (w) \int_{G/\Gamma} (I_\alpha(x:\epsilon) - K_\alpha^{\prime,c}(x:\epsilon))\, d_G(x) = 0.$$

Both $I_\alpha(?:\epsilon)$ and $K_\alpha^{\prime,c}(?:\epsilon)$ are evidently locally integrable on G/Γ, so

$$\int_{\pi(\Omega_{0t})} (I_\alpha(x:\epsilon) - K_\alpha^{\prime,c}(x:\epsilon))\, d_G(x)$$

exists and in fact vanishes, the integrand being identically zero provided ϵ is sufficiently small, as we suppose (cf. the proof of Lemma 6.2). The main point, then, is to establish the finiteness of the integrals

$$\int_{A_i(0t)} \left| \int_K \int_{M_i/Z_\Gamma} \int_{N_i/N_i \cap \Gamma} (I_\alpha(km_i a_i n_i:\epsilon) - K_\alpha^{\prime,c}(km_i a_i n_i:\epsilon))\, d_K(k)\, d_{M_i}(m_i)\, d_{N_i}(n_i) \right| e^{2|\rho|(H_i(a_i))}\, d_{A_i}(a_i)$$

and to show that they approach zero as ϵ approaches zero. This will be done as follows. On $\pi_i(\mathfrak{C}_{0t}\kappa_i)$, we shall write

$$I_\alpha(?:\epsilon) = (\mathrm{I})_\epsilon^i + (\mathrm{II})_\epsilon^i + f_\epsilon^i + F_\epsilon^i,$$
$$K_\alpha^{\prime,c}(?:\epsilon) = (\mathrm{I})_\epsilon^i + (\mathrm{II})_\epsilon^i + (\mathrm{III})_\epsilon^i + (\mathrm{IV})_\epsilon^i,$$

the terms on the right being certain functions on $\pi_i(\mathfrak{C}_{0t}\kappa_i)$ which will be defined presently. The troublesome terms are $(\mathrm{I})_\epsilon^i$ and $(\mathrm{II})_\epsilon^i$ but since they are common to both $I_\alpha(?:\epsilon)$ and $K_\alpha^{\prime,c}(?:\epsilon)$, no difficulty actually arises. The integrals of f_ϵ^i, F_ϵ^i, $(\mathrm{III})_\epsilon^i$, and $(\mathrm{IV})_\epsilon^i$ turn out to be finite (f_ϵ^i and F_ϵ^i are even integrable over $\pi_i(\mathfrak{C}_{0t}\kappa_i)$) and to approach zero as ϵ approaches zero.

It will be convenient to start off with a careful look at $K_\alpha^{\prime,c}$. Suppose that $x \in \mathfrak{C}_{0t}$—then, by definition,

$$K_\alpha^{\prime,c}(x\kappa_i:\epsilon) = \int_{\mathrm{Re}(s)=0} \left(\frac{1}{4\pi} \sum_{m,\, n \in I} \mathbf{U}_{mn}^s(\alpha) \cdot \mathbf{E}_\epsilon^\prime(e_m:s:x\kappa_i)\overline{\mathbf{E}_\epsilon^\prime(e_n:s:x\kappa_i)} \right) |ds|$$

or still (cf. the proof of Lemma 6.2),

$$K_\alpha^{\prime,c}(x\kappa_i:\epsilon) = \sum_{\vartheta} \int_{\mathrm{Re}(s)=0} \left(\frac{1}{4\pi} \sum_{m,\, n \in I(\vartheta)} \mathbf{U}_{mn}^s(\alpha) \cdot \mathbf{E}_{P_i}(e_m:s:x\kappa_i)\overline{\mathbf{E}_{P_i}(e_n:s:x\kappa_i)} \right) |ds| \times \chi_{\epsilon,i}(x\kappa_i).$$

Write

$$\mathbf{U}^{\vartheta,s}(\alpha)e_n = \sum_{m \in I(\vartheta)} (\mathbf{U}^{\vartheta,s}(\alpha)e_n, e_m)e_m.$$

Then we have

$$K_\alpha^{\prime,\mathrm{c}}(x\kappa_i:\epsilon)$$

$$= \sum_{\vartheta} \int_{\mathrm{Re}(s)=0} \left(\frac{1}{4\pi} \sum_{n\in I(\vartheta)} \mathbf{E}_{P_i}(\mathbf{U}^{\vartheta,s}(\alpha)e_n : s : x\kappa_i)\overline{\mathbf{E}_{P_i}(e_n : s : x\kappa_i)}\right)|ds| \cdot \chi_{\epsilon,i}(x\kappa_i).$$

Substitute in the usual formulas for the constant term of an Eisenstein series—then we can formally write $K_\alpha^{\prime,\mathrm{c}}(x\kappa_i:\epsilon)$ as the sum of the following four expressions:

$(\mathrm{I})_\epsilon^i$:

$$\frac{1}{4\pi} \sum_{\vartheta} \int_{\mathrm{Re}(s)=0} \left(\sum_{n\in I(\vartheta)} (\mathbf{U}^{\vartheta,s}(\alpha)e_n)_i(x\kappa_i)\overline{(e_n)_i(x\kappa_i)}\right)|ds|$$
$$\times e^{-2|\rho|(H_i(x\kappa_i))} \cdot \chi_{\epsilon,i}(x\kappa_i);$$

$(\mathrm{II})_\epsilon^i$:

$$\frac{1}{4\pi} \sum_{\vartheta} \int_{\mathrm{Re}(s)=0} \left(\sum_{n\in I(\vartheta)} (\mathbf{c}_\vartheta(s)\mathbf{U}^{\vartheta,s}(\alpha)e_n)_i(x\kappa_i)\overline{(\mathbf{c}_\vartheta(s)e_n)_i(x\kappa_i)}\right)|ds|$$
$$\times e^{-2|\rho|(H_i(x\kappa_i))} \cdot \chi_{\epsilon,i}(x\kappa_i);$$

$(\mathrm{III})_\epsilon^i$:

$$\frac{1}{4\pi} \sum_{\vartheta} \int_{\mathrm{Re}(s)=0} \left(\sum_{n\in I(\vartheta)} (\mathbf{c}_\vartheta(s)\mathbf{U}^{\vartheta,s}(\alpha)e_n)_i(x\kappa_i)\overline{(e_n)_i(x\kappa_i)}\right)e^{-2s(H_i(x\kappa_i))}|ds|$$
$$\times e^{-2|\rho|(H_i(x\kappa_i))} \cdot \chi_{\epsilon,i}(x\kappa_i);$$

$(\mathrm{IV})_\epsilon^i$:

$$\frac{1}{4\pi} \sum_{\vartheta} \int_{\mathrm{Re}(s)=0} \left(\sum_{n\in I(\vartheta)} (\mathbf{U}^{\vartheta,s}(\alpha)e_n)_i(x\kappa_i)\overline{(\mathbf{c}_\vartheta(s)e_n)_i(x\kappa_i)}\right)e^{2s(H_i(x\kappa_i))}|ds|$$
$$\times e^{-2|\rho|(H_i(x\kappa_i))} \cdot \chi_{\epsilon,i}(x\kappa_i).$$

Before justifying these formalities, we insert an ancillary result. Recalling the notations introduced in Section 3, write

$$\mathscr{E} = \sum_{i=1}^{r} \oplus \mathscr{E}_i,$$

where

$$\mathscr{E}_i = \sum_{\vartheta_i} \oplus \mathscr{E}(\vartheta_i).$$

If by $\mathbf{U}_i^s$ we understand the restriction of $\mathbf{U}^s$ to $\mathscr{E}_i$, then

$$\mathbf{U}^s = \sum_{i=1}^{r} \oplus \mathbf{U}_i^s.$$

We remark that $\mathscr{E}_i$ can be identified with the Hilbert space of those Borel functions $\Phi_i : G/(\Gamma \cap P_i) \cdot A_i \cdot N_i \to \mathbf{C}$ which have the property that for every $x \in G$, the function

$$m_i \mapsto \Phi_i(xm_i)$$

belongs to $L^2(M_i/Z_\Gamma)$, and is such that

$$\|\Phi_i\|^2 = \int_K \int_{M_i/Z_\Gamma} \Phi_i(km_i)\overline{\Phi_i(km_i)}\, d_K(k)\, d_{M_i}(m_i) < +\infty.$$

Here, in accordance with our agreement above, the invariant measure on M_i/Z_Γ is the one in which the total volume of M_i/Z_Γ is $\mathbf{c}_G^i/[Z_\Gamma]$. Given $\Phi_i \in \mathscr{E}_i$,

$$\mathbf{U}_i^s(x)\Phi_i$$

is the function

$$y \mapsto e^{(-s+|\rho|)(H_i(y))} \cdot e^{(s-|\rho|)(H_i(x^{-1}y))} \cdot \Phi_i(x^{-1}y) \qquad (x, y \in G).$$

SUBLEMMA. *Let α be a K-finite element in $C_c^\infty(G)$—then, for any s, $\mathbf{U}_i^s(\alpha)$ is of the Hilbert–Schmidt class. Put*

$$\mathbf{U}_\alpha^i(x, y : s) = e^{(-s+|\rho|)(H_i(x))} e^{(s+|\rho|)(H_i(y))}$$

$$\times \sum_{z \in Z_\Gamma} \int_{A_i} \int_{N_i} \alpha(xza_in_iy^{-1}) e^{(-s+|\rho|)(H_i(a_i))}\, d_{A_i}(a_i)\, d_{N_i}(n_i).$$

Then the kernel $\mathbf{U}_\alpha^i(x, y : s)$ represents $\mathbf{U}_i^s(\alpha)$.

Proof. Each of the representations $\mathbf{U}_i^s$ operates on the same Hilbert space, viz., $\mathscr{E}_i$. On the basis of what has been said in Section 3, it is clear that there exists an absolute constant $C > 0$ and an integer $n > 0$ such that the dimension of the δth-isotypic component of $\mathbf{U}_i^s$ (any s) cannot exceed $C \cdot (d(\delta))^n$. This being the case, it then follows from a standard generality (cf. Warner [34a, p. 336]) that $\mathbf{U}_i^s(\alpha)$ is a Hilbert–Schmidt operator, hence is represented by a kernel square integrable over

$$(K \times M_i/Z_\Gamma) \times (K \times M_i/Z_\Gamma).$$

Fix $\Phi_i \in \mathscr{E}_i$—then we have

$$
\begin{aligned}
&(\mathbf{U}_i^s(\alpha)\Phi_i)(x) \\
&= \int_G \alpha(y)(\mathbf{U}_i^s(y)\Phi_i)(x)\, d_G(y) \\
&= e^{(-s+|\rho|)(H_i(x))} \cdot \int_G \alpha(y) e^{(s-|\rho|)(H_i(y^{-1}x))} \Phi_i(y^{-1}x)\, d_G(y) \\
&= e^{(-s+|\rho|)(H_i(x))} \cdot \int_G \alpha(xy^{-1}) e^{(s-|\rho|)(H_i(y))} \Phi_i(y)\, d_G(y) \\
&= \mathbf{c}_G^i \cdot e^{(-s+|\rho|)(H_i(x))} \\
&\quad \times \int_K \int_{A_i} \int_{N_i} \alpha(xn_i^{-1}a_i^{-1}k^{-1}) e^{(s+|\rho|)(H_i(a_i))} \Phi_i(ka_in_i)\, d_K(k)\, d_{A_i}(a_i)\, d_{N_i}(n_i) \\
&= \mathbf{c}_G^i \cdot e^{(-s+|\rho|)(H_i(x))} \\
&\quad \times \int_K \int_{A_i} \int_{N_i} \alpha(xa_in_ik^{-1}) e^{(-s+|\rho|)(H_i(a_i))} \Phi_i(k)\, d_K(k)\, d_{A_i}(a_i)\, d_{N_i}(n_i) \\
&= \frac{\mathbf{c}_G^i}{[Z_\Gamma]} \cdot e^{(-s+|\rho|)(H_i(x))} \\
&\quad \times \int_K \sum_{z \in Z_\Gamma} \int_{A_i} \int_{N_i} \alpha(xza_in_ik^{-1})\ e^{(-s+|\rho|)(H_i(a_i))} \Phi_i(k)\, d_K(k)\, d_{A_i}(a_i)\, d_{N_i}(n_i) \\
&= \frac{\mathbf{c}_G^i}{[Z_\Gamma]} \cdot \int_K \mathbf{U}_\alpha^i(x, k : s)\Phi_i(k)\, d_K(k) \\
&= \frac{\mathbf{c}_G^i}{[Z_\Gamma]} \cdot \int_K \int_{M_i} \mathbf{U}_\alpha^i(x, km_i : s)\Phi_i(km_i)\, d_K(k)\, d_{M_i}(m_i) \\
&= \frac{1}{[Z_\Gamma]} \cdot \int_K \int_{M_i/Z_\Gamma} \sum_{z \in Z_\Gamma} \mathbf{U}_\alpha^i(x, km_iz : s)\Phi_i(km_iz)\, d_K(k)\, d_{M_i}(m_i) \\
&= \int_K \int_{M_i/Z_\Gamma} \mathbf{U}_\alpha^i(x, km_i : s)\Phi_i(km_i)\, d_K(k)\, d_{M_i}(m_i).
\end{aligned}
$$

Because $\mathbf{U}_\alpha^i(?, ? : s)$ is continuous on the compact space

$$(K \times M_i/Z_\Gamma) \times (K \times M_i/Z_\Gamma),$$

hence square integrable there, we are done. ■

Remark. Suppose that $\mathrm{Re}(s) = 0$—then the above result is valid for any K-finite function α in $\mathscr{C}^1(G)$.

It is now an easy matter to justify the preceding manipulations. Fix an $x \in G$—then we need only show that the following four expressions are finite:

(I_x):

$$\sum_{\vartheta} \int_{\mathrm{Re}(s)=0} \left| \sum_{n \in I(\vartheta)} (\mathbf{U}^{\vartheta,s}(\alpha)e_n)_i(x)\overline{(e_n)_i(x)} \right| |ds|;$$

(II_x):

$$\sum_{\vartheta} \int_{\mathrm{Re}(s)=0} \left| \sum_{n \in I(\vartheta)} (\mathbf{c}_\vartheta(s)\mathbf{U}^{\vartheta,s}(\alpha)e_n)_i(x)\overline{(\mathbf{c}_\vartheta(s)e_n)_i(x)} \right| |ds|;$$

(III_x):

$$\sum_{\vartheta} \int_{\mathrm{Re}(s)=0} \left| \sum_{n \in I(\vartheta)} (\mathbf{c}_\vartheta(s)\mathbf{U}^{\vartheta,s}(\alpha)e_n)_i(x)\overline{(e_n)_i(x)} \right| |ds|;$$

(IV_x):

$$\sum_{\vartheta} \int_{\mathrm{Re}(s)=0} \left| \sum_{n \in I(\vartheta)} (\mathbf{U}^{\vartheta,s}(\alpha)e_n)_i(x)\overline{(\mathbf{c}_\vartheta(s)e_n)_i(x)} \right| |ds|.$$

We have assumed that the orthonormal basis for $\mathscr{E}$ has been formed by putting together orthonormal bases from each of the $\mathscr{E}(\vartheta, \delta)$. We shall now assume in addition that the orthonormal basis for $\mathscr{E}(\vartheta, \delta)$ is made up of orthonormal bases $\{e_{n_j}\}$ from each $\mathscr{E}(\vartheta_j, \delta)$ $(1 \leqslant j \leqslant r)$ so that, in an obvious notation, $I(\vartheta) = \bigcup_{j=1}^r I(\vartheta_j)$. Write $\mathbf{P}_i(\vartheta)$ for the orthogonal projection of $\mathscr{E}(\vartheta)$ onto $\mathscr{E}(\vartheta_i)$. Consider first (I_x). We have

$$\begin{aligned}
&\sum_{n \in I(\vartheta)} (\mathbf{U}^{\vartheta,s}(\alpha)e_n)_i(x)\overline{(e_n)_i(y)} \\
&\quad = \sum_{j=1}^r \sum_{n_j \in I(\vartheta_j)} (\mathbf{U}^{\vartheta,s}(\alpha)e_{n_j})_i(x)\overline{(e_{n_j})_i(y)} \\
&\quad = \sum_{j=1}^r \sum_{n_j \in I(\vartheta_j)} (\mathbf{P}_i(\vartheta)\mathbf{U}^{\vartheta,s}(\alpha)e_{n_j})(x)\overline{(\mathbf{P}_i(\vartheta)e_{n_j})(y)} \\
&\quad = \sum_{n_i \in I(\vartheta_i)} (U^{\vartheta_i,s}(\alpha)e_{n_i})(x)\overline{e_{n_i}(y)}.
\end{aligned}$$

Therefore the function $Q_\alpha^i(x, y:s:\vartheta)$ defined by the sum

$$\sum_{n \in I(\vartheta)} (\mathbf{U}^{\vartheta,s}(\alpha)e_n)_i(x)\overline{(e_n)_i(y)}$$

is the kernel of $U^{\vartheta_i,s}(\alpha)$. Per the proof of Lemma 4.8, introduce α_{H}, β_{H}, μ_{H}, ν_{H}—then we can estimate $|Q_\alpha^i(x, x:s:\vartheta)|$ as follows:

$$\begin{aligned}
&|Q_\alpha^i(x, x:s:\vartheta)| \\
&\quad \leqslant \left|\int_{K\times M_i/Z_\Gamma} Q_\alpha^i(x, y:s:\vartheta)Q_\nu^i(y, x:s:\vartheta)\,dy\right| \\
&\qquad + \left|\int_{K\times M_i/Z_\Gamma} Q_\beta^i(x, y:s:\vartheta)Q_\mu^i(y, x:s:\vartheta)\,dy\right| \\
&\quad \leqslant \left(\int_{K\times M_i/Z_\Gamma} |Q_\alpha^i(x, y:s:\vartheta)|^2\,dy\right)^{1/2}\left(\int_{K\times M_i/Z_\Gamma} |Q_\nu^i(y, x:s:\vartheta)|^2\,dy\right)^{1/2} \\
&\qquad + \left(\int_{K\times M_i/Z_\Gamma} |Q_\beta^i(x, y:s:\vartheta)|^2\,dy\right)^{1/2}\left(\int_{K\times M_i/Z_\Gamma} |Q_\mu^i(y, x:s:\vartheta)|^2\,dy\right)^{1/2}
\end{aligned}$$

If $\sigma = \alpha$ or β, then

$$\begin{aligned}
\int_{K\times M_i/Z_\Gamma} |Q_\sigma^i(x, y:s:\vartheta)|^2\,dy &= \int_{K\times M_i/Z_\Gamma} Q_\sigma^i(x, y:s:\vartheta)\overline{Q_\sigma^i(x, y:s:\vartheta)}\,dy \\
&= \int_{K\times M_i/Z_\Gamma} Q_\sigma^i(x, y:s:\vartheta)Q_{\sigma^*}^i(y, x:s:\vartheta)\,dy \\
&= Q_{\sigma*\sigma^*}^i(x, x:s:\vartheta),
\end{aligned}$$

whereas if $\tau = \nu$ or μ, then

$$\begin{aligned}
\int_{K\times M_i/Z_\Gamma} |Q_\tau^i(y, x:s:\vartheta)|^2\,dy &= \int_{K\times M_i/Z_\Gamma} Q_\tau^i(y, x:s:\vartheta)\overline{Q_\tau^i(y, x:s:\vartheta)}\,dy \\
&= \int_{K\times M_i/Z_\tau} Q_\tau^i(y, x:s:\vartheta)Q_{\tau^*}^i(x, y:s:\vartheta)\,dy \\
&= Q_{\tau^**\tau}^i(x, x:s:\vartheta).
\end{aligned}$$

Therefore $|Q_\alpha^i(x, x:s:\vartheta)|$ is majorized by

$$(Q_{\alpha_{\mathbf{H}}}^i(x, x:s:\vartheta))^{1/2}(Q_{\nu_{\mathbf{H}}}^i(x, x:s:\vartheta))^{1/2} + (Q_{\beta_{\mathbf{H}}}^i(x, x:s:\vartheta))^{1/2}(Q_{\mu_{\mathbf{H}}}^i(x, x:s:\vartheta))^{1/2}.$$

So, thanks to the Schwarz inequality, to establish the finiteness of (I_x) it can be supposed that α has been replaced by $\phi = \psi * \psi^*$ say, ψ a K-finite compactly supported, sufficiently differentiable function. This said, let $\mathscr{F} = \{\vartheta\}$ be any finite collection of orbits—then obviously

$$\sum_{\vartheta\in\mathscr{F}} Q_\phi^i(x, x:s:\vartheta) \leqslant \mathbf{U}_\phi^i(x, x:s).$$

Because

$$|Q_\phi^i(x, y:s:\vartheta)| \leqslant (Q_\phi^i(x, x:s:\vartheta))^{1/2}\cdot(Q_\phi^i(y, y:s:\vartheta))^{1/2},$$

it follows that the series

$$\sum_\vartheta Q_\phi^i(x, y:s:\vartheta)$$

is absolutely convergent and defines a separately continuous function (cf. Lemma 4.9) which, moreover, is a kernel representing $\mathbf{U}_i^s(\phi)$. But then, by continuity,

$$\sum_{\vartheta} Q_\phi^i(x, y : s : \vartheta) = \mathbf{U}_\phi^i(x, y : s).$$

In view of the explicit formula for $\mathbf{U}_\phi^i(x, x : s)$ provided by the sublemma, it is clear that

$$\int_{\mathrm{Re}(s)=0} \mathbf{U}_\phi^i(x, x : s)\,|ds| < +\infty.$$

Therefore (I_x) is finite. The main point in verifying the finiteness of (II_x) is to note that

$$\sum_{n \in I(\vartheta)} (\mathbf{c}_\vartheta(s)\mathbf{U}^{\vartheta,s}(\alpha)e_n)_i(x)\overline{(\mathbf{c}_\vartheta(s)e_n)_i(y)}$$

is the kernel of $U^{\vartheta_i,-s}(\alpha)$. Once this has been established, one can then argue as above. Owing to Lemma 3.6, $\mathbf{c}_\vartheta(s)\mathbf{U}^{\vartheta,s}(\alpha) = \mathbf{U}^{\vartheta,-s}(\alpha)\mathbf{c}_\vartheta(s)$. Let $\mathbf{\Phi} \in \mathscr{E}(\vartheta)$—then, due to the fact that $\mathbf{c}_\vartheta(s)$ is a unitary operator for pure imaginary s, we can write

$$\Phi_i = \mathbf{P}_i(\vartheta)\mathbf{\Phi} = \sum_{n \in I(\vartheta)} (\mathbf{P}_i(\vartheta)\mathbf{\Phi}, \mathbf{c}_\vartheta(s)e_n)\mathbf{c}_\vartheta(s)e_n,$$

so

$$\begin{aligned} U^{\vartheta_i,-s}(\alpha)\Phi_i(x) &= \sum_{n \in I(\vartheta)} (\mathbf{P}_i(\vartheta)\mathbf{\Phi}, \mathbf{c}_\vartheta(s)e_n)(\mathbf{P}_i(\vartheta)\mathbf{U}^{\vartheta,-s}(\alpha)\mathbf{c}_\vartheta(s)e_n)(x) \\ &= \sum_{n \in I(\vartheta)} (\Phi_i, \mathbf{P}_i(\vartheta)\mathbf{c}_\vartheta(s)e_n)(\mathbf{P}_i(\vartheta)\mathbf{c}_\vartheta(s)\mathbf{U}^{\vartheta,s}(\alpha)e_n)(x) \\ &= \int_{K \times M_i/Z_\Gamma} \left(\sum_{n \in I(\vartheta)} (\mathbf{c}_\vartheta(s)\mathbf{U}^{\vartheta,s}(\alpha)e_n)_i(x)\overline{(\mathbf{c}_\vartheta(s)e_n)_i(y)} \right) \Phi_i(y)\,dy, \end{aligned}$$

which, of course, is equivalent to our contention. Let us turn to (III_x). We have

$$\begin{aligned} &\sum_{n \in I(\vartheta)} (\mathbf{c}_\vartheta(s)\mathbf{U}^{\vartheta,s}(\alpha)e_n)_i(x)\overline{(e_n)_i(y)} \\ &\quad = \sum_{j=1}^{r} \sum_{n_j \in I(\vartheta_j)} (\mathbf{c}_\vartheta(s)\mathbf{U}^{\vartheta,s}(\alpha)e_{n_j})_i(x)\overline{(e_{n_j})_i(y)} \\ &\quad = \sum_{j=1}^{r} \sum_{n_j \in I(\vartheta_j)} (\mathbf{P}_i(\vartheta)\mathbf{c}_\vartheta(s)\mathbf{U}^{\vartheta,s}(\alpha)e_{n_j})(x)\overline{(\mathbf{P}_i(\vartheta)e_{n_j})(y)} \\ &\quad = \sum_{n_i \in I(\vartheta_i)} (\mathbf{P}_i(\vartheta)\mathbf{c}_\vartheta(s)\mathbf{U}^{\vartheta,s}(\alpha)\mathbf{P}_i(\vartheta)e_{n_i})(x)\overline{e_{n_i}(y)}. \end{aligned}$$

Therefore the function defined by the sum

$$\sum_{n \in I(\vartheta)} (\mathbf{c}_\vartheta(s)\mathbf{U}^{\vartheta,s}(\alpha)e_n)_i(x)\overline{(e_n)_i(y)}$$

is the kernel of $\mathbf{P}_i(\vartheta)\mathbf{c}_\vartheta(s)\mathbf{U}^{\vartheta,s}(\alpha)\mathbf{P}_i(\vartheta)$. Write

$$\begin{aligned}\mathbf{P}_i(\vartheta)\mathbf{c}_\vartheta(s)\mathbf{U}^{\vartheta,s}(\alpha)\mathbf{P}_i(\vartheta) &= (\mathbf{P}_i(\vartheta)\mathbf{c}_\vartheta(s)\mathbf{U}^{\vartheta,s}(\alpha)\mathbf{P}_i(\vartheta))(\mathbf{U}^{\vartheta,s}(v)\mathbf{P}_i(\vartheta)) \\ &\quad + (\mathbf{P}_i(\vartheta)\mathbf{c}_\vartheta(s)\mathbf{U}^{\vartheta,s}(\beta)\mathbf{P}_i(\vartheta))(\mathbf{U}^{\vartheta,s}(\mu)\mathbf{P}_i(\vartheta)).\end{aligned}$$

Suppose that $\sigma = \alpha$ or β—then

$$(\mathbf{P}_i(\vartheta)\mathbf{c}_\vartheta(s)\mathbf{U}^{\vartheta,s}(\sigma)) = (\mathbf{P}_i(\vartheta)\mathbf{c}_\vartheta(s)\mathbf{U}^{\vartheta,s}(\sigma)\mathbf{P}_i(\vartheta)) + \sum_{j \neq i} (\mathbf{P}_i(\vartheta)\mathbf{c}_\vartheta(s)\mathbf{U}^{\vartheta,s}(\sigma)\mathbf{P}_j(\vartheta)),$$

$$(\mathbf{P}_i(\vartheta)\mathbf{c}_\vartheta(s)\mathbf{U}^{\vartheta,s}(\sigma))^* = (\mathbf{P}_i(\vartheta)\mathbf{c}_\vartheta(s)\mathbf{U}^{\vartheta,s}(\sigma)\mathbf{P}_i(\vartheta))^* + \sum_{j \neq i} (\mathbf{P}_i(\vartheta)\mathbf{c}_\vartheta(s)\mathbf{U}^{\vartheta,s}(\sigma)\mathbf{P}_j(\vartheta))^*,$$

so, since $\mathbf{P}_i(\vartheta)\mathbf{P}_j(\vartheta) = 0$ $(i \neq j)$, we have

$$\begin{aligned}&(\mathbf{P}_i(\vartheta)\mathbf{c}_\vartheta(s)\mathbf{U}^{\vartheta,s}(\sigma))(\mathbf{P}_i(\vartheta)\mathbf{c}_\vartheta(s)\mathbf{U}^{\vartheta,s}(\sigma))^* \\ &\quad = (\mathbf{P}_i(\vartheta)\mathbf{c}_\vartheta(s)\mathbf{U}^{\vartheta,s}(\sigma)\mathbf{P}_i(\vartheta))(\mathbf{P}_i(\vartheta)\mathbf{c}_\vartheta(s)\mathbf{U}^{\vartheta,s}(\sigma)\mathbf{P}_i(\vartheta))^* \\ &\qquad + \sum_{j \neq i} (\mathbf{P}_i(\vartheta)\mathbf{c}_\vartheta(s)\mathbf{U}^{\vartheta,s}(\sigma)\mathbf{P}_j(\vartheta))(\mathbf{P}_i(\vartheta)\mathbf{c}_\vartheta(s)\mathbf{U}^{\vartheta,s}(\sigma)\mathbf{P}_j(\vartheta))^*\end{aligned}$$

which shows that

$$\begin{aligned}&(\mathbf{P}_i(\vartheta)\mathbf{c}_\vartheta(s)\mathbf{U}^{\vartheta,s}(\sigma))(\mathbf{P}_i(\vartheta)\mathbf{c}_\vartheta(s)\mathbf{U}^{\vartheta,s}(\sigma))^* \\ &\quad - (\mathbf{P}_i(\vartheta)\mathbf{c}_\vartheta(s)\mathbf{U}^{\vartheta,s}(\sigma)\mathbf{P}_i(\vartheta))(\mathbf{P}_i(\vartheta)\mathbf{c}_\vartheta(s)\mathbf{U}^{\vartheta,s}(\sigma)\mathbf{P}_i(\vartheta))^*\end{aligned}$$

is a positive (Hilbert–Schmidt) operator on $\mathscr{E}(\vartheta_i)$. Because

$$\begin{aligned}(\mathbf{P}_i(\vartheta)\mathbf{c}_\vartheta(s)\mathbf{U}^{\vartheta,s}(\sigma))(\mathbf{P}_i(\vartheta)\mathbf{c}_\vartheta(s)\mathbf{U}^{\vartheta,s}(\sigma))^* &= \mathbf{P}_i(\vartheta)\mathbf{c}_\vartheta(s)\mathbf{U}^{\vartheta,s}(\sigma)\mathbf{U}^{\vartheta,s}(\sigma)^*\mathbf{c}_\vartheta(s)^*\mathbf{P}_i(\vartheta)^* \\ &= \mathbf{P}_i(\vartheta)\mathbf{c}_\vartheta(s)\mathbf{U}^{\vartheta,s}(\sigma * \sigma^*)\mathbf{c}_\vartheta(s)^{-1}\mathbf{P}_i(\vartheta) \\ &= \mathbf{P}_i(\vartheta)\mathbf{U}^{\vartheta,-s}(\sigma * \sigma^*)\mathbf{P}_i(\vartheta) \\ &= U^{\vartheta_i,-s}(\sigma * \sigma^*),\end{aligned}$$

it follows that the kernel of

$$(\mathbf{P}_i(\vartheta)\mathbf{c}_\vartheta(s)\mathbf{U}^{\vartheta,s}(\sigma)\mathbf{P}_i(\vartheta))(\mathbf{P}_i(\vartheta)\mathbf{c}_\vartheta(s)\mathbf{U}^{\vartheta,s}(\sigma)\mathbf{P}_i(\vartheta))^*$$

is bounded on the diagonal by the kernel of $U^{\vartheta_i,-s}(\sigma * \sigma^*)$. This being the case, if we now proceed as above, we find that the kernel of

$$\mathbf{P}_i(\vartheta)\mathbf{c}_\vartheta(s)\mathbf{U}^{\vartheta,s}(\alpha)\mathbf{P}_i(\vartheta)$$

can be estimated on the diagonal by

$$\begin{aligned}&(Q^i_{\alpha_\mathbf{H}}(x, x: -s:\vartheta))^{1/2}(Q^i_{v_\mathbf{H}}(x, x: s:\vartheta))^{1/2} \\ &\quad + (Q^i_{\beta_\mathbf{H}}(x, x: -s:\vartheta))^{1/2}(Q^i_{\mu_\mathbf{H}}(x, x: s:\vartheta))^{1/2}.\end{aligned}$$

Therefore (III_x) is finite. The reader will have no difficulty in verifying in a similar fashion that (IV_x) is also finite.

Let us go back to $K_\alpha^{\prime,\mathrm{c}}(x\kappa_i:\epsilon)$, which, as we now know, admits the decomposition

$$K_\alpha^{\prime,\mathrm{c}}(x\kappa_i:\epsilon) = (\mathrm{I})_\epsilon^i + (\mathrm{II})_\epsilon^i + (\mathrm{III})_\epsilon^i + (\mathrm{IV})_\epsilon^i.$$

On the basis of what has been said so far, $(\mathrm{I})_\epsilon^i$ can be written as

$$\frac{1}{4\pi}\int_{\mathrm{Re}(s)=0} \mathbf{U}_\alpha^i(x\kappa_i, x\kappa_i : s)\,|ds| \cdot e^{-2|\rho|(H_i(x\kappa_i))} \cdot \chi_{\epsilon,i}(x\kappa_i)$$

or still, by the sublemma,

$$\frac{1}{4\pi}\int_{\mathrm{Re}(s)=0}\left(\sum_{z\in Z_\Gamma}\int_{A_i}\int_{N_i}\alpha(x\kappa_i z a_i n_i \kappa_i^{-1}x^{-1})e^{(-s+|\rho|)(H_i(a_i))}\,d_{A_i}(a_i)\,d_{N_i}(n_i)\right)|ds|$$
$$\times\ \chi_{\epsilon,i}(x\kappa_i).$$

The integral

$$\int_{A_i}\alpha(x\kappa_i z a_i n_i \kappa_i^{-1}x^{-1})e^{(-s+|\rho|)(H_i(a_i))}\,d_{A_i}(a_i)$$

can be viewed as a Fourier transform. Therefore, by the Fourier inversion formula, our expression becomes

$$\frac{1}{2}\cdot\sum_{z\in Z_\Gamma}\int_{N_i}\alpha(x\kappa_i z n_i \kappa_i^{-1}x^{-1})\,d_{N_i}(n_i)\cdot\chi_{\epsilon,i}(x\kappa_i).$$

Since $(\mathrm{II})_\epsilon^i$ can be written in terms of $\mathbf{U}_\alpha^i(x\kappa_i, x\kappa_i : -s)$, it then follows that

$$(\mathrm{I})_\epsilon^i + (\mathrm{II})_\epsilon^i = \sum_{z\in Z_\Gamma}\int_{N_i}\alpha(x\kappa_i z n_i \kappa_i^{-1}x^{-1})\,d_{N_i}(n_i)\cdot\chi_{\epsilon,i}(x\kappa_i).$$

We shall leave $(\mathrm{I})_\epsilon^i$ and $(\mathrm{II})_\epsilon^i$ for the time being; they will be connected below with the other part of the first parabolic term. To finish our investigation of $K_\alpha^{\prime,\mathrm{c}}$, we shall establish the finiteness of the integrals

$$\int_{A_i(0t)}\left|\int_K\int_{M_i/Z_\Gamma}\int_{N_i/N_i\cap\Gamma}(\mathrm{III})_\epsilon^i(km_ia_in_i)\,d_K(k)\,d_{M_i}(m_i)\,d_{N_i}(n_i)\right|e^{2|\rho|(H_i(a_i))}\,d_{A_i}(a_i),$$

$$\int_{A_i(0t)}\left|\int_K\int_{M_i/Z_\Gamma}\int_{N_i/N_i\cap\Gamma}(\mathrm{IV})_\epsilon^i(km_ia_in_i)\,d_K(k)\,d_{M_i}(m_i)\,d_{N_i}(n_i)\right|e^{2|\rho|(H_i(a_i))}\,d_{A_i}(a_i)$$

and show that they approach zero as ϵ approaches zero. We need only discuss the first integral, the discussion of the second being similar. The first integral is evidently majorized by

$$\int_{A_i(\epsilon)}\left|\sum_{\vartheta}\sum_{n_i\in I(\vartheta_i)}\int_{\mathrm{Re}(s)=0}(\mathbf{P}_i(\vartheta)\mathbf{c}_\vartheta(s)\mathbf{U}^{\vartheta,s}(\alpha)\mathbf{P}_i(\vartheta)e_{n_i}, e_{n_i})e^{-2s(H_i(a_i))}\,|ds|\right|d_{A_i}(a_i).$$

The sums in question are, of course, finite. Let $T_i(s)$ be the adjoint of $\mathbf{P}_i(\vartheta)\mathbf{c}_\vartheta(s)\mathbf{P}_i(\vartheta)$—then we have

$$\begin{aligned}(\mathbf{P}_i(\vartheta)\mathbf{c}_\vartheta(s)\mathbf{U}^{\vartheta,s}(\alpha)\mathbf{P}_i(\vartheta)e_{n_i}, e_{n_i}) &= (U^{\vartheta_i,s}(\alpha)e_{n_i}, T_i(s)e_{n_i}) \\ &= \sum_{\tilde{n}_i \in I(\vartheta_i)} (U^{\vartheta_i,s}(\alpha)e_{n_i}, e_{\tilde{n}_i})\overline{(T_i(s)e_{n_i}, e_{\tilde{n}_i})},\end{aligned}$$

the summation being finite. It is known from the theory of the Eisenstein integral on G that the function

$$s \mapsto (U^{\vartheta_i,s}(\alpha)e_{n_i}, e_{\tilde{n}_i})$$

is uniformly rapidly decreasing on a small strip around the imaginary axis (cf. Eguchi [8, p. 169]). On the other hand, the function

$$s \mapsto (T_i(s)e_{n_i}, e_{\tilde{n}_i})$$

is bounded at infinity in any strip of the form $\{s: s_0 \leqslant \mathrm{Re}(s) \leqslant 0\}$ (cf. Harish-Chandra [13a, p. 107]). So, if $s_0 < 0$ is sufficiently close to 0, then we can apply Cauchy's theorem to find that

$$\begin{aligned}&\int_{A_i(\epsilon)} \left| \sum_{\vartheta} \sum_{n_i \in I(\vartheta_i)} \int_{\mathrm{Re}(s)=0} (\mathbf{P}_i(\vartheta)\mathbf{c}_\vartheta(s)\mathbf{U}^{\vartheta,s}(\alpha)\mathbf{P}_i(\vartheta)e_{n_i}, e_{n_i})e^{-2s(H_i(a_i))}|ds| \right| d_{A_i}(a_i) \\ &= \int_{A_i(\epsilon)} \left| \sum_{\vartheta} \sum_{n_i \in I(\vartheta_i)} \int_{\mathrm{Re}(s)=0} (\mathbf{P}_i(\vartheta)\mathbf{c}_\vartheta(s_0+s)\mathbf{U}^{\vartheta,s_0+s}(\alpha)\mathbf{P}_i(\vartheta)e_{n_i}, e_{n_i}) \right. \\ &\qquad \left. \times e^{-2s(H_i(a_i))}|ds| \right| e^{-2s_0(H_i(a_i))} d_{A_i}(a_i).\end{aligned}$$

The latter integral is in turn estimated by a finite sum of the form

$$\sum_p \int_{A_i(\epsilon)} f_{s_0}|\hat{f}_p|,$$

where $f_{s_0}(?) = e^{-2s_0?}$ and $\hat{f}_p$ is the Fourier transform of an integrable function. Because $\hat{f}_p$ vanishes at infinity, the presence of f_{s_0} forces

$$\int_{A_i(\epsilon)} f_{s_0}|\hat{f}_p| < +\infty,$$

which then approaches zero as ϵ approaches zero.

Remark. The reason for shifting the line of integration in the above verification is that an integrable C^∞ function on the line need not have an integrable Fourier transform. In fact, as was pointed out to us by D. Wick, if f is smooth and integrable and if $\hat{f}$ is also integrable, then, by Fourier inversion, f must vanish at infinity. But this certainly need not be the case in general. For example, let $\phi \in C^\infty(\mathbf{R})$ be such that $\mathrm{spt}(\phi) \subset [0, 1]$, $\phi \geqslant 0$,

$\phi(\frac{1}{2}) = 1$, and $\int_{\mathbf{R}} \phi = 1$. Put

$$f(x) = \sum_{n=1}^{\infty} \phi(2^n(x-n)) \qquad (x \in \mathbf{R}).$$

Then f is smooth and integrable but does not vanish at infinity.

For later use, it should be noted that all of the preceding discussion of $K_\alpha^{\prime,c}(?:\epsilon)$ is valid for any K-finite function α in $\mathscr{C}^1(G)$.

To complete our investigation of the first parabolic term, we have to study $I_\alpha(x\kappa_i:\epsilon)$ $(x \in \mathfrak{C}_{0t})$, which can be written as

$$\sum_{z \in Z_\Gamma} \sum_{\substack{\eta \in \Gamma \cap N_i \\ \eta \neq 1}} \alpha(x\kappa_i z\eta\kappa_i^{-1}x^{-1}) \cdot \chi_{\epsilon,i}(x\kappa_i),$$

or, equivalently, as

$$\sum_{z \in Z_\Gamma} \sum_{\eta \in \Gamma \cap N_i} \alpha(x\kappa_i z\eta\kappa_i^{-1}x^{-1}) \cdot \chi_{\epsilon,i}(x\kappa_i) - \left(\sum_{z \in Z_\Gamma} \alpha(z)\right) \cdot \chi_{\epsilon,i}(x\kappa_i).$$

Call f_ϵ^i the function on $\pi_i(\mathfrak{C}_{0t}\kappa_i)$ defined by the second expression—then we have

$$\int_{\pi_i(\mathfrak{C}_{0t}\kappa_i)} f_\epsilon^i = o(\epsilon) \qquad \text{as} \quad \epsilon \downarrow 0.$$

To treat the first expression, we shall use the Poisson summation formula. Write

$$\Gamma \cap N_i = \bigcup_{p=1}^{\iota(i)} \eta_p \Xi_i.$$

Then

$$\sum_{\eta \in \Gamma \cap N_i} \alpha(x\kappa_i z\eta\kappa_i^{-1}x^{-1}) = \sum_{p=1}^{\iota(i)} \sum_{\xi \in \Xi_i} \alpha(x\kappa_i z\eta_p\xi\kappa_i^{-1}x^{-1}).$$

Call $\hat{\mathfrak{n}}_i$ the unitary dual of $\mathfrak{n}_i$, $L_i^\perp$ the unitary dual of $\mathfrak{n}_i/L_i$. Given $y \in G$, $n_i \in N_i$, put

$$F_\alpha^i(y:z:n_i:\hat{X}) = \int_{\mathfrak{n}_i} \alpha(yzn_i\exp(X)y^{-1})\mathbf{e}(\langle X, \hat{X}\rangle)\,d_{\mathfrak{n}_i}(X) \qquad (\hat{X} \in \hat{\mathfrak{n}}_i),$$

where $\mathbf{e}(?) = e^{2\pi\sqrt{-1}?}$. Applying now the Poisson summation formula, we have

$$\sum_{\xi \in \Xi_i} \alpha(x\kappa_i z\eta_p\xi\kappa_i^{-1}x^{-1}) \cdot \chi_{\epsilon,i}(x\kappa_i) = \frac{1}{\iota(i)} \cdot F_\alpha^i(x\kappa_i:z:\eta_p:0) \cdot \chi_{\epsilon,i}(x\kappa_i)$$

$$+ \frac{1}{\iota(i)} \cdot \sum_{\substack{l^\perp \in L_i^\perp \\ l^\perp \neq 0}} F_\alpha^i(x\kappa_i:z:\eta_p:l^\perp) \cdot \chi_{\epsilon,i}(x\kappa_i).$$

The result of summing the left-hand side of the equality over p and Z_Γ is $I_\alpha(?:\epsilon) - f_\epsilon^i$. The result of summing the first term on the right-hand side of the equality over p and Z_Γ is

$$\frac{1}{\iota(i)} \cdot \sum_{z \in Z_\Gamma} \sum_{p=1}^{\iota(i)} F_\alpha^i(x\kappa_i : z : \eta_p : 0) \cdot \chi_{\epsilon,i}(x\kappa_i)$$

which, in view of the fact that the Haar measure on N_i is the exponentiation of the Haar measure on $\mathfrak{n}_i$, can be rewritten in the form

$$\sum_{z \in Z_\Gamma} \int_{N_i} \alpha(x\kappa_i z n_i \kappa_i^{-1} x^{-1})\, d_{N_i}(n_i) \cdot \chi_{\epsilon,i}(x\kappa_i)$$

and this is $(\mathrm{I})_\epsilon^i + (\mathrm{II})_\epsilon^i$ above. So, if we call the result of summing the second term on the right-hand side of the equality F_ϵ^i, then we have on $\pi_i(\mathfrak{C}_{0t}\kappa_i)$:

$$I_\alpha(?:\epsilon) = (\mathrm{I})_\epsilon^i + (\mathrm{II})_\epsilon^i + f_\epsilon^i + F_\epsilon^i.$$

The only thing left to do is to verify that F_ϵ^i is integrable on $\pi_i(\mathfrak{C}_{0t}\kappa_i)$ and that

$$\int_{\pi_i(\mathfrak{C}_{0t}\kappa_i)} F_\epsilon^i = o(\epsilon) \qquad \text{as} \quad \epsilon \downarrow 0.$$

To establish this, we can assume that $i = 1$ and then drop it from the notation. Put

$$\log^* \epsilon = (\log \epsilon)/|\lambda|.$$

Fix $z \in Z_\Gamma$ and $1 \leqslant p \leqslant \iota(i)$—then we need only consider

$$\int_K \int_{-\infty}^{\log^* \epsilon} \int_{\omega_0} \sum_{\substack{l^\perp \in L^\perp \\ l^\perp \neq 0}} |F_\alpha(ka(t)n : z : \eta_p : l^\perp)| e^{2|\rho|t}\, d_K(k)\, dt\, d_N(n).$$

There certainly is no loss of generality in assuming that $a(t)\omega_0 a(-t) \subset \omega_0$ for all $t \leqslant 0$. So, upon making the obvious change of variable, we are led to

$$\int_{K \times \omega_0} \left(\int_{-\infty}^{\log^* \epsilon} \sum_{\substack{l^\perp \in L^\perp \\ l^\perp \neq 0}} |F_\alpha(kna(t) : z : \eta_p : l^\perp)|\, dt \right) d_K(k)\, d_N(n).$$

Given $\hat{X} \in \hat{\mathfrak{n}}$ and $t \in \mathbf{R}$, let $\hat{X}_t$ be the element of $\hat{\mathfrak{n}}$ determined by the rule

$$\langle X, \hat{X}_t \rangle = \langle \mathrm{Ad}(a(t))X, \hat{X} \rangle \qquad (X \in \mathfrak{n}).$$

If $\|?\|$ is a norm on $\hat{\mathfrak{n}}$, then it is clear that there exists a constant $c > 0$ such that

$$\|\hat{X}_t\| \geqslant e^{ct}\|\hat{X}\| \qquad (\hat{X} \in \hat{\mathfrak{n}})$$

for all $t \geqslant 0$. This said, write

$$
\begin{aligned}
&F_\alpha(kna(t):z:\eta_p:l^\perp) \\
&\quad = \int_{\mathfrak{n}} \alpha(knza(t)\eta_p a(-t)\exp(\mathrm{Ad}(a(t))X)n^{-1}k^{-1})\mathbf{e}(\langle X, l^\perp\rangle)\, d_{\mathfrak{n}}(X) \\
&\quad = e^{-2|\rho|t} \cdot \int_{\mathfrak{n}} \alpha(knza(t)\eta_p a(-t)\exp(X)n^{-1}k^{-1})\mathbf{e}(\langle \mathrm{Ad}(a(-t))X, l^\perp\rangle)\, d_{\mathfrak{n}}(X) \\
&\quad = e^{-2|\rho|t} \cdot F_\alpha(kn:z:a(t)\eta_p a(-t):l^\perp_{-t}).
\end{aligned}
$$

The function $F_\alpha(kn:z:a(t)\eta_p a(-t):?)$ is rapidly decreasing on $\hat{\mathfrak{n}}$ and is continuous in the parameters k, n, and t. Accordingly, given any integer $v > 0$, there is a constant $C_v > 0$ such that

$$
|F_\alpha(kn:z:a(t)\eta_p a(-t):\hat{X})| \leqslant C_v(1 + \|\hat{X}\|)^{-v}
$$

for all $\hat{X} \in \hat{\mathfrak{n}}$, the estimate being uniform in $k \in K$, $n \in \omega_0$, $z \in Z_\Gamma$, and t provided t is bounded above. Therefore

$$
\int_{K\times\omega_0} \left(\int_{-\infty}^{\log^* \epsilon} e^{-2|\rho|t} \cdot \sum_{\substack{l^\perp \in L^\perp \\ l^\perp \neq 0}} |F_\alpha(kn:z:a(t)\eta_p a(-t):l^\perp_{-t})|\, dt \right) d_K(k)\, d_N(n)
$$

is majorized by

$$
C_v \cdot \mathrm{meas}(K \times \omega_0) \cdot \int_{-\infty}^{\log^* \epsilon} e^{-2|\rho|t} \left(\sum_{\substack{l^\perp \in L^\perp \\ l^\perp \neq 0}} (1 + \|l^\perp_{-t}\|)^{-v} \right) dt.
$$

We have

$$
\begin{aligned}
&\int_{-\infty}^{\log^* \epsilon} e^{-2|\rho|t} \left(\sum_{\substack{l^\perp \in L^\perp \\ l^\perp \neq 0}} (1 + \|l^\perp_{-t}\|)^{-v} \right) dt \\
&\qquad \leqslant \int_{-\infty}^{\log^* \epsilon} e^{(cv-2|\rho|)t} \left(\sum_{\substack{l^\perp \in L^\perp \\ l^\perp \neq 0}} (e^{ct} + \|l^\perp\|)^{-v} \right) dt \\
&\qquad \leqslant \left(\sum_{\substack{l^\perp \in L^\perp \\ l^\perp \neq 0}} \|l^\perp\|^{-v} \right) \cdot \int_{-\infty}^{\log^* \epsilon} e^{(cv-2|\rho|)t}\, dt
\end{aligned}
$$

which is finite for large enough v and approaches zero as ϵ approaches zero. Thus F^i_ϵ is indeed integrable on $\pi_i(\mathfrak{C}_{0t}\kappa_i)$ and

$$
\int_{\pi_i(\mathfrak{C}_{0t}\kappa_i)} F^i_\epsilon = o(\epsilon) \qquad \text{as} \quad \epsilon \downarrow 0,
$$

as desired.

SUMMARY (The first parabolic term). *The first parabolic term is weakly integrable over G/Γ and its weak integral tends to 0 as $\epsilon \downarrow 0$.*

We shall now investigate the second parabolic term. Write

$$\begin{aligned}
&\int_{G/\Gamma} J_\alpha(x:\epsilon)\, d_G(x) \\
&\quad = \int_{G/\Gamma} \sum_{i=1}^{r} \left(\sum_{\delta \in \Gamma/\Gamma \cap P_i} \sum_{z \in Z_\Gamma} \sum_{\substack{\eta \in \Gamma \cap N_i \\ \eta \neq 1}} \alpha(x\,\delta z\eta\,\delta^{-1}x^{-1}) \cdot (1 - \chi_{\epsilon,i}(x\,\delta)) \right) d_G(x) \\
&\quad = \sum_{i=1}^{r} \int_{G/\Gamma \cap P_i} \left(\sum_{z \in Z_\Gamma} \sum_{\substack{\eta \in \Gamma \cap N_i \\ \eta \neq 1}} \alpha(xz\eta x^{-1}) \cdot (1 - \chi_{\epsilon,i}(x)) \right) d_G(x) \\
&\quad = \sum_{i=1}^{r} \mathbf{c}_G^i \int_K \int_{A_i} \int_{N_i/N_i \cap \Gamma} \sum_{\substack{\eta \in \Gamma \cap N_i \\ \eta \neq 1}} \alpha(k a_i n_i \eta n_i^{-1} a_i^{-1} k^{-1}) \\
&\qquad \times (1 - \chi_{\epsilon,i}(k a_i n_i)) \cdot e^{2|\rho|(H_i(a_i))}\, d_K(k)\, d_{A_i}(a_i)\, d_{N_i}(n_i) \\
&\quad = \sum_{i=1}^{r} \int_{A_i} \int_{N_i/N_i \cap \Gamma} \sum_{\substack{\eta \in \Gamma \cap N_i \\ \eta \neq 1}} \alpha_K^i(a_i n_i \eta n_i^{-1} a_i^{-1}) \\
&\qquad \times (1 - \chi_{\epsilon,i}(a_i n_i)) \cdot e^{2|\rho|(H_i(a_i))}\, d_{A_i}(a_i)\, d_{N_i}(n_i).
\end{aligned}$$

Here

$$\alpha_K^i(x) = \mathbf{c}_G^i \int_K \alpha(kxk^{-1})\, d_K(k) \qquad (x \in G).$$

Therefore

$$\begin{aligned}
&\int_{G/\Gamma} |J_\alpha(x:\epsilon)|\, d_G(x) \\
&\quad \leqslant \sum_{i=1}^{r} \int_{\log^* \epsilon}^{\infty} \int_{N_i/N_i \cap \Gamma} \left(\sum_{\substack{\eta \in \Gamma \cap N_i \\ \eta \neq 1}} |\alpha_K^i(a_i(t) n_i \eta n_i^{-1} a_i(-t))| \right) \cdot e^{2|\rho|t}\, d_{N_i}(n_i)\, dt \\
&\quad \leqslant \sum_{i=1}^{r} \int_{\log^* \epsilon}^{\infty} \int_{N_i/N_i \cap \Gamma} \left(\sum_{\substack{\eta \in \Gamma \cap N_i \\ \eta \neq 1}} |\alpha|_K^i(a_i(t) n_i \eta n_i^{-1} a_i(-t)) \right) \cdot e^{2|\rho|t}\, d_{N_i}(n_i)\, dt.
\end{aligned}$$

Here

$$|\alpha|_K^i(x) = \mathbf{c}_G^i \int_K |\alpha|(kxk^{-1})\, d_K(k) \qquad (x \in G).$$

Because $\Gamma \cap N_i$ is discrete, there evidently exist $t_i(\alpha) > 0$ such that the expression majorizing

$$\int_{G/\Gamma} |J_\alpha(x:\epsilon)|\, d_G(x)$$

is equal to

$$\sum_{i=1}^{r} \int_{\log^* \epsilon}^{t_i(\alpha)} \int_{N_i/N_i \cap \Gamma} \left(\sum_{\substack{\eta \in \Gamma \cap N_i \\ \eta \neq 1}} |\alpha|_K^i(a_i(t) n_i \eta n_i^{-1} a_i(-t)) \right) \cdot e^{2|\rho|t}\, d_{N_i}(n_i)\, dt.$$

As this is a finite quantity, it follows that the parabolic term is integrable over G/Γ. Define now a function of the complex variable s by the prescription

$\Omega_\alpha(\epsilon:s)$

$$= \sum_{i=1}^{r} \int_{\log^* \epsilon}^{\infty} \int_{N_i/N_i \cap \Gamma} \left(\sum_{\substack{\eta \in \Gamma \cap N_i \\ \eta \neq 1}} \alpha_K^i(a_i(t) n_i \eta n_i^{-1} a_i(-t)) \right) \cdot e^{2|\rho|t(1+s)}\, d_{N_i}(n_i)\, dt.$$

On the basis of what has been said so far, it is clear that for any s the integral defining $\Omega_\alpha(\epsilon:s)$ is absolutely convergent. Moreover, $\Omega_\alpha(\epsilon:s)$ is an entire function of s whose value at $s = 0$ equals

$$\int_{G/\Gamma} J_\alpha(x:\epsilon)\, d_G(x).$$

To calculate the latter entity, we shall decompose $\Omega_\alpha(\epsilon:s)$ by the formula

$$\int_{\log^* \epsilon}^{\infty} = \int_{-\infty}^{\infty} - \int_{-\infty}^{\log^* \epsilon}$$

and study the resulting terms separately. Employing earlier notations, write

$$\sum_{\substack{\eta \in \Gamma \cap N_i \\ \eta \neq 1}} \alpha_K^i(a_i(t) n_i \eta n_i^{-1} a_i(-t))$$

$$= \frac{1}{\iota(i)} \cdot \sum_{p=1}^{\iota(i)} \left[\sum_{\substack{l^\perp \in L_i^\perp \\ l^\perp \neq 0}} F_{\alpha_K^i}^i(a_i(t) n_i : 1 : \eta_p : l^\perp) \right]$$

$$+ \frac{1}{\iota(i)} \cdot \sum_{p=1}^{\iota(i)} F_{\alpha_K^i}^i(a_i(t) n_i : 1 : \eta_p : 0) - \alpha_K^i(1).$$

We may argue as in the discussion of the first parabolic term and conclude that for any s the expression

$$\frac{1}{\iota(i)} \cdot \sum_{p=1}^{\iota(i)} \int_{-\infty}^{\log^* \epsilon} \int_{N_i/N_i \cap \Gamma} \sum_{\substack{l^\perp \in L_i^\perp \\ l^\perp \neq 0}} |F_{\alpha_K^i}^i(a_i(t) n_i : 1 : \eta_p : l^\perp)|$$

$$\times \left|e^{2|\rho|t(1+s)}\right| \cdot d_{N_i}(n_i)\, dt$$

is finite and approaches zero as ϵ approaches zero. Next

$$\begin{aligned}\frac{1}{\iota(i)}\cdot\sum_{p=1}^{\iota(i)} F^i_{\alpha_K}(a_i(t)n_i:1:\eta_p:0) &= \int_{N_i}\alpha^i_K(a_i(t)n_i\tilde{n}_i n_i^{-1}a_i(-t))\,d_{N_i}(\tilde{n}_i)\\ &= \int_{N_i}\alpha^i_K(a_i(t)\tilde{n}_i a_i(-t))\,d_{N_i}(\tilde{n}_i)\\ &= e^{-2|\rho|t}\int_{N_i}\alpha^i_K(\tilde{n}_i)\,d_{N_i}(\tilde{n}_i),\end{aligned}$$

and, for $\mathrm{Re}(s) > 0$,

$$\begin{aligned}&\int_{-\infty}^{\log^*\epsilon}\int_{N_i/N_i\cap\Gamma}\left(e^{-2|\rho|t}\int_{N_i}\alpha^i_K(\tilde{n}_i)\,d_{N_i}(\tilde{n}_i)\right)e^{2|\rho|t(1+s)}\,d_{N_i}(n_i)\,dt\\ &\quad=\frac{\epsilon^{(2|\rho|/|\lambda|)s}}{2|\rho|s}\cdot\int_{N_i}\alpha^i_K(n_i)\,d_{N_i}(n_i).\end{aligned}$$

Finally, for $\mathrm{Re}(s) > -1$,

$$\int_{-\infty}^{\log^*\epsilon}\int_{N_i/N_i\cap\Gamma}(-\alpha^i_K(1))e^{2|\rho|t(1+s)}\,d_{N_i}(n_i)\,dt = \frac{\epsilon^{(2|\rho|/|\lambda|)(1+s)}}{2|\rho|(1+s)}\cdot(-\alpha^i_K(1)).$$

These calculations make it clear that if we introduce $Ч_\alpha$ by the definition

$$Ч_\alpha(s) = \sum_{i=1}^{r}\int_{-\infty}^{\infty}\int_{N_i/N_i\cap\Gamma}\left(\sum_{\substack{\eta\in\Gamma\cap N_i\\ \eta\neq 1}}\alpha^i_K(a_i(t)n_i\eta n_i^{-1}a_i(-t))\right)\cdot e^{2|\rho|t(1+s)}\,d_{N_i}(n_i)\,dt,$$

then the integral defining $Ч_\alpha$ is absolutely convergent for $\mathrm{Re}(s) > 0$, and admits a continuation to the entire complex plane as a meromorphic function whose only possible singularities are simple poles at $s = 0$ and $s = -1$. Furthermore, up to a term which approaches zero as ϵ approaches zero, the integral

$$\int_{G/\Gamma} J_\alpha(x:\epsilon)\,d_G(x)$$

is equal to

$$\lim_{s\to 0}\left\{Ч_\alpha(s) - \sum_{i=1}^{r}\frac{\epsilon^{(2|\rho|/|\lambda|)s}}{2|\rho|s}\cdot\int_{N_i}\alpha^i_K(n_i)\,d_{N_i}(n_i)\right\}.$$

The constant term of the Laurent expansion about $s = 0$ of $Ч_\alpha$ is

$$\lim_{s\to 0}\frac{d}{ds}(sЧ_\alpha(s)).$$

The constant term of the Laurent expansion about $s = 0$ of

$$\sum_{i=1}^{r} \frac{\epsilon^{(2|\rho|/|\lambda|)s}}{2|\rho|s} \cdot \int_{N_i} \alpha_K^i(n_i)\, d_{N_i}(n_i)$$

is

$$\frac{\log \epsilon}{|\lambda|} \cdot \sum_{i=1}^{r} \int_{N_i} \alpha_K^i(n_i)\, d_{N_i}(n_i).$$

It is these two terms, then, which figure substantively into the evaluation of

$$\int_{G/\Gamma} J_\alpha(x:\epsilon)\, d_G(x).$$

SUMMARY (The second parabolic term). *The second parabolic term is integrable over G/Γ and, modulo an expression which approaches zero as ϵ approaches zero, its integral is equal to*

$$\lim_{s \to 0} \frac{d}{ds}(s\text{Ч}_\alpha(s)) - \frac{\log \epsilon}{|\lambda|} \cdot \sum_{i=1}^{r} \int_{N_i} \alpha_K^i(n_i)\, d_{N_i}(n_i).$$

We shall now investigate the third parabolic term. Since the first four terms of the kernel of K_α^{d} are weakly integrable over G/Γ, it follows that the fifth term, namely $-K_\alpha^{\prime\prime,\mathrm{c}}(x:\epsilon)$, is also weakly integrable over G/Γ. We can therefore turn directly to the task of computing its integral explicitly.

Let $0 < t \leqslant {}_0t$—then

$$-(w)\int_{G/\Gamma} K_\alpha^{\prime\prime,\mathrm{c}}(x:\epsilon)\, d_G(x) = -\lim_{t \downarrow 0} \int_{\pi(\Omega_t)} K_\alpha^{\prime\prime,\mathrm{c}}(x:\epsilon)\, d_G(x).$$

Owing to Lemmas 4.8 and 6.2, there is a constant $C > 0$ and an integer $n \geqslant 0$ such

$$\sum_{\vartheta} \int_{\mathrm{Re}(s)=0} \frac{1}{4\pi} \Bigg| \sum_{m,\, n \in I(\vartheta)} \mathbf{U}_{mn}^{s}(\alpha) \cdot \mathbf{E}(e_m : s : x\kappa_i)\overline{\mathbf{E}(e_n : s : x\kappa_i)}$$

$$- \sum_{m,\, n \in I(\vartheta)} \mathbf{U}_{mn}^{s}(\alpha) \cdot \mathbf{E}_\epsilon'(e_m : s : x\kappa_i)\overline{\mathbf{E}_\epsilon'(e_n : s : x\kappa_i)} \Bigg|\, |ds| \leqslant C(\xi_\lambda(a_x))^{-n}$$

for all $x \in \mathfrak{S}_{t_0,\omega_0}$, $\kappa_i \in \mathfrak{s}$. Applying Fubini's theorem, we then deduce that

$$-\lim_{t \downarrow 0} \int_{\pi(\Omega_t)} K_\alpha^{\prime\prime,\mathrm{c}}(x:\epsilon)\, d_G(x)$$

$$= -\lim_{t \downarrow 0} \sum_{\vartheta} \int_{\mathrm{Re}(s)=0} \int_{\pi(\Omega_t)} \frac{1}{4\pi} \Bigg(\sum_{m,\, n \in I(\vartheta)} \mathbf{U}_{mn}^{s}(\alpha) \cdot \mathbf{E}(e_m : s : x)\overline{\mathbf{E}(e_n : s : x)}$$

$$- \sum_{m,\, n \in I(\vartheta)} \mathbf{U}_{mn}^{s}(\alpha) \cdot \mathbf{E}_\epsilon'(e_m : s : x)\overline{\mathbf{E}_\epsilon'(e_n : s : x)} \Bigg) d_G(x)\, |ds|.$$

The last integrand is the sum of

$$\frac{1}{4\pi}\sum_{m,\, n\in I(\vartheta)} \mathbf{U}^s_{mn}(\alpha)\cdot \mathbf{E}''_\epsilon(e_m:s:x)\overline{\mathbf{E}''_\epsilon(e_n:s:x)}$$

and

$$\frac{1}{4\pi}\sum_{m,\, n\in I(\vartheta)} \mathbf{U}^s_{mn}(\alpha)\cdot (\mathbf{E}'_\epsilon(e_m:s:x)\overline{\mathbf{E}''_\epsilon(e_n:s:x)}+\overline{\mathbf{E}'_\epsilon(e_n:s:x)}\mathbf{E}''_\epsilon(e_m:s:x)).$$

We claim that if $m, n \in I(\vartheta)$, then

$$\int_{\pi(\Omega_t)} \mathbf{E}'_\epsilon(e_m:s:x)\overline{\mathbf{E}''_\epsilon(e_n:s:x)}\, d_G(x) = 0.$$

To begin with, the function in question is evidently integrable over G/Γ (cf. Lemma 2.2), so

$$\begin{aligned}\int_{G/\Gamma} &\mathbf{E}'_\epsilon(e_m:s:x)\overline{\mathbf{E}''_\epsilon(e_n:s:x)}\, d_G(x)\\ &= \int_{\pi(\Omega_t)} \mathbf{E}'_\epsilon(e_m:s:x)\overline{\mathbf{E}''_\epsilon(e_n:s:x)}\, d_G(x)\\ &\quad + \sum_{i=1}^{r}\int_{\pi_i(\mathfrak{C}_t\kappa_i)} \mathbf{E}'_\epsilon(e_m:s:x)\overline{\mathbf{E}''_\epsilon(e_n:s:x)}\, d_G(x).\end{aligned}$$

We have

$$\begin{aligned}\int_{G/\Gamma} &\mathbf{E}'_\epsilon(e_m:s:x)\overline{\mathbf{E}''_\epsilon(e_n:s:x)}\, d_G(x)\\ &= \int_{G/\Gamma}\left(\sum_{i=1}^{r}\sum_{\delta\in\Gamma/\Gamma\cap P_i} \mathbf{E}_{P_i}(e_m:s:x\,\delta)\cdot\chi_{\epsilon,i}(x\,\delta)\right)\overline{\mathbf{E}''_\epsilon(e_n:s:x)}\, d_G(x)\\ &= \sum_{i=1}^{r}\int_{G/\Gamma\cap P_i} \mathbf{E}_{P_i}(e_m:s:x)\cdot(\chi_{\epsilon,i}(x)\overline{\mathbf{E}''_\epsilon(e_n:s:x)})\, d_G(x)\\ &= \sum_{i=1}^{r}\int_{G/(\Gamma\cap P_i)\,.\,N_i} \mathbf{E}_{P_i}(e_m:s:x)\\ &\quad\times(\chi_{\epsilon,i}(x)\int_{N_i/N_i\cap\Gamma}\overline{\mathbf{E}''_\epsilon(e_n:s:xn_i)}\, d_{N_i}(n_i))\, d_G(x)\\ &= 0.\end{aligned}$$

In passing to the last equality, we use the fact that if $\chi_{\epsilon,i}(x) \neq 0$, then

$$\mathbf{E}''_\epsilon(?:s:xn_i) = \mathbf{E}(?:s:xn_i) - \mathbf{E}_{P_i}(?:s:x),$$

which, upon integration over $N_i/N_i \cap \Gamma$, gives 0. In an entirely analogous fashion, one verifies that each of the integrals

$$\int_{\pi_i(\mathfrak{C}_t\kappa_i)} \mathbf{E}'_\epsilon(e_m:s:x)\overline{\mathbf{E}''_\epsilon(e_n:s:x)}\, d_G(x)$$

also vanishes. Therefore

$$\int_{\pi(\Omega_t)} \mathbf{E}'_\epsilon(e_m:s:x)\overline{\mathbf{E}''_\epsilon(e_n:s:x)}\, d_G(x) = 0.$$

Hence the claim. It then follows that

$$-\lim_{t\downarrow 0} \int_{\pi(\Omega_t)} K''^{,c}_\alpha(x:\epsilon)\, d_G(x)$$

$$= -\lim_{t\downarrow 0} \sum_{\vartheta} \int_{\mathrm{Re}(s)=0} \int_{\pi(\Omega_t)} \frac{1}{4\pi}\left(\sum_{m,\, n\in I(\vartheta)} \mathbf{U}^s_{mn}(\alpha)\cdot \mathbf{E}''_\epsilon(e_m:s:x)\overline{\mathbf{E}''_\epsilon(e_n:s:x)}\right) d_G(x)\, |ds|.$$

Proceeding now as in the proof of Lemma 4.8, we shall show that

$$\sum_{\vartheta} \int_{\mathrm{Re}(s)=0} \int_{G/\Gamma} \frac{1}{4\pi}\left|\sum_{m,\, n\in I(\vartheta)} \mathbf{U}^s_{mn}(\alpha)\cdot \mathbf{E}''_\epsilon(e_m:s:x)\overline{\mathbf{E}''_\epsilon(e_n:s:x)}\right| d_G(x)\, |ds| < +\infty.$$

Indeed

$$\left|\sum_{m,\, n\in I(\vartheta)} \mathbf{U}^s_{mn}(\alpha)\cdot \mathbf{E}''_\epsilon(e_m:s:x)\overline{\mathbf{E}''_\epsilon(e_n:s:x)}\right|$$

is bounded by

$$\left|\sum_{m,\, n\in I(\vartheta)} \mathbf{U}^s_{mn}(\alpha_H)\cdot \mathbf{E}''_\epsilon(e_m:s:x)\overline{\mathbf{E}''_\epsilon(e_n:s:x)}\right|^{1/2}$$

$$\times \left|\sum_{m,\, n\in I(\vartheta)} \mathbf{U}^s_{mn}(\nu_H)\cdot \mathbf{E}''_\epsilon(e_m:s:x)\overline{\mathbf{E}''_\epsilon(e_n:s:x)}\right|^{1/2}$$

$$+ \left|\sum_{m,\, n\in I(\vartheta)} \mathbf{U}^s_{mn}(\beta_H)\cdot \mathbf{E}''_\epsilon(e_m:s:x)\overline{\mathbf{E}''_\epsilon(e_n:s:x)}\right|^{1/2}$$

$$\times \left|\sum_{m,\, n\in I(\vartheta)} \mathbf{U}^s_{mn}(\mu_H)\cdot \mathbf{E}''_\epsilon(e_m:s:x)\overline{\mathbf{E}''_\epsilon(e_m:s:x)}\right|^{1/2},$$

so, thanks to the Schwarz inequality, we need deal only with the case when α is replaced by ϕ, $\phi = \psi * \psi^*$, ψ subject to the usual conditions. Because $\mathbf{U}^{\vartheta,s}(\phi)$ is then a positive semidefinite operator,

$$\sum_{m,\, n\in I(\vartheta)} \mathbf{U}^s_{mn}(\phi)\cdot \mathbf{E}''_\epsilon(e_m:s:x)\overline{\mathbf{E}''_\epsilon(e_n:s:x)} \geqslant 0,$$

thus, by what has been said above,

$$\lim_{t\downarrow 0} \sum_{\vartheta} \int_{\mathrm{Re}(s)=0} \int_{\pi(\Omega_t)} \frac{1}{4\pi}\left(\sum_{m,\, n\in I(\vartheta)} \mathbf{U}^s_{mn}(\phi)\cdot \mathbf{E}''_\epsilon(e_m:s:x)\overline{\mathbf{E}''_\epsilon(e_n:s:x)}\right) d_G(x)\, |ds|$$

is actually a finite quantity. But this implies that

$$\sum_{\vartheta} \int_{\mathrm{Re}(s)=0} \int_{G/\Gamma} \frac{1}{4\pi} \left(\sum_{m,\, n \in I(\vartheta)} \mathbf{U}^s_{mn}(\phi) \cdot \mathbf{E}''_\epsilon(e_m : s : x) \overline{\mathbf{E}''_\epsilon(e_n : s : x)} \right) d_G(x)\, |ds| < +\infty,$$

as desired. Therefore

$$-(w) \int_{G/\Gamma} K''^{,c}_\alpha(x : \epsilon)\, d_G(x)$$
$$= -\frac{1}{4\pi} \sum_{\vartheta} \int_{\mathrm{Re}(s)=0} \sum_{m,\, n \in I(\vartheta)} \mathbf{U}^s_{mn}(\alpha) \left(\int_{G/\Gamma} \mathbf{E}''_\epsilon(e_m : s : x) \overline{\mathbf{E}''_\epsilon(e_n : s : x)}\, d_G(x) \right) |ds|.$$

According to Langlands [19a, p. 145], if $m, n \in I(\vartheta)$, then for any nonzero pure imaginary number s, the inner product

$$\int_{G/\Gamma} \mathbf{E}''_\epsilon(e_m : s : x) \overline{\mathbf{E}''_\epsilon(e_n : s : x)}\, d_G(x)$$

is the sum of

$$-2 \cdot \frac{\log \epsilon}{|\lambda|} \cdot (e_m, e_n), \qquad -\left(\left(\frac{d}{ds} \mathbf{c}_\vartheta(s) \right) e_m, \mathbf{c}_\vartheta(s) e_n \right),$$

and

$$-\frac{1}{2s} \{ \epsilon^{2s/|\lambda|} (e_m, \mathbf{c}_\vartheta(s) e_n) - \epsilon^{-2s/|\lambda|} (\mathbf{c}_\vartheta(s) e_m, e_n) \}.$$

To complete our calculation of

$$-(w) \int_{G/\Gamma} K''^{,c}_\alpha(x : \epsilon)\, d_G(x),$$

we shall substitute each of these three terms into the above expression on the right. The first gives

$$\frac{\log \epsilon}{|\lambda|} \cdot \frac{1}{2\pi} \cdot \sum_{\vartheta} \int_{\mathrm{Re}(s)=0} \mathrm{tr}(\mathbf{U}^{\vartheta, s}(\alpha))\, |ds|.$$

To justify this step, we remind ourselves of two points: (1) The sum over ϑ is finite; (2) as a function of s, $\mathrm{tr}(\mathbf{U}^{\vartheta,s}(\alpha))$ is integrable over the imaginary axis (cf. Warner [34a, p. 472]). We have

$$\sum_{\vartheta} \mathrm{tr}(\mathbf{U}^{\vartheta,s}(\alpha)) = \sum_{i=1}^{r} \sum_{\vartheta} \mathrm{tr}(U^{\vartheta_i, s}(\alpha))$$
$$= \sum_{i=1}^{r} \mathrm{tr}(\mathbf{U}^s_i(\alpha))$$
$$= \sum_{i=1}^{r} \int_K \int_{M_i/Z\Gamma} \mathbf{U}^i_\alpha(km_i, km_i : s)\, d_K(k)\, d_{M_i}(m_i),$$

the last step being legitimate in view of Theorem 4.10. Using now the explicit formula for $\mathbf{U}_\alpha^i(?, ?: s)$ which was developed earlier and the Fourier inversion theorem, we find that

$$\frac{\log \epsilon}{|\lambda|} \cdot \frac{1}{2\pi} \cdot \sum_{\vartheta} \int_{\mathrm{Re}(s)=0} \mathrm{tr}(\mathbf{U}^{\vartheta,s}(\alpha))\,|ds| = \frac{\log \epsilon}{|\lambda|} \cdot \sum_{i=1}^{r} \int_{N_i} \alpha_K^i(n_i)\, d_{N_i}(n_i),$$

which cancels with the substantive ϵ-dependent part of the integral of the second parabolic term. The result of inserting formally the next term is the expression

$$\frac{1}{4\pi} \sum_{\vartheta} \int_{\mathrm{Re}(s)=0} \mathrm{tr}\left(\mathbf{c}_\vartheta(-s) \cdot \left(\frac{d}{ds}\,\mathbf{c}_\vartheta(s)\right) \cdot \mathbf{U}^{\vartheta,s}(\alpha)\right) |ds|.$$

To justify this step, it is enough to show that

$$\frac{1}{4\pi} \sum_{\vartheta} \int_{\mathrm{Re}(s)=0} \left|\mathrm{tr}\left(\mathbf{c}_\vartheta(-s) \cdot \left(\frac{d}{ds}\,\mathbf{c}_\vartheta(s)\right) \cdot \mathbf{U}^{\vartheta,s}(\alpha)\right)\right| |ds| < +\infty.$$

Instead of making a direct verification, it will be more convenient to proceed to the last term and prove that

$$\frac{1}{4\pi} \int_{\mathrm{Re}(s)=0} \sum_{\vartheta} \sum_{n \in I(\vartheta)} \left| \frac{\epsilon^{2s/|\lambda|}}{2s} (\mathbf{c}_\vartheta(-s)\mathbf{U}^{\vartheta,s}(\alpha)e_n, e_n) - \frac{\epsilon^{-2s/|\lambda|}}{2s} (\mathbf{c}_\vartheta(s)\mathbf{U}^{\vartheta,s}(\alpha)e_n, e_n) \right| |ds| < +\infty.$$

But the function

$$s \mapsto \frac{\epsilon^{2s/|\lambda|}}{2s} (\mathbf{c}_\vartheta(-s)\mathbf{U}^{\vartheta,s}(\alpha)e_n, e_n) - \frac{\epsilon^{-2s/|\lambda|}}{2s} (\mathbf{c}_\vartheta(s)\mathbf{U}^{\vartheta,s}(\alpha)e_n, e_n)$$

is holomorphic at $s = 0$ and is in fact integrable over the imaginary axis, the function

$$s \mapsto (\mathbf{U}^{\vartheta,s}(\alpha)e_n, e_n)$$

being, as mentioned above, rapidly decreasing for pure imaginary s. The contribution of the last term is the sum of

$$\frac{1}{4\pi} \sum_{\vartheta} \sum_{n \in I(\vartheta)} \int_{\mathrm{Re}(s)=0} \frac{\epsilon^{2s/|\lambda|}}{2s} \cdot \{(\mathbf{c}_\vartheta(-s)\mathbf{U}^{\vartheta,s}(\alpha)e_n, e_n) - (\mathbf{c}_\vartheta(s)\mathbf{U}^{\vartheta,s}(\alpha)e_n, e_n)\}\,|ds|$$

and

$$\frac{1}{4\pi} \sum_{\vartheta} \sum_{n \in I(\vartheta)} \int_{\mathrm{Re}(s)=0} \frac{\epsilon^{2s/|\lambda|} - \epsilon^{-2s/|\lambda|}}{2s} \cdot (\mathbf{c}_\vartheta(s)\mathbf{U}^{\vartheta,s}(\alpha)e_n, e_n)\,|ds|.$$

In view of the Riemann–Lebesgue lemma, the first expression approaches zero as ϵ approaches zero. On the other hand, using the theory of the Dirichlet integral, one shows without difficulty that the second expression approaches

$$-\frac{1}{4}\sum_{\vartheta} \operatorname{tr}(\mathbf{c}_{\vartheta}(0)\mathbf{U}^{\vartheta,0}(\alpha))$$

as ϵ approaches zero.

SUMMARY (The third parabolic term). *The third parabolic term is weakly integrable over G/Γ and, modulo an expression which approaches zero as ϵ approaches zero, its weak integral is equal to the sum of*

$$\frac{\log\epsilon}{|\lambda|}\sum_{i=1}^{r}\int_{N_i}\alpha_{\mathbf{K}}^{i}(n_i)\,d_{N_i}(n_i),$$

$$\frac{1}{4\pi}\sum_{\vartheta}\int_{\mathrm{Re}(s)=0}\operatorname{tr}\left(\mathbf{c}_{\vartheta}(-s)\cdot\left(\frac{d}{ds}\mathbf{c}_{\vartheta}(s)\right)\cdot\mathbf{U}^{\vartheta,s}(\alpha)\right)|ds|,$$

and

$$-\frac{1}{4}\sum_{\vartheta}\operatorname{tr}(\mathbf{c}_{\vartheta}(0)\mathbf{U}^{\vartheta,0}(\alpha)).$$

The result of the computations made in this section is the following theorem.

THEOREM 6.3 (The Selberg trace formula). *Let α be a K-finite function in $C_c^{\infty}(G)$—then the trace of $L_{G/\Gamma}^{\mathrm{d}}(\alpha)$ is the sum of*

$$\operatorname{vol}(G/\Gamma)\left(\sum_{z\in Z_{\Gamma}}\alpha(z)\right),$$

$$(S)\sum_{\{\gamma\}_{\Gamma}}\operatorname{vol}(G_{\gamma}/\Gamma_{\gamma})\cdot\int_{G/G_{\gamma}}\alpha(x\gamma x^{-1})\,d_{G/G_{\gamma}}(x),$$

$$\lim_{s\to 0}\frac{d}{ds}(s\text{Ч}_{\alpha}(s)),$$

$$\frac{1}{4\pi}\sum_{\vartheta}\int_{\mathrm{Re}(s)=0}\operatorname{tr}\left(\mathbf{c}_{\vartheta}(-s)\cdot\left(\frac{d}{ds}\mathbf{c}_{\vartheta}(s)\right)\cdot\mathbf{U}^{\vartheta,s}(\alpha)\right)|ds|,$$

and

$$-\frac{1}{4}\sum_{\vartheta}\operatorname{tr}(\mathbf{c}_{\vartheta}(0)\mathbf{U}^{\vartheta,0}(\alpha)).$$

All the series and integrals are absolutely convergent.

The assumption that α is compactly supported has been used at a number of places in the proof of the above theorem. For the applications, it is important to know that the Selberg trace formula retains its validity for certain classes of functions whose support is no longer compact. Before taking up this question, it will be convenient to undertake some preparation which turns out to be of value in other contexts as well.

7. Zeta Functions of Epstein Type Attached to Γ

Let α be a K-finite function in $C_c^\infty(G)$—then, as we have seen, the Selberg trace formula associates with α the constant term of Ч_α around the origin:

$$\lim_{s\to 0}\frac{d}{ds}(s\text{Ч}_\alpha(s)).$$

In this section we shall make a preliminary investigation of Ч_α in the particular case when α is K-bi-invariant. It will turn out that Ч_α can be expressed in terms of zeta functions of Epstein type and certain distributions. The implication of this conclusion will be explored more fully later on. The zeta functions themselves will appear in the next section when we extend Theorem 6.3 to certain subspaces of $\mathscr{C}^1(G)$.

We shall begin by reviewing the basic facts governing the classical Epstein zeta function. Let V be an n-dimensional vector space over $\mathbf{R}$, Q a positive definite quadratic form on V. Given a lattice L in V, put

$$\zeta_Q(L:s) = \sum_{l\in L}^{*}\frac{1}{Q(l)^s} \qquad (s\in\mathbf{C}),$$

the * over the summation sign meaning that we sum over the nonzero elements of L. The series defining $\zeta_Q(L:?)$ is absolutely convergent in the half-plane $\mathrm{Re}(s) > n/2$ and can be continued to the entire complex plane as a meromorphic function with but a simple pole at $s = n/2$ where

$$\lim_{s\to n/2}\left(s-\frac{n}{2}\right)\zeta_Q(L:s) = \frac{\pi^{n/2}}{(D(L))^{1/2}\cdot\Gamma(n/2)},$$

$D(L)$ the discriminant of L with respect to Q. $\zeta_Q(L:?)$ is called the *Epstein zeta function attached to L by Q*. If $L^\perp$ is the dual lattice of L, then $\zeta_Q(L:?)$ and $\zeta_Q(L^\perp:?)$ are connected by the functional equation

$$\pi^{-s}\Gamma(s)\zeta_Q(L:s) = (D(L))^{-1/2}\cdot\pi^{s-n/2}\Gamma\left(\frac{n}{2}-s\right)\zeta_Q\left(L^\perp:\frac{n}{2}-s\right).$$

In general, $\zeta_Q(L:?)$ has no Euler product expansion.

Suppose now that α is a K-bi-invariant function in $C_c^\infty(G)$—then, by definition,

$$\text{Ч}_\alpha(s) = \sum_{i=1}^{r} \mathbf{c}_G^i \int_{-\infty}^{\infty} \int_{N_i/N_i \cap \Gamma} \left(\sum_{\substack{\eta \in \Gamma \cap N_i \\ \eta \neq 1}} \alpha(a_i(t) n_i \eta n_i^{-1} a_i(-t)) \right) \cdot e^{2|\rho|t(1+s)} \, d_{N_i}(n_i) \, dt$$

for all s such that $\operatorname{Re}(s) > 0$. We recall that the integral defining Ч_α is absolutely convergent and can be continued to the entire complex plane as a meromorphic function whose only possible singularities are simple poles at $s = 0$ and $s = -1$. In order to study the integrals

$$I_\alpha^i(s) = \int_{-\infty}^{\infty} \int_{N_i/N_i \cap \Gamma} \left(\sum_{\substack{\eta \in \Gamma \cap N_i \\ \eta \neq 1}} \alpha(a_i(t) n_i \eta n_i^{-1} a_i(-t)) \right) \cdot e^{2|\rho|t(1+s)} \, d_{N_i}(n_i) \, dt,$$

we can assume that $i = 1$ and then drop it from the notation. This said, we shall now proceed to the formulation of a technically important lemma.

Put

$$a^*(t) = \exp(tH^*) \qquad (t \in \mathbf{R}),$$

where H^* is the element in the Lie algebra of A such that $\lambda(H^*) = 1$. A given $x \in G$ always can be written in the form

$$x = k' a^*(t(x)) k'',$$

where $k', k'' \in K$ and $t(x) \geqslant 0$. We shall refer to $t(x)$, which is unique, as the *radial component* of x.

Let $m(\lambda)$ be the multiplicity of λ, $m(2\lambda)$ the multiplicity of 2λ—then, in an obvious notation, the Lie algebra $\mathfrak{n}$ of N admits the direct sum decomposition

$$\mathfrak{n} = \mathfrak{n}_\lambda + \mathfrak{n}_{2\lambda},$$

where, of course,

$$\dim(\mathfrak{n}_\lambda) = m(\lambda), \qquad \dim(\mathfrak{n}_{2\lambda}) = m(2\lambda).$$

Set

$$N_\lambda = \exp(\mathfrak{n}_\lambda), \qquad N_{2\lambda} = \exp(\mathfrak{n}_{2\lambda}),$$

so that $N = N_\lambda \cdot N_{2\lambda}$ with $N_\lambda \cap N_{2\lambda} = \{1\}$. We remark that N is a two-step nilpotent group with center $N_{2\lambda}$ (when $m(2\lambda) > 0$).

LEMMA 7.1. *Given $n \in N$, write $n = \exp(X_\lambda)\exp(X_{2\lambda})$ where $X_\lambda \in \mathfrak{n}_\lambda$, $X_{2\lambda} \in \mathfrak{n}_{2\lambda}$. Let $T(t, n)$ be the radial component of $a^*(t) n a^*(-t)$ $(t \in \mathbf{R})$—then*

$$\cosh(T(t,n)) = \left[\left(1 + e^{2t} \cdot \frac{A\|X_\lambda\|^2}{2} \right)^2 + e^{4t} \cdot A\|X_{2\lambda}\|^2 \right]^{1/2}.$$

Here

$$\Lambda^{-1} = 4(m(\lambda) + 4m(2\lambda)).$$

[$\|?\|$ is the canonical norm on $\mathfrak{n}$ associated with the usual Euclidean structure derived from the Killing form.]

Proof. The technique of proof is to reduce everything down to $\mathbf{SU}(1,1)$ or $\mathbf{SU}(2,1)$ (according to whether N is abelian or nonabelian) and then make an explicit matrix computation. The ideas here are well known (cf. Helgason [14, pp. 54–63]), so we shall omit the details. ∎

Let us agree to identify $\mathbf{R}$ with A via the map $t \mapsto a^*(t)$. If, as above, α is a compactly supported K-bi-invariant C^∞ function on G, then the restriction of α to A may be viewed as a compactly supported C^∞ function on $\mathbf{R}$ which is, moreover, even, α being invariant under the Weyl group $W(A)$. Define now a function $\overleftarrow{\alpha}: [1, +\infty[\to \mathbf{R}$ by the rule:

$$\overleftarrow{\alpha} = \alpha \circ \cosh^{-1}.$$

Then it can be shown (cf. Takahashi [29b, p. 328]) that $\overleftarrow{\alpha}$ is again compactly supported and C^∞ (where at 1, of course, it is a question of differentiability from the right). An important feature of this formalism is the following: Suppose the notations are as in Lemma 7.1—then

$$\alpha(a^*(t)na^*(-t)) = \overleftarrow{\alpha}\left(\left[\left(1 + e^{2t} \cdot \frac{\Lambda\|X_\lambda\|^2}{2}\right)^2 + e^{4t} \cdot \Lambda\|X_{2\lambda}\|^2\right]^{1/2}\right).$$

According to our conventions, $a(t) = \exp(tH)$ $(t \in \mathbf{R})$ where H has the property that $\lambda(H) = |\lambda|$. So, if we make the change of variable $t \mapsto t/|\lambda|$, then we have

$$\int_{-\infty}^{\infty} \int_{N/N \cap \Gamma} \left(\sum_{\substack{\eta \in \Gamma \cap N \\ \eta \neq 1}} \alpha(a(t)n\eta n^{-1}a(-t))\right) \cdot e^{2|\rho|t(1+s)}\, d_N(n)\, dt$$

$$= \frac{1}{|\lambda|} \cdot \int_{-\infty}^{\infty} \int_{N/N \cap \Gamma} \left(\sum_{\substack{\eta \in \Gamma \cap N \\ \eta \neq 1}} \alpha(a^*(t)n\eta n^{-1}a^*(-t))\right) \cdot e^{(2|\rho|/|\lambda|)t(1+s)}\, d_N(n)\, dt.$$

The definitions imply readily that

$$\frac{|\rho|}{|\lambda|} = \frac{m(\lambda)}{2} + m(2\lambda).$$

Moreover, it is not difficult to compute $|\lambda|$ explicitly. In fact

$$|\lambda| = \frac{1}{(2m(\lambda) + 8m(2\lambda))^{1/2}}.$$

We shall now divide the investigation of $Ч_\alpha$ into two cases, namely according to whether N is abelian or not.

To begin with, let us assume that N is abelian. Put $\Gamma_\lambda = \Gamma \cap N$—then $\log(\Gamma_\lambda)$ is a lattice in $\mathfrak{n}$, call it Γ_λ again. Bearing in mind that $\mathrm{vol}(N/N \cap \Gamma) = 1$, we have

$$I_\alpha(s) = \frac{1}{|\lambda|} \cdot \sum_{X_\lambda \in \Gamma_\lambda}^{*} \int_{-\infty}^{\infty} \overline{\alpha}\left(1 + e^{2t} \cdot \frac{\Lambda \|X_\lambda\|^2}{2}\right) e^{(2|\rho|/|\lambda|)t(1+s)}\, dt,$$

the * over the summation sign meaning that we sum over the nonzero elements of Γ_λ. Make the change of variable $y = e^{2t} \cdot \Lambda \|X_\lambda\|^2$—then, since here $|\rho|/|\lambda| = m(\lambda)/2$, we find that

$$I_\alpha(s) = \frac{1}{2|\lambda|} \cdot \frac{1}{(\sqrt{\Lambda})^{m(\lambda)(1+s)}} \cdot \sum_{X_\lambda \in \Gamma_\lambda}^{*} \frac{1}{\|X_\lambda\|^{m(\lambda)(1+s)}}$$
$$\times \int_0^\infty \overline{\alpha}\left(1 + \frac{y}{2}\right) y^{m(\lambda)(1+s)/2} \frac{dy}{y}.$$

Let

$$\zeta_{\Gamma_\lambda}(s) = \sum_{X_\lambda \in \Gamma_\lambda}^{*} \frac{1}{\|X_\lambda\|^{m(\lambda)(1+s)}}.$$

If by $\zeta(\Gamma_\lambda : s)$ we understand the Epstein zeta function attached to Γ_λ by $\|?\|$, then

$$\zeta_{\Gamma_\lambda}(s) = \zeta\left(\Gamma_\lambda : \frac{m(\lambda)(1+s)}{2}\right),$$

so, by the results mentioned earlier, the series defining ζ_{Γ_λ} is absolutely convergent for $\mathrm{Re}(s) > 0$, while, at $s = 0$, ζ_{Γ_λ} has a simple pole. On the other hand, let

$$T_\lambda(\alpha : s) = \int_0^\infty \overline{\alpha}\left(1 + \frac{y}{2}\right) y^{m(\lambda)(1+s)/2} \frac{dy}{y}.$$

Then it is clear that the integral defining $T_\lambda(\alpha : s)$ is actually absolutely convergent for $\mathrm{Re}(s) > -1$, thus represents a holomorphic function of s there. It therefore follows that one has the following identity of meromorphic functions in the half-plane $\mathrm{Re}(s) > -1$:

$$I_\alpha(s) = \frac{1}{2|\lambda|} \cdot \frac{1}{(\sqrt{\Lambda})^{m(\lambda)(1+s)}} \cdot \zeta_{\Gamma_\lambda}(s) \cdot T_\lambda(\alpha : s).$$

Call R_{Γ_λ} (respectively C_{Γ_λ}) the residue of ζ_{Γ_λ} at $s = 0$ (respectively the constant term of ζ_{Γ_λ} at $s = 0$). An easy calculation then gives

$$\lim_{s\to 0}\frac{d}{ds}(sI_\alpha(s)) = \frac{1}{2|\lambda|}\cdot\lim_{s\to 0}\left[\frac{d}{ds}\left(\frac{T_\lambda(\alpha:s)}{(\sqrt{\Lambda})^{m(\lambda)(1+s)}}\right)\cdot(s\zeta_{\Gamma_\lambda}(s)) + \frac{T_\lambda(\alpha:s)}{(\sqrt{\Lambda})^{m(\lambda)(1+s)}}\cdot\frac{d}{ds}(s\zeta_{\Gamma_\lambda}(s))\right]$$

$$= \frac{1}{2|\lambda|}\cdot\frac{1}{(\sqrt{\Lambda})^{m(\lambda)}}\left[\left(C_{\Gamma_\lambda} - R_{\Gamma_\lambda}\left(\frac{m(\lambda)}{2}\right)\log\Lambda\right)\cdot T_\lambda(\alpha:0) + R_{\Gamma_\lambda}\left(\frac{m(\lambda)}{2}\right)\cdot T'_\lambda(\alpha:0)\right],$$

where

$$T_\lambda(\alpha:0) = \int_0^\infty \overline{\alpha}\left(1+\frac{y}{2}\right)y^{m(\lambda)/2}\frac{dy}{y},$$

$$T'_\lambda(\alpha:0) = \int_0^\infty \overline{\alpha}\left(1+\frac{y}{2}\right)y^{m(\lambda)/2}\frac{\log y}{y}\,dy.$$

The assignments

$$\alpha\mapsto T_\lambda(\alpha:0), \qquad \alpha\mapsto T'_\lambda(\alpha:0)$$

are distributions which will be dealt with systematically later on.

SUMMARY (The abelian case). *Assume that* $m(2\lambda) = 0$. *Let* α *be a* K*-bi-invariant function in* $C_c^\infty(G)$*—then*

$$\lim_{s\to 0}\frac{d}{ds}(s\text{Ч}_\alpha(s))$$

$$= \frac{1}{2|\lambda|}\cdot\frac{1}{(\sqrt{\Lambda})^{m(\lambda)}}\cdot\left[\sum_{i=1}^{r}\mathbf{c}_G^i(C_{\Gamma_\lambda^i} - R_{\Gamma_\lambda^i}(m(\lambda)/2)\log\Lambda)\right]\cdot T_\lambda(\alpha:0)$$

$$+ \frac{1}{2|\lambda|}\cdot\frac{1}{(\sqrt{\Lambda})^{m(\lambda)}}\cdot\left[\sum_{i=1}^{r}\mathbf{c}_G^i R_{\Gamma_\lambda^i}(m(\lambda)/2)\right]\cdot T'_\lambda(\alpha:0).$$

Let us turn now to the case when N is nonabelian. We have then $N = N_\lambda\cdot N_{2\lambda}$ where both $N_\lambda \neq \{1\}$, $N_{2\lambda}\neq\{1\}$. It should be kept in mind that $N_{2\lambda}$ is the center of N. This said, put $\Gamma_{2\lambda} = \Gamma\cap N_{2\lambda}$—then $\log(\Gamma_{2\lambda})$ is a lattice in $\mathfrak{n}_{2\lambda}$, call it $\Gamma_{2\lambda}$ again (cf. Raghunathan [23b, p. 31]). On the other hand, set $\Gamma_\lambda = \Gamma\cap N/\Gamma_{2\lambda}$—then, under the identification $N_\lambda \sim N/N_{2\lambda}$, we have $\Gamma_\lambda \hookrightarrow N_\lambda$, and so $\log(\Gamma_\lambda)$ is a lattice in $\mathfrak{n}_\lambda$, call it Γ_λ again (cf. Raghunathan

[23b, p. 33]). Fix $t \in \mathbf{R}$—then we can write

$$\int_{N/N\cap\Gamma}\left(\sum_{\substack{\eta\in\Gamma\cap N\\ \eta\neq 1}}\alpha(a^*(t)n\eta n^{-1}a^*(-t))\right)d_N(n)$$

$$= \int_{N/N_{2\lambda}\cdot(N\cap\Gamma)}\cdot\left[\int_{N_{2\lambda}\cdot(N\cap\Gamma)/N\cap\Gamma}\left(\sum_{\substack{\eta\in\Gamma\cap N\\ \eta\neq 1}}\alpha(a^*(t)\mu\nu\eta\nu^{-1}\mu^{-1}a^*(-t))\right)d\nu\right]d\mu$$

$$= \mathrm{vol}(N_{2\lambda}/\Gamma_{2\lambda})\cdot\int_{N/N_{2\lambda}\cdot(N\cap\Gamma)}\left(\sum_{\substack{\eta\in\Gamma\cap N\\ \eta\neq 1}}\alpha(a^*(t)\mu\eta\mu^{-1}a^*(-t))\right)d\mu$$

$$= \mathrm{vol}(N_{2\lambda}/\Gamma_{2\lambda})\cdot\int_{(N/N_{2\lambda})/(N_{2\lambda}\cdot(N\cap\Gamma)/N_{2\lambda})}\left(\sum_{\substack{\eta\in\Gamma\cap N\\ \eta\neq 1}}\alpha(a^*(t)\mu\eta\mu^{-1}a^*(-t))\right)d\mu$$

$$= \mathrm{vol}(N_{2\lambda}/\Gamma_{2\lambda})\cdot\int_{N_\lambda/\Gamma_\lambda}\left(\sum_{\substack{\eta\in\Gamma\cap N\\ \eta\neq 1}}\alpha(a^*(t)n_\lambda\eta n_\lambda^{-1}a^*(-t))\right)d_{N_\lambda}(n_\lambda)$$

$$= \mathrm{vol}(N_{2\lambda}/\Gamma_{2\lambda})\cdot\int_{N_\lambda/\Gamma_\lambda}\left[\sum_{\eta\in\Gamma\cap N-\Gamma_{2\lambda}}\alpha(a^*(t)n_\lambda\eta n_\lambda^{-1}a^*(-t))\right.$$

$$\left.+\sum_{\substack{\eta\in\Gamma_{2\lambda}\\ \eta\neq 1}}\alpha(a^*(t)n_\lambda\eta n_\lambda^{-1}a^*(-t))\right]d_{N_\lambda}(n_\lambda)$$

$$= \mathrm{vol}(N_{2\lambda}/\Gamma_{2\lambda})\cdot\int_{N_\lambda/\Gamma_\lambda}\left(\sum_{\eta\in\Gamma\cap N-\Gamma_{2\lambda}}\alpha(a^*(t)n_\lambda\eta n_\lambda^{-1}a^*(-t))\right)d_{N_\lambda}(n_\lambda)$$
$$+\ \mathrm{vol}(N_\lambda/\Gamma_\lambda)\cdot\mathrm{vol}(N_{2\lambda}/\Gamma_{2\lambda})\cdot\sum_{\substack{\eta\in\Gamma_{2\lambda}\\ \eta\neq 1}}\alpha(a^*(t)\eta a^*(-t)).$$

Insert the last expression into the formula defining $I_\alpha(s)$—then we claim that

$$I_\alpha(s) = \frac{1}{|\lambda|}\cdot\mathrm{vol}\left(\frac{N_{2\lambda}}{\Gamma_{2\lambda}}\right)\cdot\int_{-\infty}^{\infty}\int_{N_\lambda/\Gamma_\lambda}\left(\sum_{\eta\in\Gamma\cap N-\Gamma_{2\lambda}}\alpha(a^*(t)n_\lambda\eta n_\lambda^{-1}a^*(-t))\right)$$
$$\times\ e^{(2|\rho|/|\lambda|)t(1+s)}\,d_{N_\lambda}(n_\lambda)\,dt + \frac{1}{|\lambda|}\cdot\mathrm{vol}\left(\frac{N_\lambda}{\Gamma_\lambda}\right)\cdot\mathrm{vol}\left(\frac{N_{2\lambda}}{\Gamma_{2\lambda}}\right)$$
$$\times\int_{-\infty}^{\infty}\left(\sum_{\substack{\eta\in\Gamma_{2\lambda}\\ \eta\neq 1}}\alpha(a^*(t)\eta a^*(-t))\right)\cdot e^{(2|\rho|/|\lambda|)t(1+s)}\,dt,$$

where both the integrals on the right-hand side are absolutely convergent for $\mathrm{Re}(s) > 0$. Since this is known to be the case for $I_\alpha(s)$, we need only deal explicitly with one of the two integrals on the right, say the second. Formally, we have

$$\int_{-\infty}^{\infty}\left(\sum_{\substack{\eta\in\Gamma_{2\lambda}\\ \eta\neq 1}}\alpha(a^*(t)\eta a^*(-t))\right)\cdot e^{(2|\rho|/|\lambda|)t(1+s)}\,dt$$
$$=\sum_{X_{2\lambda}\in\Gamma_{2\lambda}}^{*}\int_{-\infty}^{\infty}\bar{\alpha}((1+e^{4t}\cdot\Lambda\|X_{2\lambda}\|^2)^{1/2})e^{(2|\rho|/|\lambda|)t(1+s)}\,dt,$$

the * over the summation sign meaning that we sum over the nonzero elements of $\Gamma_{2\lambda}$. Make the change of variable $y = e^{2t}\cdot\sqrt{\Lambda}\|X_{2\lambda}\|$—then, since here $|\rho|/|\lambda| = m(\lambda)/2 + m(2\lambda)$, our integral becomes

$$\frac{1}{2}\cdot\left(\frac{1}{\sqrt{\Lambda}}\right)^{(m(\lambda)/2+m(2\lambda))(1+s)}\cdot\sum_{X_{2\lambda}\in\Gamma_{2\lambda}}^{*}\frac{1}{\|X_{2\lambda}\|^{(m(\lambda)/2+m(2\lambda))(1+s)}}$$
$$\times\int_0^{\infty}\bar{\alpha}((1+y^2)^{1/2})\,y^{(m(\lambda)/2+m(2\lambda))(1+s)}\frac{dy}{y}.$$

Let

$$\zeta_{\Gamma_{2\lambda}}(s)=\sum_{X_{2\lambda}\in\Gamma_{2\lambda}}^{*}\frac{1}{\|X_{2\lambda}\|^{(m(\lambda)/2+m(2\lambda))(1+s)}}.$$

If by $\zeta(\Gamma_{2\lambda}:s)$ we understand the Epstein zeta function attached to $\Gamma_{2\lambda}$ by $\|?\|$, then

$$\zeta_{\Gamma_{2\lambda}}(s)=\zeta\left(\Gamma_{2\lambda}:\frac{(m(\lambda)/2+m(2\lambda))(1+s)}{2}\right),$$

so, by the results mentioned earlier, the series defining $\zeta_{\Gamma_{2\lambda}}$ is absolutely convergent for

$$\mathrm{Re}(s) > \frac{-m(\lambda)/2}{(m(\lambda)/2+m(2\lambda))},$$

hence, in particular, is holomorphic at $s = 0$. On the other hand, let

$$T_{2\lambda}(\alpha:s)=\int_0^{\infty}\bar{\alpha}((1+y^2)^{1/2})\,y^{(m(\lambda)/2+m(2\lambda))(1+s)}\frac{dy}{y}.$$

Then it is clear that the integral defining $T_{2\lambda}(\alpha:s)$ is actually absolutely convergent for $\mathrm{Re}(s) > -1$, thus represents a holomorphic function of s

there. On the basis of these facts, it follows that

$$\int_{-\infty}^{\infty}\left(\sum_{\substack{\eta\in\Gamma_{2\lambda}\\ \eta\neq 1}}\alpha(a^*(t)\eta a^*(-t))\right)\cdot e^{(2|\rho|/|\lambda|)t(1+s)}\,dt$$

is absolutely convergent for $\mathrm{Re}(s)>0$ (and even a little bit more). Therefore

$$\int_{-\infty}^{\infty}\int_{N_\lambda/\Gamma_\lambda}\left(\sum_{\eta\in\Gamma\cap N-\Gamma_{2\lambda}}\alpha(a^*(t)n_\lambda\eta n_\lambda^{-1}a^*(-t))\right)\cdot e^{(2|\rho|/|\lambda|)t(1+s)}\,d_{N_\lambda}(n_\lambda)\,dt$$

is also absolutely convergent for $\mathrm{Re}(s)>0$. The set $\Gamma\cap N-\Gamma_{2\lambda}$ is a union of $\Gamma\cap N$-conjugacy classes. Let

$$(\lambda)\sum_{\{\eta\}_{\Gamma\cap N}}$$

stand for a sum over the $\Gamma\cap N$-conjugacy classes of the elements of $\Gamma\cap N-\Gamma_{2\lambda}$. For any $t\in\mathbf{R}$, we then have

$$\begin{aligned}&\int_{N_\lambda/\Gamma_\lambda}\left(\sum_{\eta\in\Gamma\cap N-\Gamma_{2\lambda}}\alpha(a^*(t)n_\lambda\eta n_\lambda^{-1}a^*(-t))\right)d_{N_\lambda}(n_\lambda)\\ &\quad=(\lambda)\sum_{\{\eta\}_{\Gamma\cap N}}\mathrm{vol}(N_{\lambda,\eta}/\Gamma_{\lambda,\eta})\cdot\int_{N_\lambda/N_{\lambda,\eta}}\alpha(a^*(t)n_\lambda\eta n_\lambda^{-1}a^*(-t))\,d_{N_\lambda/N_{\lambda,\eta}}(n_\lambda).\end{aligned}$$

Here $N_{\lambda,\eta}$ (respectively $\Gamma_{\lambda,\eta}$) is the centralizer of η in N_λ (respectively Γ_λ). Write $\eta=\exp(X_\lambda)\exp(X_{2\lambda})$—then $X_\lambda\neq 0$ (since $\eta\in\Gamma\cap N-\Gamma_{2\lambda}$), thus $\mathrm{ad}(X_\lambda):\mathfrak{n}_\lambda\to\mathfrak{n}_{2\lambda}$ is surjective (cf. Garland and Raghunathan [11, p. 287]), and so there is a natural identification $N_\lambda/N_{\lambda,\eta}\sim N_{2\lambda}$. We shall agree to fix the invariant measure on $N_\lambda/N_{\lambda,\eta}$ by transportation of the Haar measure on $N_{2\lambda}$ which, in turn, will be fixed below. This agreement then specifies $\mathrm{vol}(N_{\lambda,\eta}/\Gamma_{\lambda,\eta})$. Using now the fact

$$\exp(X'_\lambda)\exp(X''_\lambda)=\exp(X'_\lambda+X''_\lambda)\exp(\tfrac{1}{2}[X'_\lambda,X''_\lambda])$$

for $X'_\lambda, X''_\lambda\in\mathfrak{n}_\lambda$, we then find that

$$\begin{aligned}&\int_{N_\lambda/N_{\lambda,\eta}}\alpha(a^*(t)n_\lambda\eta n_\lambda^{-1}a^*(-t))\,d_{N_\lambda/N_{\lambda,\eta}}(n_\lambda)\\ &=\int_{\mathfrak{n}_{2\lambda}}\bar{\alpha}\left(\left(\left(1+e^{2t}\cdot\frac{\Lambda\|X_\lambda\|^2}{2}\right)^2+e^{4t}\cdot\Lambda\|Y_{2\lambda}\|^2\right)^{1/2}\right)d_{\mathfrak{n}_{2\lambda}}(Y_{2\lambda})\\ &=\left[\frac{(1+e^{2t}\cdot(\Lambda\|X_\lambda\|^2/2))}{e^{2t}\cdot\sqrt{\Lambda}}\right]^{m(2\lambda)}\\ &\quad\times\int_{\mathfrak{n}_{2\lambda}}\bar{\alpha}\left(\left(1+e^{2t}\cdot\frac{\Lambda\|X_\lambda\|^2}{2}\right)\cdot(1+\|Y_{2\lambda}\|^2)^{1/2}\right)d_{\mathfrak{n}_{2\lambda}}(Y_{2\lambda}).\end{aligned}$$

Multiply the last expression through by

$$e^{(2|\rho|/|\lambda|)t(1+s)}$$

and integrate over $\mathbf{R}$. In the resulting integral make the change of variable $y = e^{2t} \cdot \Lambda \|X_\lambda\|^2$ to get

$$\frac{1}{2} \cdot \left(\frac{\sqrt{\Lambda}}{2}\right)^{m(2\lambda)} \cdot \left(\frac{1}{\Lambda}\right)^{(m(\lambda)/2 + m(2\lambda))(1+s)}$$

$$\times \|X_\lambda\|^{2m(2\lambda)} \cdot \frac{1}{\|X_\lambda\|^{(m(\lambda) + 2m(2\lambda))(1+s)}} \cdot T_\lambda(\alpha : s),$$

where, by definition,

$$T_\lambda(\alpha : s) = \int_0^\infty \int_{\mathfrak{n}_{2\lambda}} \overline{\alpha}\left(\left(1 + \frac{y}{2}\right)(1 + \|Y_{2\lambda}\|^2)^{1/2}\right) \cdot \left(1 + \frac{2}{y}\right)^{m(2\lambda)}$$

$$\times\, y^{(m(\lambda)/2 + m(2\lambda))(1+s)} \frac{dy}{y} \, d_{\mathfrak{n}_{2\lambda}}(Y_{2\lambda}).$$

The integral defining $T_\lambda(\alpha : s)$ is evidently absolutely convergent in the region

$$\text{Re}(s) > \frac{-m(\lambda)/2}{(m(\lambda)/2 + m(2\lambda))},$$

thus represents a holomorphic function of s there. It remains to sum over the $\{\eta\}_{\Gamma \cap N}$. For this purpose, we shall make a simplifying observation. The first component of η is obviously dependent only on $\{\eta\}_{\Gamma \cap N}$. We have, therefore, a map

$$\Phi_\lambda : \{\{\eta\}_{\Gamma \cap N} : \eta \in \Gamma \cap N - \Gamma_{2\lambda}\} \to \Gamma_\lambda - \{0\},$$

namely the rule which assigns to each $\{\eta\}_{\Gamma \cap N}$ its first component. Φ_λ is obviously surjective. In addition, it has finite fibers. To see this, we notice that the second components of $\{\eta\}_{\Gamma \cap N}$ are the elements of the set

$$X_{2\lambda} + \text{ad}(X_\lambda)\Gamma_\lambda \qquad (\eta = \exp(X_\lambda)\exp(X_{2\lambda}))$$

so there is a natural way of assigning to $\{\eta\}_{\Gamma \cap N}$ a coset in $\mathfrak{n}_{2\lambda}/\text{ad}(X_\lambda)\Gamma_\lambda$. Any other conjugacy class with first component X_λ also determines a point in $\mathfrak{n}_{2\lambda}/\text{ad}(X_\lambda)\Gamma_\lambda$. Because $\text{ad}(X_\lambda)\Gamma_\lambda$ is a lattice in $\mathfrak{n}_{2\lambda}$, $\mathfrak{n}_{2\lambda}/\text{ad}(X_\lambda)\Gamma_\lambda$ is compact. The finiteness of the fibers of Φ_λ is then immediate. Given $X_\lambda \in \Gamma_\lambda - \{0\}$, call $C(X_\lambda)$ the number of elements in $\Phi_\lambda^{-1}(X_\lambda)$ times $\text{vol}(N_{\lambda,\eta}/\Gamma_{\lambda,\eta})$ where η is any element of $\Gamma \cap N - \Gamma_{2\lambda}$ such that

$$\Phi_\lambda(\{\eta\}_{\Gamma \cap N}) = X_\lambda.$$

The total volume of $N/N \cap \Gamma$ being one, specify the invariant measures on N_λ/Γ_λ and $N_{2\lambda}/\Gamma_{2\lambda}$ by demanding that

$$\mathrm{vol}(N_\lambda/\Gamma_\lambda) = 1, \qquad \mathrm{vol}(N_{2\lambda}/\Gamma_{2\lambda}) = 1.$$

The cardinality of the fiber $\Phi_\lambda^{-1}(X_\lambda)$ is evidently the number of $\Gamma \cap N$-conjugacy classes which are N-conjugate to $\{\eta\}_{\Gamma \cap N}$. Accordingly, as remarked by M. S. Osborne, elementary combinatorial considerations then readily imply that $C(X_\lambda) = 1$ (for the agreed-to normalizations of the invariant measures). Thus we have

$$\begin{aligned}\int_{-\infty}^{\infty} \int_{N_\lambda/\Gamma_\lambda} &\left(\sum_{\eta \in \Gamma \cap N - \Gamma_{2\lambda}} \alpha(a^*(t) n_\lambda \eta n_\lambda^{-1} a^*(-t)) \right) \cdot e^{(2|\rho|/|\lambda|)t(1+s)} \, d_{N_\lambda}(n_\lambda) \, dt \\ &= \frac{1}{2} \cdot \left(\frac{\sqrt{\Lambda}}{2} \right)^{m(2\lambda)} \cdot \left(\frac{1}{\Lambda} \right)^{(m(\lambda)/2 + m(2\lambda)(1+s)} \\ &\times \sum_{X_\lambda \in \Gamma_\lambda}^{*} \frac{1}{\|X_\lambda\|^{(m(\lambda) + 2m(2\lambda))s + m(\lambda)}} \, T_\lambda(\alpha : s),\end{aligned}$$

the * over the summation sign meaning that we sum over the nonzero elements of Γ_λ. Let

$$\zeta_{\Gamma_\lambda}(s) = \sum_{X_\lambda \in \Gamma_\lambda}^{*} \frac{1}{\|X_\lambda\|^{(m(\lambda) + 2m(2\lambda))s + m(\lambda)}}.$$

If by $\zeta(\Gamma_\lambda : s)$ we understand the Epstein zeta function attached to Γ_λ by $\|?\|$, then

$$\zeta_{\Gamma_\lambda}(s) = \zeta\left(\Gamma_\lambda : \frac{(m(\lambda) + 2m(2\lambda))s + m(\lambda)}{2} \right),$$

so, by the results mentioned earlier, the series defining ζ_{Γ_λ} is absolutely convergent for $\mathrm{Re}(s) > 0$, while, at $s = 0$, ζ_{Γ_λ} has a simple pole. Putting everything together then leads to the identity

$$\begin{aligned}I_\alpha(s) = \frac{1}{2|\lambda|} \Bigg[&\left(\frac{\sqrt{\Lambda}}{2} \right)^{m(2\lambda)} \cdot \left(\frac{1}{\Lambda} \right)^{(m(\lambda)/2 + m(2\lambda))(1+s)} \cdot \zeta_{\Gamma_\lambda}(s) \cdot T_\lambda(\alpha : s) \\ &+ \left(\frac{1}{\sqrt{\Lambda}} \right)^{(m(\lambda)/2 + m(2\lambda))(1+s)} \cdot \zeta_{\Gamma_{2\lambda}}(s) \cdot T_{2\lambda}(\alpha : s) \Bigg]\end{aligned}$$

valid for

$$\mathrm{Re}(s) > \frac{-m(\lambda)/2}{(m(\lambda)/2 + m(2\lambda))}.$$

Call R_{Γ_λ} (respectively C_{Γ_λ}) the residue of ζ_{Γ_λ} at $s = 0$ (respectively the constant term of ζ_{Γ_λ} at $s = 0$); call $C_{\Gamma_{2\lambda}}$ the constant term of $\zeta_{\Gamma_{2\lambda}}$ at $s = 0$. An easy calculation then gives

$$\begin{aligned}
\lim_{s\to 0} \frac{d}{ds}(sI_\alpha(s)) &= \frac{1}{2|\lambda|}\cdot\left(\frac{\sqrt{\Lambda}}{2}\right)^{m(2\lambda)}\cdot \lim_{s\to 0}\left[\frac{d}{ds}\left(\left(\frac{1}{\Lambda}\right)^{(m(\lambda)/2+m(2\lambda))(1+s)}\cdot T_\lambda(\alpha:s)\right)\right.\\
&\quad\left.\times (s\zeta_{\Gamma_\lambda}(s)) + \left(\frac{1}{\Lambda}\right)^{(m(\lambda)/2+m(2\lambda))(1+s)}\cdot T_\lambda(\alpha:s)\cdot\frac{d}{ds}(s\zeta_{\Gamma_\lambda}(s))\right]\\
&\quad+\frac{1}{2|\lambda|}\cdot\lim_{s\to 0}\left[\left(\frac{1}{\sqrt{\Lambda}}\right)^{(m(\lambda)/2+m(2\lambda))(1+s)}\cdot\zeta_{\Gamma_{2\lambda}}(s)\cdot T_{2\lambda}(\alpha:s)\right]\\
&= \frac{1}{2|\lambda|}\cdot\left(\frac{\sqrt{\Lambda}}{2}\right)^{m(2\lambda)}\cdot\left(\frac{1}{\Lambda}\right)^{(m(\lambda)/2+m(2\lambda))}\\
&\quad\times\left[\left(C_{\Gamma_\lambda} - R_{\Gamma_\lambda}\left(\frac{m(\lambda)}{2}+m(2\lambda)\right)\log\Lambda\right)\cdot T_\lambda(\alpha:0)\right.\\
&\quad\left.+R_{\Gamma_\lambda}\cdot\left(\frac{m(\lambda)}{2}+m(2\lambda)\right)\cdot T'_\lambda(\alpha:0)\right]\\
&\quad+\frac{1}{2|\lambda|}\cdot\left(\frac{1}{\sqrt{\Lambda}}\right)^{(m(\lambda)/2+m(2\lambda))}\cdot[C_{\Gamma_{2\lambda}}\cdot T_{2\lambda}(\alpha:0)]
\end{aligned}$$

where

$$\begin{aligned}
T_\lambda(\alpha:0) &= \int_0^\infty\int_{\mathfrak{n}_{2\lambda}}\overline{\alpha}\left(\left(1+\frac{y}{2}\right)(1+\|Y_{2\lambda}\|^2)^{1/2}\right)\cdot\left(1+\frac{2}{y}\right)^{m(2\lambda)}\\
&\quad\times y^{(m(\lambda)/2+m(2\lambda))}\frac{dy}{y}\,d_{\mathfrak{n}_{2\lambda}}(Y_{2\lambda}),\\
T'_\lambda(\alpha:0) &= \int_0^\infty\int_{\mathfrak{n}_{2\lambda}}\overline{\alpha}\left(\left(1+\frac{y}{2}\right)(1+\|Y_{2\lambda}\|^2)^{1/2}\right)\cdot\left(1+\frac{2}{y}\right)^{m(2\lambda)}\\
&\quad\times y^{(m(\lambda)/2+m(2\lambda))}\frac{\log y}{y}\,dy\,d_{\mathfrak{n}_{2\lambda}}(Y_{2\lambda}),\\
T_{2\lambda}(\alpha:0) &= \int_0^\infty\overline{\alpha}((1+y^2)^{1/2})y^{(m(\lambda)/2+m(2\lambda))}\frac{dy}{y}.
\end{aligned}$$

The assignments

$$\alpha\mapsto T_\lambda(\alpha:0),\qquad \alpha\mapsto T'_\lambda(\alpha:0),\qquad \alpha\mapsto T_{2\lambda}(\alpha:0)$$

are distributions which will be dealt with systematically later on.

SUMMARY (The nonabelian case). *Assume that $m(2\lambda) > 0$. Let α be a K-bi-invariant function in $C_c^\infty(G)$—then*

$$\lim_{s\to 0}\frac{d}{ds}(s\text{Ч}_\alpha(s))$$

$$= \frac{1}{2^{m(2\lambda)+1}}\cdot\frac{1}{|\lambda|}\cdot\frac{1}{(\sqrt{\Lambda})^{m(\lambda)+m(2\lambda)}}\cdot\left[\sum_{i=1}^{r}\mathbf{c}_G^i\left(C_{\Gamma_\lambda^i}-R_{\Gamma_\lambda^i}\left(\frac{m(\lambda)}{2}+m(2\lambda)\right)\log\Lambda\right)\right]$$

$$\times T_\lambda(\alpha:0)+\frac{1}{2^{m(2\lambda)+1}}\cdot\frac{1}{|\lambda|}\cdot\frac{1}{(\sqrt{\Lambda})^{m(\lambda)+m(2\lambda)}}\cdot\left[\sum_{i=1}^{r}\mathbf{c}_G^iR_{\Gamma_\lambda^i}\left(\frac{m(\lambda)}{2}+m(2\lambda)\right)\right]$$

$$\times T'_\lambda(\alpha:0)+\frac{1}{2|\lambda|}\cdot\left(\frac{1}{\sqrt{\Lambda}}\right)^{(m(\lambda)/2+m(2\lambda))}\cdot\left[\sum_{i=1}^{r}\mathbf{c}_G^iC_{\Gamma_{2\lambda}^i}\right]\cdot T_{2\lambda}(\alpha:0).$$

Suppose that N is abelian—then we put

$$\mathbf{C}_\lambda(\Gamma)=\frac{1}{2|\lambda|}\cdot\frac{1}{(\sqrt{\Lambda})^{m(\lambda)}}\cdot\left[\sum_{i=1}^{r}\mathbf{c}_G^i(C_{\Gamma_\lambda^i}-R_{\Gamma_\lambda^i}(m(\lambda)/2)\log\Lambda)\right],$$

$$\mathbf{C}'_\lambda(\Gamma)=\frac{1}{2|\lambda|}\cdot\frac{1}{(\sqrt{\Lambda})^{m(\lambda)}}\cdot\left[\sum_{i=1}^{r}\mathbf{c}_G^iR_{\Gamma_\lambda^i}(m(\lambda)/2)\right].$$

Suppose that N is nonabelian—then we put

$$\mathbf{C}_\lambda(\Gamma)=\frac{1}{2^{m(2\lambda)+1}}\cdot\frac{1}{|\lambda|}\cdot\frac{1}{(\sqrt{\Lambda})^{m(\lambda)+m(2\lambda)}}$$

$$\times\left[\sum_{i=1}^{r}\mathbf{c}_G^i\left(C_{\Gamma_\lambda^i}-R_{\Gamma_\lambda^i}\left(\frac{m(\lambda)}{2}+m(2\lambda)\right)\log\Lambda\right)\right],$$

$$\mathbf{C}'_\lambda(\Gamma)=\frac{1}{2^{m(2\lambda)+1}}\cdot\frac{1}{|\lambda|}\cdot\frac{1}{(\sqrt{\Lambda})^{m(\lambda)+m(2\lambda)}}\cdot\left[\sum_{i=1}^{r}\mathbf{c}_G^iR_{\Gamma_\lambda^i}\left(\frac{m(\lambda)}{2}+m(2\lambda)\right)\right],$$

$$\mathbf{C}_{2\lambda}(\Gamma)=\frac{1}{2|\lambda|}\cdot\left(\frac{1}{\sqrt{\Lambda}}\right)^{(m(\lambda)/2+m(2\lambda))}\cdot\left[\sum_{i=1}^{r}\mathbf{c}_G^iC_{\Gamma_{2\lambda}^i}\right].$$

In these notations, we then have the following theorem.

THEOREM 7.2. *Let α be a K-bi-invariant function in $C_c^\infty(G)$.*
(i) *Suppose that N is abelian—then*

$$\lim_{s\to 0}\frac{d}{ds}(s\text{Ч}_\alpha(s))=\mathbf{C}_\lambda(\Gamma)\cdot T_\lambda(\alpha:0)+\mathbf{C}'_\lambda(\Gamma)\cdot T'_\lambda(\alpha:0).$$

(ii) *Suppose that N is nonabelian—then*

$$\lim_{s\to 0}\frac{d}{ds}(s\text{Ч}_\alpha(s)) = \mathbf{C}_\lambda(\Gamma)\cdot T_\lambda(\alpha:0) + \mathbf{C}'_\lambda(\Gamma)\cdot T'_\lambda(\alpha:0) + \mathbf{C}_{2\lambda}(\Gamma)\cdot T_{2\lambda}(\alpha:0).$$

8. Extension to $\mathscr{C}^1_\epsilon(G)$

The objective of the present section will be to show that Theorem 6.3 is still true for functions in a class of subspaces of $\mathscr{C}^1(G)$. Such functions need not have compact support.

Let $\epsilon > 0$ be an arbitrary positive number. For any $\alpha \in C^\infty(G)$, write

$${}_{D_1}|\alpha|^{\epsilon}_{r,D_2} = \sup_{x\in G}\,(1+\sigma(x))^r \vdash\!\circ\!\dashv^{-2-\epsilon}(x)\,|\alpha(D_1 ; x; D_2)|\quad(D_1, D_2 \in \mathfrak{G};\ r\in\mathbf{R});$$

Call $\mathscr{C}^1_\epsilon(G)$ the set of all α in $C^\infty(G)$ such that ${}_{D_1}|\alpha|^{\epsilon}_{r,D_2} < +\infty$ for all $r \in \mathbf{R}$ and pairs $(D_1, D_2)\in\mathfrak{G}\times\mathfrak{G}$. Topologize $\mathscr{C}^1_\epsilon(G)$ by means of the seminorms ${}_{D_1}|?|^{\epsilon}_{r,D_2}$—then $\mathscr{C}^1_\epsilon(G)$ becomes a locally convex, complete, Hausdorff, topological vector space. The inclusions

$$\begin{array}{ccccccc} C^\infty_c(G) & \hookrightarrow & \mathscr{C}^1_\epsilon(G) & \hookrightarrow & \mathscr{C}^1(G) & \hookrightarrow & L^2(G) \\ & & \downarrow & & \downarrow & & \\ & & L^{2/(2+\epsilon)}(G) & & L^1(G) & & \end{array}$$

are continuous with dense range.

Remark. Suppose that $\text{rank}(G) = \text{rank}(K)$, so that G has a discrete series. Let $U \in \hat{G}$ be in the integrable discrete series—then it can be shown (cf. Trombi and Varadarajan [30b, p. 268]) that there exists $\epsilon_U > 0$ such that all the K-finite matrix coefficients of U are in $\mathscr{C}^1_{\epsilon_U}(G)$. Such functions, being analytic, do not have compact support. At the opposite extreme, in the class one case one can produce functions in $\mathscr{C}^1_\epsilon(G)$ which are not of compact support by imposing conditions on their spherical Fourier transforms. We shall return to this point later on.

There are two important estimates on $\vdash\!\circ\!\dashv$ which we shall need to use below.

(i) There is an integer $d > 0$ and a constant $D > 0$ such that

$$\vdash\!\circ\!\dashv(a) \leqslant D\,\frac{(1+\sigma(a))^d}{e^{|\rho|(H(a))}}$$

for all $a \in A$ such that $\xi_\lambda(a) \geqslant 1$ (cf. Warner [34b, p. 154]).

(ii) There is a constant $D > 0$ such that

$$\vdash\!\circ\!\dashv\,(an) \leqslant D(1+\sigma(an))^d \cdot e^{-|\rho|(H(a))} \cdot e^{-|\rho|(H(\theta(n^{-1})))}$$

for all $a \in A$, $n \in N$ (cf. Warner [34b, p. 239]).

[Here θ is the ambient Cartan involution associated with the pair (G, K).]

Fix now $\epsilon > 0$. Let α be a K-finite function in $\mathscr{C}_\epsilon^1(G)$—then, as we know, $L_{G/\Gamma}^{\mathrm{d}}(\alpha)$ is a trace class operator and, moreover,

$$\mathrm{tr}(L_{G/\Gamma}^{\mathrm{d}}(\alpha)) = \int_{G/\Gamma} K_\alpha^{\mathrm{d}}(x,x)\, d_G(x).$$

To calculate the integral on the right, we shall proceed exactly as in Section 6. To be in agreement with what was said there, it will be convenient to allow ourselves to denote still by ϵ a parameter varying between 0 and ${}_0t$; no confusion will arise from this double usage of the symbol. Just as before, then, $K_\alpha^{\mathrm{d}}(x,x)$ admits the decomposition

$$\sum_{z \in Z_\Gamma} \alpha(z), \qquad \sum_{\gamma \in \Gamma_S} \alpha(x\gamma x^{-1}),$$

$$I_\alpha(x:\epsilon) - K_\alpha^{\prime,\mathrm{c}}(x:\epsilon), \qquad J_\alpha(x:\epsilon), \qquad -K_\alpha^{\prime\prime,\mathrm{c}}(x:\epsilon)$$

into the central, semisimple, and first, second, and third parabolic terms, respectively. The central term offers no difficulty:

$$\int_{G/\Gamma} \left(\sum_{z \in Z_\Gamma} \alpha(z) \right) d_G(x) = \mathrm{vol}(G/\Gamma) \left(\sum_{z \in Z_\Gamma} \alpha(z) \right).$$

Turning next to the semisimple term, one makes the expected claim, namely that

$$\int_{G/\Gamma} \left(\sum_{\gamma \in \Gamma_S} \alpha(x\gamma x^{-1}) \right) d_G(x) = (S) \sum_{\{\gamma\}_\Gamma} \mathrm{vol}(G_\gamma/\Gamma_\gamma) \cdot \int_{G/G_\gamma} \alpha(x\gamma x^{-1})\, d_{G/G_\gamma}(x),$$

where the series and integrals are absolutely convergent. This is in fact an immediate consequence of the following lemma, which implies that the assignment

$$\alpha \mapsto \int_{G/\Gamma} \left(\sum_{\gamma \in \Gamma_S} |\alpha(x\gamma x^{-1})| \right) d_G(x) \qquad (\alpha \in \mathscr{C}_\epsilon^1(G))$$

is continuous in the topology of $\mathscr{C}_\epsilon^1(G)$.

LEMMA 8.1. *There is a constant $C(\epsilon) > 0$ such that*

$$\int_{G/\Gamma} \left(\sum_{\gamma \in \Gamma_S} |\alpha(x\gamma x^{-1})| \right) d_G(x) \leqslant C(\epsilon) \cdot |\alpha|_{d(2+\epsilon)}^{\epsilon}$$

for all α in $\mathscr{C}_\epsilon^1(G)$.

Proof. Fix α in $\mathscr{C}_\epsilon^1(G)$—then we have

$$\int_{G/\Gamma}\left(\sum_{\gamma\in\Gamma_S}|\alpha(x\gamma x^{-1})|\right)d_G(x)$$

$$\leqslant |\alpha|_{d(2+\epsilon)}^{\epsilon}\cdot\int_{G/\Gamma}\left(\sum_{\gamma\in\Gamma_S}\Xi^{2+\epsilon}(x\gamma x^{-1})\cdot(1+\sigma(x\gamma x^{-1}))^{-d(2+\epsilon)}\right)d_G(x)$$

$$= |\alpha|_{d(2+\epsilon)}^{\epsilon}\cdot\left[\int_{\pi(\Omega_{0t})}\left(\sum_{\gamma\in\Gamma_S}\Xi^{2+\epsilon}(x\gamma x^{-1})\cdot(1+\sigma(x\gamma x^{-1}))^{-d(2+\epsilon)}\right)d_G(x)\right.$$

$$\left.+\sum_{i=1}^{r}\int_{\pi(\mathfrak{C}_{0t}\kappa_i)}\left(\sum_{\gamma\in\Gamma_S}\Xi^{2+\epsilon}(x\gamma x^{-1})\cdot(1+\sigma(x\gamma x^{-1}))^{-d(2+\epsilon)}\right)d_G(x)\right].$$

Since the function

$$x\mapsto\sum_{\gamma\in\Gamma_S}\Xi^{2+\epsilon}(x\gamma x^{-1})\cdot(1+\sigma(x\gamma x^{-1}))^{-d(2+\epsilon)}$$

is continuous on G/Γ (cf. Section 4), its integral taken over the compact set $\pi(\Omega_{0t})$ is finite. To deal with the integrals over the $\pi(\mathfrak{C}_{0t}\kappa_i)$, we can assume that $i=1$ and then drop it from the notation. We shall establish the finiteness of the integral over $\pi(\mathfrak{C}_{0t})$ by showing that our integrand is bounded on

$$K\times M/Z_\Gamma\times A({}_0t)\times N/N\cap\Gamma.$$

Suppose that $x=kan$ where $k\in K$, $a\in A({}_0t)$, $n\in N/N\cap\Gamma$—then

$$\begin{aligned}\Xi^{2+\epsilon}(x\gamma x^{-1})&=\Xi^{2+\epsilon}(k\cdot ana^{-1}\cdot a\gamma a^{-1}\cdot an^{-1}a^{-1}\cdot k^{-1})\\&=\Xi^{2+\epsilon}(ana^{-1}\cdot a\gamma a^{-1}\cdot an^{-1}a^{-1})\\&\leqslant C\,\Xi^{2+\epsilon}(a\gamma a^{-1})\qquad(\gamma\in\Gamma_S),\end{aligned}$$

C a positive constant. In passing to the inequality, we use the fact that the $ana^{-1}(a\in A({}_0t),\ n\in N/N\cap\Gamma)$ stay in a compact subset of N (cf. Borel and Harish-Chandra [4, p. 501]) and then invoke a standard property of Ξ (cf. Warner [34b, p. 153]). In a similar way, using a standard property of σ (cf. Warner [34b, p. 67]), we find that, under the same assumption on x,

$$(1+\sigma(x\gamma x^{-1}))^{-d(2+\epsilon)}\leqslant C(1+\sigma(a\gamma a^{-1}))^{-d(2+\epsilon)}\qquad(\gamma\in\Gamma_S),$$

C a positive constant. It suffices, therefore, to show that the function

$$a\mapsto\sum_{\gamma\in\Gamma_S}\Xi^{2+\epsilon}(a\gamma a^{-1})\cdot(1+\sigma(a\gamma a^{-1}))^{-d(2+\epsilon)}$$

is bounded on $A({}_0t)$, or, since $\Gamma_S\subset\Gamma-\Gamma\cap P$, that the function

$$a\mapsto\sum_{\gamma\in\Gamma-\Gamma\cap P}\Xi^{2+\epsilon}(a\gamma a^{-1})\cdot(1+\sigma(a\gamma a^{-1}))^{-d(2+\epsilon)}$$

is bounded on $A({}_0t)$. Write

$$\sum_{\gamma\in\Gamma-\Gamma\cap P} \vdash\!\circ\!\dashv^{2+\epsilon}(a\gamma a^{-1})\cdot(1+\sigma(a\gamma a^{-1}))^{-d(2+\epsilon)}$$

$$=\sum_{\gamma\in(\Gamma-\Gamma\cap P)/\Gamma\cap P}\ \sum_{\delta\in\Gamma\cap P} \vdash\!\circ\!\dashv^{2+\epsilon}(a\gamma\,\delta a^{-1})\cdot(1+\sigma(a\gamma\,\delta a^{-1}))^{-d(2+\epsilon)}.$$

We have

$$\sum_{\delta\in\Gamma\cap P} \vdash\!\circ\!\dashv^{2+\epsilon}(a\gamma\,\delta a^{-1})\cdot(1+\sigma(a\gamma\,\delta a^{-1}))^{-d(2+\epsilon)}$$

$$=\sum_{z\in Z_\Gamma}\ \sum_{\eta\in\Gamma\cap N} \vdash\!\circ\!\dashv^{2+\epsilon}(a\gamma z\eta a^{-1})\cdot(1+\sigma(a\gamma z\eta a^{-1}))^{-d(2+\epsilon)}$$

$$=[Z_\Gamma]\sum_{\eta\in\Gamma\cap N} \vdash\!\circ\!\dashv^{2+\epsilon}(a\gamma\eta a^{-1})\cdot(1+\sigma(a\gamma\eta a^{-1}))^{-d(2+\epsilon)},$$

Z_Γ being contained in the center of G, hence in K (cf. Lemma 5.1). Fix a compact symmetric neighborhood $\mathcal{N}$ of the identity in N with the property that

$$a\mathcal{N}a^{-1}\subset\mathcal{N}$$

for all $a\in A({}_0t)$. This can always be arranged (under the supposition that ${}_0t<1$). Let $n\in\mathcal{N}$—then, for $a\in A({}_0t)$,

$$\begin{aligned}\vdash\!\circ\!\dashv^{2+\epsilon}(a\gamma\eta na^{-1}) &= \vdash\!\circ\!\dashv^{2+\epsilon}(a\gamma\eta a^{-1}\cdot ana^{-1})\\ &\geqslant R^{-1}\vdash\!\circ\!\dashv^{2+\epsilon}(a\gamma\eta a^{-1}),\\ (1+\sigma(a\gamma\eta na^{-1}))^{-d(2+\epsilon)} &= (1+\sigma(a\gamma\eta a^{-1}\cdot ana^{-1})^{-d(2+\epsilon)}\\ &\geqslant S^{-1}(1+\sigma(a\gamma\eta a^{-1}))^{-d(2+\epsilon)},\end{aligned}$$

R, S positive constants (independent of γ or η). Therefore

$$\sum_{\eta\in\Gamma\cap N} \vdash\!\circ\!\dashv^{2+\epsilon}(a\gamma\eta a^{-1})\cdot(1+\sigma(a\gamma\eta a^{-1}))^{-d(2+\epsilon)}$$

$$\leqslant\frac{RS}{\text{meas}(\mathcal{N})}\cdot\sum_{\eta\in\Gamma\cap N}\int_{\mathcal{N}} \vdash\!\circ\!\dashv^{2+\epsilon}(a\gamma\eta na^{-1})$$
$$\times(1+\sigma(a\gamma\eta na^{-1}))^{-d(2+\epsilon)}\,d_N(n).$$

There is no loss of generality in assuming that $\Gamma\cap\mathcal{N}^2=\{1\}$. So, ignoring constant factors, we see that

$$\sum_{\eta\in\Gamma\cap N} \vdash\!\circ\!\dashv^{2+\epsilon}(a\gamma\eta a^{-1})\cdot(1+\sigma(a\gamma\eta a^{-1}))^{-d(2+\epsilon)}$$

is majorized by

$$\int_N \vdash\!\circ\!\dashv^{2+\epsilon}(a\gamma na^{-1})\cdot(1+\sigma(a\gamma na^{-1}))^{-d(2+\epsilon)}\,d_N(n).$$

The last integral can be written in the form

$$e^{-2|\rho|(H(a))} \cdot \int_N \vdash\!\circ\!\dashv^{2+\epsilon}(\exp(H(a\gamma))a^{-1}n)$$
$$\times (1 + \sigma(\exp(H(a\gamma))a^{-1}n))^{-d(2+\epsilon)}\, d_N(n).$$

Owing to the estimate

$$\begin{aligned}&\vdash\!\circ\!\dashv^{2+\epsilon}(\exp(H(a\gamma))a^{-1}n)\\ &\qquad \leqslant D(1 + \sigma(\exp(H(a\gamma))a^{-1}n)^{d(2+\epsilon)} \cdot e^{-(2+\epsilon)|\rho|(H(a\gamma)-H(a))}\\ &\qquad \times e^{-(2+\epsilon)|\rho|(H(\theta(n^{-1})))},\end{aligned}$$

it follows that, up to a positive absolute constant, the series

$$\sum_{\eta \in \Gamma \cap N} \vdash\!\circ\!\dashv^{2+\epsilon}(a\gamma\eta a^{-1}) \cdot (1 + \sigma(a\gamma\eta a^{-1}))^{-d(2+\epsilon)}$$

is majorized by

$$e^{\epsilon|\rho|(H(a))} \cdot e^{(-|\rho|(1+\epsilon)-|\rho|)(H(a\gamma))} \int_{N^-} e^{-(|\rho|(1+\epsilon)+|\rho|)(H(n))}\, d_{N^-}(n),$$

where $N^- = \theta(N)$, the unipotent subgroup opposed to N. The integral over N^- is of Gindikin–Karpelevič type and, because of the presence of the term $|\rho|(1 + \epsilon)$, is finite (cf. Warner [34b, p. 323]). These estimates then imply the existence of a constant $M(\epsilon) > 0$, say, with the property that

$$\begin{aligned}&\sum_{\gamma \in \Gamma - \Gamma \cap P} \vdash\!\circ\!\dashv^{2+\epsilon}(a\gamma a^{-1}) \cdot (1 + \sigma(a\gamma a^{-1}))^{-d(2+\epsilon)}\\ &\qquad \leqslant M(\epsilon) \cdot e^{\epsilon|\rho|(H(a))} \cdot \sum_{\gamma \in (\Gamma - \Gamma \cap P)/\Gamma \cap P} e^{(-|\rho|(1+\epsilon)-|\rho|)(H(a\gamma))}.\end{aligned}$$

Write

$$\begin{aligned}&\sum_{\gamma \in (\Gamma - \Gamma \cap P)/\Gamma \cap P} e^{(-|\rho|(1+\epsilon)-|\rho|)(H(a\gamma))}\\ &\qquad = \sum_{\gamma \in \Gamma/\Gamma \cap P} e^{(-|\rho|(1+\epsilon)-|\rho|)(H(a\gamma))} - e^{(-|\rho|(1+\epsilon)-|\rho|)(H(a))}\\ &\qquad = E(P:1:-|\rho|(1+\epsilon):a) - e^{(-|\rho|(1+\epsilon)-|\rho|)(H(a))}\\ &\qquad = E(P:1:-|\rho|(1+\epsilon):a) - E_P(P:1:-|\rho|(1+\epsilon):a)\\ &\qquad\quad + c_{P|P}(w:-|\rho|(1+\epsilon))e^{\epsilon|\rho|(H(a))}.\end{aligned}$$

Here the notations are as in Section 2 (w is the nontrivial element of $W(A)$). Thanks to Lemma 2.2,

$$E(P:1:-|\rho|(1+\epsilon):a) - E_P(P:1:-|\rho|(1+\epsilon):a)$$

stays bounded as a ranges over $A({}_0t)$. Since the same is true of $e^{\epsilon|\rho|(H(a))}$, the proof of the lemma is now complete. ■

We come now to the three parabolic terms. Consider the first one:

$$I_\alpha(?:\epsilon) - K_\alpha^{\prime,c}(?:\epsilon).$$

It has already been pointed out in Section 6 that the discussion given there of $K_\alpha^{\prime,c}(?:\epsilon)$ for compactly supported K-finite α is actually valid for any K-finite α in $\mathscr{C}^1(G)$, hence, in particular, for any K-finite α in $\mathscr{C}_\epsilon^1(G)$. One cannot, however, immediately say the same about $I_\alpha(?:\epsilon)$, since the Poisson summation formula was applied to functions of the form

$$\alpha_{x,y}, X \mapsto \alpha(xy\exp(X)x^{-1}) \qquad (X \in \mathfrak{n}; x, y \in G).$$

Naturally, when α has compact support, there is no difficulty. In general, one might expect that restricting functions on G to N would take $\mathscr{C}^1(G)$ (which we regard as the integrable Schwartz space on G) to $\mathscr{C}(N)$ (the Schwartz space on N). While it is always true that

$$\mathrm{Res}_{G\to N}(\mathscr{C}^1(G)) \hookrightarrow L^1(N)$$

(cf. Warner [34b, p. 243]), it is unfortunately not the case that

$$\mathrm{Res}_{G\to N}(\mathscr{C}^1(G)) \hookrightarrow \mathscr{C}(N).$$

Therefore one must proceed with care when attempting to apply the Poisson summation formula to functions $\alpha_{x,y}$ where $\alpha \in \mathscr{C}_\epsilon^1(G)$. For the record, let us set down the following well-known criterion.

SUBLEMMA. *Let V be an n-dimensional vector space over $\mathbf{R}$, $\hat{V}$ its unitary dual. Let f be an integrable C^∞ function on V, $\hat{f}$ its Fourier transform. Suppose there exist constants $\delta > 0$, $\Delta > 0$ such that*

$$|f(X)| \leqslant \Delta(1 + \|X\|)^{-n-\delta} \qquad (X \in V),$$
$$|\hat{f}(\hat{X})| \leqslant \Delta(1 + \|\hat{X}\|)^{-n-\delta} \qquad (\hat{X} \in \hat{V}),$$

$\|?\|$ a norm. Then: The Poisson summation formula is applicable to the pair $(f, \hat{f})$.

Given a K-finite function α in $\mathscr{C}_\epsilon^1(G)$, we shall now verify that the functions

$$\alpha_{x,y}, X \mapsto \alpha(xy\exp(X)x^{-1}) \qquad (X \in \mathfrak{n}; x, y \in G)$$

do in fact meet the above criterion. Our discussion will show, moreover, that the necessary estimates are uniform for x, y ranging in compact subsets of G. Suppose that Ω is a compact subset of G and $x, y \in \Omega$—then, for any $X \in \mathfrak{n}$, we have

$$|\alpha_{x,y}(X)| \leqslant C_\Omega \cdot |\alpha|_{d(2+\epsilon)}^{\epsilon} \cdot \vdash\circ\dashv^{2+\epsilon}(\exp X) \cdot (1 + \sigma(\exp X))^{-d(2+\epsilon)},$$

C_Ω a positive constant. According to an estimate mentioned at the beginning of this section,

$$\Xi^{2+\epsilon}(\exp X) \leqslant D(1+\sigma(\exp X))^{d(2+\epsilon)} \cdot e^{-(2+\epsilon)|\rho|(H(\exp -\theta X))} \qquad (X \in \mathfrak{n}).$$

So, up to a positive absolute constant,

$$|\alpha_{x,y}(X)|$$

is majorized by

$$e^{-(2+\epsilon)|\rho|(H(\exp -\theta X))} \qquad (X \in \mathfrak{n}).$$

Given $X \in \mathfrak{n}$, write $X = X_\lambda + X_{2\lambda}$ ($X_\lambda \in \mathfrak{n}_\lambda, X_{2\lambda} \in \mathfrak{n}_{2\lambda}$)—then it is known that

$$e^{|\rho|(H(\exp -\theta X))} = [(1 + \Lambda\|X_\lambda\|^2)^2 + 4\Lambda\|X_{2\lambda}\|^2]^{(m(\lambda)+2m(2\lambda))/4}$$

(cf. Helgason [14, p. 59]). [Here, as in Lemma 7.1, $\Lambda^{-1} = 4(m(\lambda) + 4m(2\lambda))$.] Raising both sides of this expression to the $(2+\epsilon)$-power and taking inverses then leads at once to the conclusion that there exist $\delta > 0$, $\Delta > 0$ such that

$$|\alpha_{x,y}(X)| \leqslant \Delta(1 + \|X\|)^{-m(\lambda)-m(2\lambda)-\delta} \qquad (X \in \mathfrak{n}; x, y \in \Omega).$$

To finish our justification, we must show that an analogous estimate obtains for the Fourier transform $\hat{\alpha}_{x,y}$ of $\alpha_{x,y}$. For this purpose, we need only prove that the derivatives of $\alpha_{x,y}$ of order up to $m(\lambda) + m(2\lambda) + 1$ are integrable. Let $X_1, \ldots, X_d$ be a basis for $\mathfrak{g}$. Assuming first that N is abelian, one computes that for any $X_0 \in \mathfrak{n}$,

$$\begin{aligned}(X_0\alpha_{x,y})(X) &= (\mathrm{Ad}(x)X_0\alpha)(xy\exp(X)x^{-1})\\ &= \sum_{i=1}^{d} C_i(x:X_0)(X_i\alpha)_{x,y}(X) \qquad (X \in \mathfrak{n}),\end{aligned}$$

where, by definition

$$\mathrm{Ad}(x)X_0 = \sum_{i=1}^{d} C_i(x:X_0)X_i \qquad (C_i(x:X_0) \in \mathbf{R}).$$

The $X_i\alpha$ are again in $\mathscr{C}_\epsilon^1(G)$, hence the $(X_i\alpha)_{x,y}$ can be estimated as above. In addition, the $C_i(x:X_0)$ stay bounded for fixed X_0, x being in a compact subset of G. The estimates are therefore uniform in x and y. The contention in the abelian case thus follows by iteration. [In reality the discussion shows, of course, that all the derivatives of $\alpha_{x,y}$ are integrable in the abelian case, which means that $\hat{\alpha}_{x,y}$ decays faster at infinity than the inverse of any polynomial. Even so, this still does not imply that $\alpha_{x,y} \in \mathscr{C}(N)$, since $\hat{\alpha}_{x,y}$ need not be smooth. By analogy, consider, e.g., $\alpha(x) = (1 - \cos x)/x^2$ ($x \in \mathbf{R}$).] Assume now that N is not abelian—then, as we shall see, the position is more delicate.

One readily computes that for any $X_0 \in \mathfrak{n}$,

$$(X_0\alpha_{x,y})(X) = (\mathrm{Ad}(x)X_0\alpha)(xy\exp(X)x^{-1}) + (\mathrm{Ad}(x)(\tfrac{1}{2}[X_\lambda, X_{0,\lambda}])\alpha)(xy\exp(X)x^{-1}),$$

where

$$X = X_\lambda + X_{2\lambda} \qquad (X_\lambda \in \mathfrak{n}_\lambda, X_{2\lambda} \in \mathfrak{n}_{2\lambda}),$$
$$X_0 = X_{0,\lambda} + X_{0,2\lambda} \qquad (X_{0,\lambda} \in \mathfrak{n}_\lambda, X_{0,2\lambda} \in \mathfrak{n}_{2\lambda}).$$

The term $\mathrm{Ad}(x)X_0\alpha$ can be treated as in the abelian case, thus need not be considered further. Let $X_{2\lambda}^1, \ldots, X_{2\lambda}^{m(2\lambda)}$ be a basis for $\mathfrak{n}_{2\lambda}$—then we can write

$$\frac{1}{2}[X_\lambda, X_{0,\lambda}] = \sum_{j=1}^{m(2\lambda)} C_j(X_\lambda : X_{0,\lambda})X_{2\lambda}^j.$$

$C_j(? : ?)$ is linear in both variables. We have, therefore,

$$\left(\mathrm{Ad}(x)\left(\frac{1}{2}[X_\lambda, X_{0,\lambda}]\right)\alpha\right)(xy\exp(X)x^{-1})$$
$$= \sum_{i=1}^{d}\sum_{j=1}^{m(2\lambda)} C_i(x : X_{2\lambda}^j)C_j(X_\lambda : X_{0,\lambda})(X_i\alpha)_{x,y}(X).$$

Put $V = (m(\lambda) + 2m(2\lambda))/4$—then it follows by iteration that the integrability of a derivative of $\alpha_{x,y}$ of order $m(\lambda) + m(2\lambda) + m (m \geqslant 1)$ reduces to the integrability of

$$\frac{(1 + \|X_\lambda\|)^{m(\lambda)+m(2\lambda)+m}}{[(1 + A\|X_\lambda\|^2)^2 + 4A\|X_{2\lambda}\|^2]^{V(2+\epsilon)}}.$$

Passing to polar coordinates in $\mathfrak{n}_\lambda$ and $\mathfrak{n}_{2\lambda}$, the issue then becomes the finiteness of the integral

$$\int_0^\infty \int_0^\infty \frac{(1 + r)^{m(\lambda)+m(2\lambda)+m}}{[(1 + r^2)^2 + s^2]^{V(2+\epsilon)}} r^{m(\lambda)-1}s^{m(2\lambda)-1}\, dr\, ds.$$

Uncouple this integral by the change of variable $s = (1 + r^2)t$ to get

$$\int_0^\infty \frac{(1 + r)^{m(\lambda)+m(2\lambda)+m}}{(1 + r^2)^{m(\lambda)+m(2\lambda)+2V\epsilon}} r^{m(\lambda)-1}\, dr \cdot \int_0^\infty \frac{t^{m(2\lambda)-1}}{(1 + t^2)^{V(2+\epsilon)}}\, dt.$$

The first integral is convergent provided $m \leqslant m(2\lambda)$, while the second integral is convergent as it stands. Consequently, any derivative of $\alpha_{x,y}$ of order up to $m(\lambda) + 2m(2\lambda)$ is integrable. In both the abelian and nonabelian cases, therefore, there exist $\delta > 0$, $\Delta > 0$ such that

$$|\hat{\alpha}_{x,y}(\hat{X})| \leqslant \Delta(1 + \|\hat{X}\|)^{-m(\lambda)-m(2\lambda)-\delta} \qquad (\hat{X} \in \hat{\mathfrak{n}}; x, y \in \Omega).$$

There is now no difficulty in at least applying the Poisson summation formula to study $I_\alpha(?:\epsilon)$. Everything goes through just as before except at the very end of the argument where it becomes necessary to make a choice for the integer ν appearing there. In the abelian case, no problem actually arises, since as has been noted above, $\hat{\alpha}_{x,y}$ decays faster at infinity than the inverse of any polynomial. On the other hand, in the nonabelian case, one can only say, so far anyway, that

$$\hat{\alpha}_{x,y}(\hat{X}) = 0((1 + \|\hat{X}\|)^{-m(\lambda)-2m(2\lambda)}) \qquad (\hat{X} \in \hat{\mathfrak{n}}; x, y \in \Omega),$$

and, unfortunately, $\nu = m(\lambda) + 2m(2\lambda)$ does not suffice to secure convergence. What we need to do is build more decay into the estimate of $\hat{\alpha}_{x,y}$. Let us grant temporarily that there exists $\delta > 0$ such that

$$\hat{\alpha}_{x,y}(\hat{X}) = 0((1 + \|\hat{X}\|)^{-m(\lambda)-2m(2\lambda)-\delta}) \qquad (\hat{X} \in \hat{\mathfrak{n}}; x, y \in \Omega).$$

Then Lemma 8.2 provides the required convergence argument in the nonabelian case, thus serving to complete the discussion of $I_\alpha(?:\epsilon)$. The conclusions, then, about the first parabolic term are unchanged.

LEMMA 8.2. *Let $\hat{L}$ be a lattice in $\hat{\mathfrak{n}}$. Assuming that $m(2\lambda) > 0$, put*

$$\nu = m(\lambda) + 2m(2\lambda).$$

Then, for every $\delta > 0$,

$$\int_0^\infty e^{2|\rho|t}\left(\sum_{\substack{\hat{l} \in \hat{L} \\ \hat{l} \neq 0}} (1 + \|\hat{l}_t\|)^{-\nu-\delta}\right) dt < +\infty.$$

Proof. Employing obvious notations, write $\hat{\mathfrak{n}} = \hat{\mathfrak{n}}_\lambda + \hat{\mathfrak{n}}_{2\lambda}$, so that every $\hat{X} \in \hat{\mathfrak{n}}$ admits the decomposition $\hat{X} = \hat{X}_\lambda + \hat{X}_{2\lambda}$ where $\hat{X}_\lambda \in \hat{\mathfrak{n}}_\lambda$, $\hat{X}_{2\lambda} \in \hat{\mathfrak{n}}_{2\lambda}$; for the proof, it can be supposed that $\|\hat{X}\| = \|\hat{X}_\lambda\| + \|\hat{X}_{2\lambda}\|$. This said, write

$$\int_0^\infty e^{2|\rho|t}\left(\sum_{\substack{\hat{l} \in \hat{L} \\ \hat{l} \neq 0}} (1 + \|\hat{l}_t\|)^{-\nu-\delta}\right) dt$$

$$= \int_0^\infty e^{2|\rho|t}\left(\sum_{\substack{\hat{l} \in \hat{L} \\ \hat{l}_{2\lambda} = 0}} (1 + e^{|\lambda|t}\|\hat{l}_\lambda\|)^{-\nu-\delta}\right) dt$$

$$+ \int_0^\infty e^{2|\rho|t}\left(\sum_{\substack{\hat{l} \in \hat{L} \\ \hat{l}_{2\lambda} \neq 0}} (1 + e^{|\lambda|t}\|\hat{l}_\lambda\| + e^{2|\lambda|t}\|\hat{l}_{2\lambda}\|)^{-\nu-\delta}\right) dt.$$

On the one hand, we have

$$\int_0^\infty e^{2|\rho|t}\left(\sum_{\substack{\hat{l}\in\hat{L}\\ \hat{l}_{2\lambda}=0}}(1+e^{|\lambda|t}\|\hat{l}_\lambda\|)^{-\nu-\delta}\right)dt$$

$$=\sum_{\substack{\hat{l}\in\hat{L}\\ \hat{l}_{2\lambda}=0}}\int_0^\infty \frac{e^{\nu|\lambda|t}}{(1+e^{|\lambda|t}\|\hat{l}_\lambda\|)^{\nu+\delta}}\,dt$$

$$<\frac{1}{|\lambda|}\cdot\sum_{\substack{\hat{l}\in\hat{L}\\ \hat{l}_{2\lambda}=0}}\frac{1}{\|\hat{l}_\lambda\|^\nu}\cdot\int_0^\infty\frac{u^{\nu-1}}{(1+u)^{\nu+\delta}}\,du<+\infty.$$

On the other hand, we have

$$\int_0^\infty e^{2|\rho|t}\left(\sum_{\substack{\hat{l}\in\hat{L}\\ \hat{l}_{2\lambda}\neq 0}}(1+e^{|\lambda|t}\|\hat{l}_\lambda\|+e^{2|\lambda|t}\|\hat{l}_{2\lambda}\|)^{-\nu-\delta}\right)dt$$

$$=\int_0^\infty e^{2|\rho|t}\left(\sum_{\substack{\hat{l}\in\hat{L}\\ \hat{l}_{2\lambda}\neq 0}}(e^{|\lambda|t}\cdot e^{-|\lambda|t}+e^{|\lambda|t}\|\hat{l}_\lambda\|+e^{2|\lambda|t}\|\hat{l}_{2\lambda}\|)^{-\nu-\delta}\right)dt$$

$$=\int_0^\infty e^{-\delta|\lambda|t}\left(\sum_{\substack{\hat{l}\in\hat{L}\\ \hat{l}_{2\lambda}\neq 0}}(e^{-|\lambda|t}+\|\hat{l}_\lambda\|+e^{|\lambda|t}\|\hat{l}_{2\lambda}\|)^{-\nu-\delta}\right)dt$$

$$<\sum_{\substack{\hat{l}\in\hat{L}\\ \hat{l}_{2\lambda}\neq 0}}\frac{1}{\|\hat{l}\|^{\nu+\delta}}\cdot\int_0^\infty e^{-\delta|\lambda|t}\,dt<+\infty.$$

Hence the lemma. ■

There remains the problem of producing a $\delta>0$ such that

$$\hat{\alpha}_{x,y}(\hat{X})=0((1+\|\hat{X}\|)^{-m(\lambda)-2m(2\lambda)-\delta})\qquad(\hat{X}\in\hat{\mathfrak{n}};\,x,y\in\Omega).$$

For this purpose, we shall need the following distribution-theoretic result, which was kindly communicated to us by G. B. Folland.

Sublemma. *Let s be a nonzero complex number such that $-1<\mathrm{Re}(s)<n$—then the assignment T_s given by the rule*

$$T_s(f)=\int_{\|x\|\leqslant 1}\frac{f(x)-f(0)}{\|x\|^{n-s}}\,dx+\int_{\|x\|\geqslant 1}\frac{f(x)}{\|x\|^{n-s}}\,dx\qquad(f\in\mathscr{C}(\mathbf{R}^n))$$

defines a tempered distribution. The Fourier transform $\hat{T}_s$ of T_s is the function

$$\hat{T}_s(\xi)=\frac{\pi^{(n/2)-s}\Gamma(s/2)}{\Gamma((n-s)/2)}\cdot\|\xi\|^{-s}-\frac{2\pi^{n/2}}{s\Gamma(n/2)}.$$

[We adopt here the convention that the Fourier transform $\hat{f}$ of $f \in \mathscr{C}(\mathbf{R}^n)$ is to be written

$$\hat{f}(\xi) = \int e^{2\pi\sqrt{-1}x\cdot\xi} f(x)\,dx,$$

dx being Lebesgue measure. Accordingly, if T is a tempered distribution, then, by definition, its Fourier transform $\hat{T}$ is the tempered distribution: $\hat{T}(f) = T(\hat{f})\,(f \in \mathscr{C}(\mathbf{R}^n))$.]

Proof. The only issue in the temperedness of T_s is the first term in its definition and no difficulty is present since $f(x) - f(0) = 0(\|x\|)$. This said, let now $0 < \mathrm{Re}(s) < n$. Put

$$F_s(x) = \|x\|^{s-n}.$$

Then F_s is a locally integrable function on $\mathbf{R}^n$ and defines a tempered distribution. It is well known that its Fourier transform $\hat{F}_s$ is the function

$$\hat{F}_s(\xi) = \frac{\pi^{(n/2)-s}\Gamma(s/2)}{\Gamma((n-s)/2)} \cdot \|\xi\|^{-s}.$$

Write

$$\begin{aligned} T_s(f) &= \int_{\mathbf{R}^n} \frac{f(x)}{\|x\|^{n-s}}\,dx - \int_{\|x\|\leqslant 1} \frac{f(0)}{\|x\|^{n-s}}\,dx \\ &= \int_{\mathbf{R}^n} \frac{f(x)}{\|x\|^{n-s}}\,dx - f(0)\cdot\frac{2\pi^{n/2}}{s\Gamma(n/2)} \\ &= F_s(f) - \frac{2\pi^{n/2}}{s\Gamma(n/2)}\,\delta(f). \end{aligned}$$

Thus

$$T_s = F_s - \frac{2\pi^{n/2}}{s\Gamma(n/2)} \cdot \delta,$$

and so

$$\hat{T}_s = \hat{F}_s - \frac{2\pi^{n/2}}{s\Gamma(n/2)} \cdot \hat{\delta}.$$

It therefore follows that $\hat{T}_s$ is the function

$$\hat{T}_s(\xi) = \frac{\pi^{(n/2)-s}\Gamma(s/2)}{\Gamma((n-s)/2)} \cdot \|\xi\|^{-s} - \frac{2\pi^{n/2}}{s\Gamma(n/2)}.$$

To finish the proof, one need only note that the assignments

$$s \mapsto T_s, \qquad s \mapsto \hat{T}_s$$

are holomorphic in s, hence, by analytic continuation the formula for $\hat{T}_s$ still holds for s in the region $-1 < \operatorname{Re}(s) < 0$. ■

Let D be any differential operator on $\mathfrak{n}$ with constant coefficients of order not exceeding $m(\lambda) + 2m(2\lambda)$; let $\beta = D\alpha_{x,y}$—then, as has been shown above, $D\beta \in L^1(\mathfrak{n})$. Consider the functions

$$\frac{(1 + \|X_\lambda\|)^m}{[(1 + \Lambda\|X_\lambda\|^2)^2 + 4\Lambda\|X_{2\lambda}\|^2]^{V(2+\epsilon)}} \qquad (0 \leqslant m \leqslant m(\lambda) + 2m(2\lambda)).$$

Introducing polar coordinates and making some simple estimates, one finds that these functions are in $L^{1-2\delta}(\mathfrak{n})$ provided $\delta > 0$ is so chosen that

$$2\delta < \operatorname{Min}\begin{cases} \dfrac{\epsilon}{1+\epsilon} \\[2ex] 1 - \dfrac{m(2\lambda)}{2V(2+\epsilon)}. \end{cases}$$

Consequently, $\beta \in L^{1-2\delta}(\mathfrak{n})$. To produce the desired estimate on $\hat{\alpha}_{x,y}$, therefore, we need only show that $\beta * T_{-\delta} \in L^1(\mathfrak{n})$. Let $d = m(\lambda) + m(2\lambda)$, the dimension of $\mathfrak{n}$—then, by definition,

$$\beta * T_{-\delta}(X) = \int_{\|Y\| \leqslant 1} \frac{\beta(X-Y) - \beta(X)}{\|Y\|^{d+\delta}}\, dY + \int_{\|Y\| \geqslant 1} \frac{\beta(X-Y)}{\|Y\|^{d+\delta}}\, dY.$$

The function

$$X \mapsto \int_{\|Y\| \geqslant 1} \frac{\beta(X-Y)}{\|Y\|^{d+\delta}}\, dY$$

is the convolution of two integrable functions, hence is integrable. Write

$$|\beta(X-Y) - \beta(X)| = |\beta(X-Y) - \beta(X)|^{2\delta} \cdot |\beta(X-Y) - \beta(X)|^{1-2\delta}.$$

Because the first-order derivatives of β are bounded, β satisfies a uniform Lipschitz condition:

$$|\beta(X-Y) - \beta(X)| \leqslant C\|Y\| \qquad (\text{some } C > 0).$$

The function

$$X \mapsto \int_{\|Y\| \leqslant 1} \frac{\beta(X-Y) - \beta(X)}{\|Y\|^{d+\delta}}\, dY$$

is thus majorized in absolute value by the function

$$X \mapsto C \cdot \int_{\|Y\| \leqslant 1} \frac{|\beta(X - Y)|^{1-2\delta}}{\|Y\|^{d-\delta}} \, dY + C \cdot |\beta(X)|^{1-2\delta} \cdot \int_{\|Y\| \leqslant 1} \frac{dY}{\|Y\|^{d-\delta}}.$$

The first term is integrable, being the convolution of two integrable functions; the second term is integrable, β being in $L^{1-2\delta}(\mathfrak{n})$. Therefore $\beta * T_{-\delta} \in L^1(\mathfrak{n})$, as we wished to prove.

Let us proceed to the second parabolic term:

$$J_\alpha(x : \epsilon).$$

We recall that for compactly supported K-finite α, this term is actually integrable over G/Γ. The proof, however, uses in an essential way the hypothesis of compact support. Accordingly, the argument will have to be modified to treat K-finite elements α in $\mathscr{C}_\epsilon^1(G)$. The following lemma will suffice for this purpose.

LEMMA 8.3. *Let $\mathscr{S}_\epsilon$ be the strip*

$$\{s : -\epsilon/2 < \mathrm{Re}(s) < 1 + \epsilon/2\}.$$

Let $\alpha \in \mathscr{C}_\epsilon^1(G)$—then, for any real number r and all $s \in \mathscr{S}_\epsilon$, the integral

$$\int_r^\infty \int_{N/N \cap \Gamma} \left(\sum_{\substack{\eta \in \Gamma \cap N \\ \eta \neq 1}} \alpha_K(a^*(t) n \eta n^{-1} a^*(-t)) \right) \cdot e^{(2|\rho|/|\lambda|)t(1+s)} \, d_N(n) \, dt$$

is absolutely convergent, hence represents a holomorphic function of s there. Moreover, the assignment

$$\alpha \mapsto \int_r^\infty \int_{N/N \cap \Gamma} \left(\sum_{\substack{\eta \in \Gamma \cap N \\ \eta \neq 1}} \alpha_K(a^*(t) n \eta n^{-1} a^*(-t)) \right) \cdot e^{(2|\rho|/|\lambda|)t(1+s)} \, d_N(n) \, dt$$

$$(\alpha \in \mathscr{C}_\epsilon^1(G))$$

is continuous in the topology of $\mathscr{C}_\epsilon^1(G)$.

Proof. We can and will assume that s is real. Elementary considerations, which need not be detailed, then show that it is enough to establish the convergence of the integral

$$\int_0^\infty \int_{N/N \cap \Gamma} \left(\sum_{\substack{\eta \in \Gamma \cap N \\ \eta \neq 1}} \vdash\!\circ\!\dashv^{2+\epsilon}(a^*(t) n \eta n^{-1} a^*(-t)) \right.$$
$$\left. \times (1 + \sigma(a^*(t) n \eta n^{-1} a^*(-t)) \right)^{-d(2+\epsilon)} \cdot e^{(2|\rho|/|\lambda|)t(1+s)} \, d_N(n) \, dt$$

for all s lying between $-\epsilon/2$ and $1 + \epsilon/2$. Let us remind ourselves that on the set of all $a \in A$ such that $\xi_\lambda(a) \geqslant 1$, $\vdash\!\circ\!\dashv^{2+\epsilon}(a)$ is majorized, up to a positive absolute constant, by

$$\frac{(1+\sigma(a))^{d(2+\epsilon)}}{[\cosh(|\rho|(H(a)))]^{2+\epsilon}}.$$

This said, suppose first that N is abelian—then, in view of Lemma 7.1, we are led to study the convergence of

$$\sum_{X_\lambda \in \Gamma_\lambda}^{*} \int_0^\infty \frac{e^{(2|\rho|/|\lambda|)t(1+s)}}{\left[1 + e^{2t}\cdot\dfrac{\Lambda\|X_\lambda\|^2}{2}\right]^{(|\rho|/|\lambda|)(2+\epsilon)}}\,dt.$$

Ignoring the $\Lambda/2$, make the change of variable $u = e^t\|X_\lambda\|$—then we are reduced to considering

$$\sum_{X_\lambda \in \Gamma_\lambda}^{*} \frac{1}{\|X_\lambda\|^{m(\lambda)(1+s)}} \cdot \int_{\|X_\lambda\|}^\infty \frac{u^{m(\lambda)(1+s)-1}}{[1+u^2]^{m(\lambda)(1+\epsilon/2)}}\,du,$$

which, for appropriately chosen $C > 0$, is in turn majorized by

$$\sum_{X_\lambda \in \Gamma_\lambda}^{*} \frac{1}{\|X_\lambda\|^{m(\lambda)(1+s+\epsilon/2)}} \cdot \int_C^\infty \frac{u^{m(\lambda)(1+s)-1}}{[1+u^2]^{m(\lambda)(1+\epsilon/4)}}\,du.$$

The series is convergent if $s > -\epsilon/2$, while the integral is convergent if $s < 1 + \epsilon/2$. Both entities converge when $s \in \mathscr{S}_\epsilon$, thus settling the abelian case. Assume now that N is nonabelian. Proceeding as in Section 7, break up the integral over $N/N \cap \Gamma$ into two separate integrals, using Lemma 7.1 as we do so. The first integral whose convergence must be examined is then

$$\sum_{X_\lambda \in \Gamma_\lambda}^{*} \int_0^\infty \int_{\mathfrak{n}_{2\lambda}} \frac{1}{\left[\left(\left(1 + e^{2t}\cdot\dfrac{\Lambda\|X_\lambda\|^2}{2}\right)^2 + e^{4t}\cdot\Lambda\|Y_{2\lambda}\|^2\right)^{1/2}\right]^{(|\rho|/|\lambda|)(2+\epsilon)}}$$
$$\times\, e^{(2|\rho|/|\lambda|)t(1+s)}\,d_{\mathfrak{n}_{2\lambda}}(Y_{2\lambda})\,dt.$$

The double integral can be written as

$$\int_0^\infty \left[\frac{(1 + e^{2t}\cdot(\Lambda\|X_\lambda\|^2/2))}{e^{2t}\cdot\sqrt{\Lambda}}\right]^{m(2\lambda)} \cdot \frac{e^{(2|\rho|/|\lambda|)t(1+s)}}{[1 + e^{2t}\cdot(\Lambda\|X_\lambda\|^2/2)]^{(|\rho|/|\lambda|)(2+\epsilon)}}\,dt$$
$$\times \int_{\mathfrak{n}_{2\lambda}} \frac{1}{[(1+\|Y_{2\lambda}\|^2)^{1/2}]^{(|\rho|/|\lambda|)(2+\epsilon)}}\,d_{\mathfrak{n}_{2\lambda}}(Y_{2\lambda}).$$

The integral over $\mathfrak{n}_{2\lambda}$ is finite, thus may be disregarded. In the integral with respect to t, let us agree to ignore the $\Lambda/2$ and $\sqrt{\Lambda}$ and make the change of variable $u = e^t\|X_\lambda\|$—then we get

$$\frac{\|X_\lambda\|^{2m(2\lambda)}}{\|X_\lambda\|^{(m(\lambda)+2m(2\lambda))(1+s)}} \cdot \int_{\|X_\lambda\|}^{\infty} \left[\frac{1+u^2}{u^2}\right]^{m(2\lambda)} \cdot \frac{u^{(m(\lambda)+2m(2\lambda))(1+s)}}{[1+u^2]^{(m(\lambda)/2+m(2\lambda))(2+\epsilon)}} \cdot \frac{du}{u}.$$

Selecting $C > 0$ in a suitable way and making obvious estimates, it then becomes a question of determining the convergence of

$$\sum_{X_\lambda \in \Gamma_\lambda}^{*} \frac{1}{\|X_\lambda\|^{(m(\lambda)+2m(2\lambda))(s+\epsilon/2)+m(\lambda)}}$$
$$\times \int_C^{\infty} \left[\frac{1+u^2}{u^2}\right]^{m(2\lambda)} \cdot \frac{u^{(m(\lambda)+2m(2\lambda))(1+s)}}{[1+u^2]^{(m(\lambda)/2+m(2\lambda))(2+\epsilon/2)}} \cdot \frac{du}{u}.$$

Owing to what has been said in Section 7, the series is convergent if $s > -\epsilon/2$. On the other hand, the integral is convergent if $s < 1 + \epsilon/2$. Both entities therefore converge when $s \in \mathscr{S}_\epsilon$. It remains to discuss the convergence of

$$\sum_{X_{2\lambda} \in \Gamma_{2\lambda}}^{*} \int_0^{\infty} \frac{e^{(2|\rho|/|\lambda|)t(1+s)}}{[(1+e^{4t}\cdot \Lambda\|X_{2\lambda}\|^2)^{1/2}]^{(|\rho|/|\lambda|)(2+\epsilon)}}\, dt.$$

Ignoring the Λ, make the change of variable $u = e^{2t}\|X_{2\lambda}\|$—then we are led to consider

$$\sum_{X_{2\lambda} \in \Gamma_{2\lambda}}^{*} \frac{1}{\|X_{2\lambda}\|^{(m(\lambda)/2+m(2\lambda))(1+s)}} \cdot \int_{\|X_{2\lambda}\|}^{\infty} \frac{u^{(m(\lambda)/2+m(2\lambda))(1+s)}}{[(1+u^2)^{1/2}]^{(m(\lambda)/2+m(2\lambda))(2+\epsilon)}} \cdot \frac{du}{u},$$

which, for appropriately chosen $C > 0$, is in turn majorized by

$$\sum_{X_{2\lambda} \in \Gamma_{2\lambda}}^{*} \frac{1}{\|X_{2\lambda}\|^{(m(\lambda)/2+m(2\lambda))(1+s+\epsilon/2)}} \cdot \int_C^{\infty} \frac{u^{(m(\lambda)/2+m(2\lambda))(1+s)}}{[(1+u^2)^{1/2}]^{(m(\lambda)/2+m(2\lambda))(2+\epsilon/2)}} \cdot \frac{du}{u}.$$

The series is convergent if

$$s > -\frac{\epsilon}{2} - \frac{m(\lambda)/2}{(m(\lambda)/2 + m(2\lambda))}.$$

The integral is convergent if $s < 1 + \epsilon/2$. In particular, then, both entities are convergent if $s \in \mathscr{S}_\epsilon$. This takes care of the non-abelian case, thereby finishing the proof of the lemma. ■

It follows from this result that the second parabolic term formed relative to any K-finite function in $\mathscr{C}_\epsilon^1(G)$ is integrable. One can now proceed with the analysis as before, the only proviso being that s is restricted to a certain small strip ${}_\epsilon\mathscr{S} \hookrightarrow \mathscr{S}_\epsilon$ containing the imaginary axis. [In this connection, recall that one step in the analysis calls for an application of the Poisson summation formula, so, to ensure convergence, it is necessary to impose a restriction on s.] In particular, if $Ч_\alpha$ is introduced formally by the same definition, then our estimates show that the integral defining $Ч_\alpha$ is absolutely convergent for $s \in {}_\epsilon\mathscr{S}$ with $\mathrm{Re}(s) > 0$, and admits a continuation to ${}_\epsilon\mathscr{S}$ as a meromorphic function whose only possible singularity is a simple pole at $s = 0$. The assignment

$$\alpha \mapsto \lim_{s \to 0} \frac{d}{ds}(sЧ_\alpha(s)) \qquad (\alpha \in \mathscr{C}_\epsilon^1(G))$$

is easily seen to be continuous in the topology of $\mathscr{C}_\epsilon^1(G)$. Therefore, apart from the modification of working just in ${}_\epsilon\mathscr{S}$, the conclusions about the second parabolic term are unchanged.

Remark. Suppose that our K-finite function α in $\mathscr{C}_\epsilon^1(G)$ is a cusp form on G (thus is not of compact support)—then the discussion shows that $Ч_\alpha$ is actually holomorphic at $s = 0$, hence

$$\lim_{s \to 0} Ч_\alpha(s)$$

exists in this case.

To complete the extension of the trace formula to the K-finite functions in $\mathscr{C}_\epsilon^1(G)$, we have to consider the third parabolic term:

$$-K_\alpha^{\prime\prime,c}(x:\epsilon).$$

On the basis of what has been said above, this term is weakly integrable over G/Γ. Taking into account the fact that the K-finite matrix coefficients of $\mathbf{U}^{\vartheta,s}(\alpha)$ ($\mathrm{Re}(s) = 0$) are rapidly decreasing in s for any K-finite $\alpha \in \mathscr{C}_\epsilon^1(G)$ (cf. Eguchi [8, p. 169]), one sees immediately that all the conclusions about the third parabolic term remain unchanged in the present setting.

The following theorem summarizes the preceding considerations.

THEOREM 8.4 (The Selberg trace formula). *Let α be a K-finite function in $\mathscr{C}_\epsilon^1(G)$—then the trace of $L_{G/\Gamma}^{\mathrm{d}}(\alpha)$ is the sum of*

$$\mathrm{vol}(G/\Gamma)\left(\sum_{z \in Z_\Gamma} \alpha(z)\right).$$

$$(S) \sum_{\{\gamma\}_\Gamma} \mathrm{vol}(G_\gamma/\Gamma_\gamma) \cdot \int_{G/G_\gamma} \alpha(x\gamma x^{-1})\, d_{G/G_\gamma}(x),$$

$$\lim_{s \to 0} \frac{d}{ds}(sЧ_\alpha(s)),$$

$$\frac{1}{4\pi} \sum_{\vartheta} \int_{\mathrm{Re}(s)=0} \mathrm{tr}\left(\mathbf{c}_\vartheta(-s) \cdot \left(\frac{d}{ds}\mathbf{c}_\vartheta(s)\right) \cdot \mathbf{U}^{\vartheta,s}(\alpha)\right) |ds|,$$

and

$$-\frac{1}{4}\sum_{\vartheta} \operatorname{tr}(\mathbf{c}_{\vartheta}(0)\mathbf{U}^{\vartheta,0}(\alpha)).$$

All the series and integrals are absolutely convergent.

It is a point of some importance to note that each of the terms figuring in the theorem represents assignments which are continuous in the topology of $\mathscr{C}_\epsilon^1(G)$.

In Section 6, we did not commit ourselves to a choice for the Haar measure on G. For sake of definiteness, let us choose the so-called *standard Haar measure*, whose definition we shall now recall. Let d_K be the Haar measure on K which assigns to K total volume one; let d_A (respectively d_N) be the exponentiation of normalized Lebesgue measure on $\mathfrak{a}$ (respectively $\mathfrak{n}$) (relative to the Euclidean structure associated with the Killing form)—then the standard Haar measure d_G on G is determined by the requirement

$$d_G(x) = a^{2\rho}\, d_K(k)\, d_A(a)\, d_N(n) \qquad (x = kan).$$

If we apply the same agreements to the pairs (A_i, N_i) $(i = 1, \ldots, r)$, then d_G is unchanged. The constant $\mathbf{c}_G^i$ introduced in Section 6 can thus be interpreted as the volume of $N_i/N_i \cap \Gamma$, N_i being equipped with the Haar measure defined above. Of course, our specification of d_G also fixes $\operatorname{vol}(G/\Gamma)$ and $\operatorname{vol}(G_\gamma/\Gamma_\gamma)$.

Remark. Suppose that $\operatorname{rank}(G) = \operatorname{rank}(K)$—then the dimension of $K\backslash G$ is even. Assume first that Γ has no torsion—then $K\backslash G/\Gamma$ is an even-dimensional manifold. Owing to Harder's generalization of Chern's version of the Gauss–Bonnet theorem, one can proceed exactly as in Gangolli [9c, pp. 33–40] and show that $\operatorname{vol}(G/\Gamma)$ is a certain explicitly computable multiple of the Euler–Poincaré characteristic of $K\backslash G/\Gamma$. In the general case, the presence of elliptic elements in Γ implies that $K\backslash G/\Gamma$ is only a V-manifold but the same conclusion as in the torsion-free case still obtains provided the Euler–Poincaré characteristic of $K\backslash G/\Gamma$ is understood in the sense of Satake. One should also note that it is possible to compute $\operatorname{vol}(G_\gamma/\Gamma_\gamma)$, at least if γ is hyperbolic (cf. Gangolli [9b]).

EXAMPLE. Take $G = \mathbf{SL}(2, \mathbf{R})$—then the assumption on Γ made in Section 6 is always fulfilled. As an application of the Selberg trace formula in this case, consider the problem of computing the multiplicity m_U of a discrete series representation U of G in $L_d^2(G/\Gamma)$. If U is integrable, one can obtain m_U by inserting into the Selberg trace formula its "lowest weight" matrix coefficient; the result is then a formula of Shimizu's [28a, p. 63]. If U is not integrable, one can obtain m_U by employing a device of Langlands (cf. Wallach [33b, pp. 179–180]); the result is then a formula of Eichler–Saito (cf. Ishikawa [15, p. 219]). The details will be left to the reader.

9. Class One Computations

Let α be a K-bi-invariant function in $C_c^\infty(G)$—then in Section 7 we have attached to α certain distributions

$$\left.\begin{matrix} T_\lambda(\alpha:0) \\ T'_\lambda(\alpha:0) \end{matrix}\right\} \quad (N \text{ abelian}), \qquad \left.\begin{matrix} T_\lambda(\alpha:0) \\ T'_\lambda(\alpha:0) \\ T_{2\lambda}(\alpha:0) \end{matrix}\right\} \quad (N \text{ nonabelian}).$$

The objective of the present section is to compute the Fourier transforms of these distributions in the sense of Harish-Chandra. Besides giving a complete evaluation of

$$\lim_{s\to 0} \frac{d}{ds}(s\text{Ч}_\alpha(s))$$

in an important special case, the results also play a central role in other questions as well, e.g., in the theory of zeta functions of Selberg type. The latter point and related topics will be dealt with elsewhere in a forthcoming article by R. Gangolli and the author.

The key to our computations is the so-called Abel transform and its inverse. Let α be a K-bi-invariant function in $C_c^\infty(G)$—then the *Abel transform* of α is that function F_α on A defined by the rule

$$F_\alpha(a) = a^\rho \int_N \alpha(an)\, d_N(n) \qquad (a \in A),$$

d_N being the exponentiation of normalized Lebesgue measure on $\mathfrak{n}$ (relative to the Euclidean structure associated with the Killing form). Let ν be a real-valued linear function on the Lie algebra $\mathfrak{a}$ of A; let

$$\varphi_\nu(x) = \int_K a_{xk}^{(\sqrt{-1}\nu-\rho)}\, d_K(k) \qquad (x \in G)$$

be the zonal spherical function on G associated with ν—then the *Fourier transform* of α is that function $\hat{\alpha}$ on the dual of $\mathfrak{a}$ defined by the rule

$$\hat{\alpha}(\nu) = \int_G \alpha(x)\varphi_\nu(x)\, d_G(x),$$

d_G being the standard Haar measure on G (cf. Section 8). A direct computation shows that

$$\hat{\alpha}(v) = \hat{F}_\alpha(v),$$

where

$$\hat{F}_\alpha(v) = \int_A F_\alpha(a) a^{\sqrt{-1}\,v}\, d_A(a),$$

d_A being the exponentiation of normalized Lebesgue measure on $\mathfrak{a}$ (relative to the Euclidean structure associated with the Killing form). This means that the Fourier transform of α is the (Euclidean) Fourier transform of F_α. In order to explicate this relation conveniently, we shall need the following lemma, the proof of which can be omitted (cf. Lemma 7.1).

LEMMA 9.1. *Given $n \in N$, write $n = \exp(X_\lambda)\exp(X_{2\lambda})$ where $X_\lambda \in \mathfrak{n}_\lambda$, $X_{2\lambda} \in \mathfrak{n}_{2\lambda}$. Let $T(t, n)$ be the radial component of $a^*(t)n$ ($t \in \mathbf{R}$)—then*

$$\cosh(T(t,n)) = \left[\left(\cosh t + e^t \cdot \frac{\Lambda\|X_\lambda\|^2}{2}\right)^2 + e^{2t} \cdot \Lambda\|X_{2\lambda}\|^2\right]^{1/2}$$

Here

$$\Lambda^{-1} = 4(m(\lambda) + 4m(2\lambda)).$$

Using this result, we then find that

$$\begin{aligned} F_\alpha(a^*(t)) &= e^{(|\rho|/|\lambda|)t} \int_N \alpha(a^*(t)n)\, d_N(n) \\ &= e^{(|\rho|/|\lambda|)t} \int_{\mathfrak{n}_\lambda \times \mathfrak{n}_{2\lambda}} \overline{\alpha}\Bigg(\Bigg[\left(\cosh t + e^t \cdot \frac{\Lambda\|X_\lambda\|^2}{2}\right)^2 \\ &\quad + e^{2t} \cdot \Lambda\|X_{2\lambda}\|^2\Bigg]^{1/2}\Bigg) d_{\mathfrak{n}_\lambda}(X_\lambda)\, d_{\mathfrak{n}_{2\lambda}}(X_{2\lambda}). \end{aligned}$$

In the last integral, make the change of variables

$$X_\lambda \mapsto e^{t/2} \cdot \sqrt{\Lambda} \cdot X_\lambda, \qquad X_{2\lambda} \mapsto e^t \cdot \sqrt{\Lambda} \cdot X_{2\lambda}.$$

Then we get

$$F_\alpha(a^*(t)) = \frac{1}{(\sqrt{\Lambda})^{m(\lambda)+m(2\lambda)}} \cdot \mathfrak{F}_\alpha(t),$$

where

$$\mathfrak{F}_\alpha(t) = \int_{\mathfrak{n}_\lambda \times \mathfrak{n}_{2\lambda}} \overline{\alpha}\left(\left[\left(\cosh t + \frac{\|X_\lambda\|^2}{2}\right)^2 + \|X_{2\lambda}\|^2\right]^{1/2}\right) d_{\mathfrak{n}_\lambda}(X_\lambda)\, d_{\mathfrak{n}_{2\lambda}}(X_{2\lambda}).$$

Let us also observe that

$$\begin{aligned}\int_A F_\alpha(a)a^{\sqrt{-1}\nu}\, d_A(a) &= \int_{-\infty}^{\infty} F_\alpha(a(t))e^{\sqrt{-1}\nu(H)t}\, dt\\ &= \frac{1}{|\lambda|}\cdot\int_{-\infty}^{\infty} F_\alpha(a^*(t))e^{\sqrt{-1}\nu(H^*)t}\, dt\\ &= \frac{1}{|\lambda|}\cdot\frac{1}{(\sqrt{\Lambda})^{m(\lambda)+m(2\lambda)}}\cdot\int_{-\infty}^{\infty}\mathfrak{F}_\alpha(t)e^{\sqrt{-1}\nu(H^*)t}\, dt\\ &= \frac{1}{|\lambda|}\cdot\frac{1}{(\sqrt{\Lambda})^{m(\lambda)+m(2\lambda)}}\cdot\hat{\mathfrak{F}}_\alpha(\nu(H^*)).\end{aligned}$$

In what follows, we shall identify ν with $\nu(H^*)$. Adopting this convention then leads to the relation

$$\hat{\alpha}(\nu) = \frac{1}{|\lambda|}\cdot\frac{1}{(\sqrt{\Lambda})^{m(\lambda)+m(2\lambda)}}\cdot\hat{\mathfrak{F}}_\alpha(\nu).$$

Let $x \geqslant 1$—then by the above

$$\overline{\mathfrak{F}}_\alpha(x) = \int_{\mathfrak{n}_\lambda\times\mathfrak{n}_{2\lambda}}\overline{\alpha}\left(\left[\left(x+\frac{\|X_\lambda\|^2}{2}\right)^2+\|X_{2\lambda}\|^2\right]^{1/2}\right)d_{\mathfrak{n}_\lambda}(X_\lambda)\,d_{\mathfrak{n}_{2\lambda}}(X_{2\lambda}),$$

where, of course, $\overline{\mathfrak{F}}_\alpha = \mathfrak{F}_\alpha\circ\cosh^{-1}$. Our next task will be to express $\overline{\alpha}$ as an integral transform of $\overline{\mathfrak{F}}_\alpha$. In the case of the classical groups, this problem has been considered by Takahashi [29a, p. 65; 29b, p. 329] and by T. Eaton (unpublished notes); their method is matrix computation. Before stating the result, we had best remind ourselves of the possibilities for $m(\lambda)$ and $m(2\lambda)$, as can be inferred from the classification (cf. Warner [34a, pp. 30–32]).

$m(\lambda)$	$m(2\lambda)$
$m \quad (m \geqslant 1)$	0
$2(m-1) \quad (m \geqslant 2)$	1
$4(m-1) \quad (m \geqslant 2)$	3
8	7

LEMMA 9.2. *Let α be a K-bi-invariant function in $C_c^\infty(G)$.*

(i) *Suppose that N is abelian.*

(a) *If $m(\lambda) = m$ is even, then*

$$\overline{\alpha}(x) = \left(\frac{-1}{2\pi}\right)^{m/2}\cdot\overline{\mathfrak{F}}_\alpha^{(m/2)}(x) \qquad (x \geqslant 1).$$

(b) *If $m(\lambda) = m$ is odd, then*

$$\overline{\alpha}(x) = \left(\frac{-1}{2\pi}\right)^{(m+1)/2} \cdot \int_{\mathbf{R}} \overline{\mathfrak{F}}_\alpha^{((m+1)/2)}\left(x + \frac{y^2}{2}\right) dy \qquad (x \geqslant 1).$$

(ii) *Suppose that N is nonabelian.*

(a) *If $m(\lambda) = 2(m-1)$ $(m \geqslant 2)$ and $m(2\lambda) = 1$, then*

$$\overline{\alpha}(x) = \left(\frac{-1}{2\pi}\right)^{m} \cdot \int_{\mathbf{R}} \frac{1}{(x^2+y^2)^{1/2}} \cdot \overline{\mathfrak{F}}_\alpha^{(m)}((x^2+y^2)^{1/2})\, dy \qquad (x \geqslant 1).$$

(b) *If $m(\lambda) = 4(m-1)$ $(m \geqslant 2)$ and $m(2\lambda) = 3$, then*

$$\overline{\alpha}(x) = \left(\frac{-1}{2\pi}\right)^{2m} \cdot \int_{\mathbf{R}} \left(\frac{1}{x^2+y^2} \cdot \overline{\mathfrak{F}}_\alpha^{(2m)}((x^2+y^2)^{1/2}) - \frac{1}{(x^2+y^2)^{3/2}} \cdot \overline{\mathfrak{F}}_\alpha^{(2m-1)}((x^2+y^2)^{1/2})\right) dy \qquad (x \geqslant 1).$$

(c) *If $m(\lambda) = 8$ and $m(2\lambda) = 7$, then*

$$\overline{\alpha}(x) = \left(\frac{-1}{2\pi}\right)^{8} \cdot \int_{\mathbf{R}} \left(-15 \cdot \frac{1}{(x^2+y^2)^{7/2}} \cdot \overline{\mathfrak{F}}_\alpha^{(5)}((x^2+y^2)^{1/2}) + 15 \cdot \frac{1}{(x^2+y^2)^{6/2}} \times \overline{\mathfrak{F}}_\alpha^{(6)}((x^2+y^2)^{1/2}) - 6 \cdot \frac{1}{(x^2+y^2)^{5/2}} \cdot \overline{\mathfrak{F}}_\alpha^{(7)}((x^2+y^2)^{1/2}) + \frac{1}{(x^2+y^2)^{4/2}} \cdot \overline{\mathfrak{F}}_\alpha^{(8)}((x^2+y^2)^{1/2})\right) dy \qquad (x \geqslant 1).$$

SUBLEMMA. *Let $f \in C_c^\infty([1, +\infty[)$—then*

$$\int_{\mathbf{R}} \int_{\mathbf{R}} f'\left(x + \frac{a^2+b^2}{2}\right) da\, db = (-2\pi) f(x) \qquad (x \geqslant 1).$$

Proof. Passing to polar coordinates, we find that

$$\begin{aligned}\int_{\mathbf{R}} \int_{\mathbf{R}} f'\left(x + \frac{a^2+b^2}{2}\right) da\, db &= \int_0^{2\pi} \left(\int_0^\infty f'\left(x + \frac{r^2}{2}\right) r\, dr\right) d\theta \\ &= 2\pi \int_0^\infty f'(x+y)\, dy = (-2\pi) f(x),\end{aligned}$$

as desired. ■

Proof of Lemma 9.2. Suppose first that N is abelian. Let k be a non-negative integer—then

$$\overline{\mathfrak{F}}_\alpha^{(k)}(x) = \int_{\mathfrak{n}_\lambda} \overline{\alpha}^{(k)}\left(x + \frac{\|X_\lambda\|^2}{2}\right) d_{\mathfrak{n}_\lambda}(X_\lambda) \qquad (x \geqslant 1).$$

Using this fact, a direct application of the sublemma then leads at once to the stated formulas. Assume now that N is nonabelian. According to the table above, $m(\lambda)$ is even while $m(2\lambda)$ is odd. Introduce $\tilde{\alpha}$ by the definition

$$\tilde{\alpha}(x, X_{2\lambda}) = \overline{\alpha}((x^2 + \|X_{2\lambda}\|^2)^{1/2}) \qquad (x \in \mathbf{R}, X_{2\lambda} \in \mathfrak{n}_{2\lambda}).$$

Then, for all $x \geqslant 1$, we have

$$\overline{\mathfrak{F}}_\alpha^{(m(\lambda)/2)}(x) = \int_{\mathfrak{n}_\lambda \times \mathfrak{n}_{2\lambda}} \tilde{\alpha}_{(m(\lambda)/2)}\left(x + \frac{\|X_\lambda\|^2}{2}, X_{2\lambda}\right) d_{\mathfrak{n}_\lambda}(X_\lambda)\, d_{\mathfrak{n}_{2\lambda}}(X_{2\lambda}),$$

$\tilde{\alpha}_{(m(\lambda)/2)}$ being the $m(\lambda)/2$-partial derivative of $\tilde{\alpha}$ with respect to the first argument. It therefore follows from the sublemma that

$$\overline{\mathfrak{F}}_\alpha^{(m(\lambda)/2)}(x) = (-2\pi)^{m(\lambda)/2} \cdot \int_{\mathfrak{n}_{2\lambda}} \tilde{\alpha}(x, X_{2\lambda})\, d_{\mathfrak{n}_{2\lambda}}(X_{2\lambda}) \qquad (x \geqslant 1).$$

Set

$$\beta = \overline{\alpha} \circ \sqrt{}, \qquad \mathfrak{G}_\beta = \overline{\mathfrak{F}}_\alpha^{(m(\lambda)/2)} \circ \sqrt{}.$$

In these notations,

$$\mathfrak{G}_\beta(x) = (-2\pi)^{m(\lambda)/2} \cdot \int_{\mathfrak{n}_{2\lambda}} \beta(x + \|X_{2\lambda}\|^2)\, d_{\mathfrak{n}_{2\lambda}}(X_{2\lambda}) \qquad (x \geqslant 1),$$

thus

$$\begin{aligned}
&\mathfrak{G}_\beta^{(m(2\lambda)+1/2)}(x) \\
&\quad = (-2\pi)^{m(\lambda)/2} \cdot \int_{\mathfrak{n}_{2\lambda}} \beta^{(m(2\lambda)+1/2)}(x + \|X_{2\lambda}\|^2)\, d_{\mathfrak{n}_{2\lambda}}(X_{2\lambda}) \qquad (x \geqslant 1),
\end{aligned}$$

and so

$$\begin{aligned}
&\int_{\mathbf{R}} \mathfrak{G}_\beta^{(m(2\lambda)+1/2)}(x + y^2)\, dy \\
&\quad = (-2\pi)^{m(\lambda)/2} \cdot \int_{\mathbf{R}} \int_{\mathfrak{n}_{2\lambda}} \beta^{(m(2\lambda)+1/2)}(x + y^2 + \|X_{2\lambda}\|^2)\, d_{\mathfrak{n}_{2\lambda}}(X_{2\lambda})\, dy \\
&\quad = (-2\pi)^{m(\lambda)/2} \cdot \left(\frac{1}{\sqrt{2}}\right)^{m(2\lambda)+1} \\
&\qquad \times \int_{\mathbf{R}} \int_{\mathfrak{n}_{2\lambda}} \beta^{(m(2\lambda)+1/2)}\left(x + \frac{y^2 + \|X_{2\lambda}\|^2}{2}\right) d_{\mathfrak{n}_{2\lambda}}(X_{2\lambda})\, dy \\
&\quad = 2^{m(\lambda)/2} \cdot (-\pi)^{m(\lambda)+m(2\lambda)+1/2} \cdot \beta(x) \qquad (x \geqslant 1).
\end{aligned}$$

In passing to the last equality, we have used the sublemma once again. Since $\overline{\alpha}(x) = \beta(x^2)$ $(x \geqslant 1)$, we conclude that

$$\overline{\alpha}(x) = 2^{-m(\lambda)/2} \cdot (-\pi)^{-m(\lambda)+m(2\lambda)+1/2} \cdot \int_{\mathbf{R}} \mathfrak{G}_\beta^{(m(2\lambda)+1/2)}(x^2 + y^2)\, dy \qquad (x \geqslant 1).$$

To finish the proof, we have only to calculate

$$\mathfrak{G}_{\beta}^{(m(2\lambda)+1/2)}$$

per each of the three possibilities for $m(2\lambda)$. This is completely straightforward (albeit tedious in the last case), hence can be omitted. ■

We shall now take up the problem mentioned at the beginning of this section, namely, the computation of the Fourier transforms of the distributions appearing in Theorem 7.2. It will be convenient to consider first $T_{\lambda}(\alpha:0)$ and $T'_{\lambda}(\alpha:0)$ in both the abelian and nonabelian cases, reserving for later the study of $T_{2\lambda}(\alpha:0)$ (which is exclusive to the nonabelian case).

In the abelian case, we have, by definition,

$$T_{\lambda}(\alpha:0) = \int_0^{\infty} \overline{\alpha}\left(1 + \frac{y}{2}\right) y^{m(\lambda)/2}\,\frac{dy}{y}.$$

But

$$\begin{aligned}
\overline{\mathfrak{F}}_{\alpha}(1) &= \int_{\mathfrak{n}_{\lambda}} \overline{\alpha}\left(1 + \frac{\|X_{\lambda}\|^2}{2}\right) d_{\mathfrak{n}_{\lambda}}(X_{\lambda}) \\
&= A(\mathfrak{n}_{\lambda}) \cdot \int_0^{\infty} \overline{\alpha}\left(1 + \frac{r^2}{2}\right) r^{m(\lambda)-1}\,dr \\
&= \frac{A(\mathfrak{n}_{\lambda})}{2} \cdot \int_0^{\infty} \overline{\alpha}\left(1 + \frac{y}{2}\right) y^{m(\lambda)/2}\,\frac{dy}{y},
\end{aligned}$$

where $A(\mathfrak{n}_{\lambda})$ is the area of the unit sphere in $\mathfrak{n}_{\lambda}$, that is,

$$A(\mathfrak{n}_{\lambda}) = \frac{2(\sqrt{\pi})^{m(\lambda)}}{\Gamma(m(\lambda)/2)}.$$

It therefore follows that

$$\begin{aligned}
T_{\lambda}(\alpha:0) &= \frac{2}{A(\mathfrak{n}_{\lambda})} \cdot \overline{\mathfrak{F}}_{\alpha}(1) \\
&= \frac{2}{A(\mathfrak{n}_{\lambda})} \cdot \mathfrak{F}_{\alpha}(0) \\
&= \frac{2}{A(\mathfrak{n}_{\lambda})} \cdot \frac{1}{2\pi} \cdot \int_{-\infty}^{\infty} \hat{\mathfrak{F}}_{\alpha}(v)\,dv \\
&= \frac{1}{A(\mathfrak{n}_{\lambda})} \cdot |\lambda| \cdot (\sqrt{\Lambda})^{m(\lambda)} \cdot \frac{1}{\pi} \cdot \int_{-\infty}^{\infty} \hat{\alpha}(v)\,dv,
\end{aligned}$$

which gives the Fourier transform of $T_\lambda(\alpha:0)$ in the abelian case. Turning to the nonabelian case, we have, by definition,

$$T_\lambda(\alpha:0) = \int_0^\infty \int_{\mathfrak{n}_{2\lambda}} \overline{\alpha}\left(\left(1+\frac{y}{2}\right)(1+\|Y_{2\lambda}\|^2)^{1/2}\right)\cdot\left(1+\frac{2}{y}\right)^{m(2\lambda)}$$
$$\times\, y^{(m(\lambda)/2+m(2\lambda))}\frac{dy}{y}\, d_{\mathfrak{n}_{2\lambda}}(Y_{2\lambda}).$$

But

$$\begin{aligned}\mathfrak{F}_\alpha(1) &= \int_{\mathfrak{n}_\lambda\times\mathfrak{n}_{2\lambda}} \overline{\alpha}\left(\left[\left(1+\frac{\|X_\lambda\|^2}{2}\right)^2+\|X_{2\lambda}\|^2\right]^{1/2}\right) d_{\mathfrak{n}_\lambda}(X_\lambda)\, d_{\mathfrak{n}_{2\lambda}}(X_{2\lambda})\\
&= A(\mathfrak{n}_\lambda)\cdot\int_0^\infty\int_{\mathfrak{n}_{2\lambda}}\overline{\alpha}\left(\left[\left(1+\frac{r^2}{2}\right)^2+\|X_{2\lambda}\|^2\right]^{1/2}\right) r^{m(\lambda)-1}\,dr\, d_{\mathfrak{n}_{2\lambda}}(X_{2\lambda})\\
&= \frac{A(\mathfrak{n}_\lambda)}{2}\cdot\int_0^\infty\int_{\mathfrak{n}_{2\lambda}}\overline{\alpha}\left(\left[\left(1+\frac{y}{2}\right)^2+\|Y_{2\lambda}\|^2\right]^{1/2}\right) y^{m(\lambda)/2}\frac{dy}{y}\, d_{\mathfrak{n}_{2\lambda}}(Y_{2\lambda})\\
&= \frac{A(\mathfrak{n}_\lambda)}{2}\cdot\frac{1}{2^{m(2\lambda)}}\cdot\int_0^\infty\int_{\mathfrak{n}_{2\lambda}}\overline{\alpha}\left(\left(1+\frac{y}{2}\right)(1+\|Y_{2\lambda}\|^2)^{1/2}\right)\\
&\quad\times\left(1+\frac{2}{y}\right)^{m(2\lambda)}\cdot y^{(m(\lambda)/2+m(2\lambda))}\frac{dy}{y}\, d_{\mathfrak{n}_{2\lambda}}(Y_{2\lambda}).\end{aligned}$$

Here, as above, $A(\mathfrak{n}_\lambda)$ is the area of the unit sphere in $\mathfrak{n}_\lambda$. It therefore follows that

$$T_\lambda(\alpha:0) = \frac{2^{m(2\lambda)}}{A(\mathfrak{n}_\lambda)}\cdot|\lambda|\cdot(\sqrt{\Lambda})^{m(\lambda)+m(2\lambda)}\cdot\frac{1}{\pi}\cdot\int_{-\infty}^\infty \hat{\alpha}(v)\,dv,$$

which gives the Fourier transform of $T_\lambda(\alpha:0)$ in the nonabelian case. Of course the last formula makes sense in the abelian case as well, provided we take $m(2\lambda)=0$.

Before discussing $T'_\lambda(\alpha:0)$, we shall insert an ancillary result which is of considerable importance for our development.

SUBLEMMA. *Let $f\in C_c^\infty(\mathbf{R})$ be even; let $\overline{f} = f\circ\cosh^{-1}$—then*

$$\int_0^\infty \overline{f}'\left(1+\frac{y}{2}\right)\log y\,dy = -\hat{f}(0)+4\gamma\cdot f(0)+\frac{2}{\pi}\cdot\int_{-\infty}^\infty \hat{f}(v)\cdot\frac{\Gamma'(1+\sqrt{-1}\,v)}{\Gamma(1+\sqrt{-1}\,v)}\,dv,$$

γ being Euler's constant.

[$\hat{f}$ is the Fourier transform of f.]

Proof. We have

$$\int_0^\infty \overline{f}'\left(1 + \frac{y}{2}\right)\log y \, dy$$

$$= 2\int_1^\infty \overline{f}'(y)\log(2(y-1))\,dy$$

$$= 2\lim_{\epsilon \downarrow 1}\left\{\log((y-1))\cdot \overline{f}(y)\Big|_\epsilon^\infty - \int_\epsilon^\infty \frac{\overline{f}(y)}{y-1}\,dy\right\}$$

$$= 2\lim_{\epsilon \downarrow 1}\left\{-\log(2(\epsilon-1))\cdot \overline{f}(\epsilon) - \int_\epsilon^\infty \frac{\overline{f}(y)}{y-1}\,dy\right\}.$$

In the last integral on the right, make the change of variable $y = \cosh t$. Bearing in mind that

$$\overline{f}(y) = f(\cosh^{-1}(y)),$$

we then get

$$\int_\epsilon^\infty \frac{\overline{f}(y)}{y-1}\,dy = \int_{\epsilon^*}^\infty f(t)\left(\frac{e^{t/2}+e^{-t/2}}{e^{t/2}-e^{-t/2}}\right)dt$$

where $\epsilon^* = \log(\epsilon + (\epsilon^2 - 1)^{1/2})$ $(\epsilon > 1)$. Write

$$f(t) = \frac{1}{2\pi}\cdot\int_{-\infty}^\infty \hat{f}(v)e^{-\sqrt{-1}\,vt}\,dv.$$

Then

$$\int_{\epsilon^*}^\infty f(t)\left(\frac{e^{t/2}+e^{-t/2}}{e^{t/2}-e^{-t/2}}\right)dt$$

$$= \frac{1}{2\pi}\cdot\int_{\epsilon^*}^\infty\int_{-\infty}^\infty \left(\frac{e^{t/2}+e^{-t/2}}{e^{t/2}-e^{-t/2}}\right)e^{-\sqrt{-1}\,vt}\hat{f}(v)\,dv\,dt$$

$$= \frac{1}{2\pi}\cdot\int_{\epsilon^*}^\infty\int_{-\infty}^\infty \left(1 + \frac{2e^{-t}}{1-e^{-t}}\right)e^{-\sqrt{-1}\,vt}\hat{f}(v)\,dv\,dt$$

$$= \frac{1}{2\pi}\cdot\int_{\epsilon^*}^\infty\int_{-\infty}^\infty \left(e^{-\sqrt{-1}\,vt} + \frac{2e^{-t}}{t} - \frac{2e^{-t}}{t} + \frac{2e^{-t(1+\sqrt{-1}\,v)}}{1-e^{-t}}\right)\hat{f}(v)\,dv\,dt$$

$$= \frac{1}{2\pi}\cdot\int_{\epsilon^*}^\infty\int_{-\infty}^\infty \left(e^{-\sqrt{-1}\,vt} + \frac{2e^{-t}}{t}\right)\hat{f}(v)\,dv\,dt$$

$$-\frac{1}{\pi}\int_{-\infty}^\infty\left[\int_{\epsilon^*}^\infty\left(\frac{e^{-t}}{t} - \frac{e^{-t(1+\sqrt{-1}\,v)}}{1-e^{-t}}\right)dt\right]\hat{f}(v)\,dv$$

$$= \int_{\epsilon^*}^\infty f(t)\,dt + 2f(0)\left[-e^{-\epsilon^*}\cdot\log\epsilon^* + \int_{\epsilon^*}^\infty e^{-t}\cdot\log t\,dt\right]$$

$$-\frac{1}{\pi}\int_{-\infty}^\infty\left[\int_{\epsilon^*}^\infty\left(\frac{e^{-t}}{t} - \frac{e^{-t(1+\sqrt{-1}\,v)}}{1-e^{-t}}\right)dt\right]\hat{f}(v)\,dv.$$

It is well known that

$$\int_0^\infty e^{-t} \cdot \log t \, dt = -\gamma$$

and

$$\int_0^\infty \left(\frac{e^{-t}}{t} - \frac{e^{-t(1+\sqrt{-1}\,v)}}{1-e^{-t}}\right) dt = \frac{\Gamma'(1+\sqrt{-1}\,v)}{\Gamma(1+\sqrt{-1}\,v)}.$$

When ϵ is close to 1, ϵ^* is close to zero and

$$2(\epsilon - 1) \sim (\epsilon^*)^2.$$

Therefore

$$\begin{aligned}
&\int_0^\infty \overleftarrow{f}'\left(1 + \frac{y}{2}\right) \log y \, dy \\
&\quad = 2 \lim_{\epsilon \downarrow 1} \{-\log(2(\epsilon - 1)) \cdot \overleftarrow{f}(\epsilon) + 2f(0) \cdot e^{-\epsilon^*} \cdot \log \epsilon^*\} \\
&\qquad - \hat{f}(0) + 4\gamma \cdot f(0) + \frac{2}{\pi} \cdot \int_{-\infty}^{\infty} \hat{f}(v) \cdot \frac{\Gamma'(1+\sqrt{-1}\,v)}{\Gamma(1+\sqrt{-1}\,v)} \, dv \\
&\quad = -\hat{f}(0) + 4\gamma \cdot f(0) + \frac{2}{\pi} \cdot \int_{-\infty}^{\infty} \hat{f}(v) \cdot \frac{\Gamma'(1+\sqrt{-1}\,v)}{\Gamma(1+\sqrt{-1}\,v)} \, dv,
\end{aligned}$$

as we wished to prove. ■

Our goal now will be to show that

$$\begin{aligned}
T'_\lambda(\alpha:0) = \frac{2^{m(2\lambda)}}{A(\mathfrak{n}_\lambda)} \cdot |\lambda| \cdot (\sqrt{\Lambda})^{m(\lambda)+m(2\lambda)} \cdot \Bigg[& \frac{1}{\pi}\left(\frac{\Gamma'(m(\lambda)/2)}{\Gamma(m(\lambda)/2)} - \gamma\right) \\
\times \int_{-\infty}^{\infty} \hat{\alpha}(v) \, dv + \hat{\alpha}(0) - \frac{2}{\pi} \cdot & \int_{-\infty}^{\infty} \hat{\alpha}(v) \cdot \frac{\Gamma'(1+\sqrt{-1}\,v)}{\Gamma(1+\sqrt{-1}\,v)} \, dv \Bigg]
\end{aligned}$$

where, of course, in the abelian case, it is understood that $m(2\lambda) = 0$. Needless to say, this formula gives the Fourier transform of $T'_\lambda(\alpha:0)$.

Suppose first that N is abelian—then, by definition,

$$T'_\lambda(\alpha:0) = \int_0^\infty \overleftarrow{\alpha}\left(1 + \frac{y}{2}\right) y^{m(\lambda)/2} \frac{\log y}{y} \, dy.$$

If $m(\lambda) = m = 2n$ is even, then by Lemma 9.2 ((i)–(a)), we have

$$\overleftarrow{\alpha}(x) = \left(\frac{-1}{2\pi}\right)^n \cdot \overleftarrow{\mathfrak{F}}_\alpha^{(n)}(x) \qquad (x \geqslant 1).$$

Therefore

$$T'_\lambda(\alpha:0) = \left(\frac{-1}{2\pi}\right)^n \cdot \int_0^\infty \overline{\mathfrak{F}}_\alpha^{(n)}\left(1+\frac{y}{2}\right) y^n \frac{\log y}{y}\, dy$$

$$= -\left(\frac{1}{2}\right) \cdot \frac{1}{\pi^n} \cdot \int_0^\infty \overline{\mathfrak{F}}'_\alpha\left(1+\frac{y}{2}\right) \cdot \left(\frac{d}{dy}\right)^{n-1} (y^{n-1}\log y)\, dy$$

$$= -\left(\frac{1}{2}\right) \cdot \frac{1}{\pi^n} \cdot (n-1)! \cdot \int_0^\infty \overline{\mathfrak{F}}'_\alpha\left(1+\frac{y}{2}\right)\left(\log y + 1 + \frac{1}{2} + \cdots + \frac{1}{n-1}\right) dy$$

$$= -\left(\frac{1}{2}\right) \cdot \frac{1}{\pi^n} \cdot (n-1)! \cdot \int_0^\infty \overline{\mathfrak{F}}'_\alpha\left(1+\frac{y}{2}\right)\left(\log y + \frac{\Gamma'(n)}{\Gamma(n)} + \gamma\right) dy$$

$$= \frac{(n-1)!}{\pi^n} \cdot \left(\frac{\Gamma'(n)}{\Gamma(n)} + \gamma\right) \cdot \mathfrak{F}_\alpha(0)$$

$$- \left(\frac{1}{2}\right) \cdot \frac{1}{\pi^n} \cdot (n-1)! \cdot \int_0^\infty \overline{\mathfrak{F}}'_\alpha\left(1+\frac{y}{2}\right) \log y\, dy.$$

Using the sublemma and combining terms then leads at once to the claimed formula for $T'_\lambda(\alpha:0)$ in the case when $m(\lambda)$ is even. If $m(\lambda) = m = 2n+1$ is odd, then by Lemma 9.2 ((i)–(b)), we have

$$\overline{\alpha}(x) = \left(\frac{-1}{2\pi}\right)^{n+1} \cdot \int_{-\infty}^\infty \overline{\mathfrak{F}}_\alpha^{(n+1)}\left(x+\frac{y^2}{2}\right) dy \qquad (x \geqslant 1).$$

Therefore

$$T'_\lambda(\alpha:0) = \left(\frac{-1}{2\pi}\right)^{n+1} \cdot \int_0^\infty \int_{-\infty}^\infty \overline{\mathfrak{F}}_\alpha^{(n+1)}\left(1+\frac{y}{2}+\frac{x^2}{2}\right) y^{n-1/2} \log y\, dx\, dy$$

$$= 2 \cdot \left(\frac{-1}{2\pi}\right)^{n+1} \cdot \int_0^\infty \int_0^\infty \overline{\mathfrak{F}}_\alpha^{(n+1)}\left(1+\frac{y}{2}+\frac{x^2}{2}\right) y^{n-1/2} \log y\, dx\, dy.$$

In the last integral, make the change of variable $x = \sqrt{2}\cdot(v-(1+y/2))^{1/2}$ to get

$$\frac{2}{\sqrt{2}} \cdot \left(\frac{-1}{2\pi}\right)^{n+1} \cdot \int_0^\infty \int_{1+y/2}^\infty \overline{\mathfrak{F}}_\alpha^{(n+1)}(v) \cdot \frac{y^{n-1/2}\log y}{(v-(1+y/2))^{1/2}}\, dv\, dy$$

or still

$$\frac{2}{\sqrt{2}} \cdot \left(\frac{-1}{2\pi}\right)^{n+1} \cdot \int_1^\infty \left(\int_0^{2(v-1)} \frac{y^{n-1/2}\log y}{(v-(1+y/2))^{1/2}}\, dy\right) \overline{\mathfrak{F}}_\alpha^{(n+1)}(v)\, dv.$$

Put $y = u \cdot 2(v-1)$—then our integral becomes

$$\left(\frac{-1}{\pi}\right)^{n+1} \cdot \int_1^\infty \left(\int_0^1 \frac{\log(2u(v-1))}{\sqrt{1-u}} \cdot u^{n-1/2}\, du\right)(v-1)^n \overline{\mathfrak{F}}_\alpha^{(n+1)}(v)\, dv.$$

Write

$$\int_0^1 \frac{\log(2u(v-1))}{(1-u)^{1/2}} \cdot u^{n-1/2}\, du$$

$$= \log(2(v-1)) \cdot \int_0^1 \frac{u^{n-1/2}}{(1-u)^{1/2}}\, du + \int_0^1 \frac{u^{n-1/2}\log u}{(1-u)^{1/2}}\, du$$

$$= \log(2(v-1)) \cdot \frac{\Gamma(n+\frac{1}{2})\Gamma(\frac{1}{2})}{\Gamma(n+1)}$$

$$+ \frac{\Gamma(n+\frac{1}{2})\Gamma(\frac{1}{2})}{\Gamma(n+1)} \cdot \left[\frac{\Gamma'(n+\frac{1}{2})}{\Gamma(n+\frac{1}{2})} - \frac{\Gamma'(n+1)}{\Gamma(n+1)}\right].$$

Obviously

$$\int_1^\infty \overline{\mathfrak{F}}_\alpha^{(n+1)}(v)(v-1)^n\, dv = (-1)^n \cdot (-n!\,\mathfrak{F}_\alpha(0)).$$

In addition

$$\int_1^\infty \overline{\mathfrak{F}}_\alpha^{(n+1)}(v)(v-1)^n \log(2(v-1))\, dv$$

$$= \frac{1}{2^{n+1}} \cdot \int_0^\infty \overline{\mathfrak{F}}_\alpha^{(n+1)}\left(1+\frac{y}{2}\right) y^n \log y\, dy$$

$$= \frac{(-1)^n}{2} \cdot \int_0^\infty \overline{\mathfrak{F}}'_\alpha\left(1+\frac{y}{2}\right) \cdot \left(\frac{d}{dy}\right)^n (y^n \log y)\, dy$$

$$= (-1)^n \cdot \frac{n!}{2} \cdot \int_0^\infty \overline{\mathfrak{F}}'_\alpha\left(1+\frac{y}{2}\right)\left(\log y + 1 + \frac{1}{2} + \cdots + \frac{1}{n}\right) dy$$

$$= (-1)^n \cdot \frac{n!}{2} \cdot \int_0^\infty \overline{\mathfrak{F}}'_\alpha\left(1+\frac{y}{2}\right)\left(\log y + \frac{\Gamma'(n+1)}{\Gamma(n+1)} + \gamma\right) dy$$

$$= (-1)^n \cdot \left[-n!\left(\frac{\Gamma'(n+1)}{\Gamma(n+1)} + \gamma\right) \cdot \mathfrak{F}_\alpha(0) + \frac{n!}{2} \cdot \int_0^\infty \overline{\mathfrak{F}}'_\alpha\left(1+\frac{y}{2}\right) \log y\, dy\right].$$

Using the sublemma and combining terms then leads at once to the claimed formula for $T'_\lambda(\alpha:0)$ in the case when $m(\lambda)$ is odd.

Assume now that N is nonabelian—then, by definition,

$$T'_\lambda(\alpha:0) = \int_0^\infty \int_{\mathfrak{n}_{2\lambda}} \overline{\alpha}\left(\left(1+\frac{y}{2}\right)(1+\|Y_{2\lambda}\|^2)^{1/2}\right)\cdot\left(1+\frac{2}{y}\right)^{m(2\lambda)}$$

$$\times\, y^{(m(\lambda)/2+m(2\lambda))}\frac{\log y}{y}\,dy\,d_{\mathfrak{n}_{2\lambda}}(Y_{2\lambda}),$$

or still

$$T'_\lambda(\alpha:0) = 2^{m(2\lambda)}\cdot\int_0^\infty \int_{\mathfrak{n}_{2\lambda}} \overline{\alpha}\left(\left[\left(1+\frac{y}{2}\right)^2+\|Y_{2\lambda}\|^2\right]^{1/2}\right)$$

$$\times\, y^{m(\lambda)/2}\frac{\log y}{y}\,dy\,d_{\mathfrak{n}_{2\lambda}}(Y_{2\lambda}).$$

For use below, let us remind ourselves that

$$\int_0^{\pi/2}(\sin\theta)^n\,d\theta = \frac{1}{2}\cdot B\left(\frac{n+1}{2},\frac{1}{2}\right).$$

This said, suppose to begin with that $m(\lambda)=2(m-1)$ $(m\geqslant 2)$, $m(2\lambda)=1$—then, by Lemma 9.2 ((ii)–(a)), we have

$$\overline{\alpha}(x) = \left(\frac{-1}{2\pi}\right)^m\cdot\int_{-\infty}^\infty (x^2+y^2)^{-1/2}\cdot\overline{\mathfrak{F}}_\alpha^{(m)}((x^2+y^2)^{1/2})\,dy \qquad (x\geqslant 1)$$

or still

$$\overline{\alpha}(x) = 2\cdot\left(\frac{-1}{2\pi}\right)^m\cdot\int_x^\infty (y^2-x^2)^{-1/2}\cdot\overline{\mathfrak{F}}_\alpha^{(m)}(y)\,dy \qquad (x\geqslant 1).$$

Therefore

$$T'_\lambda(\alpha:0) = 2^3\cdot\left(\frac{-1}{2\pi}\right)^m$$

$$\times\int_0^\infty\int_0^\infty\left(\int_{[(1+y/2)^2+z^2]^{1/2}}^\infty \overline{\mathfrak{F}}_\alpha^{(m)}(x)\cdot\left[x^2-\left(1+\frac{y}{2}\right)^2-z^2\right]^{-1/2}dx\right)$$

$$\times\, y^{m-2}\log y\,dy\,dz$$

$$= 2^3\cdot\left(\frac{-1}{2\pi}\right)^m\cdot\int_0^\infty y^{m-2}\log y\,dy\int_{1+y/2}^\infty \overline{\mathfrak{F}}_\alpha^{(m)}(x)\,dx$$

$$\times\int_0^{[x^2-(1+y/2)^2]^{1/2}}\left[x^2-\left(1+\frac{y}{2}\right)^2-z^2\right]^{-1/2}dz$$

$$= (-1)^m \cdot \frac{1}{2^{m-2}} \cdot \frac{1}{\pi^{m-1}} \cdot \int_0^\infty y^{m-2} \log y \, dy \int_{1+y/2}^\infty \overleftarrow{\mathfrak{F}}_\alpha^{(m)}(x)\, dx$$

$$= (-1)^{m+1} \cdot \frac{1}{2^{m-2}} \cdot \frac{1}{\pi^{m-1}} \cdot \int_0^\infty \overleftarrow{\mathfrak{F}}_\alpha^{(m-1)}\left(1 + \frac{y}{2}\right) y^{m-2} \log y \, dy$$

$$= -\left(\frac{1}{\pi^{m-1}}\right) \cdot \int_0^\infty \overleftarrow{\mathfrak{F}}_\alpha'\left(1 + \frac{y}{2}\right)\left(\frac{d}{dy}\right)^{m-2} (y^{m-2} \log y)\, dy$$

$$= -\left(\frac{1}{\pi^{m-1}}\right) \cdot (m-2)!$$

$$\times \int_0^\infty \overleftarrow{\mathfrak{F}}_\alpha'\left(1 + \frac{y}{2}\right)\left(\log y + 1 + \frac{1}{2} + \cdots + \frac{1}{m-2}\right) dy$$

$$= -\left(\frac{1}{\pi^{m-1}}\right) \cdot (m-2)! \cdot \int_0^\infty \overleftarrow{\mathfrak{F}}_\alpha'\left(1 + \frac{y}{2}\right)\left(\log y + \frac{\Gamma'(m-2)}{\Gamma(m-2)} + \gamma\right) dy$$

$$= 2 \cdot \frac{(m-2)!}{\pi^{m-1}} \cdot \left(\frac{\Gamma'(m-2)}{\Gamma(m-2)} + \gamma\right) \cdot \mathfrak{F}_\alpha(0)$$

$$- \frac{(m-2)!}{\pi^{m-1}} \cdot \int_0^\infty \overleftarrow{\mathfrak{F}}_\alpha'\left(1 + \frac{y}{2}\right) \log y \, dy.$$

Using the sublemma and combining terms then leads at once to the claimed formula for $T'_\lambda(\alpha:0)$ in the case when $m(\lambda) = 2(m-1)$ $(m \geqslant 2)$, $m(2\lambda) = 1$. Let us proceed to the next possibility: $m(\lambda) = 4(m-1)$ $(m \geqslant 2)$, $m(2\lambda) = 3$. In view of Lemma 9.2 ((ii)–(b)), we have

$$\overleftarrow{\alpha}(x) = \left(\frac{-1}{2\pi}\right)^{2m} \cdot \int_{-\infty}^\infty \left(\frac{1}{x^2 + y^2} \cdot \overleftarrow{\mathfrak{F}}_\alpha^{(2m)}((x^2 + y^2)^{1/2})\right.$$

$$\left. - \frac{1}{(x^2 + y^2)^{3/2}} \cdot \overleftarrow{\mathfrak{F}}_\alpha^{(2m-1)}((x^2 + y^2)^{1/2})\right) dy \qquad (x \geqslant 1)$$

or still

$$\overleftarrow{\alpha}(x) = 2 \cdot \left(\frac{1}{2\pi}\right)^{2m} \cdot \int_x^\infty \left(\frac{1}{y(y^2 - x^2)^{1/2}} \cdot \overleftarrow{\mathfrak{F}}_\alpha^{(2m)}(y)\right.$$

$$\left. - \frac{1}{y^2(y^2 - x^2)^{1/2}} \cdot \overleftarrow{\mathfrak{F}}_\alpha^{(2m-1)}(y)\right) dy \qquad (x \geqslant 1).$$

Therefore

$$T'_\lambda(\alpha:0) = 2^4 \cdot \left(\frac{1}{2\pi}\right)^{2m} \cdot \int_0^\infty \int_{\mathbf{R}^3} \left(\int_{[(1+y/2)^2+\|Z\|^2]^{1/2}} \left\{ \frac{\mathfrak{F}_\alpha^{(2m)}(x)}{x} - \frac{\mathfrak{F}_\alpha^{(2m-1)}(x)}{x^2} \right\} \right.$$

$$\left. \times \left[x^2 - \left(1+\frac{y}{2}\right)^2 - \|Z\|^2 \right]^{-1/2} dx \right) y^{2m-3} \log y \, dy \, dZ$$

$$= 2^5 \cdot \left(\frac{1}{2\pi}\right)^{2m} \cdot \frac{\pi^{3/2}}{\Gamma(3/2)}$$

$$\times \int_0^\infty \int_0^\infty \left(\int_{[(1+y/2)^2+r^2]^{1/2}} \left\{ \frac{\mathfrak{F}_\alpha^{(2m)}(x)}{x} - \frac{\mathfrak{F}_\alpha^{(2m-1)}(x)}{x^2} \right\} \right.$$

$$\left. \times \left[x^2 - \left(1+\frac{y}{2}\right)^2 - r^2 \right]^{-1/2} dx \right) y^{2m-3} \log y \, r^2 \, dy \, dr$$

$$= 2^5 \cdot \left(\frac{1}{2\pi}\right)^{2m} \cdot \frac{\pi^{3/2}}{\Gamma(3/2)}$$

$$\times \int_0^\infty y^{2m-3} \log y \, dy \int_{1+y/2}^\infty \left\{ \frac{\mathfrak{F}_\alpha^{(2m)}(x)}{x} - \frac{\mathfrak{F}_\alpha^{(2m-1)}(x)}{x^2} \right\} dx$$

$$\times \int_0^{[x^2-(1+y/2)^2]^{1/2}} \left[x^2 - \left(1+\frac{y}{2}\right)^2 - r^2 \right]^{-1/2} r^2 \, dr$$

$$= \frac{1}{2^{2m-4}} \cdot \frac{1}{\pi^{2m-2}} \cdot \int_0^\infty y^{2m-3} \log y \, dy \int_{1+y/2}^\infty \left\{ \frac{\mathfrak{F}_\alpha^{(2m)}(x)}{x} - \frac{\mathfrak{F}_\alpha^{(2m-1)}(x)}{x^2} \right\}$$

$$\times \left[x^2 - \left(1+\frac{y}{2}\right)^2 \right] dx.$$

Observing that

$$\left\{ \frac{\mathfrak{F}_\alpha^{(2m)}(x)}{x} - \frac{\mathfrak{F}_\alpha^{(2m-1)}(x)}{x^2} \right\} = \frac{d}{dx}\left(\frac{\mathfrak{F}_\alpha^{(2m-1)}(x)}{x} \right),$$

an integration by parts gives

$$\int_{1+y/2}^\infty \left\{ \frac{\mathfrak{F}_\alpha^{(2m)}(x)}{x} - \frac{\mathfrak{F}_\alpha^{(2m-1)}(x)}{x^2} \right\} x^2 \, dx$$

$$= 2 \cdot \mathfrak{F}_\alpha^{(2m-2)}\left(1+\frac{y}{2}\right) - \left(1+\frac{y}{2}\right) \mathfrak{F}_\alpha^{(2m-1)}\left(1+\frac{y}{2}\right).$$

Inserting this, we get

$$
\begin{aligned}
T'_\lambda(\alpha:0) &= \frac{1}{2^{2m-5}}\cdot\frac{1}{\pi^{2m-2}}\cdot\int_0^\infty \overleftarrow{\mathfrak{F}}_\alpha^{(2m-2)}\left(1+\frac{y}{2}\right)y^{2m-3}\log y\,dy \\
&= -2^2\cdot\left(\frac{1}{\pi^{2m-2}}\right)\cdot\int_0^\infty \overleftarrow{\mathfrak{F}}'_\alpha\left(1+\frac{y}{2}\right)\left(\frac{d}{dy}\right)^{2m-3}(y^{2m-3}\log y)\,dy \\
&= -2^2\cdot\left(\frac{1}{\pi^{2m-2}}\right)\cdot(2m-3)! \\
&\quad\times\int_0^\infty \mathfrak{F}'_\alpha\left(1+\frac{y}{2}\right)\left(\log y + 1 + \frac{1}{2} + \cdots + \frac{1}{2m-3}\right)dy \\
&= -2^2\cdot\left(\frac{1}{\pi^{2m-2}}\right)\cdot(2m-3)! \\
&\quad\times\int_0^\infty \overleftarrow{\mathfrak{F}}'_\alpha\left(1+\frac{y}{2}\right)\left(\log y + \frac{\Gamma'(2m-3)}{\Gamma(2m-3)} + \gamma\right)dy \\
&= 2^3\cdot\frac{(2m-3)!}{\pi^{2m-2}}\cdot\left(\frac{\Gamma'(2m-3)}{\Gamma(2m-3)}+\gamma\right)\cdot\mathfrak{F}_\alpha(0) \\
&\quad - 2^2\cdot\frac{(2m-3)!}{\pi^{2m-2}}\cdot\int_0^\infty \overleftarrow{\mathfrak{F}}'_\alpha\left(1+\frac{y}{2}\right)\log y\,dy.
\end{aligned}
$$

Using the sublemma and combining terms then leads at once to the claimed formula for $T'_\lambda(\alpha:0)$ in the case when $m(\lambda) = 4(m-1)$ $(m \geqslant 2)$, $m(2\lambda) = 3$. We shall now take up the last case, namely: $m(\lambda) = 8$, $m(2\lambda) = 7$. In view of Lemma 9.2 ((ii)–(c)), we have

$$
\begin{aligned}
\overleftarrow{\alpha}(x) = \left(\frac{-1}{2\pi}\right)^8\cdot\int_{-\infty}^\infty\Bigg(&-15\cdot\frac{1}{(x^2+y^2)^{7/2}}\cdot\overleftarrow{\mathfrak{F}}_\alpha^{(5)}((x^2+y^2)^{1/2}) + 15 \\
&\times\frac{1}{(x^2+y^2)^{6/2}}\cdot\overleftarrow{\mathfrak{F}}_\alpha^{(6)}((x^2+y^2)^{1/2}) - 6\cdot\frac{1}{(x^2+y^2)^{5/2}} \\
&\times\overleftarrow{\mathfrak{F}}_\alpha^{(7)}((x^2+y^2)^{1/2}) + \frac{1}{(x^2+y^2)^{4/2}} \\
&\times\overleftarrow{\mathfrak{F}}_\alpha^{(8)}((x^2+y^2)^{1/2})\Bigg)dy \qquad (x \geqslant 1)
\end{aligned}
$$

or still

$$\overline{\alpha}(x) = 2 \cdot \left(\frac{1}{2\pi}\right)^8 \cdot \int_x^\infty \frac{1}{(y^2 - x^2)^{1/2}} \cdot \left[-\frac{15}{y^6} \cdot \overline{\mathfrak{F}}_\alpha^{(5)}(y) + \frac{15}{y^5} \cdot \overline{\mathfrak{F}}_\alpha^{(6)}(y) - \frac{6}{y^4} \cdot \overline{\mathfrak{F}}_\alpha^{(7)}(y) + \frac{1}{y^3} \cdot \overline{\mathfrak{F}}_\alpha^{(8)}(y) \right] dy \qquad (x \geqslant 1).$$

Therefore

$$\begin{aligned} T'_\lambda(\alpha:0) &= \frac{1}{\pi^8} \cdot \int_0^\infty \int_{\mathbf{R}^7} \left(\int_{[(1+y/2)^2 + \|Z\|^2]^{1/2}} \left\{ -\frac{15}{x^6} \cdot \overline{\mathfrak{F}}_\alpha^{(5)}(x) + \frac{15}{x^5} \cdot \overline{\mathfrak{F}}_\alpha^{(6)}(x) \right.\right. \\ &\quad \left.\left. - \frac{6}{x^4} \cdot \overline{\mathfrak{F}}_\alpha^{(7)}(x) + \frac{1}{x^3} \cdot \overline{\mathfrak{F}}_\alpha^{(8)}(x) \right\} \left[x^2 - \left(1 + \frac{y}{2}\right)^2 - \|Z\|^2 \right]^{-1/2} dx \right) \\ &\quad \times y^3 \log y \, dy \, dZ \\ &= 2 \cdot \frac{1}{\pi^8} \cdot \frac{(\sqrt{\pi})^7}{\Gamma(7/2)} \cdot \int_0^\infty \int_0^\infty \left(\int_{[(1+y/2)^2 + r^2]^{1/2}} \left\{ -\frac{15}{x^6} \cdot \overline{\mathfrak{F}}_\alpha^{(5)}(x) + \frac{15}{x^5} \cdot \overline{\mathfrak{F}}_\alpha^{(6)}(x) \right.\right. \\ &\quad \left.\left. - \frac{6}{x^4} \cdot \overline{\mathfrak{F}}_\alpha^{(7)}(x) + \frac{1}{x^3} \cdot \overline{\mathfrak{F}}_\alpha^{(8)}(x) \right\} \left[x^2 - \left(1 + \frac{y}{2}\right)^2 - r^2 \right]^{-1/2} dx \right) \\ &\quad \times y^3 (\log y) r^6 \, dy \, dr \\ &= 2 \cdot \frac{1}{\pi^8} \cdot \frac{(\sqrt{\pi})^7}{\Gamma(7/2)} \cdot \int_0^\infty y^3 \log y \, dy \int_{1+y/2}^\infty \left\{ -\frac{15}{x^6} \cdot \overline{\mathfrak{F}}_\alpha^{(5)}(x) + \frac{15}{x^5} \cdot \overline{\mathfrak{F}}_\alpha^{(6)}(x) \right. \\ &\quad \left. - \frac{6}{x^4} \cdot \overline{\mathfrak{F}}_\alpha^{(7)}(x) + \frac{1}{x^3} \cdot \overline{\mathfrak{F}}_\alpha^{(8)}(x) \right\} dx \\ &\quad \times \int_0^{[x^2 - (1+y/2)^2]^{1/2}} \left[x^2 - \left(1 + \frac{y}{2}\right)^2 - r^2 \right]^{-1/2} r^6 \, dr \\ &= \frac{1}{6} \cdot \frac{1}{\pi^4} \int_{1+y/2}^\infty \left\{ -\frac{15}{x^6} \cdot \overline{\mathfrak{F}}_\alpha^{(5)}(x) + \frac{15}{x^5} \cdot \overline{\mathfrak{F}}_\alpha^{(6)}(x) \right. \\ &\quad \left. - \frac{6}{x^4} \cdot \overline{\mathfrak{F}}_\alpha^{(7)}(x) + \frac{1}{x^3} \cdot \overline{\mathfrak{F}}_\alpha^{(8)}(x) \right\} \left[x^2 - \left(1 + \frac{y}{2}\right)^2 \right]^3 dx. \end{aligned}$$

Successive integration by parts gives

$$\begin{aligned} &\int_{1+y/2}^\infty \left\{ -\frac{15}{x^6} \cdot \overline{\mathfrak{F}}_\alpha^{(5)}(x) + \frac{15}{x^5} \cdot \overline{\mathfrak{F}}_\alpha^{(6)}(x) \right. \\ &\quad \left. - \frac{6}{x^4} \cdot \overline{\mathfrak{F}}_\alpha^{(7)}(x) + \frac{1}{x^3} \cdot \overline{\mathfrak{F}}_\alpha^{(8)}(x) \right\} \left[x^2 - \left(1 + \frac{y}{2}\right)^2 \right]^3 dx = 48 \cdot \overline{\mathfrak{F}}_\alpha^{(4)}\left(1 + \frac{y}{2}\right). \end{aligned}$$

Inserting this, we get

$$\begin{aligned}
T'_\lambda(\alpha:0) &= 2^3 \cdot \frac{1}{\pi^4} \cdot \int_0^\infty \overline{\mathfrak{F}}_\alpha^{(4)}\left(1+\frac{y}{2}\right) y^3 \log y \, dy \\
&= -2^6 \cdot \frac{1}{\pi^4} \cdot \int_0^\infty \overline{\mathfrak{F}}'_\alpha\left(1+\frac{y}{2}\right)\left(\frac{d}{dy}\right)^3 (y^3 \log y)\, dy \\
&= -2^6 \cdot \frac{1}{\pi^4} \cdot 3! \cdot \int_0^\infty \overline{\mathfrak{F}}'_\alpha\left(1+\frac{y}{2}\right)\left(\log y + 1 + \frac{1}{2} + \frac{1}{3}\right) dy \\
&= -2^6 \cdot \frac{1}{\pi^4} \cdot 3! \cdot \int_0^\infty \overline{\mathfrak{F}}'_\alpha\left(1+\frac{y}{2}\right)\left(\log y + \frac{\Gamma'(3)}{\Gamma(3)} + \gamma\right) dy \\
&= 2^7 \cdot \frac{3!}{\pi^4} \cdot \left(\frac{\Gamma'(3)}{\Gamma(3)} + \gamma\right) \cdot \mathfrak{F}_\alpha(0) - 2^6 \cdot \frac{3!}{\pi^4} \cdot \int_0^\infty \overline{\mathfrak{F}}'_\alpha\left(1+\frac{y}{2}\right) \log y \, dy.
\end{aligned}$$

Using the sublemma and combining terms then leads at once to the claimed formula for $T'_\lambda(\alpha:0)$ in the case when $m(\lambda) = 8$, $m(2\lambda) = 7$.

The computation of the Fourier transforms of $T_\lambda(\alpha:0)$ and $T'_\lambda(\alpha:0)$ has been made under the assumption that α is a K-bi-invariant function in $C_c^\infty(G)$. But, on the basis of what has been said in Section 8, it is clear that the computations are actually valid for any K-bi-invariant function α in $\mathscr{C}_\epsilon^1(G)$. We have therefore proved the following theorem.

THEOREM 9.3. *Let α be a K-bi-invariant function in $\mathscr{C}_\epsilon^1(G)$—then*

$$T_\lambda(\alpha:0) = \frac{2^{m(2\lambda)}}{A(\mathfrak{n}_\lambda)} \cdot |\lambda| \cdot (\sqrt{\Lambda})^{m(\lambda)+m(2\lambda)} \cdot \frac{1}{\pi} \cdot \int_{-\infty}^{\infty} \hat{\alpha}(v)\, dv,$$

and

$$T'_\lambda(\alpha:0) = \frac{2^{m(2\lambda)}}{A(\mathfrak{n}_\lambda)} \cdot |\lambda| \cdot (\sqrt{\Lambda})^{m(\lambda)+m(2\lambda)} \cdot \left[\frac{1}{\pi}\left(\frac{\Gamma'(m(\lambda)/2)}{\Gamma(m(\lambda)/2)} - \gamma\right)\right.$$
$$\left. \times \int_{-\infty}^{\infty} \hat{\alpha}(v)\, dv + \hat{\alpha}(0) - \frac{2}{\pi} \cdot \int_{-\infty}^{\infty} \hat{\alpha}(v) \cdot \frac{\Gamma'(1+\sqrt{-1}\,v)}{\Gamma(1+\sqrt{-1}\,v)} dv\right].$$

[In the abelian case, it is understood that $m(2\lambda) = 0$.]

It remains to compute the Fourier transform of $T_{2\lambda}(\alpha:0)$. For this, we shall need some preparation.

Given an even function $f \in C_c^\infty(\mathbf{R})$, write, as above, $\overline{f} = f \circ \cosh^{-1}$. Let k and l be nonnegative integers subject to the restrictions $k \geqslant 1$, $0 \leqslant l \leqslant k$. Put

$$I_{k,l}(f) = \int_1^\infty \overline{f}^{(2k+1)}(y)(y^2-1)^l\, dy, \qquad J_{k,l}(f) = \int_1^\infty \overline{f}^{(2k)}(y)(y^2-1)^{l-1/2}\, dy.$$

Straightforward integration by parts then leads to the recurrence formulas

$$I_{k,l}(f) = 2l \cdot (2l-1) \cdot I_{k-1,\,l-1}(f) + 2^2 \cdot l \cdot (l-1) \cdot I_{k-1,\,l-2}(f),$$

$$J_{k,l}(f) = (2l-1) \cdot (2l-2) \cdot J_{k-1,\,l-1}(f) + (2l-1) \cdot (2l-3) \cdot J_{k-1,\,l-2}(f),$$

where we suppose, to ensure validity of the calculations, that $l \geqslant 2$. Therefore, as was pointed out to us by D. Ragozin, $I_{k,l}(f)$ and $J_{k,l}(f)$ can be computed inductively according to the scheme

$$I_{k,l}(f) = \sum_{i=k-l+1}^{k-[(l+1)/2]} A_i(l) I_{i,0}(f) + \sum_{i=k-l+1}^{k-[l/2]} B_i(l) I_{i,1}(f),$$

$$J_{k,l}(f) = \sum_{i=k-l+1}^{k-[(l+1)/2]} C_i(l) J_{i,0}(f) + \sum_{i=k-l+1}^{k-[l/2]} D_i(l) J_{i,1}(f),$$

the $A_i(l)$, $B_i(l)$, $C_i(l)$, and $D_i(l)$ being certain positive integers, depending on i and l but not on k, which can be determined recursively but whose explicit value we shall not insist upon. Our objective now will be to express $I_{k,l}(f)$ and $J_{k,l}(f)$ in terms of $\hat{f}$, the Fourier transform of f. For this purpose, we need only focus attention on $I_{k,0}(f)$, $I_{k,1}(f)$, $J_{k,0}(f)$, and $J_{k,1}(f)$.

Consider first $I_{k,0}(f)$ and $I_{k,1}(f)$. Obviously

$$I_{k,0}(f) = -\overline{f}^{(2k)}(1), \qquad I_{k,1}(f) = 2[\overline{f}^{(2k-1)}(1) - \overline{f}^{(2k-2)}(1)].$$

Owing to Fourier transform theory, we have

$$\overline{f}^{(n)}(1) = \frac{(-1)^n}{\pi} \cdot \int_0^\infty \hat{f}(v) \cdot \frac{v^2(v^2+1)\cdots(v^2+(n-1)^2)}{1 \cdot 3 \cdots (2n-1)}\, dv \qquad (n \geqslant 0),$$

the obvious interpretation being given to the polynomial inside the integral sign when $n = 0$. Therefore $I_{k,0}(f)$ and $I_{k,1}(f)$ can indeed be written in terms of $\hat{f}$.

Sublemma. *Let N be an integer $\geqslant 2$. Given a real number v, put*

$$\Upsilon_v^N(t) = \frac{2^{(N-2)/2} \cdot \Gamma(N/2)}{(\sinh t)^{(N-2)/2}} \cdot \mathfrak{P}^{1-N/2}_{-(1/2)+\sqrt{-1}v}(\cosh t) \qquad (t \in \mathbf{R}),$$

$\mathfrak{P}^?_?$ being the Legendre function of the first kind. Υ_v^N is then a zonal spherical function on $\mathbf{SO}(N,1)$ attached to the class one principal series representation associated with v. Consequently, for any $m \geqslant 0$ there exists a constant $C_M > 0$ and an integer $M > 0$ such that

$$\left| \left(\frac{d}{dt}\right)^m \Upsilon_v^N(t) \right| \leqslant C_M \cdot (1+v^2)^M \cdot \frac{(1+t^2)^M}{e^{(N-1)t/2}}$$

for all real v *and all* $t \geqslant 0$. *Put*

$$D^N = \left(\frac{1}{\sinh t}\frac{d}{dt}\right)^N.$$

Then

$$D^N \cos(vt) = (-1)^N \frac{v^2(v^2+1)\cdots(v^2+(N-1)^2)}{1\cdot 3\cdots(2N-1)} \cdot \Upsilon_v^{2N+1}(t) \qquad (t \in \mathbf{R}).$$

Proof. The interpretation of Υ_v^N as a zonal spherical function on $\mathbf{SO}(N, 1)$ is spelled out in Takahashi [29b, p. 325]. The stated estimate on $|(d/dt)^m \Upsilon_v^N(t)|$ then follows from a well-known generality (cf. Warner [34b, p. 361]). The final assertion is easily established by induction (cf. Takahashi [29b, p. 326]). ∎

Let us proceed to $J_{k,0}(f)$:

$$J_{k,0}(f) = \int_1^\infty \frac{\mathcal{F}^{(2k)}(y)}{(y^2-1)^{1/2}}\, dy.$$

In this integral, make the change of variable $y = \cosh t$ to get

$$J_{k,0}(f) = \int_0^\infty (D^{2k}f)(t)\, dt.$$

Write

$$(D^{2k}f)(t) = \frac{1}{\pi}\cdot \int_0^\infty \hat{f}(v)\cdot D^{2k}\cos(vt)\, dv.$$

Then, thanks to the facts stated in the sublemma, it is clear that

$$\int_0^\infty \int_0^\infty |\hat{f}(v)| \cdot |D^{2k}\cos(vt)|\, dv\, dt < +\infty,$$

so

$$J_{k,0}(f) = \int_0^\infty \hat{f}(v) G_k(v)\, dv,$$

where

$$G_k(v) = \frac{1}{\pi}\cdot \int_0^\infty D^{2k}\cos(vt)\, dt,$$

a function of polynomial growth in v.

SUBLEMMA. *Let* f *be an even rapidly decreasing function on the line—then*

$$\int_{-\infty}^\infty f'(t)\cdot \frac{\cosh t}{\sinh t}\, dt = -\frac{1}{2}\cdot \int_{-\infty}^\infty \hat{f}(v)\cdot \frac{v\sinh(\pi v)}{\cosh(\pi v)-1}\, dv.$$

[This result has been stated and proved by Harish-Chandra in his lectures at the Institute for Advanced Study; it is a simple consequence of the integral formula

$$\int_{-\infty}^{\infty} \frac{e^{\mu t} + e^{-\mu t}}{e^t - e^{-t}} \cdot \sin(vt)\,dt = \frac{\pi \sinh(\pi v)}{\cos(\pi\mu) + \cosh(\pi v)} \qquad (\mu \in \mathbf{R}, |\mu| < 1),$$

valid for all real v.]

It remains to consider $J_{k,1}(f)$:

$$J_{k,1}(f) = \int_1^\infty \overline{f}^{(2k)}(y)(y^2 - 1)^{1/2}\,dy$$

or still

$$J_{k,1}(f) = -\int_1^\infty \overline{f}^{(2k-1)}(y) \cdot \frac{y}{(y^2 - 1)^{1/2}}\,dy.$$

In the latter integral, make the change of variable $y = \cosh t$ to get

$$J_{k,1}(f) = -\frac{1}{2} \cdot \int_{-\infty}^{\infty} \frac{d}{dt}(D^{2k-2}f)(t) \cdot \frac{\cosh t}{\sinh t}\,dt.$$

Since $D^{2k-2}f$ is an even rapidly decreasing function on the line, we can apply the sublemma and write

$$J_{k,1}(f) = \frac{1}{4} \cdot \int_{-\infty}^{\infty} \widehat{D^{2k-2}f}(v) \cdot \frac{v \sinh(\pi v)}{\cosh(\pi v) - 1}\,dv.$$

In order to complete the evaluation, therefore, we have only to compute $\widehat{D^{2k-2}f}$ in terms of $\hat{f}$; for this, it can be assumed that $k \geqslant 2$ ($k = 1$ gives $\hat{f}$ itself). Obviously

$$\begin{aligned}\widehat{D^{2k-2}f}(v) &= 2 \cdot \int_0^\infty (D^{2k-2}f)(t)\cos(vt)\,dt \\ &= \frac{2}{\pi} \cdot \int_0^\infty \left(\int_0^\infty \hat{f}(\mu) \cdot D^{2k-2}\cos(\mu t)\,d\mu\right)\cos(vt)\,dt\end{aligned}$$

which we claim is the same as

$$\int_0^\infty \hat{f}(\mu)F_k(\mu : v)\,d\mu,$$

where

$$F_k(\mu : v) = \frac{2}{\pi} \cdot \int_0^\infty D^{2k-2}\cos(\mu t) \cdot \cos(vt)\,dt.$$

To justify the interchange of the order of integration, it suffices to establish the finiteness of

$$\int_0^\infty \int_0^\infty |\hat{f}(\mu)| \cdot |D^{2k-2}\cos(\mu t)|\, d\mu\, dt.$$

However, for reasons mentioned earlier, this is clear. To finish up, we need to make a simple observation about the growth of $F_k(\mu : \nu)$. Suppose that $m \geqslant 0$—then we claim that there exists a constant $C_M > 0$ and an integer $M > 0$ such that

$$|F_k(\mu : \nu)| \leqslant C_M \cdot \frac{(1 + \mu^2)^M}{(1 + \nu^2)^m}.$$

To see this, write

$$F_k(\mu : \nu) = \frac{2}{\pi} \cdot \frac{\mu^2 \cdot (\mu^2 + 1) \cdots (\mu^2 + (2k-3)^2)}{1 \cdot 3 \cdots (4k-5)} \cdot \int_0^\infty \Upsilon_\mu^{4k-3}(t) \cos(\nu t)\, dt.$$

Because $1 - (d/dt)^2$ corresponds under Fourier transformation to multiplication by $(1 + \nu^2)$, the desired estimate thus follows upon making the obvious majorizations under the integral sign. It is therefore clear that

$$\int_{-\infty}^\infty \int_0^\infty |\hat{f}(\mu)| \cdot |F_k(\mu : \nu)| \cdot \left| \frac{\nu \sinh(\pi\nu)}{\cosh(\pi\nu) - 1} \right| d\mu\, d\nu < +\infty.$$

So, if we put

$$H_k(\nu) = \frac{1}{4} \cdot \int_{-\infty}^\infty F_k(\nu : \mu) \cdot \frac{\mu \sinh(\pi\mu)}{\cosh(\pi\mu) - 1}\, d\mu,$$

then we have

$$J_{k,1}(f) = \int_0^\infty \hat{f}(\nu) H_k(\nu)\, d\nu,$$

where the growth of H_k is known.

The above theory can now be applied to evaluate the Fourier transform of $T_{2\lambda}(\alpha : 0)$. We recall that, by definition,

$$T_{2\lambda}(\alpha : 0) = \int_0^\infty \overline{\alpha}((1 + y^2)^{1/2}) y^{(m(\lambda)/2 + m(2\lambda))} \frac{dy}{y}.$$

As in the discussion of $T_\lambda(\alpha:0)$ and $T'_\lambda(\alpha:0)$, it will be convenient to proceed on a case-by-case basis. Suppose to begin with that $m(\lambda) = 2(m-1)$ $(m \geqslant 2)$, $m(2\lambda) = 1$—then, thanks to Lemma 9.2 ((ii)–(a)), we have

$$T_{2\lambda}(\alpha:0) = 2\cdot\left(\frac{-1}{2\pi}\right)^m \cdot \int_0^\infty \left(\int_{(1+y^2)^{1/2}}^\infty \overline{\mathfrak{F}}_\alpha^{(m)}(x)\cdot[x^2-1-y^2]^{-1/2}\,dx\right) y^{m-1}\,dy$$

$$= 2\cdot\left(\frac{-1}{2\pi}\right)^m \cdot \int_1^\infty \overline{\mathfrak{F}}_\alpha^{(m)}(x)\,dx \int_0^{(x^2-1)^{1/2}} \frac{y^{m-1}}{(x^2-1-y^2)^{1/2}}\,dy$$

$$= \left(\frac{-1}{2\pi}\right)^m \cdot B\left(\frac{m}{2},\frac{1}{2}\right)\cdot \int_1^\infty \overline{\mathfrak{F}}_\alpha^{(m)}(y)(y^2-1)^{(m-1)/2}\,dy$$

$$= \begin{cases} \left(\dfrac{-1}{2\pi}\right)^{2k+1}\cdot B\left(k+\dfrac{1}{2},\dfrac{1}{2}\right)\cdot I_{k,k}(\mathfrak{F}_\alpha) & \text{if} \quad m = 2k+1 \\ \left(\dfrac{1}{2\pi}\right)^{2k}\cdot B\left(k,\dfrac{1}{2}\right)\cdot J_{k,k}(\mathfrak{F}_\alpha) & \text{if} \quad m = 2k. \end{cases} \qquad (k \geqslant 1)$$

Let us proceed to the next possibility, namely: $m(\lambda) = 4(m-1)$ $(m \geqslant 2)$, $m(2\lambda) = 3$. Owing to Lemma 9.2 ((ii)–(b)), we have

$$T_{2\lambda}(\alpha:0) = 2\cdot\left(\frac{1}{2\pi}\right)^{2m}$$
$$\times \int_0^\infty \left(\int_{(1+y^2)^{1/2}}^\infty \left\{\frac{\overline{\mathfrak{F}}_\alpha^{(2m)}(x)}{x} - \frac{\overline{\mathfrak{F}}_\alpha^{(2m-1)}(x)}{x^2}\right\}\right.$$
$$\left.\times [x^2-1-y^2]^{-1/2}\,dx\right) y^{2m}\,dy$$

$$= 2\cdot\left(\frac{1}{2\pi}\right)^{2m}$$
$$\times \int_1^\infty \left\{\frac{\overline{\mathfrak{F}}_\alpha^{(2m)}(x)}{x} - \frac{\overline{\mathfrak{F}}_\alpha^{(2m-1)}(x)}{x^2}\right\} dx \int_0^{(x^2-1)^{1/2}} \frac{y^{2m}}{(x^2-1-y^2)^{1/2}}\,dy$$

$$= \left(\frac{1}{2\pi}\right)^{2m}\cdot B\left(m+\frac{1}{2},\frac{1}{2}\right)\cdot\int_1^\infty \frac{d}{dy}\left(\frac{\overline{\mathfrak{F}}_\alpha^{(2m-1)}(y)}{y}\right)(y^2-1)^m\,dy$$

$$= (-2m)\cdot\left(\frac{1}{2\pi}\right)^{2m}\cdot B\left(m+\frac{1}{2},\frac{1}{2}\right)\cdot I_{m-1,m-1}(\mathfrak{F}_\alpha).$$

Finally, we have to consider the case when $m(\lambda) = 8$, $m(2\lambda) = 7$. In view of Lemma 9.2 ((ii)–(c)),

$$T_{2\lambda}(\alpha:0) = 2 \cdot \left(\frac{1}{2\pi}\right)^8 \cdot \int_0^\infty \left(\int_{(1+y^2)^{1/2}}^\infty \left\{-\frac{15}{x^6} \cdot \mathfrak{F}_\alpha^{(5)}(x) + \frac{15}{x^5} \cdot \mathfrak{F}_\alpha^{(6)}(x) - \frac{6}{x^4} \cdot \mathfrak{F}_\alpha^{(7)}(x) + \frac{1}{x^3} \cdot \mathfrak{F}_\alpha^{(8)}(x)\right\} [x^2 - 1 - y^2]^{-1/2}\, dx\right) y^{10}\, dy$$

$$= 2 \cdot \left(\frac{1}{2\pi}\right)^8 \cdot \int_1^\infty \left\{-\frac{15}{x^6} \cdot \mathfrak{F}_\alpha^{(5)}(x) + \frac{15}{x^5} \cdot \mathfrak{F}_\alpha^{(6)}(x) - \frac{6}{x^4} \cdot \mathfrak{F}_\alpha^{(7)}(x) + \frac{1}{x^3} \cdot \mathfrak{F}_\alpha^{(8)}(x)\right\} dx \int_0^{(x^2-1)^{1/2}} \frac{y^{10}}{(x^2 - 1 - y^2)^{1/2}}\, dy$$

$$= \left(\frac{1}{2\pi}\right)^8 \cdot B\left(\frac{11}{2}, \frac{1}{2}\right) \times \int_1^\infty \left\{-\frac{15}{y^6} \cdot \mathfrak{F}_\alpha^{(5)}(y) + \frac{15}{y^5} \cdot \mathfrak{F}_\alpha^{(6)}(y) - \frac{6}{y^4} \cdot \mathfrak{F}_\alpha^{(7)}(y) + \frac{1}{y^3} \cdot \mathfrak{F}_\alpha^{(8)}(y)\right\} (y^2 - 1)^5\, dy$$

$$= -2^3 \cdot 60 \cdot \frac{1}{(2\pi)^8} \cdot B\left(\frac{11}{2}, \frac{1}{2}\right) \cdot I_{2,2}(\mathfrak{F}_\alpha).$$

The preceding computations are evidently valid for any K-bi-invariant function α in $\mathscr{C}_\epsilon^1(G)$. This said, the following theorem then serves to assemble and unify our results.

THEOREM 9.4. *Let α be a K-bi-invariant function in $\mathscr{C}_\epsilon^1(G)$.*
(i) *Suppose that $m(\lambda) \equiv 0 \pmod 4$, say $m(\lambda) = 4k_\lambda$. Let*

$$\mathbf{C}_+(2\lambda) = -(2)^{(m(2\lambda)-1)/2} \cdot \prod_{i=1}^{(m(2\lambda)-1)/2} \left(\frac{m(\lambda)}{4} + i\right) \times \frac{1}{(2\pi)^{(m(\lambda)+m(2\lambda)+1)/2}} \cdot B\left(\frac{m(\lambda)}{4} + \frac{m(2\lambda)}{2}, \frac{1}{2}\right).$$

Then

$$T_{2\lambda}(\alpha:0) = \mathbf{C}_+(2\lambda) \cdot I_{k_\lambda,k_\lambda}(\mathfrak{F}_\alpha).$$

(ii) *Suppose that $m(\lambda) \equiv 2 \pmod 4$, say $m(\lambda) = 4k_\lambda - 2$. Let*

$$\mathbf{C}_-(2\lambda) = \frac{1}{(2\pi)^{(m(\lambda)+m(2\lambda)+1)/2}} \cdot B\left(\frac{m(\lambda)}{4} + \frac{m(2\lambda)}{2}, \frac{1}{2}\right).$$

Then

$$T_{2\lambda}(\alpha:0) = \mathbf{C}_{-}(2\lambda) \cdot J_{k_\lambda,k_\lambda}(\mathfrak{F}_\alpha).$$

In addition, both $I_{k_\lambda,k_\lambda}(\mathfrak{F}_\alpha)$ *and* $J_{k_\lambda,k_\lambda}(\mathfrak{F}_\alpha)$ *are directly expressible in terms of* $\hat{\alpha}$.

Taking into account what has been said in Sections 7 and 8 along with Theorems 9.3 and 9.4, it then follows that in the K-bi-invariant case the Selberg trace formula admits a third stage formulation (cf. the Introduction), thus is in about as explicit a form as can be hoped for. There is nothing to be gained by restating it. It is, however, worth reconsidering the conditions for its validity. Owing to a theorem of Trombi and Varadarajan [30a, p. 298], the condition imposed on the K-bi-invariant function α (viz., that $\alpha \in \mathscr{C}_\epsilon^1(G)$ for some $\epsilon > 0$) can alternatively be expressed directly in terms of the Fourier transform $\hat{\alpha}$ of α: $\hat{\alpha}$ is an even function admiting a holomorphic extension to the strip

$$\mathscr{S}_\epsilon = \{\nu : |\mathrm{Im}(\nu)| < (\epsilon + 1)(m(\lambda)/2 + m(2\lambda))\}$$

with continuous boundary values having the property that for every holomorphic differential operator D with polynomial coefficients

$$\sup_{\nu \in \mathscr{S}_\varepsilon} |\hat{\alpha}(\nu; D)| < +\infty.$$

EXAMPLE. Given a complex number s, let

$$\alpha_s(x) = [\cosh(\sigma(x))]^{-s} \qquad (x \in G).$$

Then α_s is a K-bi-invariant function in G which is, moreover, in $\mathscr{C}_\epsilon^1(G)$ (for some $\epsilon > 0$) provided $\mathrm{Re}(s) > m(\lambda) + 2m(2\lambda)$. Inserting α_s into the Selberg trace formula then leads to an identity which we shall term, by analogy with the compact case, the "*remarkable formula*" *of Huber* (cf. Gangolli [9b]). The details will be left to the reader.

10. Open Problems

We shall conclude with a discussion of open problems, confining our remarks only to the present case of split-rank 1 groups.

1. Compute the Fourier transform, in the sense of Harish-Chandra, of the distribution

$$\lim_{s \to 0} \frac{d}{ds} (s \text{Ч}_\alpha(s)).$$

The solution of this problem is of crucial importance for the applications, e.g., the computation of the multiplicities of the integrable discrete series in $L^2_{\tilde{d}}(G/\Gamma)$.

2. Suppose that G has a discrete series. Is it true that if a representation in the integrable discrete series occurs in $L^2_{\tilde{d}}(G/\Gamma)$, then this representation must actually occur in $L^2_{\tilde{0}}(G/\Gamma)$, the space of cusp forms? There is some evidence to indicate that the answer may be affirmative.

3. Assign a geometric or cohomological interpretation to the multiplicities of the representations which figure in the decomposition of $L^2_{\tilde{d}}(G/\Gamma)$. We have in mind here, e.g., generalizations of the work of Griffiths and Schmid.

4. Investigate the poles of $\mathbf{c}_{\vartheta,\delta}$ in the segment $[-|\rho|, 0[$. Relate these poles to the representations which occur in ${}_0L^2(G/\Gamma)$, the space of residues. Prove or disprove that the functions in $\mathscr{C}^1(G)$ are represented by trace class operators in ${}_0L^2(G/\Gamma)$.

5. Discuss the growth of the function

$$s \mapsto \mathbf{c}_{\vartheta,\delta}(-s) \cdot \left(\frac{d}{ds}\, \mathbf{c}_{\vartheta,\delta}(s)\right)$$

along the imaginary axis. One conjectures that it must be slowly increasing. This fact would have certain important consequences.

6. Develop analogues of the asymptotic formulas for the discrete spectrum along the lines of Gangolli [9a] and Wallach [33a]. For $G = \mathbf{SL}(2, \mathbf{R})$, $\Gamma = \mathbf{SL}(2, \mathbf{Z})$ such a result was obtained by Tanaka (cf. Lax and Phillips [20]); Tanaka's result can be shown to hold true when the modular group is replaced by a congruence subgroup (unpublished remark of the author).

7. Attach to the principal series representations with fixed compact parameter a zeta function in the sense of Selberg whose "nontrivial" zeros parameterize precisely those principal series representations which occur in $L^2_{\tilde{d}}(G/\Gamma)$, the multiplicity of the representation being, up to an absolute constant, the multiplicity of the associated zero of the zeta function. This has recently been done by R. Gangolli and the author in the class one case (cf. Gangolli [9a] for the class one case when Γ is uniform in G). If $G = \mathbf{SL}(2, \mathbf{R})$ and Γ is uniform in G, then A. Sitaram (University of Washington thesis) has carried out this suggestion in the nonclass one case (the class one case in this situation being, of course, the setting for Selberg's original formulation).

8. Define the notion of "Hecke operator" and carry out the computation of its trace (cf. Selberg [27a], [27c]). Special cases have been considered by Shimizu [28b] and Ishikawa [15].

9. Let χ be a finite-dimensional unitary representation of Γ, $L^{\chi}_{G/\Gamma}$ the associated unitarily induced representation of G with representation space $L^2(G/\Gamma, \chi)$. Formulate a theory for $L^2(G/\Gamma, \chi)$ analogous to that for $L^2(G/\Gamma)$ (corresponding to the case when $\chi = 1$). Before proceeding to the Selberg trace formula proper, one will have to rewrite parts of the theory of Eisenstein series. A brief discussion when $G = \mathbf{SL}(2, \mathbf{R})$ can be found in Selberg [27a].

10. Remove the restriction on Γ imposed in Section 6. In this direction, one has the examples considered by Cohn [5] and Venkov [32].

Added in Proof. Due to the length of time between the completion and actual appearance of this work, a few comments on the status of some of the preceding problems is in order. First of all, M. Scott Osborne and the author have recently made considerable progress on 1 and 2 above. In fact, 1 is more or less solved now (some as yet unpublished work of Arthur enters in here), while 2 is completely solved. For details, we refer to our forthcoming article in the *J. Functional Analysis* where, incidentally, explicit formulas for the multiplicities of the integrable discrete series in $L^2_{\mathrm{d}}(G/\Gamma)$ are given. Concerning 7, besides the work of R. Gangolli and the author already mentioned, as well as that of A. Sitaram, a study of the case $G = \mathbf{SL}(2, \mathbf{C})$ and Γ uniform in G has just been completed by D. Scott (Ph.D. Thesis, Univ. of Washington, Seattle, Washington). Scott treats the nonclass one case; it appears likely that his methods are capable of considerable generalization. Finally, P. Moore is finishing a Ph.D. Thesis at the University of Washington dealing with certain aspects of 9.

Apart from all this, we also want to mention that M. Scott Osborne and the author have prepared a lengthy article entitled: Toward the trace formula in the sense of Selberg. This work considers in complete generality the problem of the spectral decomposition of $L^2(G/\Gamma)$ (so G is now an "arbitrary" reductive group) and includes a detailed discussion of Langlands' infamous "Chapter 7."

Acknowledgments

I am indebted to G. B. Folland, R. Gangolli, M. S. Osborne, D. Ragozin, and P. C. Trombi for numerous stimulating conversations and helpful suggestions. In addition, it is my pleasant duty to point out that the work in Sections 7 and 9 is not that of the author alone but rather represents a joint effort with R. Gangolli. Lastly, I want to thank H. B. Warner for typing the manuscript with great competence and dispatch.

References

1a. J. G. Arthur, The Selberg trace formula for groups of F-rank one, *Ann. of Math.* **100** (1974), 326–385.
1b. J. G. Arthur, Some tempered distributions on semi-simple groups of real rank one, *Ann. of Math.* **100** (1974), 553–584.

2. W. L. Baily and A. Borel, Compactifications of arithmetic quotients of bounded symmetric domains, *Ann. of Math.* **84** (1966), 442–528.
3a. A. Borel, Arithmetic properties of linear algebraic groups, *Proc. Int. Congr. Math., Stockholm*, 10–22, 1962.
3b. A. Borel, Introduction aux groupes arithmétiques, "Publications de l'Institut Mathematique de l'Université de Strasbourg," Vol. XV, Hermann, Paris, 1969.
4. A. Borel and Harish-Chandra, Arithmetic subgroups of algebraic groups, *Ann. of Math.* **75** (1962), 485–535.
5. L. Cohn, The dimension of spaces of automorphic forms on a certain two dimensional complex domain, *Memoirs Amer. Math. Soc.*, No. 158, Providence, Rhode Island, 1975.
6. M. Duflo and J. P. Labesse, Sur la formule des traces de Selberg, *Ann. Sci. École Norm. Sup.* **4** (1971), 193–284.
7. T. Eaton, Thesis, Univ. of Washington, 1972.
8. M. Eguchi, The Fourier transform of the Schwartz space on a semi-simple Lie group, *Hiroshima Math. J.* **4** (1974), 133–209.
9a. R. Gangolli, Asymptotic behavior of spectra of compact quotients of certain symmetric spaces, *Acta Math.* **121** (1968), 151–192.
9b. R. Gangolli, On the length spectrum of some compact manifolds of negative curvature, *J. Differential Geometry* (to appear).
9c. R. Gangolli, Zeta functions of Selberg's type for compact space forms of symmetric spaces of rank one, *Illinois J. Math.* **21** (1977), 1–42.
10. R. Gangolli and G. Warner, On Selberg's trace formula, *J. Math. Soc. Japan* **27** (1975), 328–343.
11. H. Garland and M. S. Raghunathan, Fundamental domains for lattices in (**R**)-rank 1 semi-simple Lie groups, *Ann. of Math.* **92** (1970), 279–326.
12. S. Gelbart, Automorphic forms on adele groups, "Ann. of Math. Studies," Vol. 83, Princeton Univ. Press, Princeton, New Jersey, 1975.
13a. Harish-Chandra, Automorphic forms on semi-simple Lie groups, "Lecture Notes in Mathematics," Vol. 62, Springer-Verlag, Berlin and New York, 1968.
13b. Harish-Chandra, Harmonic analysis on semi-simple Lie groups, *Bull. Amer. Math. Soc.* **76** (1970), 529–551.
13c. Harish-Chandra, On the theory of the Eisenstein integral, "Lecture Notes in Mathematics," Vol. 266, pp. 123–149, Springer-Verlag, Berlin and New York, 1972.
14. S. Helgason, A duality for symmetric spaces with applications to group representations, *Adv. in Math.* **5** (1970), 1–154.
15. H. Ishikawa, On the trace formula for Hecke operators, *J. Fac. Sci. Univ. of Tokyo* **20** (1973), 217–238.
16. H. Jacquet and R. P. langlands, Automorphic forms on GL(2), "Lecture Notes in Mathematics," Vol. 114, Springer-Verlag, Berlin and New York, 1970.
17. T. Kubota, "Elementary theory of Eisenstein Series," Wiley, New York, 1973.
18. S. Lang, SL(2), Addison-Wesley, Reading, Massachusetts, 1975.
19a. R. P. Langlands, On the functional equations satisfied by Eisenstein series, "Lecture Notes in Mathematics," Vol. 544, Springer-Verlag, Berlin and New York, 1976.
19b. R. P. Langlands, Eisenstein series, *Proc. Symp. Pure Math.* **9**, 235–252. Amer. Math. Soc., Providence, Rhode Island, 1966.
19c. R. P. Langlands, Dimension of spaces of automorphic forms, *Proc. Symp. Pure Math.* **9**, 253–257. Amer. Math. Soc., Providence, Rhode Island, 1966.
20. P. Lax and R. S. Phillips, Scattering theory for automorphic functions, "Ann. of Math. Studies," Vol. 87, Princeton Univ. Press, Princeton, New Jersey, 1976.

21. C. C. MOORE, Decomposition of unitary representations defined by discrete subgroups of nilpotent groups, *Ann. of Math.* **82** (1965), 146–182.
22. Y. Morita, An explicit formula for the dimension of spaces of Siegel modular forms of degree two, *J. Fac. Sci. Univ. Tokyo*, **21** (1974), 167–248.
23a. M. S. RAGHUNATHAN, Cohomology of arithmetic subgroups of algebraic groups: II, *Ann. of Math.* **87** (1968), 279–304.
23b. M. S. RAGHUNATHAN, "Discrete Subgroups of Lie Groups," Springer-Verlag, Berlin and New York, 1972.
24. R. Rao, Orbital integrals in reductive groups, *Ann. of Math.* **96** (1972), 505–510.
25. J. RINGROSE, Compact non-self-adjoint operators, "Mathematical Studies," Vol. 35, Van Nostrand-Reinhold, Princeton, New Jersey, 1971.
26. P. SALLY AND G. WARNER, The Fourier transform on semi-simple Lie groups of real rank one, *Acta Math.* **131** (1973), 1–26.
27a. A. SELBERG, Harmonic analysis and discontinuous groups in weakly symmetric Riemannian spaces with applications to Dirichlet series, *J. Indian Math. Soc.* **20** (1956), 47–87.
27b. A. SELBERG, Automorphic functions and integral operators, *Seminars Anal. Functions* **2** (1957), 152–161.
27c. A. SELBERG, Discontinuous groups and harmonic analysis, *Proc. Internat. Congr. Math.*, pp. 177–189, Stockholm, 1962.
28a. H. SHIMIZU, On discontinuous groups operating on the product of upper half planes, *Ann. of Math.* **77** (1963), 33–71.
28b. H. SHIMIZU, On traces of Hecke operators, *J. Fac. Sci. Univ. of Tokyo* **10** (1963), 1–19.
29a. R. TAKAHASHI, Sur les fonctions sphériques et la formule de Plancherel dans le groupe hyperbolique, *Japan J. Math.* **31** (1961), 55–90.
29b. R. TAKAHASHI, Sur les représentations unitaires des groupes de Lorentz généralisés, *Bull. Soc. Math. France* **91** (1963), 289–433.
30a. P. TROMBI AND V. S. VARADARAJAN, Spherical transforms on semi-simple Lie groups, *Ann. of Math.* **94** (1971), 246–303.
30b. P. TROMBI AND V. S. VARADARAJAN, Asymptotic behaviour of eigenfunctions on a semi-simple Lie group, *Acta Math.* **129** (1972), 237–280.
31. V. S. VARADARAJAN, The theory of characters and the discrete series for semi-simple Lie groups, *Proc. Symp. Pure Math.* **26**, 45–99. Amer. Math. Soc., Providence, Rhode Island, 1973.
32. A. B. Venkov, Expansions in automorphic eigenfunctions of the Laplace-Beltrami operator in classical symmetric spaces of rank 1 and the Selberg trace formula, *Proc. Steklov Inst. Math.* **125** (1973), 6–55.
33a. N. WALLACH, On an asymptotic formula of Gelfand and Gangolli for the spectrum of Γ/G, *J. Differential Geometry* (to appear).
33b. N. WALLACH, On the Selberg trace formula in the case of compact quotient, *Bull. Amer. Math. Soc.* **82** (1976), 171–195.
34a. G. WARNER, "Harmonic Analysis on Semi-simple Lie Groups," Vol. I, Springer-Verlag, Berlin and New York, 1972.
34b. G. WARNER, "Harmonic Analysis on Semi-simple Lie Groups," Vol. II, Springer-Verlag, Berlin and New York, 1972.

STUDIES IN ALGEBRA AND NUMBER THEORY
ADVANCES IN MATHEMATICS SUPPLEMENTARY STUDIES, VOL. 6

On the Unitary Representation of a Semisimple Lie Group Given by the Invariant Integral on Its Lie Algebra

RONALD L. LIPSMAN

Department of Mathematics, University of Maryland, College Park, Maryland

1. INTRODUCTION

Let G be a connected noncompact semisimple Lie group and $\mathfrak{g}$ its Lie algebra. It is not hard to see that Lebesgue measure on $\mathfrak{g}$ is invariant under the adjoint action of G. Therefore there is a natural unitary representation T_G of G on $L_2(\mathfrak{g})$ given by

$$T_G(g)f(x) = f(\operatorname{Ad} g^{-1}(x)), \qquad g \in G, f \in L_2(\mathfrak{g}), x \in \mathfrak{g}.$$

The problem I am concerned with in this paper is the decomposition of T_G into its irreducible constituents.

I first learned of this problem in [17, Vol. II, p. 54]. Warner later informed me that it had arisen (in the special case $G = SL(2, \mathbb{R})$) in the work of Rallis and Schiffman on theta functions. Since then it has come up in Gelbart's work on automorphic forms on the metaplectic group [4], and in Ehrenpreis's application of partial differential equations to group representations [3].

If Z_G is the center of G, it is obvious that $T_{G|Z_G} \equiv 1$—that is, T_G is really a representation of $G/Z_G = \operatorname{Ad} \mathfrak{g}$. It is therefore appropriate to assume (as we shall do throughout this paper) that $Z_G = \{e\}$. I will give here a complete description of the continuous spectrum of T_G. It turns out to agree with the continuous spectrum of the (left) regular representation λ_G of G both in support and in multiplicity. In fact I conjecture that T_G is unitarily equivalent to λ_G, and I plan to consider the discrete spectrum in a future paper (see [19, 20]).

This paper is organized as follows. In Section 2 we study an abstract group action on a Borel space that leaves a measure quasi-invariant. We associate to that a natural unitary representation. Then we show that the technique of disintegration of measures reduces the study of that representation to examining certain induced representations. Applying that method to

ISBN 0-12-599153-3

the representation T_G, we see that our problem comes down to decomposing the representations $\mathrm{Ind}_H^G 1$, as H varies over a system of nonconjugate Cartan subgroups of G. Each noncompact Cartan subgroup H can be written $H = BA$ where B is compact and A is a vector group. In Section 3 we prove that the representations $\mathrm{Ind}_{BA}^G \sigma \times \tau$, $\sigma \in \hat{B}$, $\tau \in \hat{A}$, are (up to unitary equivalence) *independent* of τ. We then deduce that our problem is reduced to decomposing $\mathrm{Ind}_B^G 1$. It follows easily at this stage that T_G is equivalent to a subrepresentation of λ_G and has uniform infinite multiplicity. The independence result is achieved via harmonic analysis on various unipotent abelian subgroups of G. (Not unexpectedly, certain intertwining operators related to the Kunze–Stein intertwining integral operators arise here.) The idea of proving and exploiting such an independence result was pioneered in recent work on tensor products of principal series representations (see [16, 18] and [14, Chap. II, Sect. B]). It can also be shown that such techniques have application to the representation theory of nonsemisimple subgroups of G. In Section 4 we effect a further reduction from B to a still smaller compact abelian subgroup F of G. This reduction is a fairly easy consequence of the weight theory for representations of compact connected Lie groups. Finally, in Section 5 we use the Subgroup Theorem and a reciprocity theorem to show that all the principal series corresponding to the noncompact Cartan subgroups occur in T_G. We summarize our results and point out precisely what remains to be done in Section 6.

These results were announced in a talk to the Special Session on Non-Commutative Harmonic Analysis at the 1974 American Mathematical Society winter meeting (see [13]).

2. An Application of Disintegration of Measures

Let G be a separable locally compact group and suppose X is a right Borel G-space. Suppose also that X carries a quasi-invariant, σ-finite, positive Borel measure μ. Then there is a nonnegative Borel function $\alpha(x, g)$, $x \in X$, $g \in G$, such that

$$\int_X f(x)\, d\mu(x) = \int_X f(x \cdot g)\alpha(x, g)\, d\mu(x), \qquad f \in L_1(X, \mu). \tag{2.1}$$

In addition, there is a natural continuous unitary representation T_μ of G on $L_2(X, \mu)$ given by the formula

$$T_\mu(g)f(x) = f(x \cdot g)\alpha(x, g)^{1/2}, \qquad f \in L_2(X, \mu).$$

In general, one is interested in the decomposition of this representation into irreducible representations.

We now make the observation that in case the orbit space X/G is well behaved, it is possible to reduce this question to one of decomposing certain induced representations. In fact, assume that both X and X/G are standard Borel spaces. Then Theorem 2.1 of [8] can be applied (actually slightly weaker assumptions would do) to disintegrate the measure μ. Specifically, if if we choose a pseudo-image $\bar{\mu}$ of μ on X/G, then there exist quasi-invariant measures μ_x on the orbits $\mathcal{O}_x = x \cdot G$ such that for all $f \in L_1(X, \mu)$ we have

$$\begin{aligned}\int_X f(x)\,d\mu(x) &= \int_{X/G}\int_{\mathcal{O}_x} f(y)\,d\mu_x(y)\,d\bar{\mu}(\bar{x})\\ &= \int_{X/G}\int_{G/G_x} f(x\cdot g)\,d\mu_x(\bar{g})\,d\bar{\mu}(\bar{x});\end{aligned}$$

here G_x is the stability group at $x \in X$, $\bar{g}$ is the image of g in G/G_x, and $\bar{x}$ is the image of x in X/G. Applying this to functions $|f|^2$, $f \in L_2(X, \mu)$, and using Fubini's theorem, we see immediately that the Hilbert space $L_2(X, \mu)$ can be written as a direct integral

$$L_2(X, \mu) = \int_{X/G}^{\oplus} L_2(G/G_x, \mu_x)\,d\bar{\mu}(\bar{x}).$$

Now let us compute a formula for T_μ when transferred over to the space of the direct integral. First, there must be a Borel function β such that

$$\int_{\mathcal{O}_x} f(y)\,d\mu_x(y) = \int_{\mathcal{O}_x} f(y\cdot g)\beta(y,g)\,d\mu_x(y), \qquad f \in L_1(\mathcal{O}_x, \mu_x).$$

But by disintegrating the measure μ—both before and after the action of an element $g \in G$—and using (2.1), we see that

$$\int_{\mathcal{O}_x} f(y)\,d\mu_x(y) = \int_{\mathcal{O}_x} f(y\cdot g)\alpha(y,g)\,d\mu_x(y), \qquad f \in L_1(\mathcal{O}_x, \mu_x).$$

That is, $\alpha|_{\mathcal{O}_x \times G} = \beta$. Then writing an L_2 function f as a direct integral

$$\begin{aligned}&f \to \{f_x\}_{x\in X/G}\\ &f_x(y) = f(x\cdot h) \qquad \text{if} \quad y = x\cdot h,\end{aligned}$$

we have $T_\mu(g)f = f'$ where

$$\begin{aligned}(f')_x(y) &= f'(x\cdot h) \qquad \text{if} \quad y = x\cdot h\\ &= (T_\mu(g)f)(x\cdot h)\\ &= f(x\cdot h\cdot g)\alpha(x\cdot h, g)^{1/2}\\ &= f_x(y\cdot g)\beta(y,g)^{1/2}\end{aligned}$$

But the operators $f_x(y) \to f_x(y\cdot g)\beta(y,g)^{1/2}$, $f_x \in L_2(\mathcal{O}_x, \mu_x)$, are precisely those of the representation of G induced from the trivial representation of G_x.

Therefore, we conclude that

$$T_\mu \cong \int_{X/G}^{\oplus} \operatorname{Ind}_{G_x}^G 1 \, d\bar{\mu}(\bar{x}), \tag{2.2}$$

and the problem of decomposing T_μ comes down to a description of the orbit space X/G, the stability groups G_x, and the induced representations $\operatorname{Ind}_{G_x}^G 1$. (In the preceding, the symbol $\cong$ denotes unitary equivalence; in what follows, $\approx$ shall denote quasi-equivalence.)

Now let G be a connected semisimple Lie group and $\mathfrak{g}$ its Lie algebra. G acts via the adjoint representation as a group of linear automorphisms of $\mathfrak{g}$, and it is well known [7, p. 367] that

$$|\det \operatorname{Ad}(g)| = 1, \qquad g \in G.$$

In particular, G leaves Lebesgue measure on $\mathfrak{g}$ invariant. We define the natural unitary representation

$$T_G(g)f(x) = f(\operatorname{Ad} g^{-1}(x)), \qquad g \in G, f \in L_2(\mathfrak{g}), x \in \mathfrak{g}. \tag{2.3}$$

For the reason indicated in Section 1, we assume henceforth that G is centerless, i.e., $G = \operatorname{Ad} \mathfrak{g}$. Next we claim that $\mathfrak{g}/G$ is countably separated. This can be seen as follows. If $\mathfrak{g}_c$ and $\operatorname{Ad} \mathfrak{g}_c$ are the complexifications, then G is the neutral component of the group of $\mathbb{R}$-rational points G_1 of the algebraic group $\operatorname{Ad} \mathfrak{g}_c$; $G_1 = \{g \in \operatorname{Ad} \mathfrak{g}_c : g \cdot \mathfrak{g} \subseteq \mathfrak{g}\}$. But by the reasoning in the final paragraph of Theorem 3.2 in [15], the space $\mathfrak{g}/G_1$ is countably separated. In particular, the G_1-orbits in $\mathfrak{g}$ are locally closed subsets (see [2]). But G is of finite index in G_1, and so it follows easily that any G-orbit is an open and closed subset of the G_1-orbit containing it. Thus the G-orbits are also locally closed and $\mathfrak{g}/G$ is countably separated.

Next let $\mathfrak{g}'$ be the set of regular elements in $\mathfrak{g}$. $\mathfrak{g}'$ is an open, dense submanifold whose complement is of lower dimension and thus of Lebesgue measure zero. If $\mathfrak{h}_1, \ldots, \mathfrak{h}_s$ denotes a complete system of nonconjugate Cartan subalgebras of $\mathfrak{g}$, then $\mathfrak{g}' = \bigcup_{i=1}^s G \cdot \mathfrak{h}_i'$, $\mathfrak{h}_i' = \mathfrak{h}_i \cap \mathfrak{g}'$. Also if $H_1, \ldots, H_s$ denote the corresponding Cartan subgroups (i.e., H_i is the centralizer of $\mathfrak{h}_i$ in G), then [5, Sect. 20]

$$\frac{\mathfrak{g}'}{G} = \frac{\bigcup_{i=1}^s \mathfrak{h}_i'}{G} = \bigcup_{i=1}^s \left(\frac{\mathfrak{h}_i'}{W_i}\right),$$

where W_i is the finite group $\operatorname{Norm}(H_i)/H_i$. Moreover, the stability group of a point $X \in \mathfrak{h}_i'$ under the adjoint action is the Cartan subgroup H_i itself. In particular, that stability group is independent of the point $X \in \mathfrak{h}_i'$. It is clear from the Harish-Chandra–Weyl integral formula (see [5, Sect. 20 and Lemma 41]) that a cross section for $\mathfrak{h}_i'/W_i$ is a Weyl chamber, and that taking a pseudo-image of Lebesgue measure yields a continuous measure on the chamber. Therefore, applying (2.2), we obtain the following

THEOREM 1. *Let* $\mathfrak{g}$ *be a semisimple Lie algebra,* $G = \operatorname{Ad}\mathfrak{g}$, *and* T_G *the representation of* (2.3). *If* $H_1, \ldots, H_s$ *denotes a complete system of nonconjugate Cartan subgroups of* G, *then*

$$T_G \cong \sum_{i=1}^{s}{}^{\oplus} \infty \operatorname{Ind}_{H_i}^{G} 1.$$

COROLLARY 2. *The representation* T_G *has uniform infinite multiplicity.*

Note. In what follows it will be convenient to write $H_0, H_1, \ldots, H_r$ for a complete system of nonconjugate Cartan subgroups of G—where H_0 denotes a compact Cartan subgroup if one exists, and should be ignored otherwise. Likewise we write

$$T_G \cong \sum_{i=0}^{r}{}^{\oplus} \infty \operatorname{Ind}_{H_i}^{G} 1,$$

where the term corresponding to $i = 0$ is understood to be omitted if $\operatorname{rank}(G) > \operatorname{rank}(K)$.

3. THE INDEPENDENCE THEOREM

Let H be a noncompact Cartan subgroup of $G = \operatorname{Ad}\mathfrak{g}$. We now focus attention on the induced representation $\operatorname{Ind}_H^G 1$. First let K be a maximal compact subgroup of G. Let $\mathfrak{g} = \mathfrak{k} + \mathfrak{p}$ be the corresponding Cartan decomposition and θ the associated Cartan involution. We may assume (by conjugating H if necessary) that H is invariant under θ. Then H is a direct product $H = BA$, where $A = H \cap \exp\mathfrak{p}$ and $B = H \cap K$. We can also find a cuspidal parabolic subgroup P such that $P = MAN$ is its Langlands decomposition and B is a compact Cartan subgroup of M.

The next result is the key step in the computation of the continuous spectrum of T_G.

THEOREM 3. *Let* $\sigma \in \hat{B}$, $\tau \in \hat{A}$, *and write* $\sigma \times \tau$ *for the outer product* $(\sigma \times \tau)(ba) = \sigma(b)\tau(a)$, $b \in B$, $a \in A$. *Then the representation* $\operatorname{Ind}_{BA}^{G}(\sigma \times \tau)$ *is, up to unitary equivalence, independent of* τ.

Proof. Let $\mathfrak{M}$, $\mathfrak{A}$, $\mathfrak{N}$ be the Lie algebras of M, A, N respectively. We decompose $\mathfrak{g}$ according to the (restricted) root spaces of $\mathfrak{A}$. That is, for

$\alpha \in \mathfrak{A}^* = \mathrm{Hom}_{\mathbb{R}}(\mathfrak{A}, \mathbb{R})$, we set

$$\mathfrak{g}_\alpha = \{X \in \mathfrak{g} : [H, X] = \alpha(H)X, \forall H \in \mathfrak{A}\}.$$

There are only finitely many α for which $\mathfrak{g}_\alpha \neq \{0\}$, $\mathfrak{g} = \sum_\alpha \mathfrak{g}_\alpha$, $\mathfrak{g}_0 = \mathfrak{M} + \mathfrak{A}$, and there is an ordering on the roots such that $\mathfrak{N} = \sum_{\alpha > 0} \mathfrak{g}_\alpha$. Next we choose a set of simple roots $\alpha_1, \ldots, \alpha_m$ for the ordering. Every positive root is a linear combination of these α_i with nonnegative integral coefficients, and the set $\{\alpha_1, \ldots, \alpha_m\}$ forms a linear basis for $\mathfrak{A}^*$. Thus for any $\tau \in \hat{A}$ there exist unique real numbers $t_1, \ldots, t_m$ such that

$$\tau(\exp H) = e^{i(t_1\alpha_1(H) + \cdots + t_m\alpha_m(H))}, \qquad H \in \mathfrak{A}.$$

We write

$$\tau^j(\exp H) = e^{it_j\alpha_j(H)}, \qquad H \in \mathfrak{A}, 1 \leqslant j \leqslant m,$$
$$\tau^0(\exp H) = 1, \qquad H \in \mathfrak{A}.$$

It clearly suffices to show that for *any* $\sigma \in \hat{B}$ and *any* $j = 1, \ldots, m$, we have

$$\mathrm{Ind}_{BA}^G\left[\sigma \times \prod_{i=0}^{j} \tau^i\right] \cong \mathrm{Ind}_{BA}^G\left[\sigma \times \prod_{i=0}^{j-1} \tau^i\right].$$

For notational ease, let us write $\tau_+ = \prod_{i=0}^{j} \tau^i$, $\tau_- = \prod_{i=0}^{j-1} \tau^i$, so that $\tau_+ = \tau_-\tau^j$.

Now among the roots $\alpha_j, 2\alpha_j, 3\alpha_j, \ldots$ choose the largest; that is, pick the biggest positive integer k such that $\mathfrak{g}_{k\alpha_j} \neq \{0\}$. Let $\alpha = k\alpha_j$, $\mathfrak{q} = \mathfrak{g}_\alpha$, $\mathfrak{q}^* = \mathrm{Hom}_{\mathbb{R}}(\mathfrak{q}, \mathbb{R})$, and $Q = \exp \mathfrak{q}$. Q is an abelian subgroup of N which is invariant under (conjugation by elements of) MA. Thus, by the theorem on induction in stages, it suffices to show

$$\pi_+ = \mathrm{Ind}_{BA}^{BAQ}\, \sigma \times \tau_+ \cong \mathrm{Ind}_{BA}^{BAQ}\, \sigma \times \tau_- = \pi_-.$$

We shall achieve this by: realizing these representations in $L_2(Q)$; taking the logarithm to obtain a realization in $L_2(\mathfrak{q})$; then applying the Fourier transform to switch to $L_2(\mathfrak{q}^*)$; and finally exhibiting a multiplication operator that effects the equivalence. A glance at [16, Theorem 2] shows that the key point is to construct a Borel function $D: \mathfrak{q}^* - \{0\} \to \mathbb{T}$ satisfying $D((ba)^{-1} \cdot \phi) = \tau^j(a)D(\phi)$, $b \in B$, $a \in A$, $\phi \in \mathfrak{q}^*$. We attend to that now.

First let $(\cdot,\cdot)$ be an $\mathrm{Ad}(B)$-invariant inner product on $\mathfrak{q}$. Note that $(a \cdot X, Y) = (X, a \cdot Y) = e^{\alpha(H)}(X, Y)$, $a = \exp H \in A$, $X, Y \in \mathfrak{q}$. Then define the natural vector space isomorphism $\mathfrak{q}^* \to \mathfrak{q}$ via $\phi \to X_\phi$, where $\phi(X) = (X, X_\phi)$, $X \in \mathfrak{q}$. Since $(ba \cdot \phi)(X) = \phi((ba)^{-1} \cdot X) = \phi(\mathrm{Ad}(ba)^{-1}X)$, we have

$$\begin{aligned}(X, X_{ba\cdot\phi}) &= (ba \cdot \phi)(X) = \phi(b^{-1}a^{-1} \cdot X)\\ &= (b^{-1}a^{-1} \cdot X, X_\phi) = (X, ba^{-1}X_\phi), \qquad X \in \mathfrak{q}.\end{aligned}$$

That is

$$X_{ba \cdot \phi} = ba^{-1} \cdot X_{\phi}.$$

It follows that setting $(\phi_1, \phi_2) = (X_{\phi_1}, X_{\phi_2})$, we obtain a B-invariant inner product on $\mathfrak{q}^*$. Let $S = \{\phi \in \mathfrak{q}^* : \|\phi\| = (\phi, \phi)^{1/2} = 1\}$. Then given any $\phi \in \mathfrak{q}^*$, $\phi \neq 0$, we may find $a \in A$ such that $a \cdot \phi \in S$ (since $\|a \cdot \phi\| = e^{-\alpha(H)}\|\phi\|$, $a = \exp H \in A$). Set $D(\phi) = \tau^j(a)$. We show that D is a well-defined Borel function from $\mathfrak{q}^* - \{0\}$ to $\mathbb{T}$ satisfying $D((ba)^{-1} \cdot \phi) = \tau^j(a)D(\phi)$.

First of all, if $\|a \cdot \phi\| = 1$ and $\|a' \cdot \phi\| = 1$, then $e^{-\alpha(H)} = e^{-\alpha(H')}$ where $a = \exp H$, $a' = \exp H'$. Thus $\alpha(H) = \alpha(H')$ and

$$\tau^j(a) = e^{it_j\alpha(H)/k} = e^{it_j\alpha(H')/k} = \tau^j(a').$$

Hence D is well defined, and is clearly a Borel function from $\mathfrak{q}^* - \{0\}$ to $\mathbb{T}$. Also, if $\phi \in \mathfrak{q}^*$ and $\|a_1 \cdot \phi\| = 1$, then $\|a_1 a \cdot (ba)^{-1} \cdot \phi\| = \|b^{-1}a_1 \cdot \phi\| = \|a_1 \cdot \phi\| = 1$. Therefore $D((ba)^{-1} \cdot \phi) = \tau^j(a_1 a) = \tau^j(a_1)\tau^j(a) = \tau^j(a)D(\phi)$.

Next we normalize some measures in order to accommodate the transforms we wish to make. Choose Lebesgue measure dX on $\mathfrak{q}$ and Haar measure dq on Q such that

$$\int_Q f(q)\,dq = \int_{\mathfrak{q}} f(\exp X)\,dX, \qquad f \in L_1(Q).$$

The map

$$\Psi : L_2(Q) \to L_2(\mathfrak{q}), \qquad (\Psi f)(X) = f(\exp X)$$

is then a unitary operator. Next let $\Delta_Q(ba)$ denote the modulus of the automorphism $q \to (ba)q(ba)^{-1}$, $q \in Q$. Since B is compact, $\Delta_Q(b) = 1$, $b \in B$. Also if $a = \exp H$, then

$$\begin{aligned} \Delta_Q(a) &= \det \operatorname{Ad}(a)|_{\mathfrak{q}} = \det \exp \operatorname{ad}(H)|_{\mathfrak{q}} \\ &= e^{\operatorname{tr}\operatorname{ad}(H)|_{\mathfrak{q}}} = e^{c_{\mathfrak{q}}\alpha(H)}, \qquad c_\alpha = \dim \mathfrak{g}_\alpha. \end{aligned}$$

Thus

$$\int_Q f(q)\,dq = \Delta_Q(a) \int_Q f(baqb^{-1}a^{-1})\,dq, \qquad f \in L_1(Q).$$

But then we also have

$$\int_{\mathfrak{q}} f(X)\,dX = \Delta_Q(a) \int_{\mathfrak{q}} f(\operatorname{Ad}(ba)X)\,dX, \qquad f \in L_1(\mathfrak{q}).$$

Finally we normalize Lebesgue measure $d\phi$ on $\mathfrak{q}^*$ so that

$$\int_{\mathfrak{q}} |f(X)|^2\,dX = \int_{\mathfrak{q}^*} |\hat{f}(\phi)|^2\,d\phi$$

where

$$\hat{f}(\phi) = \int_{\mathfrak{q}} f(X)e^{i\phi(X)}\,dX.$$

Now the induced representations π_+, π_- can be realized on $L_2(Q)$, and as such they are given by the formulas

$$\pi_+(ban)f(q) = \sigma(b)\tau_+(a)\Delta_Q^{-1/2}(a)f(a^{-1}b^{-1}qban),$$
$$\pi_-(ban)f(q) = \sigma(b)\tau_-(a)\Delta_Q^{-1/2}(a)f(a^{-1}b^{-1}qban),$$

$ban \in BAQ$, $q \in Q$, $f \in L_2(Q)$. We transfer to $L_2(\mathfrak{q})$ via the unitary operator Ψ. Indeed, a very simple computation shows that

$$\pi_+(ban)f(X) = \sigma(b)\tau_+(a)\Delta_Q^{-1/2}(a)f(\mathrm{Ad}(ba)^{-1}X + Y),$$
$$\pi_-(ban)f(X) = \sigma(b)\tau_-(a)\Delta_Q^{-1/2}(a)f(\mathrm{Ad}(ba)^{-1}X + Y),$$

where $ban \in BAQ$, $n = \exp Y$, $Y \in \mathfrak{q}$, $X \in \mathfrak{q}$, $f \in L_2(\mathfrak{q})$. Finally we switch to $L_2(\mathfrak{q}^*)$ via the Fourier transform. If $f \in L_1(\mathfrak{q}) \cap L_2(\mathfrak{q})$, $\hat{f} \in L_2(\mathfrak{q}^*)$, then

$$\begin{aligned}(\pi_+(ban)\hat{f})(\phi) &= (\pi_+(ban)f)\hat{}\,(\phi)\\ &= \int_{\mathfrak{q}} \sigma(b)\tau_+(a)\Delta_Q^{-1/2}(a)f(\mathrm{Ad}(ba)^{-1}X + Y)e^{i\phi(X)}\,dX\\ &= \sigma(b)\tau_+(a)\Delta_Q^{+1/2}(a)\int_{\mathfrak{q}} f(X + Y)e^{i\phi(\mathrm{Ad}(ba)X)}\,dX\\ &= \sigma(b)\tau_+(a)\Delta_Q^{+1/2}(a)\int_{\mathfrak{q}} f(X)e^{i\phi\,\mathrm{Ad}(ba)(X-Y))}\,dX\\ &= \sigma(b)\tau_+(a)\Delta_Q^{+1/2}(a)e^{-i(ba)^{-1}\cdot\phi(Y)}\hat{f}((ba)^{-1}\cdot\phi),\end{aligned}$$

where $ban \in BAQ$, $n = \exp Y$, $Y \in \mathfrak{q}$, $\phi \in \mathfrak{q}^*$. Similarly,

$$(\pi_-(ban)\hat{f})(\phi) = \sigma(b)\tau_-(a)\Delta_Q^{+1/2}(a)e^{-i(ba)^{-1}\cdot\phi(Y)}\hat{f}((ba)^{-1}\cdot\phi).$$

Now define

$$B: L_2(\mathfrak{q}^*) \to L_2(\mathfrak{q}^*), \qquad F(\phi) \to D(\phi)F(\phi).$$

B is a unitary operator, and we claim it is an intertwining operator for π_+ and π_-. Indeed if $BF = F_1$, then

$$\begin{aligned}(B\pi_+(ban)F)(\phi) &= D(\phi)(\pi_+(ban)F)(\phi)\\ &= D(\phi)\sigma(b)\tau_+(a)\Delta_Q^{+1/2}(a)e^{-i(ba)^{-1}\cdot\phi(Y)}F((ba)^{-1}\cdot\phi), \qquad \phi \neq 0,\end{aligned}$$

while

$$\begin{aligned}(\pi_-(ban)BF)(\phi) &= (\pi_-(ban)F_1)(\phi)\\ &= \sigma(b)\tau_-(a)\Delta_Q^{+1/2}(a)e^{-i(ba)^{-1}\cdot\phi(Y)}F_1((ba)^{-1}\cdot\phi)\\ &= \sigma(b)\tau_-(a)\Delta_Q^{+1/2}(a)e^{-i(ba)^{-1}\cdot\phi(Y)}D((ba)^{-1}\cdot\phi)F((ba)^{-1}\cdot\phi)\\ &= \sigma(b)\tau_-(a)\Delta_Q^{+1/2}(a)e^{-i(ba)^{-1}\cdot\phi(Y)}\tau^j(a)D(\phi)F((ba)^{-1}\cdot\phi)\\ &= (B\pi_+(ban)F)(\phi), \qquad \phi \neq 0\end{aligned}$$

since $\tau_+ = \tau_-\tau^j$. This completes the proof of Theorem 3.

Remarks. (i) If $G = SL(2, \mathbb{R})$ and

$$(\sigma \times \tau)\begin{pmatrix} h & 0 \\ 0 & h^{-1} \end{pmatrix} = |h|^{it}[\operatorname{sgn}(h)]^{\epsilon}, \qquad t \in \mathbb{R}, \quad \epsilon = 0, 1,$$

then a simple computation reveals that $(Bf)(x) = |x|^{-it/2} f(x)$, $f \in L_2(\mathbb{R})$. The Fourier transforms of these operators are the same as some of the intertwining operators of Kunze and Stein [10]. In general, it is clear (at least from a formal calculation) that the operators B, when Fourier transformed back to $L_2(\mathfrak{q})$ (or perhaps $L_2(Q)$), are singular integral operators—likely related to the intertwining operators of [11] and [9] (see also [18]). We have not thoroughly investigated this connection.

(ii) Robert Martin proved a version of Theorem 3—he assumed P was minimal and he was able to replace B by M—in his thesis [16]. In fact, independence results à la Theorem 3 have arisen in three separate pursuits: originally in the question of tensor products of principal series representations (see [16, 18] and [14, Chap. II]); then in the study of T_G—incidentally, somewhat similar techniques apparently can be utilized in the search for the discrete spectrum; and finally they play a role in studying the representation theory of certain subgroups of semisimple groups. I hope to return to the latter two topics on other occasions.

Now Theorem 3 enables us to make the following computation:

$$\infty \operatorname{Ind}_{BA}^{G} \sigma \times 1 \cong \int_{A}^{\oplus} \operatorname{Ind}_{BA}^{G} \sigma \times \tau \, d\tau \cong \operatorname{Ind}_{BA}^{G} \sigma \times \lambda_A \cong \operatorname{Ind}_{B}^{G} 1.$$

Putting $\sigma = 1$, we obtain

COROLLARY 4. *Let $H = BA$ be as above. Then*

$$\operatorname{Ind}_{H}^{G} 1 \approx \operatorname{Ind}_{B}^{G} 1.$$

Since B is compact, the representation $\operatorname{Ind}_{B}^{G} 1$ is obviously unitarily equivalent to the subrepresentation of λ_G acting in the subspace

$$\{f \in L_2(G) : f(gb) = f(g),\, b \in B,\, g \in G\}.$$

We can now state

THEOREM 5. (i) *Let $H_0, H_1, \ldots, H_r$ be a complete system of nonconjugate Cartan subgroups of G. For $i \geqslant 1$, let $H_i = B_i A_i$ as above. Also set $B_0 = H_0$. Then*

$$T_G \cong \sum_{i=0}^{r}{}^{\oplus} \infty \operatorname{Ind}_{B_i}^{G} 1.$$

(ii) *T_G is unitarily equivalent to a subrepresentation of the regular representation λ_G, and has uniform infinite multiplicity.*

We digress momentarily. Suppose L is a separable locally compact type I group and $C \subseteq L$ is a compact subgroup. There is a unique measure class $\{\mu_L\}$ on $\hat{L}$ such that

$$\lambda_L \cong \int_L^{\oplus} \pi \otimes 1_{\dim \pi}\, d\mu_L(\pi).$$

Also

$$\lambda_C \cong \sum_{\sigma \in \hat{C}}^{\oplus} \sigma \otimes 1_{\dim \sigma}.$$

For any $\pi \in \hat{G}$ we decompose $\pi|_C$ into a direct sum of irreducibles with multiplicity, namely

$$\pi|_C \cong \sum_{\sigma \in \hat{C}}^{\oplus} n(\pi, \sigma)\sigma.$$

But for fixed π, the "measure" $\sigma \to n(\pi, \sigma)$ is absolutely continuous with respect to (the discrete) Plancherel measure on $\hat{C}$. Thus we may apply Anh's reciprocity theorem [1, Sect. 1] to conclude that for a.a. $\sigma \in \hat{C}$—and hence for all $\sigma \in \hat{C}$—we have

$$\mathrm{Ind}_C^L \sigma \cong \int_L^{\oplus} n(\pi, \sigma)\pi\, d\mu_L(\pi).$$

Return now to the situation $G = \mathrm{Ad}\, \mathfrak{g}$, $\mathfrak{g}$ semisimple. What we propose to show is that for μ_G – almost all $\pi \in \hat{G}$, there exists B_i, $0 \leqslant i \leqslant r$, such that $\pi|_{B_i}$ contains a fixed vector. If we can do that, it will follow from Theorem 5 and the discussion in the preceding paragraph that $T_G \cong \lambda_G$. In the remaining sections we will establish the desired fact for all representations in the various principal series corresponding to noncompact Cartan subgroups. We plan to consider the discrete series in other papers [19, 20].

4. A Further Reduction

In this section we shall effect a further contraction of the subgroups B_i; that is, we shall show

$$\mathrm{Ind}_{B_i}^G 1 \approx \mathrm{Ind}_{F_i}^G 1,$$

where F_i is a subgroup of B_i which will be defined momentarily.

We begin with a simple consequence of the highest weight theory for compact Lie groups.

LEMMA 6. *Let K be a compact connected Lie group with trivial center, and let T be a maximal torus in K. Then*

$$\mathrm{Ind}_T^K 1 \approx \lambda_K.$$

This follows from two facts: (i) when K has no center, the weight lattice for any irreducible representation includes the zero vector; and (ii) the Frobenius (or Anh) reciprocity theorem applied to compact groups. The next result is an immediate

COROLLARY 7. *If K is any compact connected Lie group with maximal torus T and center Z_K, then*

$$\mathrm{Ind}_T^K 1 \cong \mathrm{Ind}_{Z_K}^K 1.$$

Now return once again to the semisimple group $G = \mathrm{Ad}\,\mathfrak{g}$, with K a maximal compact subgroup and B_0 a compact Cartan subgroup of G contained in K. Applying induction in stages to Corollary 7, we obtain

$$\mathrm{Ind}_{B_0}^G 1 \cong \mathrm{Ind}_{Z_K}^G 1.$$

Next we would like to generalize this result from the triple (G, K, B_0) to $(G, K \cap M_i, B_i)$, $1 \leqslant i \leqslant r$. The fact that B_i and $K \cap M_i$ may fail to be connected causes some difficulty. It can be partially circumvented as follows. Let $\mathfrak{g}_c = \mathfrak{g} + i\mathfrak{g}$ be the complexification of $\mathfrak{g}$. Given a cuspidal parabolic subgroup $P = MAN$ and Cartan subgroup $H = BA$, we set $\Gamma = K \cap \exp i\mathfrak{A}$. Then it is known [12, Theorem 2.2] that: Γ is a finite subgroup of $K \cap M$; it is central in M; and $B = B^\circ \Gamma$, where B° is the neutral component of B. Set

$$F = \Gamma \cdot Z_{K \cap M^\circ}.$$

Then we have

LEMMA 8. $\mathrm{Ind}_B^G 1 \approx \mathrm{Ind}_F^G 1.$

Proof. First observe that $K \cap M^\circ$ is a maximal compact—and so connected—subgroup of M°, and that B° is a maximal torus in it. Therefore we know from Corollary 7 that

$$\mathrm{Ind}_{B^\circ}^{K \cap M^\circ} 1 \approx \mathrm{Ind}_{Z_{K \cap M^\circ}}^{K \cap M^\circ} 1.$$

By induction in stages, we deduce that

$$\mathrm{Ind}_{B^\circ}^{M^\circ} 1 \approx \mathrm{Ind}_{Z_{K \cap M^\circ}}^{M^\circ} 1.$$

Then, since the group Γ is central in M and $M^\circ \cap \Gamma = B^\circ \cap \Gamma = Z_{K \cap M^\circ} \cap \Gamma$, we have

$$\mathrm{Ind}_B^{M^\circ \Gamma} 1 = \mathrm{Ind}_{B^\circ \Gamma}^{M^\circ \Gamma} 1 \cong \mathrm{Ind}_{\Gamma Z_{K \cap M^\circ}}^{M^\circ \Gamma} 1 = \mathrm{Ind}_F^{M^\circ \Gamma} 1.$$

Finally, applying induction in stages once again, we obtain

$$\mathrm{Ind}_B^G 1 \approx \mathrm{Ind}_F^G 1.$$

We summarize our findings thus far in

THEOREM 9. *Let $H_0, H_1, \ldots, H_r$ be as usual. Set $F_0 = Z_K$ and $F_i = \Gamma_i \cdot Z_{K \cap M_i^\circ}$, $1 \leqslant i \leqslant r$. Then*

$$T_G \cong \sum_{i=0}^{r} {}^{\oplus} \infty\, \mathrm{Ind}_{F_i}^G 1.$$

Remarks. (i) If $P = MAN$ is minimal, then $M = M^\circ \Gamma$ and $F = \Gamma Z_{K \cap M^\circ} = \Gamma Z_{M^\circ} = Z_M$. Otherwise $F = \Gamma Z_{K \cap M^\circ}$ may be larger than $Z_{K \cap M}$.

(ii) If G is compact then, by Lemma 6 and Theorem 1, we get

$$T_G \cong \infty \lambda_G \ncong \lambda_G.$$

Our conjecture of course is that for *noncompact* G

$$T_G \cong \infty \lambda_G \cong \lambda_G.$$

(iii) In the next section we shall show that given any representation in the continuous spectrum of λ_G—i.e., given any principal series representation π—there exists F_i such that $\pi|_{F_i}$ contains a fixed vector. By the comments at the end of Section 3, that will establish that T_G and λ_G have the same continuous spectrum.

5. THE CONTINUOUS SPECTRUM

The continuous spectrum of λ_G is made up of the following representations. Let $H = H_i$, $1 \leqslant i \leqslant r$, be any one of the noncompact Cartan subgroups, $P = MAN$ a corresponding cuspidal parabolic subgroup, and $F \subseteq K \cap M$ as in Section 4. The set $\hat{M}_{\mathrm{d}}$ of discrete series representations of M is nonempty. For $\sigma \in \hat{M}_{\mathrm{d}}$, $\tau \in \hat{A}$, we let $\sigma \times \tau$ denote the outer product as well as its natural extension to $P = MAN$. Form $\pi(\sigma, \tau) = \mathrm{Ind}_P^G \sigma \times \tau$. These are almost always irreducible, and (ignoring equivalences, as we may) they constitute the continuous spectrum of λ_G (see [6]). We shall now show that $\pi(\sigma, \tau)|_F$ contains a fixed vector. Since $\pi(\sigma, \tau)$ is an induced representation, it is quite natural to apply Mackey's subgroup theorem. In order for that to succeed we shall need to know how to describe the action of the group F on $\bar{N} = \theta N$. In fact we have

THEOREM 10. (i) *Let $\bar{P} = \theta P = MA\bar{N}$ be the opposed parabolic. Then there is an open, dense, co-null subset $\mathscr{U} \subseteq \bar{N}$ such that the stability group F_u*

for the action of F on $\bar{N}$ (by inner automorphisms of G) is independent of $u \in \mathscr{U}$. Call it F'.

(ii) *Let $\sigma \in \hat{M}_{\mathrm{d}}$ be such that $\sigma|_{F'}$ contains the identity representation. Then for any $\tau \in \hat{A}$, $\pi(\sigma, \tau)|_{F'}$ contains a fixed vector.*

(iii) $F' = \{e\}$.

Proof. (i) The group F is clearly a *compact abelian* Lie group. Let $\mathfrak{N}$ be the Lie algebra of $\bar{N}$. Then F acts as a group of real linear transformations of the real vector space $\mathfrak{N}$; that is

$$F \to GL(\mathfrak{N}), \qquad f \to (X \to f \cdot X = \mathrm{Ad}(f)X)$$

is a real representation of F. We shall demonstrate the existence of an open, dense, co-null subset $\mathscr{V} \subseteq \mathfrak{N}$ such that $F_X = \{f \in F : f \cdot X = X\}$ is independent of $X \in \mathscr{V}$. Then $\mathscr{U} = \exp \mathscr{V}$ will satisfy the requirements of the theorem.

Now by some fairly simple linear algebra the following must be true. There exist two families $\{V_i;\ 1 \leqslant i \leqslant n\}$ and $\{W_j;\ 1 \leqslant j \leqslant m\}$ of subspaces of $\mathfrak{N}$ with the properties: $\dim_{\mathbb{R}} V_i = 1$, $1 \leqslant i \leqslant n$; $\dim_{\mathbb{R}} W_j = 2$, $1 \leqslant j \leqslant m$; $\mathfrak{N} = \sum_{i=1}^{n} V_i + \sum_{j=1}^{m} W_j$ is a vector space direct sum; and each V_i and W_j is a *minimal* F-invariant subspace of $\mathfrak{N}$. Set

$$\mathscr{V} = \left\{ \sum_{i=1}^{n} v_i + \sum_{j=1}^{m} w_j : v_i \in V_i,\ v_i \neq 0,\ w_j \in W_j,\ w_j \neq 0 \right\}.$$

$\mathscr{V}$ is actually Zariski-open, so of course is open, dense, and co-null. If we presume F-invariant inner products on the W_j, then we can write

$$\begin{aligned} f \cdot v_i &= \chi_i(f) v_i, \qquad & \chi_i &: F \to \{\pm 1\}, \qquad & 1 &\leqslant i \leqslant n, \\ f \cdot w_j &= \sigma_j(f) w_j, \qquad & \sigma_j &: F \to O(2, \mathbb{R}), \qquad & 1 &\leqslant j \leqslant m. \end{aligned}$$

Moreover, by the commutativity of F and the minimality of the W_j, we must have $\sigma_j(F) \subseteq SO(2, \mathbb{R})$. But then it is obvious that for any $v \in \mathscr{V}$,

$$F_v = \bigcap_{i=1}^{n} \operatorname{Ker} \chi_i \cap \bigcap_{j=1}^{m} \operatorname{Ker} \sigma_j.$$

This completes the proof of (i).

(ii) We now take the group F' obtained in part (i) and a representation $\sigma \in \hat{M}_{\mathrm{d}}$ such that $\sigma|_{F'}$ contains a fixed vector. We wish to show that $\pi(\sigma, \tau)|_F$ has the same property, for any $\tau \in \hat{A}$. To achieve that we apply the subgroup theorem. First we need to know that MAN and F are regularly related. But the manifold $MAN\bar{N}$ is co-null in G, and thus (up to a null set) we have $MAN \backslash G/F = MAN \backslash MAN\bar{N}/F \cong \bar{N}/F$. The latter space must be countably separated (e.g., by the compactness of F). Therefore it is legitimate to apply the subgroup theorem. Using part (i), we obtain

$$(\mathrm{Ind}_{MAN}^{G}\,\sigma \times \tau)\big|_F \cong \int_{N/F}^{\oplus} \mathrm{Ind}_{F_X}^{F}(\sigma \times \tau)\big|_{F_X}$$
$$\cong \int_{\mathscr{U}/F}^{\oplus} \mathrm{Ind}_{F'}^{F}\,\sigma\big|_{F'} \cong \infty\ \mathrm{Ind}_{F'}^{F}\,\sigma\big|_{F'}.$$

Since F' is normal in F (which is abelian), it follows immediately that if $\sigma|_{F'}$ contains a fixed vector, then $\pi(\sigma, \tau)|_F$ actually contains 1 infinitely many times. That proves (ii).

(iii) Finally we prove that $F' = \{e\}$. We begin by eliminating any compact factors. Let $\mathfrak{g} = \mathfrak{g}_1 + \mathfrak{g}_2$ be a direct sum of ideals, where $\mathfrak{g}_1$ is a compact semisimple Lie algebra and $\mathfrak{g}_2$ has no compact factors. Then $G = \mathrm{Ad}\,\mathfrak{g} \cong G_1 \times G_2$ where $G_1 = \mathrm{Ad}\,\mathfrak{g}_1$ is compact and $G_2 = \mathrm{Ad}\,\mathfrak{g}_2$ has no compact normal subgroups. A maximal compact subgroup K must be of the form $K \cong G_1 \times K_2$, where K_2 is a maximal compact subgroup of G_2. Also cuspidal parabolics look like $P = G_1 \times P_2$, where P_2 is a cuspidal parabolic of G_2. If $P_2 = M_2 A_2 N_2$, then $P = MAN$ where $M = G_1 \times M_2$, $A = A_2$, $N = N_2$. Also, $K \cap M^\circ = G_1 \times (K_2 \cap M_2^\circ)$, so that $Z_{K \cap M^\circ} = Z_{K_2 \cap M_2^\circ}$. Clearly $\Gamma = K \cap \exp i\mathfrak{A} \subseteq K_2$. Therefore $F \subseteq G_2$ and it is no loss of generality to assume that $G = G_2$ in what follows.

Now by definition any $f \in F'$ centralizes a dense open subset of $\bar{N}$. By continuity it must centralize all of $\bar{N}$. Then for $n \in N$ we have $fnf^{-1} \equiv f \cdot n = f \cdot \theta(\theta n) = \theta(f \cdot \theta n) = \theta(\theta n) = n$, since $f \in K$. Thus F' also centralizes N. The fact that F' is trivial is a consequence of the following

Lemma 11. *Let G be a connected semisimple Lie group without compact factors. Let $P = MAN$ be a (cuspidal) parabolic subgroup with $\bar{P} = MA\bar{N}$ the opposed parabolic. Let T be the subgroup of M consisting of elements that centralize both N and $\bar{N}$. Then $T \subseteq Z_G$.*

Proof. For subgroups S_1 and S_2 of G, we write

$$Z_{S_1}(S_2) = \{g \in S_1 : gs = sg,\ \forall s \in S_2\}.$$

Then $T = Z_M(N) \cap Z_M(\bar{N})$. We begin by observing that $Z_M(N)$ is normal in M—because M normalizes N. Similarly for $Z_M(\bar{N})$; therefore T is a normal subgroup of M. Since M and A commute, T is centralized by A. In fact, N also centralizes T. For if $t \in T$, $n \in N$, then $tnt^{-1} = n$ because $t \in Z_M(N)$. Similarly $\bar{N}$ centralizes T. Thus T is a *closed* subgroup of G that is normalized by every element of the dense subset $MAN\bar{N}$. It follows by continuity that T is normalized by all of G—that is, T is a closed normal subgroup of G.

Case (i): G simple. Since G is not compact, T is not G itself. Hence T is discrete and so central.

Case (ii): G semisimple. Then there are closed, mutually commuting, simple subgroups $G_1, \ldots, G_k$ such that $G = G_1 G_2 \cdots G_k$. (This is not necessarily direct—there may be some nontrivial central intersections, but that is of no matter.) Then with obvious notation

$$P = P_1 \cdots P_k, \qquad M = M_1 \cdots M_k, \qquad N = N_1 \cdots N_k,$$

and, since G contains no compact factors, no N_j is trivial. Therefore

$$Z_M(N) = Z_{M_1}(N_1) \cdots Z_{M_k}(N_k), \qquad T = T_1 \cdots T_k.$$

Hence matters reduce to case (i), and once again the conclusion is that T is central. This concludes the proofs of Lemma 11 and Theorem 10.

Remarks. (i) It is possible to avoid the subgroup theorem in part (ii) by writing down a realization of $\pi(\sigma, \tau)$ in $L_2(\bar{N})$ and explicitly producing a fixed vector. There is no advantage to that unless one happens to be allergic to the subgroup theorem.

(ii) If G has no compact factors, the argument of Theorem 10 works just as well if the group F is replaced by B (any compact abelian subgroup of $K \cap M$ would do). That is, in some sense Section 4 could be omitted. I have presented Section 4 and Theorem 10 in this way for three reasons: first, to introduce Remark (ii) at the end of Section 4; second, because F is in general a small group, often finite—thus $\mathrm{Ind}_F^G 1$ will be "close" to λ_G, and we have further evidence for the conjecture $T_G \cong \lambda_G$; and finally, because I expect to use it in later work on the discrete spectrum.

6. Conclusions

One knows [6] that the regular representation can be decomposed

$$\lambda_G \cong \lambda_G^\circ \oplus \sum_{i=1}^{r}{}^{\oplus} \infty \int^{\oplus} \mathrm{Ind}_{P_i}^G \sigma \times \tau \, d\mu_i(\sigma, \tau),$$

where $\lambda_G^\circ \cong \infty \sum_{\pi \in \hat{G}_d}^{\oplus} \pi$ is present if and only if there exists a compact Cartan subgroup. Also the continuous measures μ_i are known quite explicitly, although their exact form plays no role in our deliberations. The content of this paper may be summarized as follows:

THEOREM 12. (i) *If* rank $G >$ rank K, *then* $T_G \cong \lambda_G$.
(ii) *If* rank $G =$ rank K, *then*

$$T_G \cong T_G^\circ \oplus \sum_{i=1}^{r}{}^{\oplus} \infty \int^{\oplus} \mathrm{Ind}_{P_i}^G \sigma \times \tau \, d\mu_i(\sigma, \tau),$$

where T_G° is equivalent to a subrepresentation of λ_G° and has uniform infinite multiplicity.

Our conjecture amounts to $T_G^\circ \cong \lambda_G^\circ$, and we expect to deal with that in another paper. We close this one by listing some groups $G = \mathrm{Ad}\,\mathfrak{g}$ for which $T_G \cong \lambda_G$: (i) $\mathfrak{g}$ has a complex structure; (ii) Any $\mathfrak{g}$ which has only one conjugacy class of Cartan subalgebras, e.g., $\mathfrak{g} = so(2n+1, 1)$, $n \geqslant 1$ or $\mathfrak{g} = su^*(2n)$, $n > 1$; (iii) $\mathfrak{g} = so(2n+1, 2m+1)$, $n \geqslant m \geqslant 1$; and (iv) $\mathfrak{g} = sl(n, \mathbb{R})$, $n > 2$.

REFERENCES

1. N. ANH, Restriction of the principal series of SL(n,$\mathbb{C}$) to some reductive subgroups, *Pacific J. Math.* **38** (1971), 295–313.
2. E. EFFROS, Transformation groups and C^*-algebras, *Ann. of Math.* **81** (1965), 38–55.
3. L. EHRENPREIS, The use of partial differential equations for the study of group representations, *Proc. Symp. Pure Math.* **26** (1973), 317–320.
4. S. GELBART, Automorphic forms on the metaplectic group, preprint.
5. HARISH-CHANDRA, Invariant eigendistributions on a semi-simple Lie group, *Trans. Amer. Math. Soc.* **119** (1965), 457–508.
6. HARISH-CHANDRA, HARMONIC analysis on semisimple Lie groups, *Bull. Amer. Math. Soc.* **76** (1970), 529–551.
7. S. HELGASON, "Differential Geometry and Symmetric Spaces," Academic Press, New York, 1962.
8. A. KLEPPNER AND R. LIPSMAN, The Plancherel formula for group extensions, *Ann. Sci. Ecole Norm. Sup.* **5** (1972), 71–120.
9. A. KNAPP AND E. STEIN, Singular integrals and the principal series III, *Proc. Nat. Acad. Sci.* **71** (1974), 4622–4624.
10. R. KUNZE AND E. STEIN, Uniformly bounded representations I, *Amer. J. Math.* **82** (1960), 1–62.
11. R. KUNZE AND E. STEIN, Uniformly bounded representations III, *Amer. J. Math.* **89** (1967), 385–442.
12. R. LIPSMAN, On the characters and equivalence of continuous series representations, *J. Math. Soc. Jpn.* **23** (1971), 452–480.
13. R. LIPSMAN, The unitary representation given by the invariant integral on a semisimple Lie algebra, *Notices Amer. Math. Soc.* **21** (1974), A-110.
14. R. LIPSMAN, "Group Representations," Springer-Verlag, Berlin and New York, 1974.
15. R. LIPSMAN, Algebraic transformation groups and representation theory, *Math. Ann.* **214** (1975), 149–157.
16. R. MARTIN, On the decomposition of tensor products of principal series representations for real-rank one semisimple groups, Thesis, Univ. of Maryland, 1973; also *Trans. Amer. Math. Soc.* **201** (1975), 177–211.
17. G. WARNER, "Harmonic Analysis of Semi-Simple Lie Groups," 2 volumes, Springer-Verlag, Berlin and New York, 1972.
18. F. WILLIAMS, Reduction of tensor products of principal series representations of complex semi-simple Lie groups, Thesis, Univ. of California at Irvine, 1972; see also Springer-Verlag Lecture Notes, Vol. 358.
19. R. LIPSMAN, On the unitarized adjoint representation of a semisimple Lie group II, *Can. J. Math.* **29** (1977), 1217–1222.
20. R. LIPSMAN, Restrictions of principal series to a real form, *Pac. J. Math.*, to appear.

STUDIES IN ALGEBRA AND NUMBER THEORY
ADVANCES IN MATHEMATICS SUPPLEMENTARY STUDIES, VOL. 6

Some Recent Results on Infinite-Dimensional Spin Groups

R. J. PLYMEN

Department of Mathematics, University of Manchester, Manchester, England

INTRODUCTION

The subject of this essay can be traced back to the quaternions of Gauss and Hamilton. For the unit quaternions act by conjugation on the real three-dimensional space of pure quaternions, producing thereby a homomorphism:

$$S^3 \to SO(3),$$

where S^3 is the 3-sphere and $SO(3)$ is the rotation group in $\mathbb{R}^3$. This is the prototype of the general situation

$$\mathrm{Spin}(n) \to SO(n), \qquad n \geqslant 3,$$

where $SO(n)$ is the special orthogonal group in $\mathbb{R}^n$ and $\mathrm{Spin}(n)$ is its double cover.

The subject of spin groups has always been inseparable from applied mathematics. Consider, for example, the quaternions in algebra and the vector product in mechanics; the covering map $SU(2) \to SO(3)$ and the Cayley–Klein parameters in mechanics; the Lie algebra isomorphism $\mathbf{su}(2) \to \mathbf{o}(3)$ and the spin of the electron; the Clifford algebra of the Minkowski quadratic form and the γ-matrices in the Dirac equation of the spinning electron; the C^*-Clifford algebra and the canonical anticommutation relations in quantum theory.

In recent times, the infinite-dimensional Clifford algebra has been studied with such vigour by mathematicians and mathematical physicists that it is now possible to present infinite-dimensional versions of some classical results about the special orthogonal groups, their double covers, Lie algebras, and projective representations. In so doing, we do not get *too* far away from classical results. There are, for example, points of contact with Fourier series,

ISBN 0-12-599153-3

the stable spin group

$$\mathrm{Spin}(\infty) = \varinjlim \mathrm{Spin}(n)$$

and the direct limit $\varinjlim \Delta_n$ of the spin representations Δ_n of Cartan.

1. The C^*-Clifford Algebra

Let E be a real vector space, and let f be a symmetric nondegenerate bilinear form on E with associated quadratic form Q. Let

$$T(E) = \sum_{i \geq 0} T^i(E) = R \oplus E \oplus E \otimes E \oplus \cdots$$

be the tensor algebra of E, and let $I(Q)$ be the two-sided ideal in $T(E)$ generated by elements of the form $x \otimes x - Q(x) \cdot 1$, where $x \in E$. The Clifford algebra $\mathrm{Cl}(Q)$ of the quadratic form Q is defined to be

$$\mathrm{Cl}(Q) = T(E)/I(Q).$$

The following facts are well known (Bourbaki [6], Atiyah, *et al.* [1]).

(i) If $i_Q : E \to \mathrm{Cl}(Q)$ is the canonical map given by $E \to T(E) \to \mathrm{Cl}(Q)$ and $\phi : E \to A$ is a linear map of E into an $\mathbb{R}$-algebra A such that for all $x \in E$

$$\phi(x)^2 = Q(x) \cdot 1_A, \tag{1}$$

then there exists a unique homomorphism $\tilde{\phi} : \mathrm{Cl}(Q) \to A$ such that the diagram

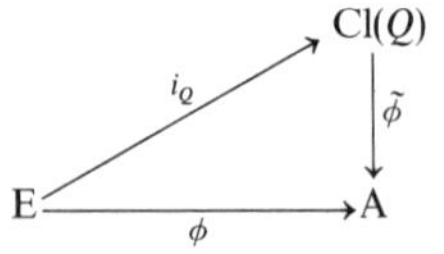

is commutative, that is, $\mathrm{Cl}(Q)$ is the universal algebra with respect to maps satisfying (1).

(ii) Let $\mathrm{Cl}^0(Q)$ be the image of $\sum_{i \geq 0} T^{2i}(E)$ and let $\mathrm{Cl}^1(Q)$ be the image of $\sum_{i \geq 0} T^{2i+1}(E)$. Then

$$\mathrm{Cl}(Q) = \mathrm{Cl}^0(Q) \oplus \mathrm{Cl}^1(Q)$$

is a $\mathbb{Z}_2$-graded algebra. That is, if $u_i \in \mathrm{Cl}^i(Q)$ and $v_j \in \mathrm{Cl}^j(Q)$, then $u_i v_j \in \mathrm{Cl}^k(Q)$ where $k = i + j \pmod 2$. The importance of the graded structure is emphasized in [1]. Define $\alpha : \mathrm{Cl}(Q) \to \mathrm{Cl}(Q)$ by

$$\alpha(x) = \begin{cases} x & \text{if } x \in \mathrm{Cl}^0(Q), \\ -x & \text{if } x \in \mathrm{Cl}^1(Q). \end{cases}$$

Then α is an automorphism of $\mathrm{Cl}(Q)$.

(iii) If $x = x_1 \otimes x_2 \otimes \cdots \otimes x_k \in T^k(E)$, then the map $x \mapsto x^*$ defined by

$$x_1 \otimes x_2 \otimes \cdots \otimes x_k \mapsto x_k \otimes \cdots \otimes x_1$$

is an anti-automorphism of $T^k(E)$ which preserves $I(Q)$, that is, it induces an anti-automorphism of $\mathrm{Cl}(Q)$, also denoted by *, called the transpose.

We shall be particularly interested in the case when E is a separable real Hilbert space, and the quadratic form Q is given by the square of the norm:

$$Q(x) = (x, x) = \|x\|^2.$$

The Clifford algebra will in this case be denoted by $\mathrm{Cl}(E)$. From now on, we shall take E to be infinite-dimensional. Such a space E is sometimes called an infinite-dimensional Euclidean space.

If x, y are vectors in E, then we have

$$xy + yx = 2(x, y).$$

Consequently

$$xy + yx = 0 \Leftrightarrow (x, y) = 0,$$

so that the geometric relation of orthogonality in E is represented in $\mathrm{Cl}(E)$ by the algebraic relation of anticommutativity.

Let $\mathbb{C}(n)$ denote the * algebra of all $n \times n$ matrices over $\mathbb{C}$. A uniformly hyperfinite (UHF) algebra of type 2^n is a C^*-algebra $\mathscr{C}$ containing a sequence M_n of * algebras such that

(1) M_n is * isomorphic to $\mathbb{C}(2^n)$;
(2) $1 \in M_n$ for each n;
(3) $M_n \subset M_{n+1}$ for each n;
(4) $\mathscr{C}$ is the norm closure of $\bigcup M_n$.

Define a linear functional τ on $\bigcup M_n$ by setting

$$\tau(A) = \mathrm{tr}_n(A)$$

if $A \in M_n$ and tr_n denotes the trace on M_n normalized so that $\mathrm{tr}_n(1) = 1$. Then τ is a state of $\bigcup M_n$, that is to say, τ is a positive linear functional on $\bigcup M_n$ such that $\tau(1) = 1$. Now τ is norm continuous, hence extends uniquely to a state τ of $\mathscr{C}$. We call τ the trace of $\mathscr{C}$. Then $\tau(AB) = \tau(BA)$ for all A, B in $\mathscr{C}$. The account of hyperfinite algebras in Topping's book [23] is well suited to our needs.

The * map on $\mathrm{Cl}(E)$ extends to the complexification $\mathrm{Cl}(E) \otimes_{\mathbf{R}} \mathbb{C}$ of $\mathrm{Cl}(E)$ according to the equation

$$(u \otimes c)^* = u^* \otimes c^*, \qquad u \in \mathrm{Cl}(E),\ c \in \mathbb{C},$$

with c^* the complex conjugate of c. The $\mathbb{Z}_2$-grading on $\mathrm{Cl}(E) \otimes_{\mathbf{R}} \mathbb{C}$ depends on the equation

$$\alpha(u \otimes c) = \alpha(u) \otimes c.$$

The notation * has been used to denote involutions on both $\mathrm{Cl}(E) \otimes_{\mathbf{R}} \mathbb{C}$ and $\bigcup M_n$. The following theorem justifies this.

THEOREM 1. There is a * monomorphism: $\bigcup M_n \to \mathrm{Cl}(E) \otimes_{\mathbf{R}} \mathbb{C}$.

Proof. Let

$$\bigotimes^n \mathbb{C}(2) = \mathbb{C}(2) \otimes \cdots \otimes \mathbb{C}(2) \qquad (n \text{ factors}).$$

Define the * monomorphism $j_{mn} : \bigotimes^m \mathbb{C}(2) \to \bigotimes^n \mathbb{C}(2)$ by

$$A_1 \otimes \cdots \otimes A_m \mapsto A_1 \otimes \cdots \otimes A_m \otimes 1 \otimes \cdots \otimes 1$$

whenever $m < n$. Then the direct limit

$$\varinjlim \bigotimes^n \mathbb{C}(2)$$

following the monomorphisms j_{mn} is well defined. For relevant material on direct limits, the reader may consult the book by Atiyah and Macdonald [2, p. 33]. Clearly

$$\varinjlim \bigotimes^n \mathbb{C}(2) \cong \bigcup M_n$$

since $\bigotimes^n \mathbb{C}(2) \cong \mathbb{C}(2^n)$.

Now let $\{e_1, e_2, \ldots\}$ be an orthonormal basis in E. Set

$$W(e_{2r-1}) = \sigma_3 \otimes \cdots \otimes \sigma_3 \otimes \sigma_1 \otimes 1 \otimes 1 \otimes \cdots,$$
$$W(e_{2r}) = \sigma_3 \otimes \cdots \otimes \sigma_3 \otimes \sigma_2 \otimes 1 \otimes 1 \otimes \cdots,$$

with σ_1, σ_2 in the rth place. Here, $\sigma_1, \sigma_2, \sigma_3$ are the Pauli spin matrices

$$\sigma_1 = \begin{pmatrix} 0 & 1 \\ 1 & 0 \end{pmatrix}, \qquad \sigma_2 = \begin{pmatrix} 0 & -i \\ i & 0 \end{pmatrix}, \qquad \sigma_3 = \begin{pmatrix} 1 & 0 \\ 0 & -1 \end{pmatrix}.$$

Now W maps $\mathrm{Cl}(E_{2n}) \otimes_{\mathbf{R}} \mathbb{C}$ onto $\bigotimes^n \mathbb{C}(2)$ where E_{2n} is the span of $\{e_1, e_2, \ldots, e_{2n}\}$. The relations between the generators $\{1, e_1, \ldots, e_{2n}\}$ are the same as the relations between the corresponding generators $\{1, W(e_1), \ldots, W(e_{2n})\}$. Further, the $\mathbb{C}$-dimension of $\mathrm{Cl}(E_{2n}) \otimes_{\mathbf{R}} \mathbb{C}$ is 2^n. Therefore W determines * isomorphisms

$$\mathrm{Cl}(E_{2n}) \otimes_{\mathbf{R}} \mathbb{C} \cong \bigotimes^n \mathbb{C}(2).$$

This concludes the proof. ■

Theorem 1 is well known. The idea of using the Pauli spin matrices goes back to Jordan and Wigner [15]. Accordingly, the mapping in Theorem 1 is sometimes called the Jordan–Wigner mapping.

We shall frequently view $\mathrm{Cl}(E) \otimes_{\mathbf{R}} \mathbb{C}$ as a * subalgebra of $\mathscr{C}$, and call $\mathscr{C}$ the C^*-Clifford algebra. It is easy to verify that, on $\mathrm{Cl}(E)$, the trace τ has the following properties:

(a) $\tau(1) = 1$;
(b) $\tau(uv) = \tau(vu)$;
(c) $\tau(\alpha(u)) = \tau(u)$.

In fact, τ is the unique linear form on $\mathrm{Cl}(E)$ satisfying (a), (b), and (c) [12, p. 248]. Note that τ annihilates $\mathrm{Cl}^1(E)$, since by (c) we have $\tau(u) = \tau(\alpha(u)) = -\tau(u)$ for all u in $\mathrm{Cl}^1(E)$. Indeed, if $u \in \mathrm{Cl}(E)$, then

$$u = a_0 1 + a_1 e_1 + \cdots + a_{12} e_1 e_2 + \cdots + a_{1 \cdots n} e_1 \cdots e_n$$

for some orthonormal set $\{e_1, \ldots, e_n\}$ and real numbers $a_1, \ldots, a_{1 \cdots n}$. In this case, we have

$$\tau(u) = a_0 .$$

This formula for the trace follows from the Jordan–Wigner mapping and the fact that the Pauli spin matrices have trace 0. Thus the trace annihilates most of $\mathrm{Cl}(E)$.

The norm on the C^*-Clifford algebra, hereafter called the C^*-norm, will be denoted $\|\cdot\|_\infty$. If we regard E as a subspace of $\mathscr{C}$, then the C^*-norm is an extension of the original norm on the Euclidean space E. To prove this, note that

$$\|u\|_\infty = \sup\{\lambda \in R : |u| - \lambda \text{ not invertible}\}$$

where $|u| = (u^*u)^{1/2}$. But $|x|^2 = x^*x = x^2 = Q(x) = \|x\|^2$ for all x in E. Therefore $|x| = \|x\|$, so that $\|x\|_\infty = \|x\|$.

There is a continuum of norms on the C^*-Clifford algebra, each of them an extension of the original norm on E. We set

$$\|u\|_p = (\tau(|u|^p))^{1/p}, \qquad 1 \leqslant p < \infty,$$

where $|u| = (u^*u)^{1/2}$. Since $|x| = \|x\|$, it follows immediately that $\|x\|_p = \|x\|$ for all vectors in the Euclidean space E.

It follows from results of Dixmier, in his fundamental paper [9] on linear forms on rings of operators, that the map $u \mapsto \|u\|_p$ is a norm, for each p with $1 \leqslant p < \infty$, and also that

$$\|u\|_1 \leqslant \|u\|_r \leqslant \|u\|_s \leqslant \|u\|_\infty$$

whenever $u \in \mathscr{C}$ and $1 \leqslant r \leqslant s \leqslant \infty$.

2. The Orthogonal and Spin Lie Algebras

Let $\mathbf{spin}(E)$ denote the linear manifold spanned by $\{xy - yx : x, y \in E\}$, and let $\mathbf{o}(E)$ denote the linear space of all antisymmetric operators of finite rank on E. Then $\mathbf{spin}(E)$ and $\mathbf{o}(E)$ are Lie algebras with Lie bracket given by

$$[X_1, X_2] = X_1X_2 - X_2X_1, \qquad X_1, X_2 \in \mathbf{o}(E),$$
$$[Y_1, Y_2] = Y_1Y_2 - Y_2Y_1, \qquad Y_1, Y_2 \in \mathbf{spin}(E).$$

The equation

$$(\psi(Y))x = Yx - xY$$

with $Y \in \mathbf{spin}(E)$ determines $\psi(Y)$ as an operator in $\mathbf{o}(E)$. Further,

$$\psi : \mathbf{spin}(E) \to \mathbf{o}(E)$$

is a Lie algebra isomorphism such that $\psi(Y^*) = \psi(Y)^*$. Here, $\psi(Y)^*$ denotes the adjoint of $\psi(Y)$ as a linear operator on E. The above statements are proved by de la Harpe in [12], and generalize some results in Jacobson's book on Lie algebras [14].

Again following Dixmier [9], we define

$$\|X\|_p = (\mathrm{Tr}(|X|^p))^{1/p}, \qquad 1 \leqslant p < \infty,$$
$$\|X\|_\infty = \text{norm of } X \text{ as an operator on } E,$$

where $|X| = (X^*X)^{1/2}$, and X lies in $\mathbf{o}(E)$.

Let $\mathbf{o}(E)_p$ (resp. $\mathbf{spin}(E)_p$) denote the completion of $\mathbf{o}(E)$ (resp. $\mathbf{spin}(E)$) in the $\|\cdot\|_p$ norm. Then $\mathbf{o}(E)_1$ may be identified with the classical Banach–Lie algebra of all antisymmetric trace-class operators on E; and $\mathbf{o}(E)_2$ may be identified with the classical Banach–Lie algebra of all antisymmetric Hilbert–Schmidt operators on E; $\mathbf{o}(E)_2$ is even a Hilbert space with inner product $\langle X_1, X_2 \rangle = \mathrm{Tr}(X_1X_2^*)$. If $\mathscr{A}$ is simultaneously a Banach space and a Lie algebra such that the multiplication is jointly continuous, then $\mathscr{A}$ is called a Banach–Lie algebra. $\mathscr{A}$ is thus an example of a topological Lie algebra. The Banach–Lie algebra $\mathscr{A}_1$ is *isomorphic* to the Banach–Lie algebra $\mathscr{A}_2$ if $\mathscr{A}_1 \cong \mathscr{A}_2$ as topological Lie algebras. For more information on such things, see [11].

Theorem 2. $\mathbf{spin}(E)_\infty \cong \mathbf{o}(E)_1$.

In detail, Theorem 2 says that the map ψ extends uniquely to an isomorphism of Banach–Lie algebras. This result is proved in [12], where it is shown that $(1/4)\psi$ is even an isometry.

Theorem 3. $\mathbf{spin}(E)_2 \cong \mathbf{o}(E)_2$.

This result is proved in [12], where it is shown that $(1/4)\sqrt{2}\cdot\psi$ extends uniquely to an isometry.

THEOREM 4. $\mathbf{spin}(E)_p \cong \mathbf{o}(E)_2$ *for all* $1 \leqslant p < \infty$.

This result, at first sight surprising, is proved in [20], and depends crucially on an earlier computation by Streater [22].

Suppose that f is a complex-valued function in $L^2(\mathbb{R})$, and that

$$\sum_{-\infty}^{\infty} a_n e^{nix}$$

is its complex Fourier series. Then it is well known that

$$\sum |a_n|^2 = (1/2\pi)\int_{-\pi}^{\pi} |f|^2\,dx,$$

the so-called Parseval's identity. This result was generalized by Young and Hausdorff. We write

$$\|a\|_p = (\sum |a_n|^p)^{1/p}, \qquad \|f\|_p = \left\{(1/2\pi)\int_{-\pi}^{\pi} |f(x)|^p\,dx\right\}^{1/p}$$

so that Parseval's identity may be written

$$\|a\|_2 = \|f\|_2.$$

Young and Hausdorff proved that if $1 < p \leqslant 2$, then

$$\|f\|_q \leqslant \|a\|_p \qquad \text{and} \qquad \|a\|_q \leqslant \|f\|_p,$$

where $1/p + 1/q = 1$. For a proof, see the classic book on inequalities by Hardy *et al.* [10].

The Hausdorff–Young inequalities survive in the context of orthogonal and spin Lie algebras in the following form.

THEOREM 5. *If* $1 \leqslant p \leqslant 2$ *and* $1/p + 1/q = 1$, *then*

$$\|Y\|_q \leqslant A_p \cdot \|\psi(Y)\|_p$$

and

$$A_q \cdot \|\psi(Y)\|_q \leqslant \|Y\|_p,$$

where $A_p = 1/(2\cdot 2^{1/p})$ *and* Y *lies in the spin algebra* $\mathbf{spin}(E)$.

Theorem 5 was announced in [22] and proved in [20], using the Riesz convexity theorem and interpolation. It is interesting to note the similarity between the two formulas:

$$\|Y\|_p = (\tau(|Y|^p))^{1/p}, \qquad \|f\|_p = (\mu(|f|^p))^{1/p},$$

where μ is the integration functional given by

$$\mu(f) = (1/2\pi) \int_{-\pi}^{\pi} f(x)\, dx.$$

Since $\mu(1) = 1$ and $\mu(|f|^2) \geqslant 0$, it follows that μ is a state. The formula for $\|Y\|_p$ is thus an example of "noncommutative integration" [9].

3. The Banach–Lie Group $\mathrm{Spin}(E)_\infty$

Let $O(E)_1$ denote the group of all orthogonal operators on E of the form $1 + A$ with $\mathrm{Tr}|A| < \infty$. Then $O(E)_1$ has a natural structure of Banach–Lie group (i.e., Lie group modeled on a Banach space) with Lie algebra $\mathbf{o}(E)_1$; the underlying topology is induced by the metric $d(1 + A_1, 1 + A_2) = \mathrm{Tr}|A_1 - A_2|$. The connected component of the identity, which is denoted by $SO(E)_1$, is of index 2 in $O(E)_1$ and has the same homotopy type as the stable classical group $SO(\infty) = \varinjlim SO(n)$. In particular, it follows from general principles that the universal covering of $SO(E)_1$ is a Banach–Lie group and that the covering map is two sheeted. Finally, the exponential map from $\mathbf{o}(E)_1$ to $SO(E)_1$ is onto. These facts are taken from [11] and [13].

Let

$$\mathrm{Spin}^c(E)_\infty = \{u \in \mathscr{C} : uu^* = 1 = u^*u, \alpha(u) = u, uEu^* \subset E\}.$$

Then $\mathrm{Spin}(E)_\infty$ has a natural structure of Banach–Lie group with Lie algebra $\mathbf{spin}(E)_\infty$; indeed the power series map $\exp: \mathbf{spin}(E)_\infty \to \mathrm{Spin}(E)_\infty$ is well defined since the spin algebra $\mathbf{spin}(E)_\infty$ lies in the C^*-Clifford algebra $\mathscr{C}$. The underlying topology on $\mathrm{Spin}(E)_\infty$ is induced by the C^*-norm.

The global analogue of Theorem 2 is

THEOREM 6. *$\mathrm{Spin}(E)_\infty$ is a model of the universal covering group of $SO(E)_1$ and the following diagram commutes.*

$$\begin{array}{ccc} \mathrm{Spin}(E)_\infty & \xrightarrow{\ \rho\ } & SO(E)_1 \\ \uparrow{\scriptstyle \exp} & & \uparrow{\scriptstyle \exp} \\ \mathbf{spin}(E)_\infty & \longrightarrow & \mathbf{o}(E)_1 \end{array}$$

Here, the covering map ρ is given by $\rho(u)x = uxu^*$ with $x \in E$.

This result is due to de la Harpe [12], and depends crucially on an earlier result of Shale and Stinespring [21].

Let now J be a complex structure in the Euclidean space E. That is,

$$J^* = -J, \qquad J^2 = -1,$$

where J^* is the adjoint of J as a linear operator on E. Then $J^*J = 1 = JJ^*$, so that J is an orthogonal operator on E. We set

$$ix = Jx, \qquad \langle x, y\rangle_J = (x, y) + i(Jx, y),$$

where $(\cdot,\cdot)$ is the original inner product on E. Then E becomes a complex Hilbert space E^J, with inner product $\langle x, y\rangle_J$ linear in the second argument.

The unique state ω_J on the C^*-Clifford algebra $\mathscr{C}$ such that

$$\omega_J(xy) = \langle x, y\rangle_J$$

for all x, y in E is the *Fock* state relative to J. It is known that ω_J is pure [4].

The Gelfand–Naimark–Segal construction yields a triple

$$(H_J, \pi_J, \Omega_J),$$

called the Fock space, the Fock representation, the Fock vacuum vector, respectively, such that

$$\omega_J(u) = \langle \Omega_J, \pi_J(u)\Omega_J\rangle_{H_J}, \qquad u \in \mathscr{C}.$$

Since ω_J is pure, π_J is an irreducible * representation of the C^*-Clifford algebra $\mathscr{C}$ on the Hilbert space H_J. Of course, Ω_J is a cyclic vector for the representation π_J.

We define

$$\Delta_J = \pi_J|\mathrm{Spin}(E)_\infty$$

and call Δ_J the spin representation relative to J.

It turns out that the Fock space splits as

$$H_J = H_J^+ + H_J^-$$

with H_J^+, H_J^- each invariant under Δ_J. Let Δ_J^+, Δ_J^- denote the subrepresentations determined by H_J^+, H_J^-. Let (e_n) be an orthonormal basis in E^J, (e_n, Je_n) be the associated basis in E, and let

$$SO(\infty) \to SO(E)_1,$$
$$\mathrm{Spin}(\infty) \to \mathrm{Spin}(E)_\infty$$

be the inclusions determined by the basis (e_n, Je_n), where $\mathrm{Spin}(\infty) = \varinjlim \mathrm{Spin}(n)$.

THEOREM 7. (i) *Δ_J^+ and Δ_J^- are irreducible (the $\frac{1}{2}$-spin representations).*
(ii) *$\Delta_J|\mathrm{Spin}(\infty) = \varinjlim \Delta_n$ where Δ_n is the spin representation of* $\mathrm{Spin}(n)$.

The spin representations Δ_n of Cartan are discussed in Cartan [7], Chevalley [8], Bott [5].

Theorem 7 is proved in [18], which is a sequel to an earlier paper of Shale and Stinespring [21].

According to a remarkable result of Manuceau and Verbeure [16], π_J is unitarily equivalent to π_K if and only if $|J - K|$ is a Hilbert–Schmidt operator on the Euclidean space E, whenever K is another complex structure on E. Consequently

$$|J - K| \text{ Hilbert–Schmidt} \Rightarrow \Delta_J \cong \Delta_K.$$

Distinct complex structures on E may give rise to distinct spin representations, so that the connection between spinors and complex structure, well known in the finite-dimensional case [1], is more subtle in the infinite-dimensional case.

By a spinor relative to J, we do of course mean a vector in H_J, regarded as a $\mathrm{Spin}(E)_\infty$-module.

Note that Δ_J is a unitary representation of $\mathrm{Spin}(E)_\infty$, hence determines a projective unitary representation of the orthogonal group $SO(E)_1$:

$$SO(E)_1 \to U(H_J)/U(1),$$

where $U(H_J)/U(1)$ is the quotient of the unitary group $U(H_J)$ by its center $U(1)$.

Set

$$a_n = (e_n + iJe_n)/2,$$
$$a_n^* = (e_n - iJe_n)/2, \qquad n = 1, 2, 3, \ldots .$$

Then a_n, a_n^* obey the canonical anticommutation relations (CAR) of quantum theory:

$$a_j a_k + a_k a_j = 0 = a_j^* a_k^* + a_k^* a_j^*,$$
$$a_j^* a_k + a_k a_j^* = \delta_{jk}.$$

Therefore a * representation of the C^*-Clifford algebra determines a representation of the CAR. For this reason, $\mathscr{C}$ is frequently called the CAR algebra: see, for example, the article by Balslev *et al.* [4].

4. THE HILBERT–LIE GROUP $\mathrm{Spin}(E)_2$

Let $O(E)_2$ denote the group of all orthogonal operators on E of the form $1 + A$ with $\mathrm{Tr}|A|^2 < \infty$. Thus an orthogonal operator in $O(E)_2$ differs from the identity by a Hilbert–Schmidt perturbation, whereas an orthogonal operator in $O(E)_1$ differs from the identity by a trace-class perturbation.

Then $O(E)_2$ has a natural structure of Hilbert–Lie group (i.e., Lie group modeled on a Hilbert space) with Lie algebra $\mathbf{o}(E)_2$; the underlying topology is induced by the metric $d(1 + A_1, 1 + A_2) = \mathrm{Tr}|A_1 - A_2|^2$. The connected component of the identity, which is denoted by $SO(E)_2$, is of index 2 in $O(E)_2$ and has $\pi_1(SO(E)_2) = \mathbb{Z}_2$. See, once again, [11] and [13].

Let $(H_\tau, \pi_\tau, \Omega_\tau)$ be the Gelfand–Naimark–Segal triple determined by the trace τ, so that

$$\tau(u) = \langle \Omega_\tau, \pi_\tau(u)\Omega_\tau \rangle_{H_\tau}$$

for all u in the C^*-Clifford algebra. Let $\mathscr{A}$ denote the von Neumann algebra on H_τ generated by the range of the representation π_τ. The unique extensions to $\mathscr{A}$ of τ, α will be denoted by τ, α. Thus $\mathscr{A}$ is $\mathbb{Z}_2$-graded. In fact, $\mathscr{A}$ is the hyperfinite II_1 factor, and τ is its trace. Let

$$\mathrm{Spin}^c(E)_2 = \{u \in \mathscr{A} : uu^* = 1 = u^*u,\ \alpha(u) = u,\ uEu^* \subset E\}$$

Blattner has shown in [3] that the following sequence is exact

$$1 \to S^1 \to \mathrm{Spin}^c(E)_2 \to SO(E)_2 \to 1$$

in the category of abstract groups and homomorphisms, where the homomorphism $\zeta\colon \mathrm{Spin}^c(E)_2 \to SO(E)_2$ is given by $\zeta(u)x = uxu^*$.

We claim that the Hilbert space H_τ has a canonical conjugation, i.e., a map $L\colon H_\tau \to H_\tau$ such that

(i) $L^2 = 1$,

(ii) $\langle L\xi, L\eta \rangle_{H_\tau} = \langle \xi, \eta \rangle^*_{H_\tau}$.

The conjugation L determines a "real" subspace, namely

$$H_\tau^r = \{\xi \in H_\tau : L\xi = \xi\}.$$

According to the Gelfand–Naimark–Segal construction, H_τ is the completion of $\mathrm{Cl}(E) \otimes_{\mathbf{R}} \mathbb{C}$ in the $\|\cdot\|_2$ norm. Then L is the unique conjugation on H_τ such that

$$L(u \otimes c) = u \otimes c^*$$

for all $u \in \mathrm{Cl}(E)$ and $c \in C$, where c^* is the complex conjugate of c. Then $\mathrm{Spin}(E)_2$ is defined to be the subgroup of $\mathrm{Spin}^c(E)_2$ which leaves invariant the real subspace H_τ^r of H_τ.

Then $\mathrm{Spin}(E)_2$ has a natural structure of Hilbert–Lie group with Lie algebra $\mathbf{spin}(E)_2$. The underlying topology on $\mathrm{Spin}(E)_2$ is induced by the strong-operator topology.

The global analogue of Theorem 3 is

THEOREM 8. $\mathrm{Spin}(E)_2$ *is a model of the universal covering group of* $SO(E)_2$ *and the following square commutes*:

$$\begin{array}{ccc} \mathrm{Spin}(E)_2 & \longrightarrow & SO(E)_2 \\ \uparrow{\scriptstyle \exp} & & \uparrow{\scriptstyle \exp} \\ \mathbf{spin}(E)_2 & \longrightarrow & \mathbf{o}(E)_2 \end{array}$$

This result is proved in [19].

Now $\mathscr{A}$ acts on H_τ. Consequently $\mathrm{Spin}(E)_2$ acts on H_τ leaving H_τ^{r} invariant. This amounts to a real representation of $\mathrm{Spin}(E)_2$ by orthogonal operators on the real Hilbert space H_τ^{r}. Let us call this real representation θ.

An orthonormal basis (e_n) determines an inclusion $SO(\infty) \to SO(E)_2$, by identifying $SO(n)$ with the subgroup of $SO(E)_2$ which fixes all basis elements save the first n. This inclusion determines an inclusion $\mathrm{Spin}(\infty) \to \mathrm{Spin}(E)_2$.

THEOREM 9. (i) $\theta|\mathrm{Spin}(\infty) = \varinjlim \theta_n$ *where* $\theta_n = 2^{[(n+1)/2]}\Delta_n$.

(ii) $\chi(u) = \pm \prod_{j=1}^{\infty} \cos \dfrac{t_j}{2}$.

Here, χ is the character of the representation θ relative to the central trace τ, i.e.,

$$\chi(u) = \tau(\theta(u)),$$

and u is one of the two elements in the spin group $\mathrm{Spin}(E)_2$ which covers the direct sum of a sequence of plane rotations through t_n radians.

With this result, proved in [17], we conclude this brief survey.

Note added in proof. $\mathrm{Spin}(E)_2$ may be defined more simply as the intersection of $\mathrm{Spin}^c(E)_2$ with the real weakly closed subalgebra of $\mathscr{A}$ generated by E. Similarly, $\mathrm{Spin}(E)_\infty$ is the intersection of $\mathrm{Spin}^c(E)_\infty$ with the real uniformly closed subalgebra of $\mathscr{C}$ generated by E.

The subject of this essay is further developed in the author's forthcoming article [24] on the Dirac operator.

REFERENCES

1. M. F. ATIYAH, R. BOTT, AND A. A. SHAPIRO, Clifford modules, *Topology* **3** (Suppl. 1) (1964), 3–38.
2. M. F. ATIYAH AND I. G. MACDONALD, "Introduction to Commutative Algebra," Addison-Wesley, London, 1969.
3. R. J. BLATTNER, Automorphic group representations, *Pacific J. Math.* **8** (1958), 665–677.

4. E. Balslev, J. Manuceau, and A. Verbeure, Representations of anti-commutation relations and Bogoliubov transformations, *Commun. Math. Phys.* **8** (1968), 315–326.
5. R. Bott, "Lectures on K(X)," Benjamin, New York, 1969.
6. N. Bourbaki, "Algèbre," Chapter 9, Hermann, Paris, 1959.
7. E. Cartan, "Lecons sur la théorie de spineurs," Hermann, Paris, 1938.
8. C. Chevalley, "The Algebraic Theory of Spinors," Columbia Univ. Press, New York, 1954.
9. J. Dixmier, Formes linéaires sur un anneau d'opérateurs, *Bull. Soc. Math. Fr.* **81** (1953), 9–39.
10. G. H. Hardy, J. E. Littlewood, and G. Polya, "Inequalities," Cambridge Univ. Press, London and New York, 1967.
11. P. de la Harpe, Classical Banach-Lie algebras and Banach-Lie groups of operators in Hilbert space, Springer Lecture Notes 285 (1972).
12. P. de la Harpe, The Clifford algebra and the spinor group of a Hilbert space, *Comp. Math.* **25** (1972), 245–261.
13. P. de la Harpe, Some properties of infinite-dimensional orthogonal groups, *in* "Global analysis and its applications," Internat. Atomic Energy Agency, Vienna, 1974.
14. N. Jacobson, "Lie Algebras," Wiley (Interscience), New York, 1962.
15. P. Jordan and E. P. Wigner, Über das Paulische Äquivalenzverbot, *Z. Phys.* **47** (1928), 631–651.
16. J. Manuceau and A. Verbeure, The theorem on unitary equivalence of Fock representations, *Ann. Inst. H. Poincaré* **16**, No. 2, (1971), 87–91.
17. R. J. Plymen, Projective representations of the infinite orthogonal group, *Mathematika* **24** (1977), 115–121.
18. R. J. Plymen, Spinors in Hilbert space, *Math. Proc. Cambridge Philos. Soc.* **80** (1976), 337–347.
19. R. J. Plymen and R. F. Streater, A model of the universal covering group of $SO(H)_2$, *Bull. London Math. Soc* **7** (1975), 283–288.
20. R. J. Plymen and R. M. G. Young, On the spin algebra of a real Hilbert space, *J. London Math. Soc.* (2), **9** (1974), 286–292.
21. D. Shale and W. F. Stinespring, Spinor representations of infinite orthogonal groups, *J. Math. Mech.* **14** (1965), 315–322.
22. R. F. Streater, Interpolating norms for orthogonal and spin Lie algebras, *Symp. Math.* **14** (1974), 173–179.
23. D. M. Topping, "Lectures on von Neumann Algebras," van Nostrand-Reinhold, Princeton, New Jersey, 1971.
24. R. J. Plymen, The Laplacian and the Dirac operator in infinitely many variables, *Compositio Math*, to appear.

STUDIES IN ALGEBRA AND NUMBER THEORY
ADVANCES IN MATHEMATICS SUPPLEMENTARY STUDIES, VOL. 6

Explicit Class Field Theory in Global Function Fields†

DAVID R. HAYES

Department of Mathematics and Statistics, University of Massachusetts, Amherst, Massachusetts

Let k be a global function field. Drinfel'd has introduced the concept of an *elliptic module*, which provides an analog in k for the classical theory of complex multiplications of elliptic functions. He has shown that the elliptic modules of rank 1 provide an explicit version of the class field theory of k. We present a simplified treatment of these rank 1 results and also generalize them somewhat, obtaining results analogous to those in the "singular" theory of imaginary quadratic number fields. However, the results are valid for arbitrary global function fields.

Contents

INTRODUCTION

Let (k, ∞) be a pair consisting of the global function field k of characteristic p together with a fixed prime divisor ∞ of k. Let the field of constants of k be $\mathbb{F}_q$ where $q = p^n$. Put $d_\infty = \deg(\infty)$. Let A be the ring of elements of k which are integral at all primes $\mathfrak{P}$ of k other than ∞. Then A is a Dedekind ring with group of units $A^* = \mathbb{F}_q \backslash \{0\}$ and with class number $h(A) = d_\infty h$, where h is the class number of k. By a *class field* of A we mean a finite abelian extension field of k on which ∞ splits completely. These class fields correspond to the generalized ideal class groups of A in the familiar way.

† Supported in part by NSF Grant MPS-75-07494.

ISBN 0-12-599153-3

Let K be a field which contains a homomorphic image of A, and let G_{aK} be the additive group scheme over K. In [3], Drinfel'd introduces the concept of an *elliptic A-module over K*, by which he means a representation of A as a ring of endomorphisms of G_{aK}. He shows [3] that the elliptic A-modules of rank 1 provide a method for constructing the class fields of A in an explicit fashion. His methods are scheme-theoretic and very elegant, and he derives further results for elliptic A-modules of higher rank. In particular, he obtains a function field analog of the theorem of Eichler–Shimura out of the theory of rank 2 elliptic A-modules.

In this paper, we develop the rank 1 applications to the class field theory of k without appealing to the scheme-theoretic machinary. In its place, we use methods which are familiar from the Deuring theory of elliptic curves with complex multiplications [1]. These methods translate into the setting of function fields in such a way that one can present the theory without any appeal to geometry at all. This is because the additive group scheme is an altogether *linear* object. In fact, the additive group scheme itself is not mentioned explicitly since we need only its endomorphism ring, which can be described independently as a ring of "twisted polynomials" (cf., Section 2). Our treatment is therefore accessible to the reader without background in algebraic geometry, although some knowledge of elliptic curves would help him understand the motivation for the methods and also help explain some of the terminology. We do assume a knowledge of basic class field theory.

The theory is developed here with the ring A replaced throughout by an arbitrary order R in A (cf., Section 1). This generalizes the Drinfel'd rank 1 theory somewhat and provides an analog over k for the "singular theory" of elliptic curves with complex multiplications. The applications to the class field theory of k appear in Sections 8 and 9, the main rank 1 theorem of Drinfel'd being essentially equal in content to Theorem 9.7.

The plan of the paper is straightforward. The general theory of elliptic R-modules is presented in Part I. It is then shown in Part II that the assumption of rank 1 yields the class field theory of k in explicit form. In Part III, we prove some special results which are useful in computing elliptic R-modules of rank 1 and present some examples. We give an especially careful introduction to the *analytic theory* in Sections 4–5 since we need it for the computations in Part III. In Section 6, we prove the existence of the *smallest field of definition* for an elliptic R-module which is "without characteristic." This is a key result. As our development diverges from that of Drinfel'd at this point, it may help to orient the reader if we explain briefly how Drinfel'd handles the matter.

The functor which associates to every ring (or scheme) X over A the isomorphism classes of elliptic A-modules of rank r over X is not representable in general. Drinfel'd solves this problem by changing the functor (cf., [2]), eventually arriving (when $r = 1$) at representing rings A_I for each sufficiently small ideal I in A. The ring A_I turns out to be the integral closure of A in the ray class field of A associated to the ideal I! Our smallest field of definition is a poor substitute for these rings but can be made to serve the same purpose. We characterize it ultimately as the *Hilbert Class Field* of A (or R). We show in an appendix that a universal rank 1 elliptic A-module does exist for fields k having a prime divisor of degree 1 if we replace *isomorphism* of elliptic A-modules by the slightly less restrictive concept of *equivalence*.

Many of the results in this paper were obtained by the author in an attempt to generalize his work on rational function fields [4] before he became aware of the work of Drinfel'd. He would like to thank J. W. S. Cassels and S. Lang for encouraging him to write up his approach. Special thanks are owed to Oxford University for generously providing facilities during the preparation of this paper.

I. Elliptic Modules

1. *Orders in A*

Let (k, ∞) and A be as indicated above. A subring R of A which contains 1 and has k as its field of fractions is called an *order* in A. For the convenience of the reader, we give in this section a quick review of the general theory of such orders.

Recall that a *fractional ideal* of R is a nonzero noetherian R-submodule $\mathfrak{A}$ of k. The *coefficient ring* $\mathrm{End}(\mathfrak{A})$ of $\mathfrak{A}$ is the set of all $x \in k$ such that $x\mathfrak{A} \subseteq \mathfrak{A}$. It is also an order in A as $A \supseteq \mathrm{End}(\mathfrak{A}) \supseteq R$. If $\mathrm{End}(\mathfrak{A}) = R$, then we say that $\mathfrak{A}$ is *proper*. We put

$$\mathfrak{A}^* = \{x \in k \mid x\mathfrak{A} \subseteq R\},$$

and if $\mathfrak{A}\mathfrak{A}^* = R$, we say that $\mathfrak{A}$ is *invertible*. Invertible ideals are proper. We say ideals $\mathfrak{A}$ and $\mathfrak{B}$ are *isomorphic* if $\mathfrak{A} = x\mathfrak{B}$ for some $x \in k$.

Notation

(1.1) $\mathscr{I}(R)$ is the monoid of all fractional ideals of R.
(1.2) $\mathscr{I}^*(R)$ is the group of invertible ideals in $\mathscr{I}(R)$.
(1.3) $\mathscr{P}(R) \subseteq \mathscr{I}^*(R)$ is the group of nonzero principal ideals.
(1.4) $\mathscr{M}(R) \subseteq \mathscr{I}(R)$ is the monoid of integral ideals.
(1.5) $\mathscr{K}(R) = \{\mathfrak{A} \in \mathscr{I}(R) \mid \mathfrak{A}A = A\}$.
(1.6) $\mathscr{K}^*(R)$ is the set of invertible ideals in $\mathscr{K}(R)$.
(1.7) $\mathbf{Pic}(R) = \mathscr{I}^*(R)/\mathscr{P}(R)$ is the ideal class group of R.

(1.8) $\mathbf{M}_1(R)$ is the set of isomorphism classes of the ideals in $\mathscr{I}(R)$.
(1.9) $\mathbf{PM}_1(R)$ is the set of isomorphism classes of the *proper* ideals in $\mathscr{I}(R)$.

Note that both $\mathbf{M}_1(R)$ and $\mathbf{PM}_1(R)$ admit the ideal class group as a group of operators in a natural way.

The factorization of an ideal $\mathfrak{A} \in \mathscr{I}(A)$ into primes determines a *divisor* of k which is prime to ∞. For $\mathfrak{A} \in \mathscr{I}(R)$, let $\partial(\mathfrak{A})$ be the divisor determined by $\mathfrak{A}A$; and put $\deg \mathfrak{A} = \deg \partial(\mathfrak{A})$ and $\deg^* \mathfrak{A} = n \cdot \deg \mathfrak{A}$. (Recall that $q = p^n$.) Then for integral ideals $\mathfrak{A}$, we have

$$\deg \mathfrak{A} = \dim_{\mathbb{F}_q}(A/\mathfrak{A}A) \tag{1.10}$$

and

$$\deg^* \mathfrak{A} = \dim_{\mathbb{F}_p}(A/\mathfrak{A}A). \tag{1.11}$$

If $\mathfrak{A} = xR$, $x \in R$, then $\partial(\mathfrak{A})$ is the divisor of zeros of x because x is integral at every prime divisor of k other than ∞. Therefore, we have

$$\deg xR = -d_\infty \cdot v_\infty(x) \qquad (x \in R). \tag{1.12}$$

It is convenient also to introduce the multiplicative homomorphism $N\colon \mathscr{I}(R) \to \mathbb{Q}$ defined by $N(\mathfrak{A}) = q^{\deg \mathfrak{A}}$. We call $N(\mathfrak{A})$ the *norm* of $\mathfrak{A}$. When $\mathfrak{A} \in \mathscr{M}(R)$, we have

$$N(\mathfrak{A}) = \#(A/\mathfrak{A}A). \tag{1.13}$$

Now, given an order R in A, let

$$\mathfrak{C} = \{x \in k \mid xA \subseteq R\} \tag{1.14}$$

be the *conductor* of R. The conductor is the largest ideal in A which is also an ideal in R. It is never zero. We write $\mathscr{M}_{\mathfrak{C}}(R)$ (resp. $\mathscr{M}_{\mathfrak{C}}(A)$) for the monoid of ideals in R (resp. A) which are prime to $\mathfrak{C}$. One checks easily that the maps $\mathfrak{A} \mapsto \mathfrak{A}A$ and $\mathfrak{A} \mapsto R \cap \mathfrak{A}$ are inverse isomorphisms between the monoids $\mathscr{M}_{\mathfrak{C}}(R)$ and $\mathscr{M}_{\mathfrak{C}}(A)$. Thus, $\mathscr{M}_{\mathfrak{C}}(R)$ is a free monoid, freely generated by its prime ideals. One also checks easily that the natural map $R/\mathfrak{A} \mapsto A/\mathfrak{A}A$ is an isomorphism for $\mathfrak{A} \in \mathscr{M}_{\mathfrak{C}}(R)$. Thus,

$$N(\mathfrak{A}) = \#(R/\mathfrak{A}) \tag{1.15}$$

for $\mathfrak{A} \in \mathscr{M}_{\mathfrak{C}}(R)$.

LEMMA 1.1. *Suppose R' is a finite direct sum of local rings. Let A' be an R'-algebra which is faithful and finitely generated as an R'-module. Call an R'-submodule P of A'* invertible *in A' if $PP^* = R'$ where*

$$P^* = \{x \in A' \mid xP \subseteq R'\}.$$

Then, every invertible submodule has the form $R'u$, where u is an invertible element of A'.

Proof. By an easy reduction, we may assume that R' is local. Let $\mathfrak{M}'$ be its maximal ideal. Let P be an invertible R'-submodule of A'. Then P is projective and hence free over R'. Let $u_1, \ldots, u_s$ be a basis for P over R'. We show that $s \geqslant 2$ leads to a contradiction. Assume $s \geqslant 2$. Since P is invertible, there is an element $x \in P^*$ which does not map each basis element into $\mathfrak{M}'$. We may assume $xu_1 = r_1 \notin \mathfrak{M}'$. Let $xu_2 = r_2$. Then $x(r_2u_1 - r_1u_2) = 0$, which implies a nonzero linear relation between u_1 and u_2 as r_1 and x are invertible in A'. Thus, $P = R'u_1$; and u_1 is necessarily invertible. ■

PROPOSITION 1.2. *We have*: (1) $\mathfrak{A} \in \mathscr{K}(R) \Rightarrow A \supseteq \mathfrak{A} \supseteq \mathfrak{C}$. (2) $\mathscr{K}(R)$ *is a finite monoid.* (3) $\mathfrak{A} \in \mathscr{K}(R)$ *is invertible* $\Leftrightarrow \mathfrak{A} = Ru + \mathfrak{C}$ *for some element* $u \in A$ *which is prime to* $\mathfrak{C}$.

Proof. If $\mathfrak{A}A = A$, then $\mathfrak{C} = \mathfrak{C}A = \mathfrak{C}\mathfrak{A}A = \mathfrak{C}\mathfrak{A} \subseteq \mathfrak{A}$. Thus, the elements of $\mathscr{K}(R)$ correspond one-to-one with the R-submodules of the finite ring $A' = A/\mathfrak{C}$. This proves (1) and (2). Suppose $\mathfrak{A} \in \mathscr{K}(R)$ is invertible. Then $\mathfrak{A}\mathfrak{A}^* = R$ implies $\mathfrak{A}^* \in \mathscr{K}(R)$ also. Thus, the canonical image of $\mathfrak{A}$ in the quotient ring A' is invertible over $R' = R/\mathfrak{C}$ in the sense of Lemma 1.1. As R' is finite, it is artinian and hence a finite direct sum of primary rings. So (3) follows from Lemma 1.1. ■

COROLLARY 1.3. *Every* $\mathfrak{A} \in \mathscr{I}^*(R)$ *is isomorphic to an ideal in* $\mathscr{M}_{\mathfrak{C}}(R)$.

Proof. Suppose $\mathfrak{A}$ is an invertible ideal of R. Since A is a Dedekind ring, there is an element $x \in k$ such that $x\mathfrak{A}A$ is contained in $\mathscr{M}_{\mathfrak{C}}(A)$. Choose $\mathfrak{B} \in \mathscr{M}_{\mathfrak{C}}(R)$ so that $x\mathfrak{A}A = \mathfrak{B}A$. Then $x\mathfrak{A}\mathfrak{B}^* \in \mathscr{K}^*(R)$ so that, by Proposition 1.2, $x\mathfrak{A}\mathfrak{B}^* = Ru + \mathfrak{C}$ for some element $u \in A$ which is prime to $\mathfrak{C}$. Choose $v \in A$ such that $1 - uv \in \mathfrak{C}$. Then $vx\mathfrak{A}\mathfrak{B}^* = \mathfrak{D}$ is in $\mathscr{M}_{\mathfrak{C}}(R)$, and $vx\mathfrak{A} = \mathfrak{D}\mathfrak{B}$. ■

COROLLARY 1.4. *We have*

$$N(\mathfrak{A}) = \#(R/\mathfrak{A}) \tag{1.16}$$

for all invertible ideals in $\mathscr{M}(R)$.

Proof. Choose $z \in R$, $z \notin \mathbb{F}_q$, so that k is separable over the rational function field $F = \mathbb{F}_p(z)$. All fractional ideals in $\mathscr{I}(R)$ are free of rank $[k:F]$ over the polynomial ring $\mathbb{F}_p[z]$. For $\mathfrak{A} \in \mathscr{I}(R)$, put $N_1(\mathfrak{A}) = |\det M|$, where M is the matrix of transition from an $\mathbb{F}_p[z]$-basis of R to an $\mathbb{F}_p[z]$-basis of $\mathfrak{A}$ and where the absolute value is the normalized absolute value at ∞ on F ($|z| = p$). We observe in the usual way that $N_1(\mathfrak{A})$ is independent of the choice of bases and that for $x \in k$ and $\mathfrak{A} \in \mathscr{I}(R)$ we have: (1) $N_1(x\mathfrak{A}) = N_1(x) \cdot N_1(\mathfrak{A})$. (2) $N_1(x) = |\mathrm{Norm}_{k/F}(x)|$. (3) $N_1(\mathfrak{A}) = \#(R/\mathfrak{A})$ if $\mathfrak{A} \subseteq R$.

Now suppose $\mathfrak{A} \in \mathscr{M}(R)$ is invertible. It suffices to prove that $N(\mathfrak{A}) = N_1(\mathfrak{A})$. By Corollary 1.3, we can choose $x \in k$ such that $x\mathfrak{A} \subseteq \mathscr{M}_{\mathfrak{C}}(R)$. Then by (1.15), we have

$$N_1(x) \cdot N_1(\mathfrak{A}) = N_1(x\mathfrak{A}) = N(x\mathfrak{A}) = N(x) \cdot N(\mathfrak{A}).$$

But $N_1(x) = |\mathrm{Norm}_{k/F}(x)| = N(x)$. ■

It is well known that $h(A) = \#\mathbf{Pic}(A) = hd_\infty$ is finite. We show next that $h(R) = \#\mathbf{Pic}(R)$ is also finite for any order R in A, and we compute $h(R)$ in terms of the class number h of k. We call $h(R)$ the *class number* of R.

THEOREM 1.5. *Let R be an order in A with conductor $\mathfrak{C}$. Then the ideal class group* $\mathbf{Pic}(R)$ *is finite; and*

$$h(R) = h \cdot d_\infty \cdot \frac{\Phi_A(\mathfrak{C})}{e_R \cdot \Phi_R(\mathfrak{C})}, \tag{1.17}$$

where $\Phi_A(\mathfrak{C})$ (resp. $\Phi_R(\mathfrak{C})$) is the number of elements in the group of units of the ring $A/\mathfrak{C}$ (resp. $R/\mathfrak{C}$) and where e_R is the index of the group of units of R in $\mathbb{F}_q^*$. The representation spaces $\mathbf{M}_1(R)$ *and* $\mathbf{PM}_1(R)$ *of* $\mathbf{Pic}(R)$ *are also finite. In fact, each orbit of* $\mathbf{Pic}(R)$ *in* $\mathbf{M}_1(R)$ *contains an isomorphism class which intersects the finite monoid $\mathscr{K}(R)$ nonvacuously.*

Proof. The map $\mathfrak{A} \mapsto \mathfrak{A}A$ from $\mathscr{I}^*(R)$ to $\mathscr{I}(A)$ induces a homomorphism $\psi: \mathbf{Pic}(R) \to \mathbf{Pic}(A)$. This homomorphism is surjective since $\mathscr{M}_{\mathfrak{C}}(A)$ contains a representative of every ideal class of A. Suppose $\mathfrak{A}$ belongs to a class in $\ker(\psi)$. Then $\mathfrak{A}A$ is principal—say $\mathfrak{A}A = xA$—and $x^{-1}\mathfrak{A} \in \mathscr{K}^*(R)$. Therefore, every ideal class in $\ker(\psi)$ intersects $\mathscr{K}^*(R)$ nonvacuously. Certainly then, $\mathbf{Pic}$ (R) is finite, and we have $h(R) = h(A) \cdot s = h \cdot d_\infty \cdot s$ where $s = \#\ker(\psi)$. We next determine s. Suppose $\mathfrak{A} = x\mathfrak{B}$ with $\mathfrak{A}$ and $\mathfrak{B}$ both in $\mathscr{K}^*(R)$. Then $A = xA$, which implies that x is a unit in A; i.e., $x \in \mathbb{F}_q^*$. Therefore,

$$s \cdot e_R = \#\mathscr{K}^*(R).$$

Now, it is clear from Proposition 1.2 that $\#\mathscr{K}^*(R)$ is the number of R-submodules of $A/\mathfrak{C}$ of the form Ru, u a unit in $A/\mathfrak{C}$. This number is easily computed, and it is found to yield the required value (1.17) for $h(R)$.

The last assertion can be proved as follows: Given $\mathfrak{A} \in \mathscr{I}(R)$, there is an invertible ideal $\mathfrak{B}$ in $\mathscr{I}^*(R)$ such that $\mathfrak{A}A$ is isomorphic to $\mathfrak{B}A$. Thus, there is an element $x \in k$ such that $x\mathfrak{A}\mathfrak{B}^* \in \mathscr{K}(R)$ as required. ■

THEOREM 1.6. *Suppose* $\mathfrak{A} \in \mathscr{I}(R)$. *Then* $\mathfrak{A}$ *is invertible* $\Leftrightarrow \mathrm{End}(\mathfrak{A}^*) = R$.

Proof. Certainly $\mathrm{End}(\mathfrak{A}^*) = R$ if $\mathfrak{A}$ is invertible. Conversely, suppose that $\mathrm{End}(\mathfrak{A}^*) = R$. We have to show that $\mathfrak{A}$ is invertible. Now, multiplying $\mathfrak{A}$ by an invertible ideal leaves $\mathrm{End}(\mathfrak{A}^*)$ invariant. Therefore, by the last assertion of Theorem 1.5, we may assume $\mathfrak{A} \in \mathscr{K}(R)$ without loss of generality. Since $\mathscr{K}(R)$ is a finite monoid, there are integers $r > 0$ and $s \geqslant 0$ such that $\mathfrak{A}^r \cdot \mathfrak{A}^s = \mathfrak{A}^s$. Successive multiplications by $\mathfrak{A}^*$ leads us to the conclusion $\mathfrak{A}^r \cdot \mathfrak{B}^s \subseteq R$ where $\mathfrak{B} = \mathfrak{A}^*\mathfrak{A}$. The cancellation rule $\mathfrak{D}\mathfrak{B} \subseteq R \Rightarrow \mathfrak{D} \subseteq R$ is valid for $\mathfrak{B}$ as follows: $\mathfrak{D}\mathfrak{A}^*\mathfrak{A} \subseteq R \Rightarrow \mathfrak{D}\mathfrak{A}^* \subseteq \mathfrak{A}^* \Rightarrow \mathfrak{D} \subseteq \mathrm{End}(\mathfrak{A}^*) = R$. Thus, $\mathfrak{A}^r \subseteq R$. We claim in fact that $\mathfrak{A}^r = R$. For if not, then $\mathfrak{A}^r$ is contained in a prime ideal of R which is itself contained in a prime ideal of A by the "going-up" theorem, contrary to our assumption that $\mathfrak{A}A = A$. Thus $\mathfrak{A}^r = R$, and $\mathfrak{A}$ is invertible. ■

Finally, let us note that every ideal $\mathfrak{C}$ in A such that $\deg^* \mathfrak{C} > 1$ is the conductor of at least one order in A. In fact, the minimal order with conductor $\mathfrak{C}$ is just $R = \mathbb{F}_p + \mathfrak{C}$.

2. *Elementary Theory of Elliptic R-Modules*

Let K be any field of characteristic p. Let $K[\varphi]$ be the $\mathbb{F}_p$-algebra generated by the elements of K and another generator φ subject only to the relations in K and the additional relations

$$\varphi w = w^p \varphi \tag{2.1}$$

for all $w \in K$. Each element of $K[\varphi]$ can be written in a unique way as a polynomial in φ with *left* coefficients from K. Therefore, $K[\varphi]$ is a *twisted polynomial ring* in the sense of [5]. There is a right division algorithm for $K[\varphi]$, and so each of its left ideals is principal. Further, any K-algebra B becomes a $K[\varphi]$-module if we map φ to the Frobenius endomorphism $b \mapsto b^p$, $b \in B$, as is immediately clear from the definitions. This construction of $K[\varphi]$-modules out of K-algebras is functorial.

Let $i: K \to K[\varphi]$ be the canonical *inclusion* which maps each element $w \in K$ onto the constant polynomial w; and let $D: K[\varphi] \to K$ be the canonical *augmentation* which maps each polynomial onto its constant coefficient.

Suppose now that (k, ∞) and A are as indicated in the introduction, and let R be an order in A. The idea of Drinfel'd is that for suitable K (e.g., K algebraically closed), there should exist nontrivial embeddings of R into $K[\varphi]$. Thus, each order in A should be a "ring of polynomials in φ." (Actually, Drinfel'd considers only $R = A$, but the generalization to an arbitrary order is both natural and obvious.)

DEFINITION 2.1. Let K be a field of characteristic p. An *elliptic R-module over K* is a ring homomorphism $\rho: R \to K[\boldsymbol{\varphi}]$ such that $\rho \neq i \circ D \circ \rho$.

We write ρ_x for the polynomial which is the image of $x \in R$ under ρ, and we write $\rho_x(t)$ for the action of ρ_x on a given element t in some K-algebra B. We note that $\rho_x(t)$ is an additive polynomial in t of degree p^d where $d = \deg \rho_x$.

PROPOSITION 2.2. *An elliptic R-module ρ is an injection of R into $K[\boldsymbol{\varphi}]$.*

Proof. As $K[\boldsymbol{\varphi}]$ contains no zero divisors, $\ker(\rho)$ is a prime ideal. Now, every nonzero prime ideal in R is maximal. Thus, $\ker(\rho) \neq 0$ implies that each ρ_x, $x \in R$, is a constant polynomial, contrary to the requirement $\rho \neq i \circ D \circ \rho$. ■

Consider the map $x \mapsto -\deg \rho_x$ for $x \in R$. As one easily checks, this map is a nontrivial valuation on R, and its unique extension to k defines the prime divisor ∞. Thus, there is a rational number r_ρ such that

$$\deg \rho_x = -n \cdot d_\infty \cdot r_\rho \cdot v_\infty(x) \tag{2.2}$$

for all $x \in R$. We call r_ρ the *rank* of ρ. We prove below that r_ρ is actually an integer.

For $x \in R$, let $j(x)$ denote the smallest integer s such that the monomial $\boldsymbol{\varphi}^s$ appears in ρ_x with nonzero coefficient. The map $x \mapsto j(x)$ is also a valuation on R. When $\ker(D \circ \rho) = 0$, this valuation is trivial, and we say that ρ is "without characteristic." Otherwise, the valuation determines a unique maximal ideal $\mathfrak{P}_0$ in A such that $\mathfrak{P}_0 \cap R = \ker(D \circ \rho)$. We call $\mathfrak{P}_0$ the "characteristic" of ρ. Let v_0 be the normalized valuation at $\mathfrak{P}_0$. Then there is a rational number h_ρ such that

$$j(x) = \deg^* \mathfrak{P}_0 \cdot h_\rho \cdot v_0(x) \tag{2.3}$$

for all $x \in R$. We put $\mathfrak{P}_0 = 0$ and $h_\rho = 0$ when ρ is "without characteristic." The invariant h_ρ is called the *height* of ρ. It too is an integer, as we prove below.

Do elliptic R-modules over K of every rank exist? This question is answered affirmatively for the case when K is a universal domain for k in Section 4. For the moment, we continue to study the elementary aspects of the theory.

Given an elliptic R-module ρ over K and a K-algebra B, let R act on B through ρ. Call the resulting R-module B_ρ. In particular, consider K_ρ^{ac} where K^{ac} is the algebraic closure of K. Since the action of $x \in R$ on $t \in K_\rho^{\text{ac}}$ is given by the polynomial $\rho_x(t)$, K_ρ^{ac} is a *divisible* R-module. Let T_ρ be the torsion submodule of K_ρ^{ac}, and for every ideal $\mathfrak{A} \in \mathscr{M}(R)$, let $T_\rho(\mathfrak{A})$ be the submodule of $\mathfrak{A}$-torsion. We shall determine the structure of the $\mathfrak{P}$-primary submodule $T_{\rho,\mathfrak{P}}$ of T_ρ for every *invertible* prime ideal $\mathfrak{P}$ in $\mathscr{M}(R)$.

Let $R_\mathfrak{P}$ be the local ring of R at $\mathfrak{P}$. Since $\mathfrak{P}$ is invertible, $R_\mathfrak{P}$ is a discrete valuation ring, and so

$$T_{\rho,\mathfrak{P}} = T_\rho \otimes_R R_\mathfrak{P}.$$

One knows that a divisible torsion module over the valuation ring $R_\mathfrak{P}$ is a direct sum of submodules each isomorphic to $k/R_\mathfrak{P}$. Now, some power of $\mathfrak{P}$ is principal—say $\mathfrak{P}^e = xR$—so that $T_\rho(\mathfrak{P}^e)$ is just the set of roots of $\rho_x(t)$ in K^{ac}. Thus, recalling (1.12), we compute

$$\begin{aligned}
\dim_{\mathbb{F}_p} T_\rho(\mathfrak{P}^e) &= \deg \rho_x - j(x)\\
&= -r_\rho \cdot d_\infty \cdot n \cdot v_\infty(x) - h_\rho \cdot \deg^* \mathfrak{P}_0 \cdot v_0(x)\\
&= r_\rho \cdot \deg^* x - h_\rho \cdot \deg^* \mathfrak{P}_0 \cdot v_0(x)\\
&= \begin{cases} r_\rho \cdot e \cdot \deg^* \mathfrak{P} & \text{if} \quad \mathfrak{P} \neq \mathfrak{P}_0 \cap R\\ (r_\rho - h_\rho) \cdot e \cdot \deg^* \mathfrak{P} & \text{if} \quad \mathfrak{P} = \mathfrak{P}_0 \cap R. \end{cases}
\end{aligned}$$

A simple count allows us to conclude from this result that

$$T_{\rho,\mathfrak{P}} \cong (k/R_\mathfrak{P})^{r_\rho} \qquad (\mathfrak{P} \neq \mathfrak{P}_0 \cap R) \tag{2.4}$$

and

$$T_{\rho,\mathfrak{P}} \cong (k/R_\mathfrak{P})^{r_\rho - h_\rho} \qquad (\mathfrak{P} = \mathfrak{P}_0 \cap R) \tag{2.5}$$

provided that $\mathfrak{P}$ is invertible. In particular, we have proved

PROPOSITION 2.3. *The rank of an elliptic R-module ρ over K is a positive integer. If $\mathfrak{P}_0 \cap R$ is invertible, then the height of ρ is also an integer.*

3. *The Action of* **Pic**(R)

In this section, we show that the ideal class group of R acts in a natural way on the isomorphism classes of elliptic R-modules over K.

DEFINITION 3.1. Let ρ and ρ' be elliptic R-modules over K. A polynomial $\sigma \in K[\boldsymbol{\varphi}]$ such that $\sigma \cdot \rho_x = \rho'_x \cdot \sigma$ for all $x \in R$ is called an *isogeny from ρ to ρ'*.

Let $\mathscr{E}\ell\ell_R(K)$ be the category whose objects are the elliptic R-modules over K and whose morphisms are the isogenies. Composition of morphisms is defined to be the ordinary product. We will work in this category. Thus, two elliptic R-modules ρ and ρ' over K are *isomorphic* if there is an element $w \neq 0$ in K such that $w \cdot \rho_x = \rho'_x \cdot w$ for all $x \in R$.

The set $\mathbf{Isog}(\rho, \rho')$ of isogenies from ρ to ρ' is obviously an additive group. Note that $\mathbf{Isog}(\rho, \rho')$ contains a nonzero polynomial only if $r_\rho = r_{\rho'}$ and $h_\rho = h_{\rho'}$. Thus, the rank and height are invariant under nonzero isogenies.

We now show how one can generate isogenies using a given ideal $\mathfrak{A} \in \mathscr{M}(R)$. For each elliptic R-module ρ over K, let $I_{\mathfrak{A},\rho}$ be the left ideal generated in $K[\boldsymbol{\varphi}]$ by the polynomials ρ_a, $a \in \mathfrak{A}$. As left ideals are principal, $I_{\mathfrak{A},\rho} = K[\boldsymbol{\varphi}] \cdot \rho_{\mathfrak{A}}$ for a uniquely determined monic polynomial $\rho_{\mathfrak{A}}$ in $K[\boldsymbol{\varphi}]$; and we have

$$\rho_{\mathfrak{A}} = \tau_1 \cdot \rho_{a_1} + \tau_2 \cdot \rho_{a_2} + \cdots + \tau_m \cdot \rho_{a_m} \tag{3.1}$$

for suitable elements $a_1, \ldots, a_m$ in $\mathfrak{A}$ and polynomials $\tau_1, \ldots, \tau_m$ in $K[\boldsymbol{\varphi}]$. Because $\mathfrak{A}$ is an ideal, $I_{\mathfrak{A},\rho}$ is carried into itself by multiplication on the *right* by the polynomials ρ_x, $x \in R$. Therefore, for every $x \in R$, there is a unique polynomial ρ'_x in $K[\boldsymbol{\varphi}]$ such that

$$\rho_{\mathfrak{A}} \cdot \rho_x = \rho'_x \cdot \rho_{\mathfrak{A}}. \tag{3.2}$$

One checks easily that the map $\rho': R \to K[\boldsymbol{\varphi}]$ defined by $x \mapsto \rho'_x$ is also an elliptic R-module over K. We introduce the notation

$$\rho' = \mathfrak{A} * \rho. \tag{3.3}$$

Thus, we have an operation $*$ of the ideals in $\mathscr{M}(R)$ on $\mathscr{E}\ell\ell_R(K)$ which we can characterize as follows: *$\mathfrak{A} * \rho$ is the unique elliptic R-module ρ' over K such that $\rho_{\mathfrak{A}} \in \mathbf{Isog}(\rho, \rho')$*

Proposition 3.2. *Let $\mathfrak{A} \in \mathscr{M}(R)$, $\rho \in \mathscr{E}\ell\ell_R(K)$, and $R_1 = \mathrm{End}(\mathfrak{A})$. Then, there is a unique elliptic R_1-module over K whose restriction to R is $\mathfrak{A} * \rho$.*

Proof. Given $z \in R_1$, choose elements $x, y \in R$, $y \neq 0$, such that $zy = x$. For all $a \in \mathfrak{A}$, we have $az \in \mathfrak{A}$ which allows us to write

$$\rho_a \cdot \rho_x = \rho_{azy} = \rho_{az} \cdot \rho_y. \tag{3.4}$$

We now multiply both sides of (3.1) by ρ_x on the right and use (3.4) to factor out ρ_y. We conclude that

$$\rho_{\mathfrak{A}} \cdot \rho_x = \tau \cdot \rho_y \tag{3.5}$$

for some polynomial $\tau \in I_{\mathfrak{A},\rho}$. Let $\tau = \tau_z \cdot \rho_{\mathfrak{A}}$, $\tau_z \in K[\boldsymbol{\varphi}]$. Then (3.2) and (3.3) show that (3.5) can be rewritten as

$$(\mathfrak{A} * \rho)_x \cdot \rho_{\mathfrak{A}} = \tau_z \cdot (\mathfrak{A} * \rho)_y \cdot \rho_{\mathfrak{A}}$$

which implies that

$$(\mathfrak{A} * \rho)_x = \tau_z \cdot (\mathfrak{A} * \rho)_y.$$

It is easily verified that τ_z is independent of the choice of x and y and that $z \mapsto \tau_z$ is an elliptic R_1-module which coincides with $\mathfrak{A} * \rho$ on R. ■

PROPOSITION 3.3. *The height of an elliptic R-module ρ over K is a nonnegative integer.*

Proof. Let $\mathfrak{C}$ be the conductor of R. By the above proposition, $\mathfrak{C} * \rho$ can be regarded as an elliptic A-module. It therefore has integral height by Proposition 2.3 as every ideal in A is invertible. But ρ and $\mathfrak{C} * \rho$ have the same height since they are isogenous. ■

For $\mathfrak{A} \in \mathscr{M}(R)$ and $\rho \in \mathscr{E}\ell\ell_R(K)$, let $j(\mathfrak{A})$ be the least integer s such that φ^s appears in $\rho_{\mathfrak{A}}$ with nonzero coefficient. Put $v_0(\mathfrak{A}) = \min\{v_0(a)\}$, $a \in \mathfrak{A}$. Then, as one easily checks,

$$j(\mathfrak{A}) = \deg^* \mathfrak{P}_0 \cdot h_\rho \cdot v_0(\mathfrak{A}), \tag{3.6}$$

a natural extension of (2.3).

Now, each elliptic R-module determines a natural homomorphism from R into K—namely, $D \circ \rho$. Proposition 3.3 and (3.6) together imply

COROLLARY 3.4. *The homomorphism $D \circ \rho$ is invariant under the operation $*$.*

Proof. From the definitions, for given $\mathfrak{A} \in \mathscr{M}(R)$ we have

$$D((\mathfrak{A} * \rho)_x) = [D(\rho_x)]^{p^{j(\mathfrak{A})}}$$

for all $x \in R$. If ρ is "without characteristic," then $j(\mathfrak{A}) = 0$, and we are done. Otherwise, each element $D(\rho_x)$ is contained in a finite subfield of K which is isomorphic to $A/\mathfrak{P}_0$. By (3.6) and Proposition 3.3, $p^{j(\mathfrak{A})}$ is a power of the order of this subfield, and so multiplying an element of this subfield by itself $p^{j(\mathfrak{A})}$ times leaves it invariant. ■

We next compute the degree of $\rho_{\mathfrak{A}}$ for *invertible* ideals $\mathfrak{A}$. First, we require two lemmas.

LEMMA 3.5. *Let $\mathfrak{A}$ be a principal ideal—say $\mathfrak{A} = zR$, $z \in R$. Let w be the leading coefficient of ρ_z. Then $\rho_{\mathfrak{A}} = w^{-1} \cdot \rho_z$, and $(\mathfrak{A} * \rho)_x = w^{-1} \cdot \rho_x \cdot w$ for all $x \in R$.*

LEMMA 3.6. *For any $\mathfrak{A} \in \mathscr{M}(R)$, the roots of $\rho_{\mathfrak{A}}(t)$ in a K-algebra B are precisely the points of $\mathfrak{A}$-torsion in the R-module B_ρ.*

The proofs are easy and will be omitted. As an immediate corollary of the second lemma, we deduce the fundamental formula

$$\deg \rho_{\mathfrak{A}} = j(\mathfrak{A}) + \dim_{\mathbb{F}_p} T_\rho(\mathfrak{A}), \tag{3.7}$$

where $\mathfrak{A} \in \mathscr{M}(R)$ and $T_\rho(\mathfrak{A})$ is the $\mathfrak{A}$-torsion submodule of K_ρ^{ac}.

PROPOSITION 3.7. *Let $\rho \in \mathscr{E}\ell\ell_R(K)$. For every invertible ideal $\mathfrak{A}$ in R,*

$$\deg \rho_{\mathfrak{A}} = r_\rho \cdot \deg^* \mathfrak{A}. \tag{3.8}$$

Proof. First, assume $\mathfrak{A} \in \mathscr{M}_{\mathfrak{C}}(R)$, where $\mathfrak{C}$ is the conductor of R. Let

$$\mathfrak{A} = \prod_{\mathfrak{P}} \mathfrak{P}^{e_{\mathfrak{P}}}$$

be the prime factorization of $\mathfrak{A}$ in the free monoid $\mathscr{M}_{\mathfrak{C}}(R)$. Then by (2.4) and (2.5),

$$\dim_{\mathbb{F}_p} T_\rho(\mathfrak{A}) = \sum_{\mathfrak{P}} \dim_{\mathbb{F}_p} T_\rho(\mathfrak{P}^{e_{\mathfrak{P}}}) = \sum_{\mathfrak{P}} e_{\mathfrak{P}} \cdot \deg^* \mathfrak{P} \cdot m_{\mathfrak{P}},$$

where $m_{\mathfrak{P}} = r_\rho$ for $\mathfrak{P} \neq \mathfrak{P}_0 \cap R$ and $m_{\mathfrak{P}} = r_\rho - h_\rho$ for $\mathfrak{P} = \mathfrak{P}_0 \cap R$. Consider the cases (1) $\mathfrak{P}_0 \notin \mathscr{M}_{\mathfrak{C}}(R)$ and (2) $\mathfrak{P}_0 \in \mathscr{M}_{\mathfrak{C}}(R)$ separately. In the first case, each $m_{\mathfrak{P}} = r_\rho$ so that $\dim_{\mathbb{F}_p} T_\rho(\mathfrak{A}) = r_\rho \cdot \deg^* \mathfrak{A}$. Since $j(\mathfrak{A}) = 0$ in this case, (3.8) follows from (3.7). In the second case, we have

$$\begin{aligned}\dim_{\mathbb{F}_p} T_\rho(\mathfrak{A}) &= r_\rho \cdot \deg^* \mathfrak{A} - h_\rho \cdot e_{\mathfrak{P}_0} \cdot \deg^* \mathfrak{P}_0 \\ &= r_\rho \cdot \deg^* \mathfrak{A} - j(\mathfrak{A})\end{aligned}$$

by (3.6) as $v_0(\mathfrak{A}) = e_{\mathfrak{P}_0}$. Thus, (3.8) follows from (3.7) in this case also.

Now let $\mathfrak{A}$ be any invertible ideal. We know from Corollary 1.3 that there is an element $z \in k$ such that $z\mathfrak{A} \in \mathscr{M}_{\mathfrak{C}}(R)$. Choose $x, y \in R$, $y \neq 0$, such that $zy = x$. From the definitions,

$$\rho_{\mathfrak{A}x} = w_1 \cdot \rho_{\mathfrak{A}} \cdot \rho_x = w_2 \cdot \rho_{\mathfrak{A}z} \cdot \rho_y$$

for suitable constants w_1 and $w_2 \in K$. Since we know $\deg \rho_{\mathfrak{A}z}$, we can easily compute $\deg \rho_{\mathfrak{A}}$. We find $\deg \rho_{\mathfrak{A}} = r_\rho \cdot \deg^* \mathfrak{A}$ as required. ∎

Suppose $\mathfrak{P}_0 \cap R$ is invertible in R. Then $\mathfrak{P}_0$ is the only prime ideal of A which sits over $\mathfrak{P}_0 \cap R$, so it is sensible (if abusive of language) to call $\mathfrak{P}_0 \cap R$ also "$\mathfrak{P}_0$."

COROLLARY 3.8. *Suppose ρ has rank 1 and "characteristic" $\mathfrak{P}_0 \in \mathscr{I}^*(R)$. If $\mathfrak{A} = \mathfrak{P}_0^e$, $e \geqslant 1$, then*

$$\rho_{\mathfrak{A}} = \varphi^{\deg^* \mathfrak{A}}. \tag{3.9}$$

Proof. This follows from (3.6) and (3.8). ∎

COROLLARY 3.9. *Suppose ρ has rank 1. Let d be the greatest common divisor of the exponents s of all those monomials $\boldsymbol{\varphi}^s$ which appear in some ρ_x, $x \in R$, with nonzero coefficient. Let f_R be the $\mathbb{F}_p$-dimension of $R \cap \mathbb{F}_q$. Then $f_R | d | n$.*

Proof. Since the elements in $R \cap \mathbb{F}_q$ map onto constant polynomials in $K[\boldsymbol{\varphi}]$ which must commute with each ρ_x, $x \in R$, it is clear that f_R divides d. Also, from the definitions each $T_\rho(\mathfrak{P})$, $\mathfrak{P}$ an invertible prime ideal in R, $\mathfrak{P} \neq \mathfrak{P}_0$, is an $\mathbb{F}_{p^d}$-vector space. So, by (2.4), d divides $\deg^* \mathfrak{P}$ for all such $\mathfrak{P}$. Now, since k has a divisor of degree 1, the greatest common divisor of the degrees of the prime divisors outside any finite set is 1. As $\deg^* \mathfrak{P} = n \cdot \deg \mathfrak{P}$, we must have d dividing n.

Proposition 3.7 has important consequences for the operation $*$ defined by (3.3).

THEOREM 3.10. *Let $\mathfrak{A}$, $\mathfrak{B}$ be invertible ideals in $\mathscr{M}(R)$. Let ρ be an elliptic R-module over K. Then*

$$\rho_{\mathfrak{A}\mathfrak{B}} = (\mathfrak{B} * \rho)_{\mathfrak{A}} \cdot \rho_{\mathfrak{B}} \tag{3.10}$$

and

$$\mathfrak{A} * (\mathfrak{B} * \rho) = (\mathfrak{A}\mathfrak{B}) * \rho. \tag{3.11}$$

Proof. For $a \in \mathfrak{A}$ and $b \in \mathfrak{B}$, we have

$$\rho_{ab} = \rho_b \cdot \rho_a = \tau_1 \cdot \rho_{\mathfrak{B}} \cdot \rho_a = \tau_1 \cdot (\mathfrak{B} * \rho)_a \cdot \rho_{\mathfrak{B}} = \tau_1 \cdot \tau_2 \cdot (\mathfrak{B} * \rho)_{\mathfrak{A}} \cdot \rho_{\mathfrak{B}}$$

for suitable polynomials τ_1 and τ_2 in $K[\boldsymbol{\varphi}]$. This implies that $(\mathfrak{B} * \rho)_{\mathfrak{A}} \cdot \rho_{\mathfrak{B}}$ is a right divisor of $\rho_{\mathfrak{A}\mathfrak{B}}$. But by Proposition 3.7, both these polynomials have the same degree. Since both are monic, the equality (3.10) must hold. Now, (3.11) is a formal consequence of (3.10). ■

Given a ring homomorphism δ from R into the field K and a positive integer r, let $\mathscr{E}_R(\delta, r)$ denote the set of isomorphism classes of elliptic R-modules ρ over K such that $r_\rho = r$ and $D \circ \rho = \delta$. We leave to the reader the simple task of checking that $*$ induces an operation (also called "$*$") of the ideals in $\mathscr{I}^*(R)$ on $\mathscr{E}_R(\delta, r)$ and that the principal ideals operate trivially. In particular, therefore, we conclude from (3.11) that $\mathscr{E}_R(\delta, r)$ admits the ideal class group $\mathbf{Pic}(R)$ as a group of operators in a natural way. This fact, for $r = 1$, is of central importance for the applications to the class field theory of k given in Part II.

4. *Analytic Theory*

Let k_∞ be the completion of k at ∞, and let Ω be the completion of the algebra closure of k_∞. The valuation v_∞ extends to a nondiscrete valuation on Ω, which we denote by the same symbol. In this section, we demonstrate the existence of elliptic R-modules over Ω for any order R in A. These appear as "complex multiplications" satisfied by certain analytic functions associated to "lattices" in Ω which are also R-modules.

As usual, for any divisor $\mathfrak{A}$ of k, $L(\mathfrak{A})$ denotes the $\mathbb{F}_q$-vector space of all elements $x \in k$ such that $v_{\mathfrak{P}}(x) \geqslant v_{\mathfrak{P}}(\mathfrak{A})$ for all prime divisors $\mathfrak{P}$ of k. We note that

$$A = \bigcup_{\mu=0}^{\infty} L(\infty^{-\mu}). \tag{4.1}$$

DEFINITION 4.1. A *lattice* in the k_∞-vector space Ω is a nonzero $\mathbb{F}_p$-vector subspace Γ satisfying:

Lat 1. Γ is discrete in the valuation topology.
Lat 2. $k_\infty \Gamma$ is finite dimensional over k_∞.
Lat 3. $x\Gamma \subseteq \Gamma$ for at least one element $x \in k_\infty$ which is not a k_∞-constant.

EXAMPLE 4.2. Let R be an order in A. Then any rank 1 R-module Γ in Ω is a lattice. Certainly **Lat** 2 and **Lat** 3 hold for such a Γ. Also, since Γ has rank 1 over R, there is an element $w \in \Omega$ such that $w\Gamma \subseteq R \subseteq A$. Now, A is discrete since each set $L(\infty^{-\mu})$ in the union (4.1) is finite. Thus, **Lat** 1 also holds for Γ.

We note that $v_\infty(x) < 0$ for any $x \in k_\infty$ satisfying **Lat** 3: If $v_\infty(x) \geqslant 0$, then $y = x^s - x$, $s = q^{d_\infty}$, also satisfies **Lat** 3. Now $v_\infty(y) > 0$ so that the sequence $y^\nu \gamma$, $\nu \geqslant 0$, converges to zero for every $\gamma \in \Gamma$. If $\gamma \neq 0$, this contradicts **Lat** 1.

The following characterization of lattices is proved in the same way as its analog for euclidean spaces.

THEOREM 4.3. *Let Γ be a lattice, and let x satisfy* **Lat** 3. *Then Γ is an $\mathbb{F}_p[x]$-module of finite rank. Further, every $\mathbb{F}_p[x]$-basis of Γ is linearly independent over the closure of $\mathbb{F}_p(x)$ in k_∞.*

DEFINITION 4.4. A polynomial or power series $F(u)$ with coefficients in an extension field of $\mathbb{F}_p$ is *linear* if only monomials with exponent a power of p appear in $F(u)$ with nonzero coefficient. Let $D(F)$ denote the coefficient of u in $F(u)$. We call $D(F)$ the *tangent vector* of F. We say that F is *separable* if $D(F) \neq 0$.

The following easily proved lemma gives a useful criterion for a polynomial to be linear.

LEMMA 4.5. *Let $F(u)$ be a polynomial without multiple roots. Then $F(u)$ is linear of degree p^s if and only if its set of roots is an s-dimensional vector space over $\mathbb{F}_p$.*

It is well known [8] that an entire function on Ω (i.e., a function defined by an everywhere convergent power series) is determined up to a multiplicative constant by its roots, multiplicities being counted. In particular, we have

LEMMA 4.6. *If $\lambda_1(u)$ and $\lambda_2(u)$ are everywhere convergent linear power series with $D(\lambda_1) = D(\lambda_2) \neq 0$ and if $\lambda_1(u)$ and $\lambda_2(u)$ have the same set of roots, then $\lambda_1(u) = \lambda_2(u)$.*

Now, let Γ be any lattice in Ω. By **Lat** 1 and **Lat** 2, the series

$$\sum \frac{1}{\gamma} \qquad (\gamma \in \Gamma, \gamma \neq 0)$$

converges absolutely. Therefore, the infinite product

$$\lambda_\Gamma(u) = u \cdot \prod_\gamma \left(1 - \frac{u}{\gamma}\right) \qquad (\gamma \in \Gamma, \gamma \neq 0) \tag{4.2}$$

converges for all $u \in \Omega$. We can multiply out the product in (4.2) into an everywhere convergent linear power series

$$\lambda_\Gamma(u) = u + \sum_{\mu=1}^{\infty} c_\mu u^{p^\mu} \tag{4.3}$$

with $D(\lambda_\Gamma) = 1$.

THEOREM 4.7. *The function $\lambda_\Gamma(u)$ is an entire function on Ω with properties: (1) $\lambda_\Gamma(u)$ is a surjective $\mathbb{F}_p$-linear endomorphism of Ω, and (2) $\lambda_\Gamma(u)$ is periodic with Γ as its group of periods.*

Proof. The theorem is a trivial consequence of (4.2) and (4.3) if we recall that an entire function is necessarily surjective. ■

We now show that $\lambda_\Gamma(u)$ admits "complex multiplications" by the elements $x \in k_\infty$ satisfying **Lat** 3. Let $\Gamma' \supseteq \Gamma$ be a pair of lattices with Γ of finite index $|\Gamma'/\Gamma|$ in Γ'. As λ_Γ induces an isomorphism between the $\mathbb{F}_p$-vector spaces Γ'/Γ and $\lambda_\Gamma(\Gamma')$, $\lambda_\Gamma(\Gamma')$ is a finite set. Put

$$P(\Gamma'/\Gamma; t) = t \cdot \prod_\sigma \left(1 - \frac{t}{\sigma}\right) \qquad (\sigma \in \lambda_\Gamma(\Gamma'), \sigma \neq 0). \tag{4.4}$$

Then $P(\Gamma'/\Gamma; t)$ is a linear polynomial of degree $|\Gamma'/\Gamma|$ which is associated in a canonical way to the pair $\Gamma' \supseteq \Gamma$.

THEOREM 4.8. *Let $\Gamma'' \supseteq \Gamma' \supseteq \Gamma$ be lattices with $|\Gamma''/\Gamma|$ finite. Then*

$$\lambda_{\Gamma'}(u) = P(\Gamma'/\Gamma; \lambda_\Gamma(u)) \tag{4.5}$$

for all $u \in \Omega$, and

$$P(\Gamma''/\Gamma; t) = P(\Gamma''/\Gamma'; P(\Gamma'/\Gamma; t)). \tag{4.6}$$

Proof. From the definitions, both sides of (4.5) have the same set of roots. Since each is defined by an everywhere convergent power series with tangent vector 1, equality follows by Lemma 4.6. Equation (4.6) is a formal consequence of (4.5). ■

Now given a lattice Γ and any element $x \in k_\infty$ satisfying $x\Gamma \subseteq \Gamma$, we have $x^{-1}\Gamma \supseteq \Gamma$, and the index $|x^{-1}\Gamma/\Gamma|$ is finite by Theorem 4.3. Put

$$\rho_x^\Gamma(t) = x \cdot P(x^{-1}\Gamma/\Gamma; t) \tag{4.7}$$

so that ρ_x^Γ is a linear polynomial of degree $|x^{-1}\Gamma/\Gamma|$ and with tangent vector x. From the definitions, if $\Gamma' = x^{-1}\Gamma$, then $\lambda_{\Gamma'}(u) = x^{-1} \cdot \lambda_\Gamma(xu)$. Therefore, we see from (4.5) that $\lambda_\Gamma(u)$ admits the "complex multiplication"

$$\lambda_\Gamma(xu) = \rho_x^\Gamma(\lambda_\Gamma(u)). \tag{4.8}$$

The polynomial ρ_x^Γ is unique and canonically defined.

From (4.8) and the uniqueness,

$$\rho_{x+y}^\Gamma(t) = \rho_x^\Gamma(t) + \rho_y^\Gamma(t), \tag{4.9}$$

$$\rho_{xy}^\Gamma(t) = \rho_x^\Gamma(\rho_y^\Gamma(t)) \tag{4.10}$$

for all x, y with $x\Gamma \subseteq \Gamma$, $y\Gamma \subseteq \Gamma$.

In particular, if R is an order in A and if Γ is a discrete R-module of finite rank r_Γ over R, then the polynomials $\rho_x^\Gamma(t)$ provide a faithful representation of R as a ring of linear endomorphisms of the $\mathbb{F}_p$-vector space Ω. It is clear that this representation factors through the twisted polynomial ring $\Omega[\varphi]$. Thus, $x \mapsto \rho_x^\Gamma$, $x \in R$, is an elliptic R-module over Ω. Let us compute $\deg \rho_x^\Gamma(t)$ in terms of $v_\infty(x)$:

$$\deg \rho_x^\Gamma(t) = |x^{-1}\Gamma/\Gamma| = |\Gamma/x\Gamma| = p^s$$

by Theorem 4.3 where

$$s = \operatorname{rank}_{\mathbb{F}_p[x]}\Gamma = r_\Gamma \cdot \operatorname{rank}_{\mathbb{F}_p[x]}R = r_\Gamma \cdot [k : \mathbb{F}_p(x)].$$

Now, as $x \in R$, ∞ is the only pole of x. Therefore, the divisor of poles of x has degree $-d_\infty \cdot v_\infty(x) = [k : \mathbb{F}_p(x)]$. Thus, we have proved

THEOREM 4.9. *If Γ is a discrete module of finite rank r_Γ over the order R, then the complex multiplications* (4.8) *for $x \in R$ define an elliptic R-module ρ^Γ of rank r_Γ.*

We show now that the functions $\lambda_\Gamma(u)$ associated to lattices Γ in Ω are the only entire linear functions satisfying complex multiplications of type (4.8). This is a key fact for the applications to the class field theory of k.

PROPOSITION 4.10. *Let $\lambda(u)$ be an everywhere convergent linear power series with tangent vector* 1. *Suppose there is a linear polynomial $\rho(t)$ and an element $x \in k_\infty$ which is not a k_∞ constant such that*

$$\lambda(xu) = \rho(\lambda(u)). \tag{4.11}$$

Then the set of roots of $\lambda(u)$ is a lattice Γ with $x\Gamma \subseteq \Gamma$, and $\lambda = \lambda_\Gamma$, $\rho = \rho_x^\Gamma$.

Proof. Let Γ be the set of roots of λ. From the general theory of entire functions on Ω, we know that $\sum \gamma^{-1}$ $(\gamma \in \Gamma, \gamma \neq 0)$ is absolutely convergent; and $\lambda(u) = \lambda_\Gamma(u)$, as both are linear with tangent vector 1. Thus, we have only to show that Γ is a lattice. Now, **Lat** 1 holds for Γ since $\sum \gamma^{-1}$ is absolutely convergent, and x satisfies **Lat** 3 by (4.11). So we have only to check **Lat** 2. Let $\xi_1, \ldots, \xi_s$ be any elements of Γ which are linearly independent over the closure E of $\mathbb{F}_p(x)$ in k_∞, and let V be the E-vector space spanned by these elements. Since k_∞ is finite dimensional over E, $V \cap \Gamma$ is a lattice, and therefore (Theorem 4.3) there is an $\mathbb{F}_p[x]$-basis $\eta_1, \ldots, \eta_s$ of $V \cap \Gamma$ which is linearly independent over E. Now, the elements of the $\mathbb{F}_p$-vector space spanned by $\lambda(\eta_1/x), \ldots, \lambda(\eta_s/x)$ are all roots of $\rho(t)$ by (4.11). If we show that these values of λ are $\mathbb{F}_p$-linearly independent, then we shall have the bound $p^s \leqslant \deg \rho(t)$ for s, which is enough to establish **Lat** 2 for Γ. Therefore, assume

$$\alpha_1 \cdot \lambda(\eta_1/x) + \cdots + \alpha_s \cdot \lambda(\eta_s/x) = 0$$

for elements $\alpha_1, \ldots, \alpha_s$ in $\mathbb{F}_p$. Then, as λ is linear, $\alpha_1\eta_1 + \cdots + \alpha_s\eta_s \in x \cdot (V \cap \Gamma)$. Since $\eta_1, \ldots, \eta_s$ is an $\mathbb{F}_p[x]$-basis for $V \cap \Gamma$, this is impossible unless $\alpha_1 = \cdots = \alpha_s = 0$. ∎

5. *Elliptic R-Modules "Without Characteristic"*

Let ρ be an elliptic R-module "without characteristic." We show in this section that ρ can be defined analytically as in Section 4, and we interpret the operator $*$ defined in Section 3 in terms of lattices.

Let $K[[\varphi]]$ be the ring of *left twisted power series* generated over K by φ subject to the relations (2.1). We define the inclusion map $i: K \to K[[\varphi]]$ and the augmentation map $D: K[[\varphi]] \to K$ as we did for twisted polynomials in Section 2.

DEFINITION 5.1. Any monomorphism $\psi: k \to K[[\varphi]]$ will be called a *formal k-module over K*. If $\psi = i \circ D \circ \psi$, then we say that ψ is *trivial*.

Since ρ is "without characteristic," each ρ_x, $x \in R$, is invertible in $K[[\boldsymbol{\varphi}]]$, and therefore ρ extends to a formal k-module over K which we also call ρ.

LEMMA 5.2. *Suppose $\sigma \in K[[\boldsymbol{\varphi}]]$ has constant term $D(\sigma) = y$ with y transcendental over $\mathbb{F}_p$. Then there is a unique power series*

$$\lambda_\sigma = \sum_{\mu=0}^{\infty} c_\mu \boldsymbol{\varphi}^\mu$$

in $K[[\boldsymbol{\varphi}]]$ with $c_0 = 1$ such that $\lambda_\sigma \cdot y = \sigma \cdot \lambda_\sigma$.

Proof. If

$$\sigma = \sum_{\nu=0}^{\infty} a_\nu \boldsymbol{\varphi}^\nu,$$

then $\lambda_\sigma \cdot y = \sigma \cdot \lambda_\sigma$ is equivalent to the recurrence

$$(y^{p^\mu} - y) \cdot c_\mu = \sum_{\nu=1}^{\mu} a_\nu \cdot c_{\mu-\nu}^{p^\nu} \tag{5.1}$$

for all $\mu \geqslant 1$. Since y is not algebraic over $\mathbb{F}_p$, the coefficient of c_μ on the left-hand side is never zero. Thus, given the initial value $c_0 = 1$, (5.1) is uniquely solvable for the c_μ, $\mu \geqslant 1$. ■

COROLLARY 5.3. *If $D(\sigma)$ is transcendental over $\mathbb{F}_p$, then $\lambda_\sigma \cdot K \cdot \lambda_\sigma^{-1}$ is the centralizer of σ in $K[[\boldsymbol{\varphi}]]$.*

Proof. Clearly, K is the centralizer of $D(\sigma)$ in $K[[\boldsymbol{\varphi}]]$. Since σ is the image of $D(\sigma)$ under the inner automorphism through λ_σ, the centralizer of σ is the image of K under the same automorphism. ■

COROLLARY 5.4. *If $D(\sigma)$ is transcendental over $\mathbb{F}_p$, then for each $x \in K$,*

$$\tau_x = \lambda_\sigma \cdot x \cdot \lambda_\sigma^{-1} \tag{5.2}$$

is the unique power series in $K[[\boldsymbol{\varphi}]]$ with constant term x which commutes with σ.

PROPOSITION 5.5. *Let ψ be a formal k-module over K. Then there is a unique monomorphism $\delta\colon k \to K$ and a unique power series*

$$\lambda_\psi = \sum_{\mu=0}^{\infty} c_\mu \boldsymbol{\varphi}^\mu \tag{5.3}$$

such that $c_0 = 1$ and $\psi_x = \lambda_\psi \cdot \delta(x) \cdot \lambda_\psi^{-1}$ for all $x \in k$.

Proof. Let $\delta = D \circ \psi$. Clearly, there exists an element $z \in k$ such that $\delta(z) = y$ is transcendental over $\mathbb{F}_p$. Let λ_ψ be the power series λ_σ associated to $\sigma = \psi_z$ by Lemma 5.2. Now if $x \in k$, then $\psi_x \cdot \psi_z = \psi_{xz} = \psi_z \cdot \psi_x$, which implies $\psi_x = \lambda_\psi \cdot \delta(x) \cdot \lambda_\psi^{-1}$ by Corollary 5.4. ■

For the remainder of this section, we will assume that $k \subseteq K$ and that for every formal k-module ψ, the isomorphism $D \circ \psi$ is just the inclusion of k into K. There is no loss of generality and considerable gain in convenience in making this assumption.

Let ψ be a nontrivial formal k-module having the property that

$$R_\psi = \{x \in k \mid \psi_x \in K[\boldsymbol{\varphi}]\} \tag{5.4}$$

has k as its field of fractions. Assume that the map $x \to -\deg \psi_x$, $x \in R_\psi$, determines the prime divisor ∞. We now show that these assumptions imply that R_ψ is an order in A so that ψ is the extension to k of an elliptic R_ψ-module.

LEMMA 5.6. *With assumptions as above, for each integer $d > 0$, there are only finitely many elements $y \in R_\psi$ such that $\deg \psi_y < d$.*

Proof. The power series (5.3) associated to ψ is not a polynomial since otherwise from $\lambda_\psi \cdot x = \psi_x \cdot \lambda_\psi$ we could deduce $\deg \psi_x = 0$ for all $x \in R_\psi$. Thus, for every $d > 0$, there is a coefficient $c_\delta \neq 0$ of λ_ψ with $\delta \geqslant d$. Fix $d > 0$ for the moment, and let $y \in R_\psi$ be such that $\deg \psi_y < d$. Then the coefficients a_ν, $\nu \geqslant 0$, of ψ_y satisfy the recurrence (5.1), which shows that each such coefficient a_ν is formally a polynomial in y of degree $\leqslant p^\nu$ and in fact is a polynomial of *exact* degree p^ν in y if $c_\nu \neq 0$. Now, if $\delta \geqslant d$ and $c_\delta \neq 0$, we have $a_\delta = 0$, and y is the root of a polynomial of degree p^δ over K. Thus, y is limited to a set of at most p^δ elements. ■

THEOREM 5.7. *With assumptions as above, the ring R_ψ is an order in A.*

Proof. Since $-v_\infty(x)$ is a positive multiple of $\deg \psi_x$ for every $x \in R_\psi$, it follows from Lemma 5.6 that R_ψ is a lattice. According to Theorem 4.3, if $x \in R_\psi$ is not a constant, then the rank of R_ψ as an $\mathbb{F}_p[x]$-module is less than or equal to the degree n_∞ of k_∞ over the closure of $\mathbb{F}_p(x)$ in k_∞. Thus, $[k : \mathbb{F}_p(x)] \leqslant n_\infty$, which implies that the local degree at ∞ of the extension $k/\mathbb{F}_p(x)$ equals the global degree. Therefore, ∞ is the unique prime divisor of k sitting over its restriction to $\mathbb{F}_p(x)$. As $v_\infty(x) < 0$ (by the remark after Example 4.2), that restriction must be the unique pole of x in $\mathbb{F}_p(x)$. This proves that $R_\psi \subset A$. Since R_ψ has field of fractions k by assumption, R_ψ is an order in A. ■

We return now to the elliptic R-module ρ which is "without characteristic." Since R is a finitely generated $\mathbb{F}_p$-algebra, the smallest subfield of K which contains the coefficients of the polynomials ρ_x, $x \in R$, is finitely generated. We can therefore map that subfield into Ω in such a way that each element of k is mapped onto itself. Thus, there is no loss of generality in studying ρ under the assumption $K = \Omega$. In any case, $K = \Omega$ suffices for our applications in Part II.

Suppose then that $K = \Omega$. We will prove that $\rho = \rho^\Gamma$ for a suitable lattice Γ in Ω. Consider the ordinary linear power series

$$\lambda_\rho(u) = u + \sum_{\mu=1}^{\infty} c_\mu u^{p^\mu}, \tag{5.5}$$

where λ_ρ is the twisted power series (5.3) associated to ρ when it is viewed as a formal k-module over Ω. We show below that $\lambda_\rho(u)$ converges for all $u \in \Omega$. Assume this fact for the moment. Then the formal identity $\lambda_\rho \cdot x = \rho_x \cdot \lambda_\rho$, $x \in R$, in $\Omega[[\boldsymbol{\varphi}]]$ implies that the entire function $\lambda_\rho(u)$ admits the complex multiplications

$$\lambda_\rho(xu) = \rho_x(\lambda_\rho(u)) \tag{5.6}$$

for all $x \in R$. Applying Proposition 4.10, we conclude that $\rho = \rho^\Gamma$, where Γ is the lattice of roots of $\lambda_\rho(u)$.

The convergence of (5.5) for all $u \in \Omega$ is an immediate consequence of

LEMMA 5.8. *There are positive constants N and e such that*

$$v_\infty(c_\mu) \geqslant \left(\frac{\mu}{e} - N\right) \cdot p^\mu \tag{5.7}$$

for all $\mu \geqslant 0$.

Proof. Choose any $y \in R$ with $\deg \rho_y = d > 0$, and let

$$\rho_y = y + a_1\boldsymbol{\varphi} + \cdots + a_d\boldsymbol{\varphi}^d. \tag{5.8}$$

For the purposes of this proof, we renormalize v_∞ if necessary so that $v_\infty(y) = 1$. We also replace ρ by $w^{-1} \cdot \rho \cdot w$, where $w \in \Omega$ is chosen so that $a_d = 1$ in (5.8). Since this replacement means that each c_μ is multiplied by the factor $w^{p^\mu - 1}$, the validity of (5.7) is not affected. This being done, we prove (5.7) with $e = d$.

Choose $\delta \geqslant d$ so large that $p^\delta + d \cdot v_\infty(a_\nu) > 0$ for $1 \leqslant \nu \leqslant d - 1$, and then choose N so large that

$$v_\infty(c_\mu) \geqslant ((\mu/d) - N) \cdot p^\mu$$

for all $\mu \leqslant \delta$. We verify (5.7) by induction on $\mu \geqslant \delta$. In the recurrence (5.1), we have

$$v_\infty((y^{p^\mu} - y) \cdot c_\mu) = -p^\mu + v_\infty(c_\mu)$$

and

$$v_\infty\left(\sum_{\nu=1}^{d} a_\nu c_{\mu-\nu}^{p^\nu}\right) \geqslant \min_{1\leqslant \nu\leqslant d}\{v_\infty(a_\nu) + p^\nu \cdot v_\infty(c_{\mu-\nu})\}.$$

Thus, it suffices to show that

$$p^\mu + v_\infty(a_\nu) + p^\nu \cdot v_\infty(c_{\mu-\nu}) \geqslant ((\mu/d) - N)\cdot p^\mu \tag{5.9}$$

for $1 \leqslant \nu \leqslant d$. Now, by the induction hypothesis

$$v_\infty(c_{\mu-\nu}) \geqslant \left(\frac{\mu-\nu}{d} - N\right)\cdot p^{\mu-\nu}$$

so that the left-hand side of (5.9) exceeds or equals

$$\left(\frac{\mu}{d} - N\right)\cdot p^\mu + \left(1 - \frac{\nu}{d}\right)\cdot p^\mu + v_\infty(a_\nu).$$

When $\nu = d$, this last expression is just the right-hand side of (5.9) as $a_d = 1$. When $1 \leqslant \nu \leqslant d-1$, it is larger than the right-hand side of (5.9) because

$$\left(1 - \frac{\nu}{d}\right)\cdot p^\mu + v_\infty(a_\nu) \geqslant \frac{1}{d}\cdot p^\mu + v_\infty(a_\nu) > 0$$

by choice of δ. Thus, by induction, (5.7) is valid with $e = d$ for all $\mu \geqslant 0$. ■

We have now proved

THEOREM 5.9. *Let ρ be an elliptic R-module over Ω with $D(\rho_x) = x$ for all $x \in R$. Then $\rho = \rho^\Gamma$, where Γ is the lattice of roots of the entire function $\lambda_\rho(u)$ defined by* (5.5).

Now let ρ and ρ' be a pair of elliptic R-modules over Ω. We will show that the additive group **Isog**(ρ, ρ') is isomorphic to

$$(\Gamma' : \Gamma) = \{z \in \Omega \mid z\Gamma \subseteq \Gamma'\},$$

where Γ' (resp. Γ) is the lattice of roots of $\lambda_{\rho'}(u)$ (resp. $\lambda_\rho(u)$) in Ω. Let $\sigma \in$ **Isog**(ρ, ρ'). Then from $\sigma\cdot\rho_x = \rho'_x\cdot\sigma$ and Proposition 5.5, we conclude that $\lambda_{\rho'}^{-1}\cdot\sigma\cdot\lambda_\rho$ commutes with x for every $x \in k$. This implies that

$$\sigma = \lambda_{\rho'}\cdot z\cdot\lambda_\rho^{-1}, \tag{5.10}$$

where $z = D(\sigma)$. The map $\sigma \mapsto D(\sigma)$ is therefore a monomorphism from the group **Isog**(ρ, ρ') into the additive group of Ω. We can rewrite (5.10) as the identity

$$\lambda_{\rho'}(zu) = \sigma(\lambda_\rho(u)) \qquad (u \in \Omega), \tag{5.11}$$

from which we infer that $z \in (\Gamma' : \Gamma)$ and that

$$\sigma(t) = z \cdot P(z^{-1}\Gamma'/\Gamma; t), \tag{5.12}$$

where P is the polynomial defined by (4.4). Conversely, if $z \in (\Gamma' : \Gamma)$, then (5.12) defines an isogeny from ρ to ρ', as one easily checks using Theorem 4.8. Thus,

$$\mathbf{Isog}(\rho, \rho') \cong (\Gamma' : \Gamma). \tag{5.13}$$

PROPOSITION 5.10. *Let $\mathfrak{A} \in \mathcal{M}(R)$, and let Γ be an R-submodule of Ω which is a lattice. Put $\rho' = \mathfrak{A} * \rho^\Gamma$, and let Γ' be the lattice of roots of ρ'. Then*

$$z^{-1}\Gamma' = (\Gamma : \mathfrak{A}), \tag{5.14}$$

where $z = D(\rho^\Gamma_\mathfrak{A})$.

Proof. By Lemma 3.6,

$$\rho^\Gamma_\mathfrak{A}(\lambda_\Gamma(u)) = 0 \Leftrightarrow \rho^\Gamma_a(\lambda_\Gamma(u)) = \lambda_\Gamma(au) = 0, \qquad a \in \mathfrak{A},$$
$$\Leftrightarrow u \in (\Gamma : \mathfrak{A}).$$

On the other hand, by (5.11)

$$\rho^\Gamma_\mathfrak{A}(\lambda_\Gamma(u)) = 0 \Leftrightarrow \lambda_{\rho'}(zu) = 0 \Leftrightarrow u \in z^{-1}\Gamma'.$$

Thus, $(\Gamma : \mathfrak{A}) = z^{-1}\Gamma'$ as required. ■

For future reference, we note the following immediate consequences of (5.14):

$$\mathfrak{A} * \rho^\Gamma \quad \text{is isomorphic to} \quad \rho^{(\Gamma : \mathfrak{A})}. \tag{5.15}$$

$$\rho^\Gamma_\mathfrak{A}(t) = \prod(t - \gamma), \qquad \gamma \in \lambda_\rho((\Gamma : \mathfrak{A})). \tag{5.16}$$

$$\deg \rho^\Gamma_\mathfrak{A} = \dim_{\mathbb{F}_p}[(\Gamma : \mathfrak{A})/\Gamma]. \tag{5.17}$$

If $\mathfrak{A} \in \mathcal{M}(R)$ is invertible, then $(\Gamma : \mathfrak{A}) = \mathfrak{A}^{-1}\Gamma$ so that (5.14) can be rewritten as

$$z^{-1}\Gamma' = \mathfrak{A}^{-1}\Gamma. \tag{5.18}$$

Now, $\dim_{\mathbb{F}_p}(\mathfrak{A}^{-1}\Gamma/\Gamma) = \dim_{\mathbb{F}_p}(\Gamma/\mathfrak{A}\Gamma) = r_\Gamma \cdot \deg^* \mathfrak{A}$ by the usual arguments based on Theorem 4.3. Therefore, from (5.17) we recover Proposition 3.7 for elliptic R-modules ρ over Ω which are "without characteristic."

Call lattices Γ and Γ' *equivalent* if $\Gamma = w\Gamma'$ for some element $w \in \Omega$. Given an integer $r > 0$, let $\mathscr{L}_r(R)$ be the set of equivalence classes of the R-submodules Γ of Ω which are lattices of rank r over R. Let $\mathbf{Pic}(R)$ act on $\mathscr{L}_r(R)$ via the operation induced by

$$(\mathfrak{A}, \Gamma) \mapsto \mathfrak{A}^{-1}\Gamma$$

for $\mathfrak{A} \in \mathscr{I}^*(R)$. Then, we have

THEOREM 5.11. *Let $\delta: R \to \Omega$ be the inclusion map. Then the representation spaces $\mathscr{E}_R(\delta, r)$ (see Section 3) and $\mathscr{L}_r(R)$ of the group* **Pic**(R) *are canonically isomorphic for every positive integer r.*

Proof. This is an immediate consequence of (5.15) and Theorem 5.9. ■

COROLLARY 5.12. *The sets $\mathbf{M}_1(R)$ (see Section 1) and $\mathscr{E}_R(\delta, 1)$ for δ the inclusion of R in Ω are in a natural one-to-one correspondence. In particular, $\mathscr{E}_R(\delta, 1)$ is a finite set.*

COROLLARY 5.13. *There are exactly $h(A)$ isomorphism classes of elliptic A-modules ρ of rank 1 over Ω such that $D(\rho_x) = x$ for all $x \in k$.*

6. *The Field of Invariants*

Throughout this section ρ is an elliptic R-module over Ω such that $D(\rho_x) = x$ for all $x \in R$.

DEFINITION 6.1. Let K be a subfield of Ω containing k. We say that ρ is *defined over* K or that K is a *field of definition* for ρ if ρ is isomorphic over Ω to an elliptic R-module ρ' such that ρ'_x has coefficients belonging to K for every $x \in R$.

PROPOSITION 6.2. *If ρ has rank 1, then k_∞ is a field of definition for ρ.*

Proof. By Theorem 5.9, $\rho = \rho^\Gamma$ for a suitable lattice Γ which is an R-module of rank 1. Choose $w \in \Omega$ such that $\Gamma' = w\Gamma \subset R \subset k_\infty$. Since k_∞ is complete, $\lambda_{\Gamma'}$ maps k_∞ into itself. Thus, by (4.7), the coefficients of $\rho_x^{\Gamma'}$ belong to k_∞ for every $x \in R$. ■

Our aim now is to show that there is a smallest field of definition for ρ. Let $k[\mathbf{X}]$ be the ring of polynomials in the indeterminates $\mathbf{X} = \{X_\nu\}_{\nu \geqslant 1}$, and let $k(\mathbf{X})$ be the field of rational functions in these same indeterminates. There is a unique doubly infinite graduation grad on $k(\mathbf{X})$ such that

$$\operatorname{grad}(X_\nu) = m_\nu = p^\nu - 1 \tag{6.1}$$

for all $\nu \geqslant 1$. The field $k(\mathbf{X})_0$ of the homogeneous elements of grade zero for this graduation is called the *field of formal invariants*. Thus, a formal invariant is a rational function f of the form

$$f = g/h, \tag{6.2}$$

where g and h are polynomials such that $\operatorname{grad} g = \operatorname{grad} h$.

Given an element w in Ω, let

$$w * \mathbf{X} = \{w^{m_\nu} \cdot X_\nu\}_{\nu \geqslant 1}. \tag{6.3}$$

It is immediate from the definitions that if g is a polynomial of grade d, then

$$g(w * \mathbf{X}) = w^d \cdot g(\mathbf{X}). \tag{6.4}$$

Therefore, if $w \neq 0$, then

$$f(w * \mathbf{X}) = f(\mathbf{X}) \tag{6.5}$$

for every formal invariant f.

Now, for $y \in k$, we write

$$\rho_y = y + \sum_{\nu=1}^{\infty} c_\nu(\rho, y) \cdot \boldsymbol{\varphi}^\nu, \tag{6.6}$$

and we let $S_{y,\rho}: k[\mathbf{X}] \to \Omega$ be the substitution homomorphism which maps X_ν onto $c_\nu(\rho, y)$, $\nu \geqslant 1$. Let $V_{y,\rho}$ be the local ring in $k(\mathbf{X})_0$ consisting of those formal invariants which can be written in the form (6.2) with $S_{y,\rho}(h) \neq 0$. The same substitution induces a homomorphism from $V_{y,\rho}$ into Ω which we also call $S_{y,\rho}$.

DEFINITION 6.3. For each nonconstant element $y \in R$, we put

$$I_y(\rho) = S_{y,\rho}(V_{y,\rho}) \subseteq \Omega.$$

We call $I_y(\rho)$ the *field of invariants* of ρ at y.

PROPOSITION 6.4. *Given y, the field $I_y(\rho)$ depends only on the isomorphism class of ρ. It is contained in every field of definition for ρ.*

Proof. For $w \in \Omega$, $w \neq 0$, let $\rho'_x = w^{-1} \cdot \rho_x \cdot w$ for every $x \in k$. Then by (6.6), we have

$$c_\nu(\rho', y) = w^{m_\nu} \cdot c_\nu(\rho, y)$$

for all $\nu \geqslant 1$. It is clear from (6.4) and (6.5) that $V_{y,\rho} = V_{y,\rho'}$ and that $S_{y,\rho} = S_{y,\rho'}$ on this local ring. This proves the first assertion of the proposition. The second is an immediate consequence of the first. ■

THEOREM 6.5. *Let y be a nonconstant element of R. Then $I_y(\rho)$ is a field of definition for ρ.*

Proof. Let $\nu_1, \ldots, \nu_s$ be the indices ν such that $c_\nu(\rho, y) \neq 0$, $\nu \geqslant 1$, and put $Y_i = X_{\nu_i}$ and $b_i = m_{\nu_i}$ for $1 \leqslant i \leqslant s$. Let d be the greatest common divisor of $b_1, \ldots, b_s$, and choose integers $e_1, \ldots, e_s$ such that $d = e_1 b_1 + \cdots + e_s b_s$. Now for each $1 \leqslant i \leqslant s$, the formal invariant

$$f_i = Y_i \cdot \left(\prod_{j=1}^{s} Y_j^{e_j} \right)^{-b_i/d} \tag{6.7}$$

belongs to $V_{y,\rho}$. Choose $w \in \Omega$ so that

$$w^{-d} = \prod_{j=1}^{s} [c_{v_j}(\rho, y)]^{e_j}.$$

Then for $1 \leqslant i \leqslant s$,

$$c_{v_i}(\rho, y) \cdot w^{b_i} = S_{y,\rho}(f_i) \in I_y(\rho)$$

so that $\rho'_y = w^{-1} \cdot \rho_y \cdot w$ has coefficients in $I_y(\rho)$. In fact, $\rho'_x = w^{-1} \cdot \rho_x \cdot w$ has coefficients in $I_y(\rho)$ for all $x \in k$. For this, it is enough to show that the power series $\lambda_{\rho'}$ of Proposition 5.5 has coefficients in $I_y(\rho)$. But from $\lambda_{\rho'} \cdot y = \rho'_y \cdot \lambda_{\rho'}$ and the corresponding recurrence (5.1), we see that $\lambda_{\rho'}$ has coefficients belonging to the overfield of k generated by the coefficients of ρ'_y. Thus, $I_y(\rho)$ is indeed a field of definition for ρ. ■

This theorem together with Proposition 6.4 shows in particular that $I_x(\rho) \subseteq I_y(\rho)$ for every nonconstant $x \in R$. Thus, the field $I_y(\rho)$ depends only on ρ and not on the choice of y. We therefore write

$$I(\rho) = I_y(\rho) \tag{6.8}$$

for any nonconstant $y \in R$, and we call $I(\rho)$ the *field of invariants* of ρ.

THEOREM 6.6. *The field $I(\rho)$ is the smallest field of definition for ρ.*

Proof. The theorem is an immediate corollary of Theorem 6.5. ■

COROLLARY 6.7. *If $\mathfrak{A} \in \mathcal{M}(R)$, then*

$$I(\mathfrak{A} * \rho) \subseteq I(\rho). \tag{6.9}$$

If $\mathfrak{A}$ is invertible, then

$$I(\mathfrak{A} * \rho) = I(\rho). \tag{6.10}$$

DEFINITION 6.8. For $y \in R$, y not a constant, we introduce the s-tuple of invariants

$$F(y, \rho) = (S_{y,\rho}(f_1), \ldots, S_{y,\rho}(f_s)), \tag{6.11}$$

where f_i for $1 \leqslant i \leqslant s$ is the formal invariant defined by (6.7).

Now let ρ' be also an elliptic R-module over Ω such that $D(\rho'_x) = x$ for all $x \in R$. We say that ρ and ρ' are *patterned alike* if

$$c_v(\rho, x) = 0 \Leftrightarrow c_v(\rho', x) = 0$$

for all $v \geqslant 1$ and all $x \in R$. Then, as an immediate consequence of the proof of Theorem 6.5, we have

PROPOSITION 6.9. *If ρ and ρ' are patterned alike, then ρ and ρ' are isomorphic over Ω if and only if $F(y, \rho) = F(y, \rho')$ for some nonconstant element $y \in R$.*

II. GENERATION OF THE CLASS FIELDS OF A

7. *Reduction of Elliptic Modules*

Let ρ be an elliptic R-module over K. A discrete valuation ring $\mathcal{O}$ in K is called *finite* with respect to ρ if $D(\rho_x) \in \mathcal{O}$ for all $x \in R$. Let $\mathfrak{p}$ be the maximal ideal of $\mathcal{O}$, and let $v_\mathfrak{p}$ be the normalized discrete valuation on K determined by $\mathcal{O}$. We form the twisted polynomial ring $\mathcal{O}[\boldsymbol{\varphi}] \subset K[\boldsymbol{\varphi}]$ with coefficients restricted to $\mathcal{O}$, and we introduce the reduction homomorphism $\mathrm{rd}_\mathfrak{p}: \mathcal{O}[\boldsymbol{\varphi}] \to (\mathcal{O}/\mathfrak{p})[\boldsymbol{\varphi}]$ induced by the operation of the canonical homomorphism $\mathcal{O} \to \mathcal{O}/\mathfrak{p}$ on the coefficients of the polynomials in $\mathcal{O}[\boldsymbol{\varphi}]$.

DEFINITION 7.1. We say that ρ has a *stable reduction modulo* $\mathfrak{p}$ if there is an element $w \in K$, $w \neq 0$, such that $\rho'_x = w^{-1} \cdot \rho_x \cdot w$ belongs to $\mathcal{O}[\boldsymbol{\varphi}]$ for all $x \in R$ and such that $\mathrm{rd}_\mathfrak{p} \circ \rho'$ is an elliptic R-module over $\mathcal{O}/\mathfrak{p}$. We say that ρ has a *good reduction modulo* $\mathfrak{p}$ if in addition the rank of $\mathrm{rd}_\mathfrak{p} \circ \rho'$ equals the rank of ρ.

For elliptic R-modules over K of rank 1, every stable reduction is good.

Suppose there is a finite extension field K' of K containing a valuation ring $\mathcal{O}'$ with maximal ideal $\mathfrak{p}'$ such that (a) $\mathcal{O}'$ sits over $\mathcal{O}$ (i.e., $K \cap \mathfrak{p}' = \mathfrak{p}$), and (b) ρ when considered as an elliptic R-module over K' has a stable (resp. good) reduction modulo $\mathfrak{p}'$. Then we say that ρ has a *potential stable* (resp. *good*) *reduction modulo* $\mathfrak{p}$.

PROPOSITION 7.2. *The elliptic R-module ρ has a potential stable reduction at every finite discrete valuation ring $\mathcal{O}$ in K.*

Proof. Let $y_1, \ldots, y_s$ generate R as an $\mathbb{F}_p$-algebra, and put

$$m = \max\left\{\frac{-v_\mathfrak{p}(c_\nu(\rho, y_i))}{p^\nu - 1}\right\},$$

where the max is over $1 \leqslant i \leqslant s$ and $1 \leqslant \nu \leqslant \deg \rho_{y_i}$. We can construct a finite extension field K' of K which has a valuation extending $v_\mathfrak{p}$ with ramification number equal to the denominator of m. Clearly, ρ has a stable reduction over K'. ■

Given an elliptic R-module ρ, we put

$$\eta_\rho(x) = c_d(\rho, x) \qquad (x \in R, d = \deg \rho_x) \tag{7.1}$$

so that $\eta_\rho(x)$ is the leading coefficient of ρ_x. From $\rho_{xy} = \rho_x \cdot \rho_y$, we infer the "cocycle condition"

$$\eta_\rho(xy) = \eta_\rho(x) \cdot [\eta_\rho(y)]^{p^{r_\rho \cdot \deg^* x}} \tag{7.2}$$

for all $x, y \in R$.

PROPOSITION 7.3. *Let ρ have rank 1 over K. Then for every finite discrete valuation $v_\mathfrak{p}$ on K, we have*

$$v_\mathfrak{p}(c_\nu(\rho, x)) \geqslant \left(\frac{p^\nu - 1}{p^{\deg^* x} - 1}\right) \cdot v_\mathfrak{p}(\eta_\rho(x)) \tag{7.3}$$

for all $x \in R$ and $1 \leqslant \nu \leqslant \deg \rho_x$.

Proof. By Proposition 7.2, there exists an element w in some extension field K' of K and an extension of $v_\mathfrak{p}$ to K' such that

$$v_\mathfrak{p}(w^{p^\nu - 1} \cdot c_\nu(\rho, x)) \geqslant 0$$

and

$$v_\mathfrak{p}(w^{p^{\deg^* x} - 1} \cdot \eta_\rho(x)) = 0$$

for all $x \in R$ and $1 \leqslant \nu \leqslant \deg^* x$. The inequality (7.3) is an immediate consequence of these two relations. ■

Let B be a subring of K. We say that the elliptic R-module ρ over K has *coefficients in B* if for all $x \in R$, $\rho_x \in B[\boldsymbol{\varphi}]$ and $\eta_\rho(x)$ is a unit in B.

COROLLARY 7.4. *Suppose ρ has rank 1, and suppose $\eta_\rho(x)$ is a unit of $\mathcal{O}$ for all $x \in R$. Then ρ has coefficients in $\mathcal{O}$.*

PROPOSITION 7.5. *Suppose ρ has coefficients in $\mathcal{O}$. Let $\mathfrak{A} \in \mathcal{M}(R)$. Then $\rho_\mathfrak{A} \in \mathcal{O}[\boldsymbol{\varphi}]$, and $\mathfrak{A} * \rho$ has coefficients in $\mathcal{O}$ also.*

Proof. The roots of $\rho_\mathfrak{A}(t)$ are integral over $\mathcal{O}$ since they are also roots of the polynomials $\rho_a(t)$, $a \in \mathfrak{A}$. Since $\mathcal{O}$ is integrally closed in K, we conclude that $\rho_\mathfrak{A} \in \mathcal{O}[\boldsymbol{\varphi}]$. That $\mathfrak{A} * \rho$ is defined over $\mathcal{O}$ is now a simple consequence of the defining relation

$$\rho_\mathfrak{A} \cdot \rho_x = (\mathfrak{A} * \rho)_x \cdot \rho_\mathfrak{A}$$

for $x \in R$. ■

PROPOSITION 7.6. *Suppose ρ has rank 1 and coefficients in $\mathcal{O}$. Suppose that the morphism* Spec $\mathcal{O} \to$ Spec R *induced by $D \circ \rho$ is unramified at $\mathfrak{p}$ and that the image $\mathfrak{P}$ of $\mathfrak{p}$ in* Spec R *is invertible. For given $e \geqslant 1$, let $\mathfrak{A} = \mathfrak{P}^e$ and $\mathfrak{B} = \mathfrak{P}^{e-1}$. Then $\rho_{\mathfrak{B}}(t)$ divides $\rho_{\mathfrak{A}}(t)$ in $\mathcal{O}[t]$, and the quotient is Eisenstein at $\mathfrak{p}$.*

Proof. We first settle the case $e = 1$. Reduction of ρ modulo $\mathfrak{p}$ produces an elliptic R-module whose "characteristic" contains $\mathfrak{P}$. Therefore, by Corollary 3.8, every coefficient of $\rho_{\mathfrak{P}}(t)$ other than its leading coefficient must belong to $\mathfrak{p}$. Thus, $\rho_{\mathfrak{P}}(t)/t$ will be Eisenstein at $\mathfrak{p}$ if we show that

$$v_{\mathfrak{p}}(D(\rho_{\mathfrak{P}})) \leqslant 1. \tag{7.4}$$

Choose $x \in R$ such that $v_{\mathfrak{P}}(x) = v_{\mathfrak{p}}(D(\rho_x)) = 1$, and let $xR = \mathfrak{D}\mathfrak{P}$. Then by (3.10)

$$\rho_x = \eta_\rho(x) \cdot (\mathfrak{P} * \rho)_{\mathfrak{D}} \cdot \rho_{\mathfrak{P}}$$

so that

$$1 = v_{\mathfrak{p}}(D(\rho_x)) = v_{\mathfrak{p}}(D((\mathfrak{P} * \rho)_{\mathfrak{D}})) + v_{\mathfrak{p}}(D(\rho_{\mathfrak{P}})) \geqslant v_{\mathfrak{p}}(D(\rho_{\mathfrak{P}}))$$

by Proposition 7.5. Thus, (7.4) must hold.

Next, for $e > 1$, we have

$$\rho_{\mathfrak{A}}(t) = f_{\mathfrak{P}}(\rho_{\mathfrak{B}}(t)) \cdot \rho_{\mathfrak{B}}(t)$$

by (3.10) where $f_{\mathfrak{P}}(t)$ is $(\mathfrak{B} * \rho)_{\mathfrak{P}}(t)$ divided by t. Now $f_{\mathfrak{P}}(t)$ has coefficients in $\mathcal{O}$ by Proposition 7.5, and so $\rho_{\mathfrak{B}}(t)$ divides $\rho_{\mathfrak{A}}(t)$ in $\mathcal{O}[t]$. Further, we observe that the constant coefficient of the quotient is $D((\mathfrak{B} * \rho)_{\mathfrak{P}})$, which has valuation 1 at $\mathfrak{p}$ by (7.4). As above, we conclude from reduction modulo $\mathfrak{p}$ and Corollary 3.8 that both $f_{\mathfrak{P}}(t)$ and $\rho_{\mathfrak{B}}(t)$ and hence $f_{\mathfrak{P}}(\rho_{\mathfrak{B}}(t)) = \rho_{\mathfrak{A}}(t)/\rho_{\mathfrak{B}}(t)$ have coefficients in $\mathfrak{p}$, except that each leading coefficient is 1. Therefore, the quotient is indeed Eisenstein at $\mathfrak{p}$. ∎

8. *The "Hilbert Class Field" of R*

Let R be a fixed order in A. By Theorem 5.11 and Corollary 6.7, if $\mathfrak{A}$ and $\mathfrak{B}$ are invertible ideals in $\mathcal{M}(R)$, then the fields of invariants $I(\rho^{\mathfrak{A}})$ and $I(\rho^{\mathfrak{B}})$ are equal. Our aim in this section is to identify this common field of invariants for the $\rho^{\mathfrak{A}}$, $\mathfrak{A}$ invertible, in terms of the class field theory of k. Let us denote this field by H_R. We will show that H_R/k is a finite abelian extension whose Galois group is isomorphic to **Pic** (R). We call H_R the *Hilbert Class Field of R*.

Let $\delta : R \to \Omega$ be the inclusion homomorphism. In this section, all elliptic R-modules ρ over Ω are assumed to satisfy $D \circ \rho = \delta$.

Let G_∞ be the Galois group of Ω/k. For $g \in G_\infty$ and $\tau = \sum a_\nu \boldsymbol{\varphi}^\nu \in \Omega[[\boldsymbol{\varphi}]]$, put $g(\tau) = \sum g(a_\nu)\boldsymbol{\varphi}^\nu$. We define the action $g\rho$ of $g \in G_\infty$ on an elliptic

R-module ρ over Ω by

$$(g\rho)_x = g(\rho_x) \tag{8.1}$$

for all $x \in R$. Of course, $g\rho$ is also an elliptic R-module over Ω of the same rank as ρ. As one easily checks from the definitions,

$$g(\mathfrak{A} * \rho) = \mathfrak{A} * (g\rho) \tag{8.2}$$

for all ideals $\mathfrak{A} \in \mathscr{M}(R)$. As this Galois action also respects isomorphism, we have the following

Proposition 8.1. *The Galois group G_∞ acts naturally on $\mathscr{E}_R(\delta, r)$ for each $r > 0$, and the Galois action commutes with the action of* $\mathbf{Pic}(R)$.

In view of the isomorphism between $\mathscr{E}_R(\delta, 1)$ and $\mathscr{L}_1(R)$ as $\mathbf{Pic}(R)$ spaces (Theorem 5.11), the group G_∞ acts on $\mathscr{L}_1(R)$ by "transport of structure" and commutes with the action of $\mathbf{Pic}(R)$. Can we describe this action *internally* (e.g., in terms of Frobenius elements)? This question seems to be of central importance for the class field theory of k. We have a good description for the action of G_∞ on the invertible ideal classes (Theorem 8.5 below), but no satisfactory answer to the general question is known. However, we can describe some subsets of $\mathscr{L}_1(R)$ which are invariant under both G_∞ and $\mathbf{Pic}(R)$. We write $\rho^{\mathfrak{A}} \sim \rho^{\mathfrak{B}}$ to indicate that $\rho^{\mathfrak{A}}$ is isomorphic to $g\rho^{\mathfrak{B}}$ for some $g \in G_\infty$.

Proposition 8.2. *Suppose $\rho^{\mathfrak{A}} \sim \rho^{\mathfrak{B}}$ where $\mathfrak{A}, \mathfrak{B} \in \mathscr{M}(R)$. Then* $\mathrm{End}(\mathfrak{A}) = \mathrm{End}(\mathfrak{B})$. *Further, if $\mathfrak{A}$ and $\mathfrak{B}$ are proper, then there is an invertible ideal $\mathfrak{D}$ such that $\mathfrak{A}^* = \mathfrak{B}^*\mathfrak{D}$.*

Proof. Let $\rho^{\mathfrak{B}}$ be isomorphic to $g\rho^{\mathfrak{A}}$, where $g \in G_\infty$. From the natural isomorphism of (5.13) and the definitions, it is clear that $g(\mathrm{End}(\mathfrak{A})) = \mathrm{End}(\mathfrak{B})$. Since $\mathrm{End}(\mathfrak{A}) \subseteq A$, we have proved the first assertion. For the second assertion, we deduce first that $\mathfrak{A} * \rho^{\mathfrak{A}} \sim \mathfrak{A} * \rho^{\mathfrak{B}}$ since the operator $*$ commutes with the Galois action. By (5.15), this means

$$\rho^R \sim \rho^{(\mathfrak{B}:\mathfrak{A})} \tag{8.3}$$

as $\mathfrak{A}$ is proper. Thus, $\mathrm{End}((\mathfrak{B}:\mathfrak{A})) = R$. Now, when $\mathfrak{B}$ is invertible, $(\mathfrak{B}:\mathfrak{A}) = \mathfrak{A}^*\mathfrak{B}$ and $R = \mathrm{End}(\mathfrak{A}^*\mathfrak{B}) = \mathrm{End}(\mathfrak{A}^*)$. We conclude from Theorem 1.6 that $\mathfrak{B}$ invertible implies $\mathfrak{A}$ invertible. Applying this fact to (8.3), we see that $(\mathfrak{B}:\mathfrak{A}) = \mathfrak{D}$ is invertible; and this easily implies that $\mathfrak{A}^* = \mathfrak{B}^*\mathfrak{D}$. ■

Now let $\mathscr{E}_R^*(\delta, 1) \subseteq \mathscr{E}_R(\delta, 1)$ be the set of isomorphism classes containing an elliptic R-module $\rho^{\mathfrak{A}}$, $\mathfrak{A} \in \mathscr{I}^*(R)$. By Theorem 5.11, $\mathscr{E}_R^*(\delta, 1)$ is a *principal homogeneous space* over $\mathbf{Pic}(R)$. We have also the following important corollary of Proposition 8.2.

PROPOSITION 8.3. *The subset $\mathscr{E}_R^*(\delta, 1) \subseteq \mathscr{E}_R(\delta, 1)$ is invariant under the action of G_∞.*

We turn now to the field H_R. Since it is the field of invariants for each $\rho^{\mathfrak{A}}$, $\mathfrak{A} \in \mathscr{I}^*(R)$, H_R is generated as a field over k by a finite number of invariants, each of which has only finitely many conjugates under the action of G_∞ by Corollary 5.12. By Proposition 8.3, these conjugates also belong to H_R. We conclude that H_R/k is a finite extension which is invariant under the action of G_∞. Now, since each $\rho^{\mathfrak{A}}$, $\mathfrak{A} \in \mathscr{I}^*(R)$, has rank 1, $H_R \subset k_\infty$ by Proposition 6.2. As is well known, the elements of k_∞ which are algebraic over k are all separable over k. Thus, we have proved

PROPOSITION 8.4. *The extension H_R/k is finite and Galois. Further, ∞ splits completely in H_R/k.*

Let G_R be the Galois group of H_R/k. We will use the natural action of G_R on $\mathscr{E}_R^*(\delta, 1)$ as a means for studying the deeper properties of H_R (cf. [1, 9]). Since each element of G_R induces an automorphism of $\mathscr{E}_R^*(\delta, 1)$ as a principal homogeneous space over $\mathbf{Pic}(R)$, we have a natural monomorphism

$$\psi : G_R \to \mathbf{Pic}(R). \tag{8.4}$$

We see in particular that H_R/k is abelian.

THEOREM 8.5. *Suppose $\mathfrak{P}$ is a nonzero prime ideal of R which does not contain the conductor $\mathfrak{C}$ and which does not ramify in H_R/k. Let $\mathbf{Frob}_{\mathfrak{P}} \in G_R$ be the Frobenius automorphism associated to $\mathfrak{P}$. Let ρ be an elliptic R-module which is defined over H_R and which represents a class in $\mathscr{E}_R^*(\delta, 1)$. Then*

$$\mathbf{Frob}_{\mathfrak{P}}(\rho) \cong \mathfrak{P} * \rho. \tag{8.5}$$

This key result is the analog of Hasse's theorem (cf. [9]). Before presenting the proof, we introduce some convenient notation and prove some lemmas. We work with an elliptic R-module ρ of the type mentioned in the theorem.

Let S be a finite set of prime divisors of H_R which contains at least: (1) all divisors of the conductor $\mathfrak{C}$ of R, (2) all prime divisors sitting over ∞, (3) any pole or zero of the $\eta_\rho(x)$, $x \in R$ (see (7.1)), and (4) any prime divisor which ramifies in H_R/k. From the cocycle condition (7.2), we infer

$$[\eta_\rho(x)]^{p^{\deg^* y} - 1} = [\eta_\rho(y)]^{p^{\deg^* x} - 1} \tag{8.6}$$

for all x, $y \in R$ so that the set of primes of type (3) above constitute a finite set. We note that by Corollary 7.4, ρ has coefficients in the valuation ring at $\mathfrak{p}$ for all $\mathfrak{p} \notin S$.

LEMMA 8.6. *Suppose $\mathfrak{p} \notin S$ is a prime divisor of H_R which sits over the prime ideal $\mathfrak{P}$ of R. Then*

$$\mathbf{Frob}_{\mathfrak{P}}(c_v(x;\rho)) \equiv c_v(x;\mathfrak{P} * \rho) \pmod{\mathfrak{p}} \tag{8.7}$$

for all $x \in R$ and $v \geqslant 1$ (see (6.6)).

Proof. From the reduction modulo $\mathfrak{p}$ of the relation

$$\rho_{\mathfrak{P}} \cdot \rho_x = (\mathfrak{P} * \rho)_x \cdot \rho_{\mathfrak{P}}$$

we conclude that

$$\boldsymbol{\varphi}^{\deg^* \mathfrak{P}} \cdot \rho_x \equiv (\mathfrak{P} * \rho)_x \cdot \boldsymbol{\varphi}^{\deg^* \mathfrak{P}} \pmod{\mathfrak{p}}$$

for all $x \in R$ by Corollary 3.8. In terms of the coefficients, this last congruence means that

$$[c_v(x;\rho)]^{p^{\deg^* \mathfrak{P}}} \equiv c_v(x;\mathfrak{P} * \rho) \pmod{\mathfrak{p}}$$

for all $x \in R$, $v \geqslant 0$, which is equivalent to (8.7) by definition of $\mathbf{Frob}_{\mathfrak{P}}$. ∎

PROPOSITION 8.7. *The elliptic R-modules $\rho^{\mathfrak{A}}$, $\mathfrak{A} \in \mathscr{I}^*(R)$, are patterned alike (see Section 6).*

Proof. Let ρ, ρ' be elliptic R-modules which are defined over H_R and each of which is isomorphic to some $\rho^{\mathfrak{A}}$, $\mathfrak{A} \in \mathscr{I}^*(R)$. Suppose for some $x \in R$ and $v \geqslant 1$ we have $c_v(x;\rho) = 0$. Consider the invariant

$$J_v(x;\rho) = \frac{[c_v(x;\rho)]^{p^{\deg^* x} - 1}}{[\eta_\rho(x)]^{p^v - 1}} = 0.$$

By Theorem 5.11, there are infinitely many invertible prime ideals $\mathfrak{P}$ in R such that ρ' is isomorphic to $\mathfrak{P} * \rho$. By virtue of (8.7), this implies that for any such $\mathfrak{P}$ and any $\mathfrak{p} \notin S$ sitting over $\mathfrak{P}$, we have

$$J_v(x;\rho') = J_v(x;\mathfrak{P} * \rho) \equiv 0 \pmod{\mathfrak{p}}.$$

Thus, $J_v(x;\rho') = 0$ as it is divisible by infinitely many prime divisors of H_R. We have now shown that $c_v(x;\rho) = 0 \Rightarrow c_v(x;\rho') = 0$. Since the roles of ρ and ρ' can be interchanged in this argument, the proposition is proved. ∎

We are now in a position to prove Theorem 8.5. Let T be a set of representatives for the classes in $\mathscr{E}_R^*(\delta, 1)$ such that each rank 1 elliptic R-module in T is defined over H_R. Fix arbitrarily a nonconstant element $y \in R$. For each pair ρ', ρ'' of elements of T, consider the difference

$$F(y,\rho') - F(y,\rho''), \tag{8.8}$$

where $F(y,\rho')$ and $F(y,\rho'')$ are the vectors of invariants (6.11) associated to ρ' and ρ''. Let S be the set of prime divisors of H_R which contains all those prime divisors required by (1)–(4) above and in addition contains: (5) all prime divisors which divide some nonzero coefficient $c_\nu(y;\rho')$ for $\rho' \in T$ and $\nu \geqslant 1$, and (6) all prime divisors which divide some nonzero coordinate of the difference vectors (8.8) for all pairs ρ', ρ'' in T. Then S is a finite set. By the usual arguments, it suffices to prove (8.5) for the prime ideals $\mathfrak{P}$ in R which do not lie under any prime divisor in S. For such a $\mathfrak{P}$, we have

$$F(y, \mathbf{Frob}_{\mathfrak{P}}(\rho)) \equiv F(y, \mathfrak{P} * \rho) \pmod{\mathfrak{p}}$$

for every $\mathfrak{p}$ lying over $\mathfrak{P}$ by Lemma 8.6. By choice of S and $\mathfrak{P}$, this congruence must imply *equality* of the two vectors of invariants. We can now apply Proposition 6.9 to conclude that $\mathbf{Frob}_{\mathfrak{P}}(\rho)$ and $\mathfrak{P} * \rho$ are isomorphic over Ω. ∎

THEOREM 8.8. *The map* ψ *of* (8.4) *is a natural isomorphism. A prime ideal* $\mathfrak{P}$ *of* A *which does not contain the conductor* $\mathfrak{C}$ *of* R *splits completely in* H_R/k *if and only if* $\mathfrak{P} \cap R$ *is principal.*

Proof. Since the classes of the primes not containing $\mathfrak{C}$ generate $\mathbf{Pic}(R)$, ψ is surjective and hence an isomorphism by Theorem 8.5. A prime $\mathfrak{P}$ splits completely if and only if $\psi(\mathbf{Frob}_{\mathfrak{P}}) = 1$. By (8.5), this is equivalent to $\mathfrak{P}$ being principal in R. ∎

DEFINITION 8.9. A finite abelian extension of k on which ∞ splits completely is called a *class field of* A.

The class fields of A correspond to the generalized ideal class groups of A in the familiar way.

We can now identify H_R as a class field of A. Let $\mathscr{P}_R(\mathfrak{C})$ denote the group of principal ideals of A generated by the ideals xA with $x \in R$ and x prime to $\mathfrak{C}$. Let $\mathscr{P}_1(\mathfrak{C})$ be the subgroup consisting of the ideals xA with $x \in k$ and $x \equiv 1 (\bmod\, \mathfrak{C})$.

THEOREM 8.10. *The extension* H_R/k *is class field to the* $\mathfrak{C}$*-ideal group* $\mathscr{P}_R(\mathfrak{C})$, *and* $\mathbf{Pic}(R)$ *is isomorphic to the* $\mathfrak{C}$*-ideal class group* $\mathscr{I}(\mathfrak{C})/\mathscr{P}_R(\mathfrak{C})$ *where* $\mathscr{I}(\mathfrak{C})$ *is the group of ideals of* A *which are prime to* $\mathfrak{C}$. *Only primes dividing the conductor* $\mathfrak{C}$ *can ramify in* H_R/k. *The field of constants of* H_R *has degree* d_∞ *over* $\mathbb{F}_q$.

Proof. This theorem follows from Theorem 8.8 by class field theory. ∎

COROLLARY 8.11. *If* $R = \mathbb{F}_{p^s} + \mathfrak{C}$ *where* $1 \leqslant s \leqslant n$, *then* H_R/k *is the* ray class field *associated to* $\mathscr{P}_1(\mathfrak{C})$.

COROLLARY 8.12. *The field* H_A *is unramified of degree* hd_∞ *over* k, *and its field of constants has degree* d_∞ *over* $\mathbb{F}_q$.

Finally, let us consider the extension H_∞/k which is the union of the Hilbert Class Fields H_R for all orders R in A. We easily deduce the following theorem from the above results and class field theory.

THEOREM 8.13. *The extension H_∞/k is the maximal abelian extension of k on which ∞ splits completely.*

9. *Another Generation of Class Fields*

In this section, we restrict ourselves to the case $R = A$. Let ρ be a rank 1 elliptic A-module over H_A such that $D(\rho_x) = x$ for all $x \in A$. We will show that for any given ideal $\mathfrak{A}$ in A, the roots of $\rho_{\mathfrak{A}}(t)$ generate an abelian extension of H_A with Galois group isomorphic to the group of units $(A/\mathfrak{A})^*$ of the residue class ring $A/\mathfrak{A}$. From this extension, we recover the ray class field of A associated to $\mathfrak{A}$. Although these results can be generalized so as to hold for an arbitrary order, they seem more natural and meaningful when stated for A.

We work with a fixed ρ. Let $\mathfrak{A}$ be an ideal of A, and let $\Lambda_{\mathfrak{A}}$ be the set of roots of $\rho_{\mathfrak{A}}(t)$. We know from Section 3 that $\Lambda_{\mathfrak{A}}$ is an $\mathbb{F}_q$-vector space of dimension $\deg \mathfrak{A}$. From Sections 2 and 3, if A acts on $\Lambda_{\mathfrak{A}}$ through ρ, then $\Lambda_{\mathfrak{A}}$ becomes a cyclic A-module isomorphic to $A/\mathfrak{A}$. Now, the polynomial $\rho_{\mathfrak{A}}(t)$ is separable since its derivative is the constant $D(\rho_{\mathfrak{A}})$. The extension $H_A(\Lambda_{\mathfrak{A}})/H_A$ is therefore Galois. Let $G_{\mathfrak{A}}$ be its Galois group. Since A acts on $\Lambda_{\mathfrak{A}}$ by polynomials with coefficients from H_A, the action of $G_{\mathfrak{A}}$ on $\Lambda_{\mathfrak{A}}$ commutes with that of A. This gives a faithful representation of $G_{\mathfrak{A}}$ as a group of A-automorphisms of $\Lambda_{\mathfrak{A}}$. As the group of all A-automorphisms of $\Lambda_{\mathfrak{A}}$ is isomorphic to $(A/\mathfrak{A})^*$ in a natural way, we have a natural monomorphism

$$\psi : G_{\mathfrak{A}} \to (A/\mathfrak{A})^*. \tag{9.1}$$

We show now that ψ is actually an isomorphism. We recall from Section 1 the notation $\Phi_A(\mathfrak{A}) = \#(A/\mathfrak{A})^*$.

PROPOSITION 9.1. *Suppose $\mathfrak{A}$ is a prime power—say $\mathfrak{A} = \mathfrak{P}^e$. Let $\mathfrak{p}$ be a prime divisor of H_A which sits over $\mathfrak{P}$. Then $H_A(\Lambda_{\mathfrak{A}})/H_A$ is totally ramified at $\mathfrak{p}$, and its degree equals $\Phi_A(\mathfrak{A})$.*

Proof. By the proof of Proposition 7.2, there exists an extension field K' of H_A which is totally ramified over $\mathfrak{p}$ and which contains an element w such that the elliptic A-module ρ' over K' defined by $\rho'_x = w^{-1} \cdot \rho_x \cdot w$ for all $x \in A$ has coefficients in the valuation ring of the unique extension $\mathfrak{p}'$ of $\mathfrak{p}$ to K'. Let $\Lambda'_{\mathfrak{A}} = w^{-1} \cdot \Lambda_{\mathfrak{A}}$ be the set of roots of $\rho'_{\mathfrak{A}}(t)$. Now, the field $K'(\Lambda_{\mathfrak{A}}) = K'(\Lambda'_{\mathfrak{A}})$ can be generated over K' by any generator of the cyclic A-module $\Lambda_{\mathfrak{A}}$. If $\mathfrak{B} = \mathfrak{P}^{e-1}$, then these generators are exactly the roots of the quotient $\rho'_{\mathfrak{A}}(t)/\rho'_{\mathfrak{B}}(t)$, which by Proposition 7.6 and a simple calculation is a polynomial of degree $\Phi_A(\mathfrak{A})$ in $K'[t]$ which is Eisenstein at $\mathfrak{p}'$. The extension $K'(\Lambda_{\mathfrak{A}})/H_A$ is totally ramified at $\mathfrak{p}$ therefore, and the degree of $K'(\Lambda_{\mathfrak{A}})/K'$ equals $\Phi_A(\mathfrak{A})$. We conclude at once that $H_A(\Lambda_{\mathfrak{A}})/H_A$ is totally ramified at $\mathfrak{p}$ and that its degree is greater than or equal to $\Phi_A(\mathfrak{A})$. But from the monomorphism (9.1), its degree is also less than or equal to $\Phi_A(\mathfrak{A})$. ■

We can now compute the degree of $H_A(\Lambda_{\mathfrak{A}})/H_A$ for an arbitrary ideal $\mathfrak{A}$.

THEOREM 9.2. *The degree of $H_A(\Lambda_{\mathfrak{A}})/H_A$ is $\Phi_A(\mathfrak{A})$, and the map* (9.1) *is an isomorphism which identifies $G_{\mathfrak{A}}$ with $(A/\mathfrak{A})^*$ in a natural way.*

Proof. Let $\mathfrak{A} = \prod \mathfrak{P}^e$ be the canonical decomposition of $\mathfrak{A}$ into a product of prime ideals. By the results in Sections 2 and 3, $\Lambda_{\mathfrak{A}}$ is the direct sum of the A-modules $\Lambda_{\mathfrak{P}^e}$ associated to the prime powers appearing in this decomposition. Thus, $H_A(\Lambda_{\mathfrak{A}})$ is the composite of the fields $H_A(\Lambda_{\mathfrak{P}^e})$. We deduce from Proposition 9.1 that the extensions $H_A(\Lambda_{\mathfrak{P}^e})/H_A$ are linearly disjoint in pairs. The degree of $H_A(\Lambda_{\mathfrak{A}})/H_A$ is therefore the product of the numbers $\Phi_A(\mathfrak{P}^e)$, and this product equals $\Phi_A(\mathfrak{A})$. The fact that the map ψ of (9.1) is surjective now follows immediately. ■

Let S be a finite set of prime divisors of H_A which contains at least (1) any zero of $D(\rho_{\mathfrak{A}})$, (2) all prime divisors sitting over ∞, (3) any pole or zero of the $\eta_\rho(x)$, $x \in A$, and (4) any prime divisor which ramifies in $H_A(\Lambda_{\mathfrak{A}})/H_A$. As in Section 8, we conclude that such an S exists and that ρ has coefficients in the valuation ring at $\mathfrak{p}$ for all $\mathfrak{p} \notin S$. Our aim now is to compute the images under ψ of the Frobenius automorphisms associated to the prime divisors $\mathfrak{p} \notin S$.

LEMMA 9.3. *Suppose $\mathfrak{p}'$ is a prime divisor of $H_A(\Lambda_{\mathfrak{A}})$ which sits over a prime divisor $\mathfrak{p} \notin S$ of H_A. Then the elements of $\Lambda_{\mathfrak{A}}$ are distinct modulo $\mathfrak{p}'$.*

Proof. We have

$$\rho_{\mathfrak{A}}(t) = \prod (t - \lambda) \qquad (\lambda \in \Lambda_{\mathfrak{A}}). \tag{9.2}$$

Fix $\lambda_1 \in \Lambda_{\mathfrak{A}}$. Taking the derivative of both sides of (9.2) and then substituting $t = \lambda_1$, we find that

$$D(\rho_{\mathfrak{A}}) = \prod (\lambda_1 - \lambda) \qquad (\lambda \in \Lambda_{\mathfrak{A}};\ \lambda \neq \lambda_1).$$

Since $\mathfrak{p}$ does not divide $D(\rho_{\mathfrak{A}})$, $\mathfrak{p}'$ cannot divide any of the differences $\lambda_1 - \lambda$, $\lambda \neq \lambda_1$. ■

Let the prime divisor $\mathfrak{p}$ of H_A sit over the prime ideal $\mathfrak{P}$ of A, and let f be the inertial degree of $\mathfrak{p}$ over $\mathfrak{P}$. From the definitions, the Frobenius automorphism associated to the ideal $\text{Norm}(\mathfrak{p}) = \mathfrak{P}^f$ is the identity, which implies that $\mathfrak{P}^f$ is a principal ideal.

LEMMA 9.4. *Suppose* $\mathfrak{p} \notin S$, *and let* $\text{Norm}(\mathfrak{p}) = xA$, $x \in A$. *Then* $\eta_\rho(x)$ *belongs to* $\mathbb{F}_q^*$ *modulo* $\mathfrak{p}$.

Proof. We reduce ρ modulo $\mathfrak{p}$. By Corollary 3.8,

$$\rho_x \equiv \eta_\rho(x) \cdot \boldsymbol{\varphi}^{\deg^* x} \pmod{\mathfrak{p}}. \tag{9.3}$$

As the order of the residue class field at $\mathfrak{p}$ is

$$(N\mathfrak{P})^f = Nx = p^{\deg^* x},$$

the power of $\boldsymbol{\varphi}$ on the right in (9.3) commutes with ρ_y modulo $\mathfrak{p}$ for all $y \in A$; and the relation $\rho_x \cdot \rho_y \equiv \rho_y \cdot \rho_x (\text{mod}\, \mathfrak{p})$ is equivalent to

$$[\eta_\rho(x)]^{p^s} \equiv \eta_\rho(x) \pmod{\mathfrak{p}} \tag{9.4}$$

for every exponent s such that $\boldsymbol{\varphi}^s$ appears in ρ_y with nonzero coefficient modulo $\mathfrak{p}$. By Corollary 3.9, the greatest common divisor of all such s for all $y \in A$ is n. Therefore, (9.4) holds with $s = n$. ■

Let $\mathfrak{p} \notin S$ be a prime divisor of H_A, and let x generate the ideal $\text{Norm}(\mathfrak{p})$ of A. By Lemma 9.4, there is a unique element $\gamma(\mathfrak{p}, x)$ in $\mathbb{F}_q^*$ such that

$$\gamma(\mathfrak{p}, x) \equiv [\eta_\rho(x)]^{-1} \pmod{\mathfrak{p}}. \tag{9.5}$$

Let $x(\mathfrak{p})$ be the unique generator x of $\text{Norm}(\mathfrak{p})$ such that $\gamma(\mathfrak{p}, x) = 1$. Thus, having fixed ρ, we have associated an element of A to each $\mathfrak{p} \notin S$ in a canonical way.

THEOREM 9.5. *Suppose* $\mathfrak{p} \notin S$ *is a prime divisor of* H_A. *Then in the abelian extension* $H_A(\Lambda_{\mathfrak{A}})/H_A$ *we have*

$$\psi(\mathbf{Frob}_\mathfrak{p}) = \text{can}(x(\mathfrak{p})), \tag{9.6}$$

where $\text{can}\colon A \to A/\mathfrak{A}$ *is the canonical homomorphism.*

Proof. For brevity write $x = x(\mathfrak{p})$. Let $\mathfrak{p}'$ be an extension of $\mathfrak{p}$ to $H_A(\Lambda_{\mathfrak{A}})$, and let $\lambda \in \Lambda_{\mathfrak{A}}$. By definition,

$$\mathbf{Frob}_\mathfrak{p}(\lambda) \equiv \lambda^{N\mathfrak{p}} \pmod{\mathfrak{p}'}$$

and by Corollary 3.8

$$\rho_x(\lambda) \equiv \eta_\rho(x) \cdot \lambda^{Nx} \equiv \lambda^{Nx} \pmod{\mathfrak{p}'}$$

by choice of x. Since $Nx = (N\mathfrak{P})^f = N\mathfrak{p}$, we conclude that

$$\mathbf{Frob}_\mathfrak{p}(\lambda) \equiv \rho_x(\lambda) \pmod{\mathfrak{p}'}$$

for all $\lambda \in \Lambda_{\mathfrak{A}}$. Now since both sides of this last congruence are themselves elements of $\Lambda_{\mathfrak{A}}$, we infer via Lemma 9.1 that actually $\mathbf{Frob}_{\mathfrak{p}}(\lambda) = \rho_x(\lambda)$ for all $\lambda \in \Lambda_{\mathfrak{A}}$. Interpreting these equalities in terms of the isomorphism ψ, we arrive at (9.6). ■

COROLLARY 9.6. *Suppose $\mathfrak{P}$ is a prime ideal of A which does not sit under any prime divisor belonging to S. Then $\mathfrak{P}$ splits completely in $H_A(\Lambda_{\mathfrak{A}})/k$ if and only if $\mathfrak{P} = xA$ with* (1) $x \equiv 1(\text{mod}\,\mathfrak{A})$ *and* (2) $\eta_\rho(x) \equiv 1(\text{mod}\,\mathfrak{p})$ *for every extension $\mathfrak{p}$ of $\mathfrak{P}$ to H_A.*

Proof. Since it is exactly the principal prime ideals which split completely in the subextension H_A/k, it suffices to prove that conditions (1) and (2) on $\mathfrak{P} = xA$ are equivalent to $\mathbf{Frob}_{\mathfrak{p}}(\lambda) = \lambda$ for every $\lambda \in \Lambda_{\mathfrak{A}}$ and every extension $\mathfrak{p}$ of $\mathfrak{P}$ to H_A. By (9.6), if $x = x(\mathfrak{p})$, then

$$\mathbf{Frob}_{\mathfrak{p}}(\lambda) = \lambda \Leftrightarrow \rho_x(\lambda) = \lambda \Leftrightarrow \rho_{x-1}(\lambda) = 0 \tag{9.7}$$

for all $\lambda \in \Lambda_{\mathfrak{A}}$. But as $\Lambda_{\mathfrak{A}}$ is isomorphic to $A/\mathfrak{A}$ as an A-module, the last proposition in (9.7) is equivalent to $x \equiv 1(\text{mod}\,\mathfrak{A})$. ■

If $c \in \mathbb{F}_q^*$, then the element $\psi^{-1}(c)$ of $G_{\mathfrak{A}}$ maps each $\lambda \in \Lambda_{\mathfrak{A}}$ onto $c\lambda$. Therefore, the fixed field $E_{\mathfrak{A}}$ of $\psi^{-1}(\mathbb{F}_q^*)$ in $H_A(\Lambda_{\mathfrak{A}})$ is generated over H_A by $(q-1)$st powers of the elements of $\Lambda_{\mathfrak{A}}$.

THEOREM 9.7. *The field $E_{\mathfrak{A}}$ is the ray class field of A associated to the ideal $\mathfrak{A}$. In particular, $E_{\mathfrak{A}}/k$ is Galois with abelian Galois group.*

Proof. Let $\mathfrak{P}$ be a prime ideal of A which does not sit under any prime divisor in S. Proceeding as in the proof of Corollary 9.6 and using (9.6), we deduce that $\mathfrak{P}$ splits completely in $E_{\mathfrak{A}}/k$ if and only if $\mathfrak{P}$ is principal with $\text{can}(x(\mathfrak{p})) \in \mathbb{F}_q^*$ for every extension $\mathfrak{p}$ of $\mathfrak{P}$ to H_A. Thus, $\mathfrak{P}$ splits completely in $E_{\mathfrak{A}}/k$ if and only if $\mathfrak{P} = xA$ for an element $x \in A$ such that $x \equiv 1(\text{mod}\,\mathfrak{A})$. By class field theory, therefore, the Galois closure of $E_{\mathfrak{A}}/k$ is the ray class field of A associated to the ideal $\mathfrak{A}$. We complete the proof by observing that the degree $h(A) \cdot (\Phi_A(\mathfrak{A})/(q-1))$ of $E_{\mathfrak{A}}/k$ is equal to that of this ray class field. ■

III. EXAMPLES OF ELLIPTIC MODULES

10. *Preliminaries*

We first introduce a concept which is somewhat less restrictive than the concept of isomorphism of elliptic A-modules. Let us note first that if $\rho \in \mathscr{E}\ell\ell_A(K)$, then $w^{-1} \cdot \rho \cdot w \in \mathscr{E}\ell\ell_A(K)$ also for any element $w \in K^{\text{ac}}$ (the algebraic closure of K) such that $w^{q-1} \in K$. This follows from the fact that each ρ_x, $x \in A$ must be a polynomial in φ^n since it commutes with the elements in $\mathbb{F}_q$.

DEFINITION 10.1. Two elliptic A-modules ρ and ρ' over K are said to be *equivalent* if there is an element $w \in K^{\mathrm{ac}}$ such that $w^{q-1} \in K$ and $\rho' = w^{-1} \cdot \rho \cdot w$.

Let B be the integral closure of A in H_A. We show in this section that the hypothesis "k has a prime divisor of degree 1" implies that every rank 1 elliptic A-module ρ over H_A is equivalent over H_A to one which has coefficients in B (see Section 7). This result will be helpful in computing the examples in Section 11. We work with a fixed ρ of rank 1 over H_A. For brevity, we write $\eta(x)$ instead of $\eta_\rho(x)$ for $x \in A$, and we put $d = d_\infty$.

LEMMA 10.2. *Suppose $x, z \in A$. Then for all prime divisors $\mathfrak{p}$ of H_A, we have*

$$v_{\mathfrak{p}}(\eta(x)) = \left(\frac{q^{\deg x} - 1}{q^{\deg z} - 1}\right) \cdot v_{\mathfrak{p}}(\eta(z)). \tag{10.1}$$

Proof. This lemma is an immediate consequence of (8.6). ■

LEMMA 10.3. *There is an element $w \in H_A^{\mathrm{ac}}$ whose $(q^d - 1)$st power belongs to H_A such that*

$$\eta'(z) = \eta(z) \cdot w^{q^{\deg z} - 1} \in H_A \tag{10.2}$$

is a constant in H_A for every $z \in A$.

Proof. By the Riemann–Roch theorem, there are elements x and y in A such that $\deg y = \deg x + d$. Let w be a $(q^d - 1)$st root of the element $\eta(x)^{q^d}/\eta(y)$. Then for this w and any prime divisor $\mathfrak{p}$ of H_A, a simple calculation using (10.1) shows that $v_{\mathfrak{p}}(\eta'(z)) = 0$. Thus, $\eta'(z)$ has no poles or zeros in H_A and must therefore be a constant. ■

PROPOSITION 10.4. *If $d = d_\infty = 1$, then ρ is equivalent over H_A to an elliptic A-module ρ' with coefficients in B such that $\eta_{\rho'}(z)$ is a constant in H_A for all $z \in A$.*

Proof. By the previous lemma, there is an element w with $w^{q-1} \in H_A$ such that the element $\eta'(z)$ of (10.2) is a constant for all $z \in A$. Put $\rho' = w^{-1} \cdot \rho \cdot w$. By Corollary 7.4, if $\mathcal{O}$ is any finite valuation ring in H_A, then ρ' has coefficients in $\mathcal{O}$. But B is the intersection of the finite valuation rings in H_A. Therefore, ρ' actually has coefficients in B. ■

The following little lemma is required for the proof of our main result (Theorem 10.6).

LEMMA 10.5. *Let $\rho \in \mathcal{E}\ell\ell_A(K)$, and let $x, y \in A$. Then x is a divisor of y in A if and only if ρ_x is a right divisor of ρ_y in $K[\boldsymbol{\varphi}]$.*

Proof. If $y = ax$, $a \in A$, then $\rho_y = \rho_a \rho_x$. Conversely, suppose $\rho_y = \tau\rho_x$ for some $\tau \in K[\boldsymbol{\varphi}]$. Proceeding by contradiction, we assume y is not a multiple of x. Let $\mathfrak{B} = Ax + Ay$. Then $\mathfrak{B} \supset Ax$ but $\mathfrak{B} \neq Ax$. Since ρ_x is a right divisor of ρ_y, $I_{\rho,\mathfrak{B}} = I_{\rho,Ax}$ which is impossible by Proposition 3.7. ■

THEOREM 10.6. *Suppose k has a prime divisor of degree 1. Then any rank 1 elliptic A-module over H_A is equivalent to an elliptic A-module with coefficients in B.*

Proof. If ∞ has degree 1, then the result follows from Proposition 10.4 above. We may assume, therefore, that A has a prime ideal $\mathfrak{P}$ with $\deg \mathfrak{P} = 1$. Fix a nonzero element x in $\mathfrak{P}$, and let $\mathfrak{A}$ be the (integral) ideal such that $\mathfrak{A}\mathfrak{P} = xA$. Then

$$\deg \mathfrak{A} + 1 = \deg x. \tag{10.3}$$

Choose $y \in \mathfrak{A}$ such that the quotient y/x does not belong to A. By the right division algorithm in $H_A[\boldsymbol{\varphi}^n]$, there are polynomials τ and γ in $H_A[\boldsymbol{\varphi}^n]$ such that $\deg \gamma < \deg x$ and $\rho_y = \tau\rho_x + \gamma$, and we conclude from Lemma 10.5 that $\gamma \neq 0$. Now $\rho_{\mathfrak{A}}$ is a right divisor of γ since both x and y belong to $\mathfrak{A}$. But $\deg \gamma \leqslant \deg x - 1 = \deg \mathfrak{A}$ by (10.3). Therefore, there is an element $t \in H_A$ such that $\gamma = t\rho_{\mathfrak{A}}$. We claim now that

$$v_{\mathfrak{p}}(t) = \left(\frac{q^{\deg \mathfrak{A}} - 1}{q^{\deg z} - 1}\right) \cdot v_{\mathfrak{p}}(\eta(z)) \tag{10.4}$$

for all *finite* prime divisors $\mathfrak{p}$ of H_A and all $z \in A$. To prove this for fixed $\mathfrak{p}$ and z, choose u in an extension field K' of H_A so that $\rho' = u^{-1} \cdot \rho \cdot u$ has a good reduction modulo $\mathfrak{p}'$ where $\mathfrak{p}'$ is a prime divisor of K' which sits over $\mathfrak{p}$. Let $v_{\mathfrak{p}}$ also denote that extension of $v_{\mathfrak{p}}$ to K' which is strictly positive on $\mathfrak{p}'$. Then

$$0 = v_{\mathfrak{p}}(\eta_{\rho'}(z)) = v_{\mathfrak{p}}(\eta(z) \cdot u^{q^{\deg z} - 1}). \tag{10.5}$$

From $\rho'_y = (u^{-1}\tau u)\rho'_x + u^{-1}t\rho_{\mathfrak{A}}u$ and Lemma 10.5 again, we see that $u^{-1}t\rho_{\mathfrak{A}}u$ reduces modulo $\mathfrak{p}'$ to a nonzero multiple of $(\mathrm{rd}_{\mathfrak{p}'} \circ \rho')_{\mathfrak{A}}$, which also has degree equal to $\deg \mathfrak{A}$. Thus,

$$t \cdot u^{q^{\deg \mathfrak{A}} - 1}$$

is a unit modulo $\mathfrak{p}'$. Using this result to compute $v_{\mathfrak{p}}(u)$ in terms of $v_{\mathfrak{p}}(t)$ and then substituting from (10.5), we arrive at (10.4).

Now, let w be a $(q-1)$st root of $t^q/\eta(x)$. A simple calculation based on (10.4) and (10.1) shows us that

$$\eta'(z) = \eta(z) \cdot w^{q^{\deg z} - 1} \in H_A$$

is a unit modulo $\mathfrak{p}$ for all finite prime divisors $\mathfrak{p}$ of H_A and all $z \in A$. We conclude as in the proof of Proposition 10.4 that $w^{-1} \cdot \rho \cdot w$ has coefficients in B. ∎

As a corollary of Theorem 10.6, we recover the Principal Ideal Theorem for A in the case when k has a prime of degree 1.

PROPOSITION 10.7. *Suppose there is a rank* 1 *elliptic A-module ρ over H_A which has coefficients in B and is such that $D(\rho_x) = x$ for all $x \in A$. Then every integral ideal in A generates a principal ideal in B.*

Proof. It suffices to prove that $\mathfrak{P}B$ is principal for every prime ideal $\mathfrak{P}$ in A. We show in fact that $\mathfrak{P}B = uB$, where $u = D(\rho_{\mathfrak{P}})$. We know from Proposition 7.6 that $v_{\mathfrak{p}}(u) = 1$ for all prime divisors $\mathfrak{p}$ of H_A which divide $\mathfrak{P}$. Thus, it suffices to show that $v_{\mathfrak{p}}(u) = 0$ for all finite $\mathfrak{p}$ which do not divide $\mathfrak{P}$. To that end, choose $e > 0$ so that $\mathfrak{P}^e = xA$ is principal, and put $\mathfrak{A} = \mathfrak{P}^{e-1}$. Then by (3.10) for $\mathfrak{p}$ finite, $\mathfrak{p} \nmid \mathfrak{P}$,

$$v_{\mathfrak{p}}(D(\mathfrak{P} * \rho)_{\mathfrak{A}}) + v_{\mathfrak{p}}(u) = v_{\mathfrak{p}}(\eta_\rho(x)^{-1} \cdot x) = 0. \tag{10.6}$$

By Proposition 7.5, $\mathfrak{P} * \rho$ also has coefficients in B. Thus, both summands on the left in (10.6) are nonnegative, and therefore both are zero. ∎

11. *Computation of Examples*

Let there be given a pair (k, ∞) as above, an order R in A, a positive integer r, and an algebraically closed extension field K of k. Let $R \cap \mathbb{F}_q = \mathbb{F}_{p^m}$, $m \leqslant n$. Then one can effectively compute an elliptic R-module of rank r over K as follows: Choose elements $x_1, \ldots, x_s$ in R which generate R as an $\mathbb{F}_{p^m}$-algebra, and write ρ_{x_i} for $1 \leqslant i \leqslant s$ as a polynomial of degree $-r \cdot n \cdot d_\infty \cdot v_\infty(x_i)/m$ in $\psi = \varphi^m$ with constant coefficient x_i but with indeterminate coefficients in degrees greater than zero. Let $\mathscr{S}$ be the system of k-algebraic relations among these coefficients determined by the commutativity conditions $\rho_{x_i}\rho_{x_j} = \rho_{x_j}\rho_{x_i}$ for $1 \leqslant i < j \leqslant s$. By Corollary 5.3, any nontrivial solution of this system in K provides an example of an elliptic R-module of rank $\leqslant r$. We can get an elliptic R-module of rank equal to r by solving the system $\mathscr{S}^*$ obtained from $\mathscr{S}$ by specializing the leading coefficient of (say) ρ_{x_1} to be 1. The analytic results in Section 4 show that a solution to $\mathscr{S}^*$

always exists in K. Further, a solution can be computed effectively by the well-known algorithms of elimination theory.

We limit ourselves to examples of rank 1. By Corollary 5.12, the system $\mathscr{S}^*$ has only finitely many solutions when $r = 1$, and in a few cases these can be found without undue effort by elimination. However, it is usually more practicable to use congruence conditions to determine a solution. This works as follows. Let us assume for simplicity that $R = A$ and $d_\infty = 1$. Then by Proposition 10.4, there is a solution to $\mathscr{S}^*$ which has coefficients in B, the integral closure of A in $H_R \subseteq K$, provided that x_1 is suitably normalized. The analytic results in Section 4 enable one to compute the valuations at ∞ of the coefficients of ρ_{x_1} for the various k-embeddings of H_A into Ω. As these coefficients lie in B, this information restricts each of them to a computable finite set. We may then eliminate all but $h(A)$ of these finitely many possibilities for ρ_{x_1} by computing the elliptic A-modules of rank 1 over the homomorphic images of B modulo some of its prime ideals. Once we have determined ρ_{x_1}, the ρ_{x_i}, $2 \leqslant i \leqslant s$, can be computed in a straightforward manner (cf., Corollary 5.4).

In the examples which follow, we specify only A and R since A determines the pair (k, ∞) uniquely.

EXAMPLE 11.1. Take $A = \mathbb{F}_4[x]$ and $R = \mathbb{F}_2 + xA$. Since k is a rational function field and $d_\infty = 1$, we compute $h(R) = 1$ from (1.17). Now, $R = \mathbb{F}_2[x, \theta x]$ where $\theta^2 + \theta + 1 = 0$. The system $\mathscr{S}$ is easily solved by elimination and yields the solution

$$\begin{aligned} \rho_x &= x + x(x+1)\varphi + x^2\varphi^2, \\ \rho_{\theta x} &= \theta x + \theta x(\theta x + 1)\varphi + \theta^2 x^2 \varphi^2. \end{aligned}$$

Since $\rho_{\theta x} + \theta^2 \rho_x = x + x\varphi$, $\rho_{\mathfrak{C}} = 1 + \varphi$ where $\mathfrak{C} = xA$ is the conductor of R. The elliptic A-module $\rho' = \mathfrak{C} * \rho$ is determined by $\rho'_x = x + x^4\varphi^2$.

EXAMPLE 11.2. Take $A = \mathbb{F}_3[x, y]$ where $y^2 = x - x^2$ and $R = A$. Again k is a rational function field, but $d_\infty = 2$. Therefore, $h(A) = 2$, and $H_A = k(c)$ where $c^2 = -1$ by Corollary 8.12. The fundamental unit of B is $\eta = 1 + x + cy$, as one easily checks. We know by Theorem 10.6 that there is a rank 1 elliptic A-module ρ with coefficients in B, and we find upon solving $\mathscr{S}$ by elimination that we can determine ρ by

$$\rho_x = x + y\eta\varphi + \bar{\eta}\varphi^2, \qquad \rho_y = y + (1 + x)\eta\varphi + c\bar{\eta}\varphi^2,$$

where $\bar{\eta} = 1 + x - cy$.

According to MacRae [6] there are exactly four "imaginary quadratic" function fields with class number 1. These lead to four examples of elliptic

A-modules with $H_A = k$. We have computed the essentially unique rank 1 elliptic A-module for each field using the method of congruence conditions outlined above. These computations are much facilitated by the fact that $B = A$ has unique factorization.

EXAMPLE 11.3. Take $A = \mathbb{F}_2[x, y]$ with $y^2 + y = x^3 + x + 1$. Then a rank 1 elliptic A-module ρ is determined by

$$\rho_x = x + (x^2 + x)\varphi + \varphi^2,$$
$$\rho_y = y + (y^2 + y)\varphi + x(y^2 + y)\varphi^2 + \varphi^3.$$

EXAMPLE 11.4. Take $A = \mathbb{F}_4[x, y]$ with $y^2 + y = x^3 + \theta, \theta^2 + \theta + 1 = 0$. Then ρ is determined by

$$\rho_x = x + (x^8 + x^2)\varphi^2 + \varphi^4,$$
$$\rho_y = y + (x^{10} + x)\varphi^2 + (x^{32} + x^8 + x^2)\varphi^4 + \varphi^6.$$

We note the factorizations $x^{10} + x = (x^4 + x)(y^4 + y)$ and

$$x^{32} + x^8 + x^2 = [x(y^4 + y)(x^3 + x + 1)(x^3 + \theta x + 1)(x^3 + \theta^2 x + 1)]^2.$$

EXAMPLE 11.5. Take $A = \mathbb{F}_3[x, y]$ with $y^2 = x^3 - x - 1$. Then ρ is determined by

$$\rho_x = x + a\varphi + \varphi^2, \qquad \rho_y = y + b\varphi + c\varphi^2 + \varphi^3,$$

where $a = y(x^3 - x)$, $b = y^4 - y^2$,
and $c = (y^3 - y)(y(x^3 - x + 1) - 1)(y(x^3 - x + 1) + 1)$.

EXAMPLE 11.6. Take $A = \mathbb{F}_2[x, y]$ with $y^2 + y = x^5 + x^3 + 1$. Then ρ is determined by

$$\rho_x = x + (x^2 + x)^2\varphi + \varphi^2,$$
$$\rho_y = y + b_1\varphi + b_2\varphi^2 + b_3\varphi^3 + b_4\varphi^4 + \varphi^5,$$

where

$$b_1 = (y^2 + y)(x^2 + x),$$
$$b_2 = x^2(x + 1)(y^2 + y)(x^3 + y)(x^3 + y + 1),$$
$$b_3 = y(y + 1)(x^5 + x^3 + x^2 + x + 1)((1 + x^2 + x^3)y + x^2 + x^4 + x^7)((1 + x^2 + x^3)y + 1 + x^3 + x^4 + x^7)$$
$$b_4 = [x(y^2 + y)(x^5 + x^2 + 1)(x + y)(x + 1 + y)]^2.$$

The factorization of b_3 is a prime factorization in A.

Finally, we present two examples in which $h(A) = 2$ and 3, respectively. In both examples, $d_\infty = 1$, and there is also a prime ideal $\mathfrak{P}$ of A such that $\deg^* \mathfrak{P} = 1$. We determine a rank 1 elliptic A-module ρ by writing down only the value of d such that

$$\rho_{\mathfrak{P}} = d + \boldsymbol{\varphi}.$$

If $\mathfrak{P}^e = xA$ ($e = 2$ or 3), then ρ_x can be computed from $\rho_{\mathfrak{P}}$ by (3.10) and (8.5). It would be worth writing down the coefficients of ρ_x if one could factor them into prime ideals in H_A. However, these factorizations are not available.

EXAMPLE 11.7. Take $A = \mathbb{F}_3[x, y]$ with $y^2 = x(x^2 - x - 1)$. One calculates $h(A) = 2$ and $B = \mathbb{F}_3[x, u, z]$ with $u^2 = x^2 - x - 1$, $z^2 = x$ and $uz = y$. And one proves in the usual way that $\eta = 1 + x + u$ is a fundamental unit in B. Let $\mathfrak{P}$ be the ideal in A generated by x and y. Then $\mathfrak{P}$ has degree 1 and $\mathfrak{P}^2 = xA$. We conclude that $\mathfrak{P}B = zB$. We know from the proof of Proposition 10.7 that $d = \epsilon\eta^s z$ where s is an integer and $\epsilon = \pm 1$. A value for s can be determined by analysis from (5.16). We find $s = -1$. Since we can change the sign of d by making an equivalence using $w^2 = -1$, we may assume that $\epsilon = 1$. Thus

$$\rho_{\mathfrak{P}} = \eta^{-1} z + \boldsymbol{\varphi}$$

determines a rank 1 elliptic A-module.

EXAMPLE 11.8. Take $A = \mathbb{F}_3[x, y]$ with $y^2 = x^3 + x^2 - 1$. One calculates $h(A) = 3$ and $B = \mathbb{F}_3[x, y, z]$ where $z^3 - z = y$. The ideal $\mathfrak{P}$ generated by $x - 1$ and $y - 1$ has degree 1, and $\mathfrak{P}^3 = (y - x)A$. We find that $\mathfrak{P}B = wB$ where $w = (x - 1)^2 + (y - 1)z + (x - 1)z^2$. Let σ be the k-automorphism of H_A which maps z to $z + 1$, and put $\eta = -x - z + z^2$. Then we compute in the usual way that η and η^σ constitute a basis for the group of units of B. Therefore, as in the above example, $d = \epsilon\eta^{s + t\sigma} w$ where s and t are integers and $\epsilon = \pm 1$. We find using (5.16) and analysis that $s = -1$, $t = 0$. From (3.10) and (8.5), we learn that $d \cdot d^\sigma \cdot d^{\sigma^2} = y - x$, which easily implies that $\epsilon = 1$. Thus,

$$\rho_{\mathfrak{P}} = \eta^{-1} w + \boldsymbol{\varphi}$$

determines a rank 1 elliptic A-module.

APPENDIX. THE UNIVERSAL ELLIPTIC A-MODULE OF RANK 1

In this appendix, we release the symbol K from its previous role as a field and allow it to range freely over the category of (commutative) A-algebras. In other words, K now denotes a commutative ring with identity for which there is given a fixed ring morphism $\delta_K: A \to K$. As in Section 2 above, we form the twisted polynomial ring $K[\boldsymbol{\varphi}]$ with left coefficients from K subject to the relations (2.1). We have the inclusion $i: K \to K[\boldsymbol{\varphi}]$ and the augmentation $D: K[\boldsymbol{\varphi}] \to K$ defined exactly as before. If $\mu: K \to K'$ is a morphism of A-algebras, then the operation of μ on the coefficients of the polynomials in $K[\boldsymbol{\varphi}]$ induces a ring morphism $K[\boldsymbol{\varphi}] \to K'[\boldsymbol{\varphi}]$, which we also call μ. The formation of $K[\boldsymbol{\varphi}]$ from K is thus *functorial* in K.

For the purposes of this appendix, we work not with $K[\boldsymbol{\varphi}]$ but rather with the subring $K[\boldsymbol{\psi}]$ generated over K by the element $\boldsymbol{\psi} = \boldsymbol{\varphi}^n$, where $n = \log_p q$. The ring $K[\boldsymbol{\psi}]$ is also a twisted polynomial ring over K with the relations

$$\boldsymbol{\psi} w = w^q \boldsymbol{\psi} \tag{A1}$$

for all $w \in K$.

Definition A1. Given an A-algebra K, an *elliptic A-module over* K is a ring morphism $\rho: A \to K[\boldsymbol{\psi}]$ such that (1) $D \circ \rho = \delta_K$, (2) $\rho \neq i \circ \delta_K$, and (3) the leading coefficient of each nonzero polynomial in the image of ρ is a unit in K.

As before, we write ρ_x for the image of $x \in A$ under ρ. We define the *rank* of ρ to be the rational number r_ρ such that

$$\deg \rho_x = -r_\rho \cdot d_\infty \cdot v_\infty(x) \tag{A2}$$

for all $x \in A$. If $\mu: K \to K'$ is a morphism from K into some field K', then we see from Proposition 2.3 applied to $\mu \circ \rho$ that r_ρ is actually an integer.

Let r denote a fixed positive integer, and put

$$s_r(x) = -r \cdot d_\infty \cdot v_\infty(x) \tag{A3}$$

for all $x \in A$. If ρ is an elliptic A-module of rank r over K, then we write

$$\rho_x = \delta_K(x) + \sum_{v=1}^{s_r(x)} c_v(\rho, x) \cdot \boldsymbol{\psi}^v$$

for all $x \in A$.

Definition A2. Let ρ and ρ' be two elliptic A-modules of the same rank r over K. We say that ρ and ρ' are *equivalent* if there is a unit $w \in K$ such that

$$c_v(\rho', x) = c_v(\rho, x) \cdot w^{(q^v - 1)/(q - 1)} \tag{A4}$$

for all $x \in A$ and all $1 \leqslant v \leqslant s_r(x)$.

Now, let F_r be the functor which associates to every A-algebra K the set of equivalence classes of the elliptic A-modules of rank r over K. We will show that the functor F_1 is representable provided that k has a prime divisor of degree 1. This result has much the same content as Theorem 10.6, and the proof we give is based on the same idea.

We first introduce the A-algebra $E_A^r = A[\mathbf{C}]/I$ where $\mathbf{C} = \{c_v(x) | x \in A, 1 \leqslant v \leqslant s_r(x)\}$ is a "triangular" array of indeterminates and where I is the ideal generated by the relations among these indeterminates which determine a ring morphism $\rho^*: A \to E_A^r[\psi]$ with

$$\rho_x^* = x + \sum_{v=1}^{s_r(x)} c_v(x) \cdot \psi^v \tag{A5}$$

for all $x \in A$. If we put

$$\operatorname{grad} c_v(x) = q^v - 1 \tag{A6}$$

for all $x \in A$ and $1 \leqslant v \leqslant s_r(x)$, then I is a homogeneous ideal in the grading induced on $A[\mathbf{C}]$, and so E_A^r is a *graded* A-algebra. Let S be the monoid generated by the leading coefficients of the polynomials ρ_x^*, $x \in A$, of (A5). Then $S^{-1}E_A^r$ is graded (in both positive and negative dimensions), and there is a canonical elliptic A-module ρ^* over $S^{-1}E_A^r$ determined by (A5). We see that $S^{-1}E_A^r$ represents the functor which associates to every A-algebra K the set of elliptic A-modules of rank r over K.

Let P_A^r be the set of homogeneous elements of grade 0 in $S^{-1}E_A^r$. If ρ is an elliptic A-module of rank r over K and if $\mu^\rho: S^{-1}E_A^r \to K$ is the canonical morphism such that $\rho = \mu^\rho \circ \rho^*$, then the restriction μ_0^ρ of μ^ρ to P_A^r depends only on the equivalence class of ρ.

THEOREM A3. *If k has a prime divisor of degree* 1, *then the A-algebra P_A^1 represents the functor F_1.*

Proof. We have just seen that for each elliptic A-module ρ of rank 1 over K, we can construct a morphism $\mu_0^\rho: P_A^1 \to K$ in a canonical way. To complete the proof, therefore, it suffices to show that for every morphism $\mu: P_A^1 \to K$ there is an equivalence class M of elliptic A-modules of rank 1 over K such that $\mu = \mu_0^\rho$ for every $\rho \in M$. This last assertion means that there is an elliptic A-module of rank 1 over P_A^1 which is equivalent in $S^{-1}E_A^1$ to ρ^*; i.e., that $S^{-1}E_A^1$ has a unit of grade $q - 1$. We now prove the existence of such a unit from the hypothesis of the theorem.

For brevity put $\eta(x) = \eta_{\rho^*}(x)$ (cf. (7.1)) for all $x \in A$. If $d_\infty = 1$, then there are elements x and y in A such that $\deg y = \deg x + 1$ by the Riemann–Roch theorem; and then $w = \eta(y)/[\eta(x)]^q$ is a unit of grade $q - 1$. Therefore, we may assume that there is a prime ideal $\mathfrak{P}$ of A such that $\deg \mathfrak{P} = 1$.

As in the proof of Theorem 10.6, we fix $x \in \mathfrak{P}$, $x \neq 0$, and we let $\mathfrak{A}$ be the ideal such that $\mathfrak{A}\mathfrak{P} = xA$. We choose $y \in \mathfrak{A}$ such that $y/x \notin A$. Since the leading coefficient of ρ_x^* is a unit in $S^{-1}E_A^1$, we may divide ρ_y^* by ρ_x^* in

$S^{-1}E_A^1[\psi]$ on the right, obtaining polynomials τ and γ such that $\rho_y^* = \tau\rho_x^* + \gamma$ with $\deg\gamma < \deg\rho_x^*$. Further, the coefficients of τ and γ will be homogeneous elements of $S^{-1}E_A^1$.

From Lemma 10.5 and (3.8) applied to some homomorphic image of ρ^* over a field K, we see that $\deg\gamma = \deg\mathfrak{A}$, as in the proof of Theorem 10.6. Let t be the leading coefficient of γ. Now, t cannot be contained in any prime ideal $\mathfrak{p}$ of $S^{-1}E_A^1$ since otherwise we would contradict (3.8) modulo $\mathfrak{p}$. Therefore, t is a unit. As grad $t = q^{\deg\mathfrak{A}} - 1$, we easily compute using (10.3) that $w = \eta(x)/t^q$ is a unit of grade $q - 1$. ■

One can show by the methods (or results) of Drinfel'd [3] that in fact P_A^1 is isomorphic to the integral closure B of A in H_A. We then easily identify $S^{-1}E_A^1$ as the ring $B[w, w^{-1}]$ of Laurent polynomials in w over B.

References

1. A. Borel *et al.*, Seminar on complex multiplications, *in* "Lecture Notes in Mathematics," No. 21, Springer-Verlag, Berlin and New York, 1966.
2. P. Deligne, Formes modulaires et représentations ℓ-adique, "Lecture Notes in Mathematics," No. 179, Springer-Verlag, Berlin and New York, 1971.
3. V. G. Drinfel'd, Elliptic modules (Russian), *Math. Sb.* **94** (1974), 594–627; *Math. USSR Sb.* **23** (1974), 561–592.
4. D. R. Hayes, Explicit class field theory for rational function fields, *Trans. Amer. Math. Soc.* **189** (1974), 77–91.
5. N. Jacobson, "The Theory of Rings," Amer. Math. Soc. Mathematical Surveys, Vol. II, New York, 1943.
6. R. E. MacRae, On unique factorization in certain rings of algebraic functions, *J. Algebra* **17** (1971), 243–261.
7. I. I. Piateckii-Shapiro, "Zeta-functions of Modular Curves," Lecture Notes in Mathematics No. 349, Springer-Verlag, Berlin and New York, 1973.
8. W. Schöbe, "Beiträge zur Funktionentheorie in nichtarchimedisch bewerten Körpern," Inaugural-Dissertation, Helios-Verlag, Halle, 1930.
9. J.-P. Serre, Complex multiplication, *in* 'Algebraic Number Theory" (J. W. S. Cassels and A. Frolich, eds.), Thompson, Washington, 1967.

STUDIES IN ALGEBRA AND NUMBER THEORY
ADVANCES IN MATHEMATICS SUPPLEMENTARY STUDIES, VOL. 6

Some Diophantine Equations Related to the Quadratic Form $ax^2 + by^2$

EDWARD A. BENDER

University of California at San Diego,
La Jolla, California

NORMAN P. HERZBERG

Institute for Defense Analyses,
Princeton, New Jersey

Given positive integers a and D, with a square-free and $D > 1$, we are interested in finding all the integers x, p, and n for which $ax^2 + D = p^n$ and p is prime to $2D$. We generally assume that if $a = 1$, then D is not a square. The following are typical results: (i) If D is not a square then there is an effectively computable function Γ of a and D, such that there are no solutions with $n > 2$ and $p > \Gamma$. (ii) When p is given, we show how to get bounds on the number of solutions; for example, when D is even and square-free there can be no more than $2p/3$ solutions. (iii) When n is given, we give a complete solution of $ax^2 + D = z^n$ for various classes of a and D. In particular we explicitly show how to solve this equation when $(z, 2D) = 1$ and the class number of $\mathbb{Q}(\sqrt{-aD})$ is not divisible by n, and give conditions on a and D which imply $(z, 2D) = 1$. (iv) The solutions of the equation $ax^2 + Dq^z = p^n$ in the positive integer variables x, z, and n, are described.

Contents

1. *Introduction*

Given positive integers a and D with a square-free and $D > 1$, we are interested in studying all the integers x, p, n for which $ax^2 + D = p^n$ and p is prime to $2D$. These results are then applied to the equations $ax^2 + D = N^z$,

ISBN 0-12-599153-3

$ax^2 + D = z^n$, and $ax^2 + Dq^z = p^n$. We also discuss some cases of $ax^2 + D = 2p^n$ and $4p^n$ briefly. We generally assume that if $a = 1$, then D is not a square. Although these equations have been studied by various authors (see the references), many have assumed that $a = 1$ or have restricted D. Our general discussion contains no restrictions on a and D except for $a > 0$, $D > 1$, and, usually, aD not a square. (The case $D = 1$ has been treated by Nagell [27, Sect. 12] when n is not a power of 2. Ljunggren [16] considers $aD = p^2$.) We provide a unified treatment which allows us to survey most of the known results, generalize many of them, and derive new ones.

Despite our care in preparing this survey, the careful reader is certain to find (easily corrected, we hope) errors. We ask the reader's pardon and hope he will inform us of those he notices.

Let us write $D = bf^2$, where b is square-free. We are then led to the study of the integer solutions (x, y, n) of the equation

$$ax^2 + by^2 = p^n, \tag{1}$$

where a and b are relatively prime, square-free, positive integers, and p is odd. To avoid trivial solutions, we assume that $xy \neq 0$ and identify the solutions obtained by changing the signs of x and y. Since p is prime to D, we may assume without loss of generality that p is prime to $abxy$. When $(p, D) \neq 1$ the situation is more complicated, because, after removing common factors, we are left with equations of the form $ax^2 + D = cp^n$.

The next three sections discuss some basic arithmetic properties of the solutions of (1). After treating prime p, we show how to extend the results to general p by introducing the notion of "families" of solutions. In Section 5 we show that the values of n in a family of solutions to (1) for which $|y| \leqslant C$ are widely spaced and that there is at most one such n for each p greater than an effectively computable function of b and C. In Sections 6 through 8 we discuss the solutions of $ax^2 + D = p^n$ in a single family. In Section 9 we apply these results and bounds on the number of families with solutions to the equation $ax^2 + D = N^z$. To illustrate, we provide bounds on the number of solutions when $aD < 25$. In Section 10 we discuss the equation $ax^2 + D = z^n$, when n does not divide the class number of $\mathbb{Q}(\sqrt{-aD})$. In Part III the more general equation $ax^2 + Dq^z = p^n$ is discussed. In most cases there are no more than five distinct values of z: three even and two odd. We also discuss the two equations $ax^2 + D = 2p^n$ and $ax^2 + D = 4p^n$, which can be handled by simple modifications of the ideas in Parts I and II.

The following notation will be used. The letters a, b, and p will always refer to the parameters in (1), and we assume they satisfy the conditions listed after (1). The highest power of q dividing an integer c is denoted by $\operatorname{ord}_q(c)$ and the greatest common divisor of r and s by (r, s). Set $\omega = \sqrt{-ab}$.

Equivalence in the ideal class group of the imaginary quadratic number field $\mathbb{Q}(\omega)$ is denoted by $\sim$.

Illustrative special results are set off as observations.

I. $ax^2 + by^2 = p^n$

2. *Solutions with p Prime*

Throughout this section we will assume that p is an odd prime. In Theorem 1 we establish that (1) has many solutions if it has any, and that the exponents n are precisely all the multiples, or all the odd multiples, of a number l defined below. In Theorems 2 and 3 we obtain some arithmetic properties of the solutions that will be useful later.

In order to give a unified treatment we must establish some notational conventions.

Suppose $a \neq 1$. Since $a \mid ab$, it ramifies in $\mathbb{Q}(\omega)$ and we may write $(a) = \mathfrak{a}^2$, thus defining the ideal $\mathfrak{a}$. If $a = 1$ let $\mathfrak{a} = (1)$. Note that in both cases $(a) = \mathfrak{a}^2$.

The proliferation of cases is due to two simple arithmetic facts. First of all, if either of a or b equals 1 the ideal $\mathfrak{a}$ is principal, while otherwise $\mathfrak{a}$ is not principal, since a is not a norm from $\mathbb{Q}(\omega)$ to $\mathbb{Q}$ (Lemma 2). Secondly, when $ab \equiv 3$ modulo 4 there are integers of $\mathbb{Q}(\omega)$ that are of the form $(r + s\sqrt{\omega})/2$, where r and s are odd integers. Hence there may be "half-integral solutions" to (1).

Since we will use half-integers below, certain statements need to be clarified. The statement "v divides y" for any v and y means "y/v is a rational integer." Two numbers are "relatively prime" if their greatest common divisor is either $\frac{1}{2}$ or 1.

Suppose (x, y, n) is a "solution" of (1) where x and y may be half integers if $ab \equiv 3$ modulo 4. We refer to such a "solution" as a *possibly half-integral* solution of (1). The word "solution" is reserved for *integral* solutions. We require the theory of half-integral solutions in some applications. You may find it convenient to ignore these solutions in a first reading.

Let $\lambda = ax \pm y\omega$. Note that λ is an algebraic integer. The norm of λ (from $\mathbb{Q}(\omega)$ to $\mathbb{Q}$) is $a^2x^2 + aby^2 = ap^n$. Since $p \nmid 2ab$, it is unramified in $\mathbb{Q}(\omega)$. If p remained prime in $\mathbb{Q}(\omega)$ we'd have $p \mid \lambda$, contradicting $p \nmid ax$. *Therefore p splits* and we can write

$$(p) = \mathfrak{p}\bar{\mathfrak{p}}.$$

Furthermore, since $p \nmid \lambda$, while $\lambda\bar{\lambda} = ap^n$ we see that with an appropriate choice of sign in λ

$$(\lambda) = \mathfrak{a}\mathfrak{p}^n.$$

The above discussion shows that when a possibly half-integral solution of (1) exists it makes sense to talk about μ, an algebraic integer of $\mathbb{Q}(\omega)$ satisfying $(\mu) = \mathfrak{a}\mathfrak{p}^L$, *where L is the smallest positive integer for which* $\mathfrak{a}\mathfrak{p}^L$ *is principal.* If $ab \neq 1$ or 3, then μ is unique up to sign and conjugation (i.e., sign changes of x and y). When $ab = 1$ or 3 there are units other than ± 1 in $\mathbb{Q}(\omega)$. If $ab = 1$, we may multiply μ by i (i.e., interchange x and y). Suppose $ab = 3$. Let $\zeta = (1 + \sqrt{-3})/2$. Precisely one of μ, $\zeta\mu$, and $\zeta^2\mu$ lies in the order $\mathbb{Z}[\sqrt{-3}]$ because the norm of μ is odd. Without loss of generality, we call it μ. Hence μ is again unique up to sign change and conjugation.

Denote the *order of* $\mathfrak{p}$ *in the ideal class group by H*. We define the integers h and l as follows:

$$h = H \quad \text{and} \quad l = L \qquad \text{if} \quad \mu \in \mathbb{Z}[\omega],$$

and

$$h = 3H \quad \text{and} \quad l = 3L \qquad \text{if} \quad \mu \notin \mathbb{Z}[\omega].$$

Remark. If $\mu \notin \mathbb{Z}[\omega]$, then $ab \equiv 3$ modulo 4 and $\mu = (r + s\omega)/2$.

Let E denote the exponent of the ideal class group of $\mathbb{Q}(\omega)$—that is, the smallest integer such that $\mathfrak{a}^E \sim 1$ for all ideals $\mathfrak{a}$ of $\mathbb{Q}(\omega)$. Set $E^* = E$ if $a = 1$ or $b = 1$ and set $E^* = E/2$ otherwise. (Theorem 1 will show that E^* is an integer when (1) has solutions.)

The following lemma explains the somewhat strange case $h = 3H$.

LEMMA 1. *If* μ *does not lie in the order* $\mathcal{O} = \mathbb{Z}[\omega]$, *then* $ab \equiv 3$ modulo 8 *and precisely those powers of* μ *of the form* μ^{3t} *lie in* $\mathcal{O}$.

Proof. Suppose $\mu \notin \mathcal{O}$. Then the trace of μ is odd. Since $(\mu) = \mathfrak{a}\mathfrak{p}^L$, the norm of μ is also odd. Thus μ^2 is the sum of an odd integer and an odd multiple of μ. Multiplying by μ, it is easy to see that $\mu^3 \in \mathbb{Z}[2\mu] \pm \mathcal{O}$. The norm of μ is $(r^2 + abs^2)/4$ where r and s are odd. Hence $ap^L \equiv (1 + ab)/4$ modulo 8. Since ap^L is odd, $ab \equiv 3$ modulo 8.

Suppose $\mu^m, \mu^n \in \mathcal{O}$ where $m \leqslant n$. Since $(ap^L)^m \mu^{n-m} = \mu^n \overline{\mu^m} \in \mathcal{O}$, μ^{n-m} is an algebraic integer, and $(ap^L)^m$ is odd, it follows that μ^{n-m} belongs to $\mathcal{O}$. Thus the set of n for which $\mu^n \in \mathcal{O}$ is closed under taking the greatest common divisor. Since 3 is in this set, the proof is complete. ■

LEMMA 2. *If* $a > 1$ *and* $b > 1$, *then a is not a norm from* $\mathbb{Q}(\omega)$ *to* $\mathbb{Q}$ *of an integer, and so* $\mathfrak{a} \not\sim (1)$.

Proof. Suppose that a is a norm. Then we have $\mathfrak{a}^2 = (\nu\bar{\nu})$ for some ν. Since $\mathfrak{a}$ is square-free, it contains no rational prime factors. Since we also have $\mathfrak{a} \sim \bar{\mathfrak{a}}$, $\mathfrak{a}$ is a product of ramified primes. Thus $(\nu) = \mathfrak{a}$. Hence $4a =$

$w^2 + aby^2$ for some integers w and y. Since a is square-free, $a \mid w$; say $w = ax$. The only solution with $a > 1$ and $b > 1$ is $a = b = 2$, which contradicts $(a, b) = 1$. ■

We can now state and prove the folk theorem.

THEOREM 1. *If $a = 1$ or $b = 1$, Equation (1) has solutions if and only if (i) p splits and (ii) n is a multiple of $l = h$. There is exactly one solution for each such n, unless $a = b = 1$, in which case there are two. In either case, E is the least common multiple of the L's of all the primes not dividing $2aD$.*

If $a > 1$ and $b > 1$, Equation (1) has solutions if and only if (i) p splits, (ii) n is an odd multiple of $l = h/2$, and (iii) $\mathfrak{a} \sim \mathfrak{p}^l$. There is exactly one solution for each such n. (Note that (ii) implies that H is even.)

The above statements are true for possibly half-integral solutions when l is replaced by L and h by H.

Proof. Let (x, y, n) be a solution of (1), and $\lambda = ax \pm y\omega$. We have already observed that p must split, and that the sign in λ may be chosen so that

$$(\lambda) = \mathfrak{a}\mathfrak{p}^n. \tag{2}$$

If $ab \neq 1$ or 3, the only units of $\mathbb{Q}(\omega)$ are ± 1, and so there is only one solution for each n. If $ab = 3$ and $\zeta = (1 + \sqrt{-3})/2$, only one of the three associates λ, $\zeta\lambda$, $\zeta^2\lambda$ lies in the order $\mathbb{Z}[\sqrt{-3}]$. Therefore the solution is unique for each n in these cases. If $ab = 1$, replacing λ by $i\lambda$ interchanges x and y, and so there are two solutions in this case.

Suppose $a = 1$ or $b = 1$. Then $\mathfrak{a}$ is principal and so $H = L$. By the definition of l, $h = l$. By (2), $\mathfrak{p}^n$ is principal and so H divides n. By Lemma 1, h divides n. Conversely, if p splits and h divides n, there is some algebraic integer $\lambda \in \mathbb{Z}[\omega]$ such that $(\lambda) = a\mathfrak{p}^n$. Write $\lambda = z + y\omega$. Since a divides $z^2 + aby^2 = ap^n$ and is square-free, we may write $z = ax$. Clearly x and y satisfy (1) and are integers. Finally, E is the least common multiple of all the H's since the primes dividing $2aD$ can be ignored. (Each ideal class contains an infinite number of primes [11, Theorem 9-2-6].)

Suppose $a > 1$ and $b > 1$. Since $(\lambda)^2$ and $\mathfrak{a}^2$ are principal ideals, $\mathfrak{p}^{2n} \sim 1$ by (2). By Lemma 2, $\mathfrak{a}$ is not principal, and so $\mathfrak{p}^n$ is not principal. We have shown that H divides $2n$ but not n. By Lemma 1, h divides $2n$ but not n. Hence h is even, n is an odd multiple of $h/2$, and $l = h/2$. That $\mathfrak{a} \sim \mathfrak{p}^l$ follows from (2).

Conversely, suppose p splits, $\mathfrak{a} \sim \mathfrak{p}^l$, and n is an odd multiple of l. Then there is some algebraic integer $\lambda \in \mathbb{Z}[\omega]$ such that $(\lambda) = \mathfrak{a}\mathfrak{p}^n$. Write $\lambda = ax + y\omega$. As in the previous case, x and y are integers and satisfy (1).

The modifications for half-integral solutions are trivial. ■

Remarks. It is well known that p splits if and only if $(-ab|p) = +1$, where $(|)$ is the Legendre symbol. This condition is often substituted for (i) in Theorem 1.

If $a > 1$ and $b > 1$, it is not necessarily true that E^* is the least common multiple of the L's of all the primes not dividing $2aD$. To see this, construct an imaginary quadratic extension whose class group is $\mathbb{Z}_2 \oplus \mathbb{Z}_4$ and such that $\mathfrak{a}$ lies in $\mathbb{Z}_2$. Since $\mathfrak{a}$ is not principal and since, by (iii), $\mathfrak{a} \sim \mathfrak{p}^l$, $\mathfrak{p}$ lies in $\mathbb{Z}_2$.

THEOREM 2. *Suppose ab is positive, square-free, and prime to p. Assume that $au^2 + bv^2 = p^m$ for some $m > 0$ and some relatively prime (half) integers u and v. Let*

$$\beta = au + v\omega. \tag{3}$$

When r is odd define $x(rm)$ and $y(rm)$ by

$$ax(rm) + y(rm)\omega = \beta^r/a^{(r-1)/2}. \tag{4}$$

Then

$$v \sum_{j=0}^{(r-1)/2} \binom{r}{2j} (-bv^2)^{(r-2j-1)/2}(au^2)^j = y(rm). \tag{5}$$

Suppose n and k are odd multiples of m. Then:

(i) *$x(n)$ and $y(n)$ are relatively prime (possibly half) integers,*
(ii) *$ax(n)^2 + by(n)^2 = p^n$,*
(iii) *if u and v are integers, $\mathrm{ord}_2(y(n))$ is constant,*
(iv) *$(y(k), y(n)) = \pm y((k, n))$, and*
(v) *we can replace y by x in* (iii) *and* (iv).

Proof. By equating the coefficients of ω in (4) after expanding the right-hand side, we obtain (5).

Let

$$\gamma = \beta^2/a = (au^2 - bv^2) + 2uv\omega. \tag{6}$$

It is easily seen that $(\beta) = \mathfrak{a}\mathfrak{p}^m$ and $(\gamma) = \mathfrak{p}^{2m}$. Hence $(ax(n) + y(n)\omega) = \mathfrak{a}\mathfrak{p}^n$. Thus $(ax(n), y(n)) = 1$. Taking norms we obtain (ii) and we see that, since $\beta\gamma^{(r-1)/2}$ is an algebraic integer, $(ax(n))^2$ is a (quarter) integer divisible by a (or $a/4$). Since a is square-free, $x(n)$ is a (half) integer. Thus (i) is true.

Working with the coefficients s and t in $s + t\omega$, we have $\gamma \equiv p^m$ modulo $2v$ when u and v are integers and so $y(n) \equiv v$ modulo $2v$. Hence $\mathrm{ord}_2(v) = \mathrm{ord}_2(y(n))$, proving (iii), and also that v divides $y(n)$. Even if u and v are half integers, $\gamma \equiv p^m$ modulo v, and so v divides $y(n)$ in any case. Setting $m = (n, k)$, we see that $y(m)$ divides $y(k)$ and $y(n)$. Hence (iv) will be proved if we show that $(y(k), y(n))$ divides $y(m)$. Since $m = (n, k)$, there are positive integers s and t such that $sk - tn = m$. Define i and j by $k = (2i + 1)m$ and $n = (2j + 1)m$, then

$$\beta = (\beta^{2i+1})^s(\beta^{2j+1})^{-t} = (\beta\gamma^i)^s(\beta\gamma^j)^{-t}a^{si-tj}$$
$$= (\beta\gamma^i)^s(\overline{\beta\gamma^j})^t a^{si-tj}/p^{nt}.$$

By the definition of β, and the fact that $(ap, y(k)) = 1$ for all k and both $\beta\gamma^i$ and $\beta\gamma^j$ are real modulo $(y(k), y(n))$, it follows that $y(m)$ is zero modulo $(y(k), y(n))$.

The last assertion in the theorem follows by symmetry. ■

THEOREM 3. *We use the notation of Theorem 2. Suppose that* $a = 1$. *Then for all* r *we can define* $x(rm)$ *and* $y(rm)$ *by*

$$x(rm) + y(rm)\omega = \beta^r. \tag{7}$$

Let n *be an even multiple of* m. *Then*:

(i) $2uv$ *divides* $y(n)$ *and*
(ii) $2uv$ *is prime to* $x(n)$.

Let n *and* k *be any multiples of* m. *Then*

(iii) $x(n)$ *and* $y(n)$ *are relatively prime (possibly half) integers.*
(iv) $(y(n), y(k)) = \pm y(n, k))$.

Proof. β^2 is real modulo $2uv$ and prime to $2uv$. Thus (i) and (ii) hold. The proof of (iii) and (iv) parallels the proof of the corresponding facts in Theorem 2, but for (iv) use $\beta^{k/m}$ and $\beta^{n/m}$ in place of $\beta\gamma^i$ and $\beta\gamma^j$. We do not need to define i and j since $a = 1$. ■

Remarks. If $b = 1$, we can still use Theorem 3 by interchanging the roles of x and y. Since the solutions of (1) are unique for each n, (4) *and* (7) *generate all solutions to* (1) *when* $m = l$.

Theorems 2(iv) and 3(iv) imply that, when $b \neq 1$ and m divides n, $y(m)$ divides $y(n)$. Setting $m = l$ and $y(n) = 1$ we see that

Observation 1. If D is square-free and $ax^2 + D = p^n$ has a solution, then it has one with $n = l$, a number dividing $3E^*$ and hence not exceeding 3 times the class number of $\mathbb{Q}(\omega)$. When $ax^2 + D = p^n$ is known to have at most one solution, this provides a practical method for finding it.

Alter and Kubota [1] have shown that the factor of 3 can be eliminated unless $ab \equiv 3$ modulo 8 and $p^L = (D \pm 2)/3$. The condition that D be square-free can be weakened considerably: It suffices to have $p^L > D/4$. See Theorem 8 in Section 5. The condition that p be prime can be replaced by $(p, 2D) = 1$. See Section 3.

COROLLARY 3.1. *If $b \neq 1$ or if $y(n)$ is even, then $y(m) \mid y(n)$.*

Proof. If $b = 1$, Theorem 3(ii) implies that n/m is odd. If $a = 1$, apply Theorem 3(iv). If n/m is odd, apply Theorem 2(iv). ■

Theorem 2(iii) can be generalized to all primes:

LEMMA 3. *We use the notation of Theorems 2 and 3. Suppose (1) has solutions at the exponents n and m, $n = rm$, and q is a prime dividing $abuv$ but not r. Then*

$$\operatorname{ord}_q(x(n)\,y(n)) = \operatorname{ord}_q(uv).$$

Proof. Suppose r is odd. By symmetry we may suppose $q \mid bv$ and so $q \nmid au$. By (5),

$$y(n) = r(au^2)^{(r-1)/2}v + Cbv^2$$

for some integer C. Since $qv \mid bv^2$, $\operatorname{ord}_q(y(n)) = \operatorname{ord}_q(v)$. Since $q \mid bv$, $q \mid by(n)$ and so $q \nmid ax(n)$. This completes the proof for r odd.

Suppose r is even. By Theorem 1, one of a, b must equal 1. By symmetry we may suppose $a = 1$. If $q \mid bv$, an argument like that in the previous paragraph completes the proof. Suppose $q \mid u$. By (7), $y(n) = rv^{(r-1)}(-b)^{(r-1)/2}u + Cu^2$. Hence $\operatorname{ord}_q(y(n)) = \operatorname{ord}_q(u) > 0$. ■

3. *General p*

When p is an arbitrary *odd number prime to abxy*, the results in the preceding section can easily be generalized by introducing the notion of "families of solutions."

We first prove that if there are any solutions (x, y, n) to (1), then all the primes dividing p split. Let $\lambda = ax \pm y\omega$. The norm of λ is $a^2x^2 + aby^2 = ap^n$. Since $(p, 2ab) = 1$, every prime q dividing p is unramified in $\mathbb{Q}(\omega)$. If a prime q remained prime in $\mathbb{Q}(\omega)$ we'd have $q \mid \lambda$, contradicting $(p, ax) = 1$. Therefore every prime q dividing p splits. This proof is essentially the same as the one in Section 2.

Let τ be the number of distinct prime factors of p. We may write

$$(p) = \prod_{j=1}^{\tau} \mathfrak{q}_j \bar{\mathfrak{q}}_j,$$

where each $\mathfrak{q}_j$ and $\bar{\mathfrak{q}}_j$ is a distinct prime power ideal in $\mathbb{Q}(\omega)$. Finally let

$$\mathfrak{p}_i = \prod_{j=1}^{\tau} \hat{\mathfrak{q}}_j,$$

where $\hat{\mathfrak{q}}_j$ is either $\mathfrak{q}_j$ or $\bar{\mathfrak{q}}_j$. There are 2^τ distinct values of $\mathfrak{p}_i$, and for each one $\mathfrak{p}_i\bar{\mathfrak{p}}_i = (p)$. It will be convenient to assume that the numbering is such that the conjugate of $\mathfrak{p}_i$ is numbered $i + 2^{\tau-1}$, for $i = 1, \ldots, 2^{\tau-1}$. Recalling the definition of the ideal $\mathfrak{a}$, we define the $2^{\tau-1}$ algebraic integers μ_i by

$$(\mu_i) = \mathfrak{a}\mathfrak{p}_i^{L_i} \qquad \text{for} \quad i = 1, \ldots, 2^{\tau-1},$$

where L_i is the smallest positive integer for which $\mathfrak{a}\mathfrak{p}_i^{L_i}$ is principal. Similarly we define the integers H_i, h_i, and l_i by mimicking Section 2.

THEOREM 4. *Suppose* (x, y, n) *is a solution to* (1). *Set* $\lambda = ax \pm y\omega$. *Then there is an* i, $1 \leqslant i \leqslant 2^{\tau-1}$, *such that, with an appropriate choice of sign in* λ,

$$(\lambda) = \mathfrak{a}\mathfrak{p}_i^n,$$

where n *is a multiple of* l_i, (*and* n *is an odd multiple of* l_i *if* $a > 1$ *and* $b > 1$). *Such a solution will be referred to as belonging to the ideal* $\mathfrak{p}_i$. *For each* (*odd*) *multiple* n *of* l_i, *there is a unique solution of* (1) *belonging to* $\mathfrak{p}_i$.

Proof. The proof is similar to the proof of the corresponding facts in Theorem 1. The only change is that we must consider the prime factors π of p one at a time to show that the ideal (λ) is divisible by only one of the two unequal conjugate primes into which π splits, and hence only one of $\mathfrak{q}_j$ or $\bar{\mathfrak{q}}_j$. ■

The following lemma generalizes part of the second half of Theorem 1.

LEMMA 4. *If* $a > 1$ *and* $b > 1$, *and we restrict our attention to a single family, then the value of* $\operatorname{ord}_2(n)$ *is constant for those exponents* n *at which* (1) *has solutions.*

Proof. Suppose solutions occur at the exponents $n = 2^k r$ and $n' = 2^{k'} r'$ where $k < k'$. By Theorem 4 there exist λ and λ' with norms ap^n and $ap^{n'}$. Set $t = 2^{k'-k}r'$. Since the norm of λ'^r/λ^t is a^{r-t}, an odd power of a, it follows that a is a norm from $\mathbb{Q}(\omega)$ to $\mathbb{Q}$, contradicting Lemma 2. ■

Theorem 4 tells us that the solutions of (1) can be divided into $2^{\tau-1}$ *families*, each family consisting of all the solutions belonging to one $\mathfrak{p}_i$. Therefore, *we limit our attention to one family at a time in the remainder of the paper unless we explicitly state otherwise.* The results in Sections 4 and 5 are for solutions in a single family. The subscript associated with the family can thus be dropped without confusion in the remainder of Part I.

It is possible to construct examples that not only have arbitrarily many families of solutions, but also have arbitrarily many different L_i. The following example is interesting because both families have solutions with

$y = 1$:

$$2 \cdot 1^2 + 13 \cdot 1^2 = 15, \qquad 2 \cdot 41^2 + 13 \cdot 1^2 = 15^3.$$

4. *Some Congruents*

We retain the notation of Section 2. In particular we assume that p is prime to $2abxy$, that (1) has a (possibly half-integral) solution at the exponent m, that β, u, and v are as in (3), and that $x(n)$ and $y(n)$ are defined by (4) or (7). Thus $(x(n), y(n), n)$ is in the same family as (u, v, m).

LEMMA 5. *Suppose $z + w\omega$ is an algebraic integer. Let $\varphi = \text{ord}_q(abw^2/2)$ and $\delta = \text{ord}_q(d)$. Suppose that $\varphi > 0$ and $q \nmid z$. If $q = 3$ suppose that $3z^2 \not\equiv abw^2$* modulo 9. *Let A stand for any number prime to q. Then*

$$(z + w\omega)^d \equiv z^d + q^\delta A w\omega \text{ modulo } q^{\delta+\varphi}.$$

Suppose that $q \mid bv^2$, that r is odd if $a \neq 1$, and that $a \not\equiv bv^2/3$ modulo 3 *if $q = 3$. Then* $\text{ord}_q(y(rm)/v) = \text{ord}_q(r)$.

Proof. Since $\varphi > 0$, $q \neq 2$ when z and w are half-integers. It is easy to prove by induction on δ that

$$(z + w\omega)^{q^\delta} \equiv z^{q^\delta} + q^\delta A w\omega \text{ modulo } q^{\delta+\varphi}.$$

(When $\delta = 1$, the special assumption for $q = 3$ is needed to derive the above.) By raising this to the power d/q^δ, a number not divisible by q, we obtain the desired result.

To see the last result, set $z = au$, $w = v$, and $d = r$; then use (4) or (7). The condition $3z^2 \not\equiv abw^2$ becomes $3a \not\equiv bv^2$ since $3 \nmid au$. Unless $q = 2$ and $\varphi = 1$, we have $\text{ord}_q(w) < \varphi$ and so $\text{ord}_q(q^\delta A w) < \delta + \varphi$. If $q = 2$, the result follows from Theorem 2(iii) with "r" $= d/2$ plus the formula $y(2k) = 2x(k)y(k)$. ■

LEMMA 6. *Let $\varphi = \text{ord}_q(au^2bv^2)$. Suppose that $\varphi > 0$. If $q = 3$ suppose that $au^2 \not\equiv 3bv^2$* modulo 9 *and $bv^2 \not\equiv 3au^2$* modulo 9. *Suppose k and n are odd multiples of m. Let $t = (n - k)/m$, $\tau = \text{ord}_q(t)$, and $\sigma = \text{ord}_q(v)$. Let A stand for any number prime to q.*

(i) *If $q \mid au^2$, then*

$$y(n) \equiv y(k)(-p^m)^{t/2} \text{ modulo } q^{\varphi+\tau}.$$

(ii) *If on the other hand $q \mid bv^2$, then*

$$y(n) \equiv y(k)(p^m)^{t/2} + Aq^{\tau+\sigma} \text{ modulo } q^{\tau+\sigma+1}.$$

(ii′) *If* $q = 2$, bv^2 *is even, and* $\tau \geqslant 2$, *then*

$$y(n) \equiv y(k) + 2^{\tau+\sigma} \text{ modulo } 2^{\tau+\sigma+1}.$$

(ii″) *If* q *is odd,* $q \mid bv^2$, *and* $p^{tm} \equiv 1$ *modulo* q, *then*

$$y(n) \equiv \pm y(k) + Aq^{\tau+\sigma} \text{ modulo } q^{\tau+\sigma+1}.$$

Proof. Again note that $q \neq 2$ when u and v are half-integers. The proof depends on the observations, made in proving Theorem 2, that $y(n)$ is the coefficient of ω in $(ax(k) + y(k)\omega)\gamma^t$, and that $\gamma = au^2 - bv^2 + 2uv\omega = p^m - 2bv^2 + 2uv\omega = 2au^2 - p^m + 2uv\omega$. Let $d = t/2$, $w = 2uv$, and $z = au^2 - bv^2$. Note that $q \nmid z$. By Lemma 5,

$$\gamma^{t/2} \equiv z^{t/2} + q^\tau Auv\omega \text{ modulo } q^{\varphi+\tau},$$

provided $3z^2 \not\equiv 4au^2bv^2$ modulo 9 when $q = 3$. If $3 \nmid au^2$, this is equivalent to $3au^2 \not\equiv bv^2$ modulo 9; if $3 \mid au^2$, it is equivalent to $3bv^2 \not\equiv au^2$ modulo 9. Multiplying the congruence for $\gamma^{t/2}$ by $ax(k) + y(k)\omega$ and equating the coefficients of ω we get

$$y(n) \equiv z^{t/2}y(k) + ax(k)q^\tau Auv \text{ modulo } q^{\varphi+\tau}.$$

Suppose first that $q \mid au^2$. Then when q is odd, $z \equiv -p^m$ modulo q^φ, while when $q = 2$, $z \equiv -p^m$ modulo $2^{\varphi+1}$. Hence $z^{t/2} \equiv (-p^m)^{t/2}$ modulo $q^{\varphi+\tau}$ for all q. Since $u \mid x(k)$ by Theorem 2(v, iv), $\operatorname{ord}_q(aux(k)) \geqslant \varphi$, and (i) follows from (8).

Suppose $q \mid bv^2$. Then $\varphi = \operatorname{ord}_q(bv^2) \geqslant \sigma + 1$, and so $z^{t/2} \equiv p^{mt/2}$ modulo $q^{\tau+\sigma+1}$ for all q. To deduce (ii) from (8) we must check that $\operatorname{ord}_q(ax(k)Auv) = \sigma$. Now by definition $q \nmid A$, while $q \nmid au$, since $q \mid bv^2$. If $\sigma = 0$, then $q \mid b$ and so $q \nmid x(k)$. If $\sigma > 0$, then $q \mid v \mid y(k)$ and so $q \nmid x(k)$. Thus $\operatorname{ord}_q(ax(k)Auv) = \operatorname{ord}_q(v) = \sigma$.

If $q = 2$ and $\tau \geqslant 2$, then since $z^2 \equiv 1$ modulo 8, $z^{t/2} \equiv 1$ modulo $2^{\tau+1}$. If q is odd and $p^{tm} \equiv 1$ modulo q, then $z^{s/2} \equiv \pm 1$ modulo q for some $s \mid t$ and prime to q. Hence $z^{t/2} \equiv \pm 1$ modulo $q^{\tau+1}$. Since $v \mid y(k)$ by Theorem 2(iv), $\operatorname{ord}_q(y(k)) \geqslant \sigma$. Hence when $q = 2$, $z^{t/2}y(k) \equiv y(k)$ modulo $2^{\tau+\sigma+1}$, while when q is odd $z^{t/2}y(k) \equiv \pm y(k)$ modulo $q^{\tau+\sigma}$. Thus (ii′) and (ii″) follow from (ii). ■

The following lemma, by giving many congruences for the same number, provides a means for obtaining contradictions, particularly if $r \equiv 3$ modulo 4. In Part II we apply it with $\epsilon = \pm 1$, while in Section 12 we apply it with $\epsilon = \pm q$.

LEMMA 7. *Suppose r is odd and v is an integer. Let $\epsilon = y(rm)/v$. Then*

$$\epsilon \equiv rp^{m(r-1)/2} \text{ modulo } bv^2, \tag{9a}$$

$$\epsilon \equiv (-1|r)p^{m(r-1)/2} \text{ modulo } au^2, \tag{9b}$$

$$\epsilon \equiv (-2|r)p^{m(r-1)/2} \text{ modulo } au^2 - bv^2, \tag{9c}$$

$$\epsilon \equiv (-1|r)p^{m(r-1)/2} \text{ modulo } au^2 - 3bv^2 \qquad \text{if } 3 \nmid r, \tag{9d}$$

$$\epsilon \equiv (-3|r)p^{m(r-1)/2} \text{ modulo } 3au^2 - bv^2 \qquad \text{if } 3 \nmid r, \tag{9e}$$

$$\epsilon \equiv (-4bv^2)^{(r-1)/2} \equiv (4au^2)^{(r-1)/2} \text{ modulo } p^m. \tag{9f}$$

If r is a prime, then

$$\epsilon = (-bv^2|r). \tag{9g}$$

The above results remain true for half-integral solutions provided all factors of 2 are removed from the moduli. Equation (9g) *is unaltered.*

Proof. By dividing (5) by v and taking the result to various moduli we obtain (9). We will illustrate this by proving (9c). From (5),

$$\epsilon \equiv (bv^2)^{(r-1)/2} \sum \binom{r}{2j} (-1)^{(r-2j-1)/2} \text{ modulo } au^2 - bv^2.$$

The summation is the imaginary part of $(1+i)^r$ and therefore equals $2^{(r-1)/2}$ in magnitude. Its sign is positive if and only if r is congruent to 1 or 3 modulo 8. Thus the sign is $(-2|r)$. Since $p^m \equiv 2bv^2$ modulo $au^2 - bv^2$, the result follows.

To prove (9d) and (9e) consider the real and imaginary parts of $(1+\sqrt{-3})^r$.

Equation (9g) comes from $\epsilon \equiv (-bv^2)^{(r-1)/2}$ modulo r. ■

LEMMA 8. *If r is odd and q is an odd prime dividing $y(rm)$ but not bv, then r and $q - (-ab|q)$ have a common factor.*

Proof. We cannot have $q|au^2$, for then $q|ax(rm)^2$ as well as $by(rm)^2$.

We apply (5). Since $q \nmid bv^2$, there is a $\lambda \neq 0$ such that $au^2 \equiv -\lambda bv^2$ modulo q. Substituting in (5) and noting that $q|\epsilon$, we obtain

$$\begin{aligned} 0 &\equiv (-bv^2)^{(r-1)/2} \sum \binom{r}{2j} \lambda^j \\ &\equiv (-bv^2)^{(r-1)/2}((1+\sqrt{\lambda})^r + (1-\sqrt{\lambda})^r)/2 \quad \text{modulo } q. \end{aligned}$$

Since q and r are odd,

$$(\sqrt{\lambda}+1)^r \equiv (\sqrt{\lambda}-1)^r \quad \text{modulo } q.$$

Hence, in the multiplicative group of $GF(q)(\sqrt{\lambda})$ the order of $\mu = (\sqrt{\lambda}+1)/(\sqrt{\lambda}-1)$ is a divisor of r. Since $\mu \neq 1$, r and $(q-1)(q+1)$ have a common factor. If $\mu \in GF(q)$, then r and $q-1$ have a common factor; otherwise, r and $q+1$ have a common factor. Now $\mu \in GF(q)$ is equivalent to $\sqrt{\lambda} \in GF(q)$, which is equivalent to $(\lambda|q) = +1$. Multiplying the definition of λ by $-bv^2$ we see that λ is a square if and only if $-ab$ is. ■

5. *Solutions of* (1) *with* $|y| \leqslant C$

Here we prove that the successive values of n in the solutions of $ax^2 + D = p^n$ are widely spaced. We actually prove more because we only need assume that $|y|$ is bounded. In Theorem 8 we show that, roughly speaking, (1) has at most one small value of y, namely $y(l)$. First a lemma.

LEMMA 9. *Let w and $z = x + yi$ be complex numbers such that $0 < |\arg(z)| < \pi/2$ and $|z| > |w| > 0$. Set $t = |y/z|$. Then*

$$t < |\arg(z)| < t/(1-t^2)^{1/2}$$

and

$$|\arg(z+w) - \arg(z)| < |w|(|z(z+w)| - |w|^2)^{-1/2}.$$

Proof. Let $f(t) = \arcsin(t)$. Then $f'(t) = (1-t^2)^{-1/2}$. By the mean value theorem we have $1 < f(t)/t < f'(t)$ when $t \neq 0$. This establishes the inequality for $\arg(z)$.

Let $v = z + w$ and let ϑ be the angle between v and z. Since $|z| > |w|$, $|\vartheta| < \pi/2$. By the law of cosines and the arithmetic–geometric mean inequality,

$$2|vz|\cos\vartheta = v\bar{v} + z\bar{z} - w\bar{w} \geqslant 2|vz| - w\bar{w}.$$

Since $|\sin| = (1-\cos^2)^{1/2}$,

$$|\sin\vartheta| \leqslant (\delta(1-\delta/4))^{1/2} < \delta^{1/2},$$

where $\delta = |w|^2/|vz|$. Combined with the first half of the lemma, this proves the second half. ■

In the theorem which follows we continue to use the notation of Section 2.

THEOREM 5. *Let $C > 0$ and $k > 1$ be given. Suppose that p is so large that*

$$p^{mk/2} > (2C/v)^2 + 2C\sqrt{b}. \tag{10}$$

Let $(x(n), y(n), n)$ and $(x(n'), y(n'), n')$ be two solutions in the same family as (u, v, m). Suppose $n' > n$ and that both n and n' are multiples of m—and both odd multiples if $a \neq 1$. Finally suppose $|y(n)| \leqslant C$ and $|y(n')| \leqslant C$. Then either

$m < n < mk$ or $m < n' < mk$ or

$$n' - n > m(\pi(p^m/bv^2 - 1)^{1/2} - 3/2). \tag{11}$$

If $b = 1$ and n and n' are not both odd multiples of m, then the conclusion still holds if π is replaced by $\pi/2$ in (11).

By symmetry the same results hold for $x(n)$ and $x(n')$ when "$a \neq 1$" is replaced by "$b \neq 1$."

Proof. We may assume $u > 0$ and $v > 0$. Define $\eta = u\sqrt{a} + v\sqrt{b}i$. By Theorems 2 and 3, $y(jm)$ is the imaginary part of $\eta^j/\sqrt{b}$. Suppose $|y(jm)| \leqslant C$. Apply Lemma 9 with $z = \eta^j$ and $w = -y(jm)\sqrt{b}i$. Since $z + w$ is real,

$$|J\pi - j\arg(\eta)| < |w|(|z(z+w)| - |w|^2)^{-1/2} \tag{12}$$

for some integer $J \geqslant 0$. When (12) holds we say that j is associated with J, except that we insist that 1 is associated with 0.

When $j \geqslant k$, $|z(z+w)| - |w|^2 \geqslant |z|(|z| - 2|w|) \geqslant p^{mj/2}(p^{mk/2} - 2C\sqrt{b})$. Combining this with (10) and (12) we obtain

$$|J\pi - j\arg(\eta)| < v\sqrt{b}/2p^{mj/4} \qquad \text{when} \quad j \geqslant k. \tag{13}$$

Since $p^{mj/4} \geqslant |\eta|$, the first part of Lemma 9 combined with (13) yields

$$|J\pi - j\arg(\eta)| < \arg(\eta)/2 \qquad \text{when} \quad j \geqslant k. \tag{14}$$

Note that $J = 0$ is impossible in (14). It follows that there is at most one j associated with each J.

Let $j' > j$ be given. Suppose $j' \geqslant k$, and either $j \geqslant k$ or $j = 1$. By adding two versions of (14) together, or using (14) and $|\arg(\eta)| \leqslant \arg(\eta)$ if $j = 1$, we obtain

$$|(J' - J)\pi - (j' - j)\arg(\eta)| < 3\arg(\eta)/2.$$

Since $j' \neq j$, we have $J' \neq J$ and so $(j' - j + 3/2)\arg(\eta) > \pi$. Multiplying this last inequality by m, and substituting the upper bound for $\arg(\eta)$ provided by Lemma 9 completes the proof of (11).

If $b = 1$ and j is even, then $y(jm)$ is the real part of η^j. As a result, we may have $w = -y(jm)\sqrt{b}$ and the value of J in (12) will then be a half-integer. The rest of the proof is unchanged. ■

Remarks. When $k = 2$ in the theorem, (10) implies (11) since $m < n < 2m$ and $m < n' < 2m$ are both impossible.

As long as (4) or (7) defines x and y, there is no reason to assume that x and y are integers; *Theorem 5 is valid when x and y are any real numbers.*

If $ab = 1$, the theorem still applies provided we think of x and y as being generated by (7) with either $x(n) = \pm f$ or $y(n) = \pm f$, but not both; i.e., we ignore the additional solutions of (1) which are obtained by interchanging x and y.

Persson [29] uses another argument to show that when r is odd and $p^m \geqslant bv^2(\operatorname{cosec} \pi/r)^2 + 1$, then $|y(rm)/v| \geqslant r$. When applicable this result improves the "3/2" in (11) to slightly more than 1.

THEOREM 6. *Let $C \geqslant 0$ be given. Then*:

(i) *There is an effectively computable function $\Pi(b, C)$ such that for all $p > \Pi(b, C)$, Equation* (1) *has at most one solution per family with $|y| \leqslant C$. If $b \neq 1$, this solution occurs at $n = l$ if at all.*

(ii) *If $b \neq 1$, there is an effectively computable function $\Gamma(a, b, C)$ such that* (1) *has no solutions with $n > 2$, $p > \Gamma(a, b, C)$ and $|y| \leqslant C$. There is also an effectively computable function $\Gamma'(a, b, C)$ such that for any $p > \Gamma'(a, b, C)$ there cannot be solutions of* (1) *at both the exponents $n = 1$ and $n = 2$ with $|y| \leqslant C$.*

Proof. We use the notation established above. Suppose there is at least one solution with $|y| \leqslant C$. If $b \neq 1$, then $y(l)$ divides $y(n)$ for all n by Theorems 2(iv) and 3(i). Hence $|y(l)| \leqslant C$. In this case set $m = l$. Suppose $b = 1$. Let k be a multiple of l such that $|y(k)| \leqslant C$ and $\operatorname{ord}_2(k) = \epsilon$ is a minimum. Set $m = 2^\epsilon l$. By Theorem 2(iv), $y(m)$ divides $y(k)$ and thus cannot exceed C in absolute value. Since ϵ is a minimum, every n such that $|y(n)| \leqslant C$ is a multiple of m.

Suppose now that there are two solutions $|y(m)| \leqslant C$ and $|y(jm)| \leqslant C$. We shall apply Baker's result [3] to (13) with π replaced by $\pi/2$ as noted in Theorem 5. The right-hand side of (13) is less than $C\sqrt{b}\,p^{-mj/4}$, which is less than $e^{-\delta j}$ for some $0 < \delta \leqslant 1$. Obviously δ is a computable function of b and C, independent of p since $p > e$. Since $\arg(\eta)$ is less than $\pi/2$, $J \leqslant j$. Using Baker's notation from [3], set

$$\begin{aligned} n &= 2, & \delta &= \delta, \\ b_1 &= J, & H &= -b_2 = j, \\ \alpha_1 &= i, & \alpha_2 &= e^{i \arg(\eta)} = \eta/|\eta|. \end{aligned}$$

It is easily shown that α_2 is a root of the quartic equation

$$(au^2 - bv^2 - p^m x^2)^2 = (2iuv\sqrt{ab})^2 = -4ab(uv)^2.$$

The maximum of the absolute values of the coefficients in this equation is thus less than $2p^{2m}$. Set $d = 4$ and $A = 2p^{2m}$. By [3] mj is less than $\Phi = m(2^{16}\delta^{-1} \log 2p^{2m})^{25}$. On the other hand, since we may replace $|v|$ by C in (11), there is an effectively computable function $\Pi(b, C)$ such that mj exceeds Φ whenever $p^m > \Pi$. This proves (i).

To prove (ii) we apply Baker's result [4] to the equations $ax^2 + bk^2 = z^l$ with $k = 1, 2, \ldots, C$, and l a divisor of $3E^*$, but not 1 or 2. (Note that $3E^*$ is bounded by 3 times the class number of $\mathbb{Q}(\omega)$, an effectively computable function of a and b, and that the roots of $az^l - abk^2$ are all distinct.) Thus none of these equations have a solution if z exceeds some computable function depending only on a, b, and C. By (i), when $p > \Phi(a, b, C)$, the only possible solutions of (1) are at $n = l$. Since the hypothesis excludes $n = 1$ or 2, the first part of (ii) follows. Now suppose $p = ax^2 + bk^2$ and $p^2 = aX^2 + bK^2$, where $0 < k$, $K \leqslant C$. Then $aX^2 = (ax^2 + bk^2)^2 - bK^2$. Since $b \neq 1$ by hypothesis, the right-hand side of this equation has 4 unequal roots, and so by [4], there is a computable function $\Delta(a, b, k, K) = \Delta(a, b, C)$ such that $|x| \leqslant \Delta$. Setting Γ' equal to the maximum of Δ and Φ, the last part of (ii) follows. ■

In the rest of this section we use Theorem 5 to establish nonexistence criteria for solutions of (1) for various types of y. Theorem 7 is stated in a somewhat peculiar form to facilitate proving later results.

THEOREM 7. *Suppose that* (1) *has a possibly half-integral solution at the exponent M and a possibly half-integral solution in the same family at the exponent rM; that $st \mid r$ with $s \geqslant t > 1$; and, if $b = 1$, that s and t are odd. Let $f = |y(rM)|$ and let $\Lambda = 4$ if the solution of* (1) *at M is integral and $\Lambda = 16$ otherwise. Then either*

(i) $p^{rM/2} \leqslant \Lambda f^2 + 2f\sqrt{b}$, *or*
(ii) $p^{kM} \leqslant (1 + (2k + 1)^2/4\pi^2)bf^2$ *whenever* $1 \leqslant k \leqslant s$.

Proof. It suffices to prove the theorem with M replaced by Mr/st and r replaced by st. By Theorems 2(iv) and 3(i), $y(kM) \mid y(rM) = \pm f$ and so $|y(kM)| \leqslant f$.

By taking derivatives of both sides of $p^{xM} \leqslant (1 + (2x + 1)/4\pi^2)bf^2$, it is easy to see that (ii) need only be proved for $k = s$.

Suppose (i) and (ii) are false. We will apply Theorem 5 with $C = f$ to the pair of exponents sM and rM with the values of "n" and "m" in Theorem 4 equal to sM and "k" $= t$. We must show that (10) holds. Since "v" $= |y(sM)|$, it follows that $v \geqslant 1$ when $\Lambda = 4$, and $v \geqslant \frac{1}{2}$ when $\Lambda = 16$. Thus the negation of (i) implies (10). Hence (11) is true. Dividing by sM and adding 1 we obtain

$$t > F(s, v) = \pi(p^{sM}/bv^2 - 1)^{1/2} - 1/2.$$

We showed above that $v \leqslant f$. Hence $F(s, v) \geqslant F(s, f)$. By the negation of (ii), $F(s, f) \geqslant s$. Thus $t > s$, a contradiction. ■

COROLLARY 7.1. *Suppose that* $L|M|n$ *and* n/M *is odd if* $b = 1$. *Suppose further that* $y(n)$ *is an integer and* $|y(n)| \leqslant C$. *Let* $D = bC^2$. *Suppose either*

(i) $p^M > D/4$;
(ii) $p^{2M} > 18D$;
(iii) $p^{3M} > 18D$ *and either* $a \neq 1$ *or* $y(n)$ *is odd.*

Then n/M *equals* 1 *or a prime.*

If we have $y(n) = \pm f$ *for more than one* n, *then* $y(M) = \pm f$. (*Thus, if* $l \neq L = M$, *there is at most one* n.)

If $b \neq 1$, $y(n)$ *is a half-integer,* $|y(n)| \leqslant C$, *and* $p^M > D$, *then* n/M *equals* 1 *or a prime.*

Proof. We may suppose $C = |y(n)|$. Suppose $n = stM$ where $s \geqslant t > 1$ (i.e., n/M is not a prime). It follows from Theorem 3 that when n/l is even, $y(n)$ is even. We will first show that when (i) holds n/M is odd. Hence $t \geqslant 3$ whenever (i) or (iii) holds.

Suppose n/M is even and (i) holds. Set $n = 2k$ and $u = x(k)$. Note that $l|k$ and that $y(k) = \pm y(n)/2u$. It follows that $p^k = u^2 + D/4u^2$. This is a convex function of u^2 and so its maximum occurs at $u = \pm 1$ or $u = \pm y(n)/2$. The former gives the maximum. Since $4|D$, $p^M \geqslant 1 + D/4 \geqslant p^k$. Since $k \geqslant M$, $p^M = p^k = 1 + D/4$. Hence $M = k = n/2$, contradicting the assumption that n/M is not a prime.

Conclusion (ii) of Theorem 7 cannot hold since with $k = 2$ in case (ii) nor with $k = 3$ in case (iii) because $p^{2m} \geqslant 9$ and so $p^{3M} > 9D/4 > (1 + 49/4\pi^2)D$. To complete the proof it suffices to show that Theorem 7(i) is false. This follows immediately if (ii) or (iii) holds.

Suppose $p^M > D/4$. We have shown that n/M must be odd. If $p^{7M/2} \geqslant 72$, then $p^{stM/2} > 72D/4 = 18D$ and so 7(i) is false. Thus we need only consider $p^M \leqslant 3$. This implies that $p^M = 3$ and $D < 12$. Hence either D is square-free or D equals 4 or 8. Suppose D is square-free. Then $y(n) = \pm 1$ and 7(i) is false since $p^{3M} > 16 + 4p^{M/2}$ when $p^M = 3$. If $D = 4$ or 8, then $y(n) = \pm 2$ and so 7(i) is false because $p^{9M/2} > 81 > 16 \cdot 2^2 + 2\sqrt{8}$.

Suppose $y(n) = \pm y(k) = \pm f$ and $k \neq n$. Since n/M and k/M are 1 or primes, $(k, n) = M$. By Theorems 2(iv) and 3(iv), $\pm y(M) = (y(k), y(n)) = f$.

Suppose $y(n)$ is a half-integer. Suppose n/M is even. As in the integral case, $p^k = u^2 + D/4u^2$; however, the maximum now occurs at $u = \frac{1}{2}$. Hence $p^M \geqslant D + \frac{1}{4} \geqslant p^k$. Thus n/M again equals 2. If n/M is odd, Theorem 7(ii) is false and 7(i) will be false whenever $p^{7M/2} > 8$ because $b \geqslant 3$ by assumption and this implies that $\Lambda f^2 < 6D$. Since $2^{7/2} > 8$, we are done. ∎

The next theorem is a surprising generalization of the assertion that if $y(n) = \pm 1$, then $y(l) = \pm 1$.

THEOREM 8. *Suppose that $y(n)$ is an integer, $|y(n)| < 2\sqrt{p^l/b}$, and, if $b = 1$, $(y(n), 2y(l)) \neq 1$. Then either $y(l) = \pm y(n)$ or $a = 1$, $n = 2l$, $y(l) = \pm y(n)/2$, and $x(l) = 1$.*

Suppose that $y(n)$ is a half-integer, $|y(n)| < \sqrt{p^L/b}$, and $b \neq 1$. Then $y(L) = \pm y(n)$.

Proof. By Theorem 3(ii), n/l is odd if $b = 1$. Suppose $y(l) \neq \pm y(n)$. Set $D = by(n)^2$. By assumption, $p^l > D/4$. By Corollary 7.1(i) with "M" $= l$, $r = n/l$ is a prime. By the second paragraph in the proof of Corollary 7.1(i), $r = 2$ implies $a = 1$, $y(l) = \pm y(n)/2$, and $x(l) = \pm 1$. Thus we may suppose that r is an odd prime.

Without loss of generality, $y(l) > 0$. Let $v = y(l)$ and $f = |y(n)|$. Since r is odd and $v \neq f$, it follows from Theorem 2 that f/v is odd and greater than 1. Thus $1 \leqslant v \leqslant f/3$. By (9g), $\epsilon = f/v \equiv 0, \pm 1$ modulo r. If $\epsilon = 0$, then $r \leqslant f/v$. If $\epsilon = \pm 1$, $r \leqslant (f/v \pm 1)/2$; since r and f/v are odd, $r \leqslant f/v$.

Suppose $r = 3$. Then $\pm f = y(3l) = v(3p^l - 4bv^2)$. If $y(3l) = -f$, it follows that $3p^l < 4bv^2 \leqslant 4D/9$, a contradiction. Thus $+f = v(3p^l - 4bv^2)$, a concave function of v. Since $1 \leqslant v \leqslant f/3$,

$$f \geqslant \min(3p^l - 4b, f(3p^l - 4D/9)/3). \tag{15}$$

Since $f/3 \geqslant 1$, it follows that $D = bf^2 \geqslant 9$. We cannot have $D = 9$ for then $bv^2 = 1$, contradicting Lemma 8 with "q" $= 3$. Since $f \geqslant f(3p^l - 4D/9)/3$ is equivalent to $p^l \leqslant 1 + 4D/27$, which is less than $D/4$ when $D > 9$, it follows from (15) that $f \geqslant 3p^l - 4b$. Hence

$$p^l \leqslant (f + 4b)/3. \tag{16}$$

Since $D \neq 9$ and f/v is odd, either $f \geqslant 5$ or $b \geqslant 2$. In the former case, $f \leqslant D/5$ and $b \leqslant D/25$. Thus (16) implies $p^l \leqslant 3D/25$, a contradiction. In the latter case, $f \leqslant D/6$ and $b \leqslant D/9$. Thus (16) implies $p^l \leqslant 11D/54$, a contradiction. Hence $r \neq 3$ and so $r \geqslant 5$.

We will now derive a contradiction from Theorem 5 with "m" $=$ "n" $= l$, "k" $= r$, "n'" $= rl$, and "C" $= f$. Since f/v is odd, $f/v \geqslant r \geqslant 5$, and D is not an odd square, we have $D \geqslant 50$. Hence $p^l > D/4 > 12$ and so

$$p^{rl/2} \geqslant p^{5l/2} > 12^{3/2} D/4 > 6D,$$

which implies (10) since the right-hand side of (10) is less than $4D + 2D$. By (11) we have, after some rearranging,

$$p^l < \{(1 + (2r+1)^2/4\pi^2)/(f/v)^2\} D.$$

Since $f/v \geqslant r$, the expression multiplying D is bounded by a decreasing function of r which has the value $(1 + (11/2\pi)^2)/25$ at $r = 5$. Since this factor is less than $\frac{1}{4}$, the proof is complete for integral $y(n)$.

Now suppose $y(n)$ is half-integral. The first two paragraphs of the proof carry over when l is replaced by L except that $p^L > D$ and the case $n = 2L$, by the last paragraph of the proof of Corollary 7.1, leads to $x(n) = \pm\frac{1}{2}$ and $y(n) = \pm y(L)$. In particular, $r \leqslant f/v$.

Suppose $r = 3$. Since $\frac{1}{2} \leqslant v \leqslant f/3$, we have

$$f \geqslant \min((3p^L - b)/2, f(3p^L - 4D/9)/3).$$

Since $f \geqslant (3p^L - b)/2$ implies the contradiction $3p^L \leqslant 2f + b < 3D$, we have $f \geqslant f(3p^L - 4D/9)/3$, which is equivalent to $p^L \leqslant 1 + 4D/27$. Since $p^L > D$, it follows that $D < 27/23$. Since $f > v \geqslant \frac{1}{2}$, $D \geqslant b$. By assumption $b \neq 1$ and so $D \geqslant 2$, a contradiction.

Finally suppose $r \geqslant 5$. The proof proceeds as in the integral case except that we obtain (10) as follows. Again $f/v \geqslant 5$. Since $b \neq 1$, $D \geqslant 2(5/2)^2 > 12$. Thus

$$p^{rL/2} > 12^{3/2}D > 16D + 2D. \quad ■$$

COROLLARY 8.1. *Suppose that* $y(m)$ *is an integer,* $|y(rm)| < 2\sqrt{p^m/b}$, *and, if* $b = 1$, $(y(rm), 2y(m)) \neq 1$. *Then either* $y(m) = \pm y(rm)$ *or* $a = 1$, $r = 2$, $y(m) = \pm y(rm)/2$, *and* $x(m) = 1$.

Suppose that $y(rm)$ *is a half-integer,* $b \neq 1$, *and* $|y(rm)| < \sqrt{p^m/b}$. *Then* $y(m) = \pm y(rm)$.

Proof. We can replace l by m in the theorem, provided $y(n)$ is generated from $ax(m) + y(m)\omega$ by (4) or (7). For half-integral solutions, the same remark applies to L. ■

II. $ax^2 + D = p^n$

We shall now study the equation

$$ax^2 + D = p^n \qquad (D = bf^2) \tag{17}$$

where $D \neq 1$ ($D \neq \frac{1}{4}$ for half-integral solutions), *a and D are relatively prime, a and b are square-free positive integers, and p is odd and prime to D.* As remarked in the introduction, we identify the solutions (x, n) and $(-x, n)$. Except for the constraint that p be odd and prime to D, which is relaxed in Section 10, *the above assumptions apply for the remainder of Part II.*

In Theorem 9 we apply Part I in a simple fashion to (17). In the remainder of Sections 6, 7, and 8 we devote most of our effort to bounding the number of solutions per family and obtaining criteria for the existence of more than one solution. In Section 9 we fix p and bound the number of families containing solutions, while in Section 10 we fix n and study all families. In

Section 11 we construct imaginary quadratic number fields all of whose class numbers are divisible by some specified integer.

In the literature, the equation $ax^2 + D = z^n$ with fixed n is usually considered instead of (17). However, the proofs give information about solutions in a family, which can be obtained by a simple transformation of the statements of the theorems. For example:

Replace "If n does not divide the class number of $\mathbb{Q}(\omega)$ . . . , then there are no solutions."

by "If . . . , then in each family the only possible solution is at $n = l$."

6. *General Results*

Solutions of (17) are the solutions of (1) for which $y(n) = \pm f$. Thus the results of Part I are immediately applicable to (17). For example Corollary 7.1 with "M" $= L$ tells us that when $l = 3L$, $p^L > D/4$, and D is not an odd square, if there is a solution in a family, there is one at the exponent $n = l$ itself. (This generalizes a result of Alter and Kubota [1].) Other simple consequences of Part I are collected in the following theorem. Note that the assumption that p^L or p^l exceed $D/4$ is weaker than the assumption that D be square-free since $p^l \geqslant p^L \geqslant (a+b)/4 > D/4$ when D is square-free.

THEOREM 9. *Suppose that* (17) *has a solution. Then every prime divisor of* p *splits in* $\mathbb{Q}(\sqrt{-ab})$. *Let the smallest solution in a particular family occur at the exponent* $n = m$. *If another solution in the family occurs at* $n = m'$, *then* m' *is an odd multiple of* m *and solutions occur at all* m'' *such that* $m \mid m'' \mid m'$.

If $p^m \geqslant 5$, *then the gaps between values of* n *for which solutions exist in the family exceed*

$$m(\pi(p^m/D - 1)^{1/2} - 3/2). \tag{18}$$

For each family there are unique positive integers u *and* v *prime to* p *such that*

$$au^2 + bv^2 = p^l. \tag{19}$$

If $p^l > D/4$ *and* D *is not an odd square, then either* $m = l$ *or* $a = 1$, $p^l = 1 + D/4$, *and* $m = 2l$.

If m/l *is odd, then* $v \mid f$, $\mathrm{ord}_2(v) = \mathrm{ord}_2(f)$, *and* $(u, f) = 1$.

If $a > 1$ *and* $b > 1$, *then* m/l *is odd.*

Suppose $a = 1$. *Then* $v \mid f$. *If* $2uv \nmid f$, *then* m/l *is odd.*

Suppose $b = 1$. *If* $2uv$ *is not prime to* f, *then* m/l *is odd.*

For half-integral solutions, l *must be replaced by* L *and* $p^l > D/4$ *by* $p^L > D$. *All the results are then true except that* (i) *when* $a = 1$ *and* $p^m = D + \frac{1}{4}$, *a solution occurs at the exponent* $n = 2m$ *as well, and* (ii) *when* $p^L > D$, *the exception* $m = 2L$ *cannot occur.*

Proof. We limit our attention to the solutions in one family. The proof of the gap result will be given last. We will begin by showing that $\mathrm{ord}_2(n)$ is constant for those n for which solutions exist.

Suppose $a > 1$ and $b > 1$. By Theorem 1, (1) and a fortiori (17) have solutions only when $\mathrm{ord}_2(n) = \mathrm{ord}_2(l)$.

Suppose $a = 1$ and that solutions exist for the exponents rt and $2st$ where r is odd and t is a multiple of l. By Theorem 2(iii) $\mathrm{ord}_2(y(rt))$ equals $\mathrm{ord}_2(y(t))$, while by Theorem 3(i) $2y(t)$ divides $y(2st)$; a contradiction since both $y(rt)$ and $y(2st)$ equal $\pm f$. (The same argument shows m is an odd multiple of l if $2uv \nmid f$.)

Suppose $b = 1$. Rewrite (17) as $f^2 + ax^2 = p^n$ so it is in the form given in Theorem 3. (Here f corresponds to "$x(n)$" and x to "$y(n)$.") Suppose that solutions exist for the exponents rt and $2st$ where r is odd and t is a multiple of l. By Theorem 2(v, iv), $x(2st) = \pm f$ divides $x(2rst)$. By Theorem 3(ii) with "m" $= rt$, $x(rt) = \pm f$ is prime to $x(2rst)$; a contradiction since $f \neq 1$ in this case. Hence $\mathrm{ord}_2(n)$ is constant.

To prove the assertions about m' and m'', apply Theorem 2(iv) to see that the set of n's for which solutions exist is closed under the operation of greatest common divisor and that $m|m''|m'$ implies $y(m)\,|\,y(m'')|\,y(m')$.

Suppose n is an odd multiple of l. By Theorem 2(v, iv), $u|x(n)$. Hence $(u, f) = 1$. The other results except for (18) follow from Theorems 4 and 8.

We now prove (18). Apply Theorem 5 with "m" as given here, "C" $=$ "v" $= f$, and "k" $= 3$. The requirement in (10) then becomes $p^{3m/2} > 4 + 2\sqrt{D}$. Since $p^m > D$, (10) will hold if $D > 4 + 2\sqrt{D}$, regardless of the value of p. This is the same as $\sqrt{D} \geqslant 1 + \sqrt{5}$. But if $\sqrt{D} < 1 + \sqrt{5}$, then $4 + 2\sqrt{D} < 11 < 5^{3/2}$, and so (10) will hold provided $p^m \geqslant 5$. Thus (18) follows from (11).

Suppose that we are considering half-integral solutions. The entire proof carries over with two exceptions. First, the proof of (18) breaks down if half-integral solutions occur at $n = m$ and $2m$; but then $p^m = D + \frac{1}{4}$, and so (18) is less than m.

Second, the proof that $\mathrm{ord}_2(n)$ is a constant for half-integral solutions when $a = 1$ must be modified. By Theorem 3(iv), the first paragraph in the statement of the theorem is true, except that half-integral solutions may occur at even multiples of m. Suppose a half-integral solution occurs at the exponent $2km$. Then $x^2 + D = p^{2km}$ and so $x < p^{km}$. Since x is a half-integer, $x \leqslant p^{km} - \frac{1}{2}$. Hence

$$D = p^{2km} - x^2 \geqslant p^{km} - \tfrac{1}{4}.$$

Since there is a half-integral solution at the exponent m, $p^m \geqslant D + \frac{1}{4}$. Thus $k = 1$ and $p^m = D + \frac{1}{4}$. ■

COROLLARY 9.1. *Let u and v be given in the notation of* (19). *Suppose* $v \nmid f$ *and, if* $b = 1$, $(2uv, f) \neq 1$, *then* (17) *has no solutions.*

Proof. This follows immediately from the theorem. ■

The results in Theorem 9 can be applied to the case $a = b = 1$ as follows: To begin with, combine the assertions about the solutions obtained by assuming $a = 1$ with those obtained by assuming $b = 1$, then interchange u and v in (19) and repeat the process. Clearly $l = 1$. Since (4) does not distinguish $a = 1$ and $b = 1$:

(i) The cases $a = 1$, $(x(1), y(1)) = (x, y)$ and $b = 1$, $(x(1), y(1)) = (x, y)$ give the same solutions when n is odd.

When n is even, we can analyze (7) and see that:

(ii) The cases $a = 1$, $(x(1), y(1)) = (x, y)$ and $a = 1$, $(x(1), y(1)) = (y, x)$ give the same solutions when n is even.
(iii) The cases $b = 1$, $(x(1), y(1)) = (x, y)$ and $b = 1$, $(x(1), y(1)) = (y, x)$ give the same solutions when n is even.

Since u and v have opposite parities, when n is odd Theorem 2(iii) implies that we cannot have solutions with both $(x(1), y(1)) = (u, v)$ and (v, u). Combining this with (ii) and (iii):

(iv) We need consider only one of the cases $a = 1, (u, v)$ and $a = 1, (v, u)$, and only one of the cases $b = 1, (u, v)$ and $b = 1, (v, u)$.

Suppose f is odd. By Theorem 3(i), n is odd when $a = 1$ and so by (i) and (iv) we are reduced to considering only one case. On the other hand, if f is even, Theorem 3(ii) implies that n is odd when $b = 1$ and (i) and (iv) apply again. By the last remark after Theorem 5, equation (18) is correct as it stands because the above discussion shows that all solutions of (17) have n an odd multiple of m. In summary, *Theorem 9 applies to the case $ab = 1$ but we may have to interchange the role of u and v in* (19); *furthermore, we must use "$a = 1$" when f is even and "$b = 1$" when f is odd.* Finally note that by Corollary 9.1, there are no solutions to $x^2 + f^2 = p^n$ if $u \nmid f$, $v \nmid f$, and, when f is odd, $(uv, f) = 1$.

We now present an algorithm for finding the smallest solution of (17) in a family *when D is not an odd square*. The various families (possibly with some repetition) can be found by letting l run over all divisors of E^*. The first step is initialization. The validity of Step 2 follows from Theorem 3 when $q = 2$ and congruence (9a) modulo q otherwise. Steps 3 to 5 are based on Lemma 8. The fact that we actually obtain the smallest solution follows from the divisibility of the n's which was proved in Theorem 9.

The algorithm uses a LIST to keep track of potential smallest solutions and a STACK to implement a branching process associated with Lemma 8.

Step 1. Choose a family and let $ax^2 + by^2 = p^l$ be the smallest solution of (1) in the family. If $y \nmid f$ or $a \neq 1$ and f/y is even, STOP: there is no solution; otherwise set $n = l$ and continue.

Step 2. Let s be the product of the distinct prime factors of $(f/v, 2by^2)$. If $s = 1$, go to Step 3; otherwise replace n by sn and x, y by x', y' where $ax' + y'\omega = (ax + y\omega)^s/a^{(s-1)/2}$. If $y \nmid f$ STOP: there is no solution; otherwise repeat Step 2.

Step 3. Choose a prime $q \mid (f/y)$, set $s = 1$, set $r = q - (-ab \mid q)$, and go to Step 5. If there is no such q, LIST (x, y, n) and continue.

Step 4. Remove (x, y, n, r, s) from the top of the STACK and go to Step 5. If the STACK is empty, choose the triple (x, y, n) in the LIST having the smallest n and STOP: (x, y, n) is the smallest solution in the family.

Step 5. Replace s by the next prime factor of r exceeding the old value of s. If there is no s, go to Step 4. Set $ax' + y'\omega = (ax + y\omega)^s/a^{(s-1)/2}$. If $y' \mid f$, put (x, y, n, r, s) on the top of the STACK, set $x = x'$, $y = y'$, $n = ns$, and go to Step 2; otherwise repeat Step 5.

The following is a slight generalization of a theorem of Apery [2]. It tells us that we almost never have more than two solutions to (17). *We are not able to verify Apery's claim* that the conclusion holds when $a = 1$ and $\operatorname{ord}_2(w^2) = 2$.

THEOREM 10. *Suppose that some family contains more than one (possibly half-integral) solution. Let $(x, n) = (w, m)$ be the smallest and let another occur at $n = rm$. Fix a prime q dividing aw^2. Let $\operatorname{ord}_q(aw^2) = \lambda$. If $q = 2$ or 3, suppose that $\lambda \geqslant 2$. Let s be the smallest positive integer for which $D^{2s} \equiv 1$* modulo q, *and let $\eta = \operatorname{ord}_q(D^s \pm 1)$, where the sign is chosen to maximize η. Then*

(i) $r = 2st + 1$;
(ii) $\eta = \lambda + \operatorname{ord}_q(r) \leqslant \lambda + 1$.

If $q = 2$, suppose further that $\lambda \geqslant 3$. Then there are only two (possibly half-integral) solutions to (17), and for the second $n = rm$, where r is an odd prime.

Proof. Note that since $\lambda > 0$, $q \neq 2$ when we are dealing with half-integral solutions. If a second (possibly half-integral) solution to (17) exists, then by Theorem 9, $n = rm$ where r is an odd integer. (If a half-integral solution occurs at the exponent $2m$, then $aw^2 = \frac{1}{4}$, and so $\lambda = 0$.) Hence we can use (5). Since q divides aw^2, the left-hand side of (5) is $(-D)^{(r-1)/2}$ modulo q. Hence $s|(r-1)/2$, proving the necessity of (i). We can assume $t \neq 0$, since $t = 0$ gives the solution $r = 1$.

Dividing (5) by $(-D)^{st}$ we get

$$\sum_{j=1}^{\infty} \binom{2st+1}{2j} \left(-\frac{aw^2}{D}\right)^j = \pm\left(-\frac{1}{D}\right)^{st} - 1. \tag{20}$$

We can allow the index of summation go to ∞ because $\binom{2st+1}{2j} = 0$ when $j > st$. *We will regard this equation and all subsequent equations in the remainder of the proof as equations involving q-adic numbers.* It clearly suffices to bound the number of q-adic integers $t \neq 0$ satisfying (20).

We can write $(-1/D)^s = \pm(1 + \gamma q^\eta)$, where γ is a q-adic unit. Note that when $q = 2$, $\eta \geqslant 2$ and $s = 1$. The divisibility of the left side of (20) by q, or by 4 when $q = 2$, then implies that $\pm(-1/D)^{st} = (1 + \gamma q^\eta)^t$, the choice of sign agreeing with that in (20). Substituting this in (20) we obtain

$$\sum_{j=1}^{\infty} \binom{2st+1}{2j} \left(-\frac{aw^2}{D}\right)^j = \sum_{i=1}^{\infty} \binom{t}{i} (\gamma q^\eta)^i. \tag{21}$$

We view (21) as an equation between two power series in the variable t over the q-adic numbers. Since

$$\operatorname{ord}_q(N!) = \sum [N/q^i] < N/(q-1)$$

and $\eta > 1/(q-1)$, the series on the right of (21) is well defined, and the coefficient of t^k goes to zero q-adically as $k \to \infty$. The terms of the sum on the left of (21) are polynomials in t, and the coefficients of the jth polynomial all have $\operatorname{ord}_q \geqslant \delta_j$, where $\delta_j \geqslant j\lambda - \operatorname{ord}_q((2j)!) > j\lambda - 2j/(q-1)$ when q is odd, and $\delta_j \geqslant j + j\lambda - 2j$ when $q = 2$. (The latter expression follows upon noting that every other factor in $(2st+1)(2st)\cdots(2st-2j+2)$ contains a factor of 2.) By the hypotheses of the theorem, $\delta_j > j/2$. Hence the series on the left of (21) is well defined and the coefficient of t^k goes to zero q-adically as $k \to \infty$. Equation (21) can be written as

$$q^\lambda st(2st+1)(c_0 + q^\lambda F(t)/6) = t\gamma q^\eta(1 + q^\eta G(t)/2), \tag{22}$$

where $c_0 = -aw^2/Dq^\lambda$ is a q-adic unit and F and G are power series in t with q-adic integer coefficients. Since $s|(q-1)$, s is a q-adic unit. Equating the q-adic values of the two sides of (22) gives $\lambda + \operatorname{ord}_q(r) = \eta$. If $q|r$, we may assume $r = q$ by Theorem 9. Hence $\eta \leqslant \lambda + 1$.

We now assume that if $q = 2$, then $\lambda \geqslant 3$. Dividing (22) by $2tq^{\lambda}$ and rearranging we get

$$(sc_0 - \gamma q^{\eta - \lambda})/2 + s^2 c_0 t - q^{\lambda} \sum_{i=1}^{\infty} c_i t^i/12 = 0, \tag{23}$$

where the c_i are q-adic integers and, as already noted, c_0 and s are q-adic units. Hence the coefficient of the linear term is a q-adic unit. By the hypotheses on λ, the coefficients of the higher-order terms are q-adic integers but not units. Thus (23) has at most one integral solution by Strassman's Theorem. (See e.g. [6, Problem 7, p. 301], or [30].)

By Theorem 2(iv), there is a solution to (17) with $n = r'm$ whenever $r' \mid r$. Since the only values for r' are 1 and r, r is necessarily a prime. ■

Condition (ii) is particularly useful when $q = 2$ and $4 \mid aw^2$ because $s = 1$ and, if there are two solutions, r is odd and so $\lambda = \eta$. Thus, with an appropriate choice of sign, $\operatorname{ord}_2(p^m \pm 1) = \operatorname{ord}_2(aw^2 + (D \pm 1)) > \lambda$. Thus

Observation 2. Suppose D is odd and $(x, n) = (w, m)$ is the smallest solution in a family. If

$$2 \leqslant \operatorname{ord}_2(p^m \pm 1) \leqslant \operatorname{ord}_2(aw^2),$$

then there is no other solution in the family.

Since one of $\operatorname{ord}_2(D \pm 1)$ equals 1 when D is odd, $\operatorname{ord}_2(D \pm 1)$ and $\operatorname{ord}_2(D^2 - 1)$ differ by 1. Thus we have:

Observation 3. If a is odd and $\operatorname{ord}_2(D^2 - 1)$ is even, then (17) has at most one solution per family.

Observation 4. If a is even, $\operatorname{ord}_2(D^2 - 1)$ is odd, and $p^m \equiv D$ modulo 4, then (17) has at most one solution per family.

Observation 2 was made by Alter and Kubota [1] and Observation 3 by Ljunggren [17].

If $q = 3$, then $s = 1$. By forcing 3 to divide w, we can apply the same parity argument provided we are careful about $r = 3$. By using (5) one easily sees that $r = 3$ if and only if $p^m = (4D \pm 1)/3$. Since this determines m and since $y(m) = \pm f$, the family is determined uniquely. Thus

Observation 5. If $3 \nmid p$, $a \equiv -D$ modulo 3, and $\operatorname{ord}_3(D \pm 1)$ is odd, then (17) has at most one solution per family, except that if $p^m = (4D \pm 1)/3$ and a equals the square-free part of $(D \pm 1)/3$, then one family has solutions at $n = m$ and $n = 3m$.

We used $D \neq 1$ in Theorem 10 only to insure that r is odd. Since $\eta = \infty$ when $D = 1$, we have from Theorem 10(ii) and Theorem 2(iv):

Observation 6. If $a \neq 1$ and $(a, p) \neq (2, 3)$ or $(6, 7)$, then $ax^2 + 1 = p^n$ has at most one solution per family with n an odd multiple of l, namely $n = l$.

Nagell [27, Sect. 12] studied the exceptional cases and showed that the conclusion is true except for $2 \cdot 11^2 + 1 = 3^5$.

COROLLARY 10.1. *Let the assumptions and notation be as given in Theorem 10. Suppose q is odd. Then*

(i) $D^{q-1} \equiv 1$ modulo q^λ.

If $q \mid w$, then

(ii) $D^{q-1} \equiv 1$ modulo q^2.

If $D = B^c$ and $\text{ord}_q(c) \leqslant \lambda - 2$, then

(iii) $B^{q-1} \equiv 1$ modulo q^2.

Proof. By the definition of η, $(-1/D)^s \equiv \pm 1$ modulo q^η, and by the definition of s, $2s \mid (q-1)$. Hence $D^{q-1} \equiv 1$ modulo q^η. By Theorem 7(ii), $\lambda \leqslant \eta$, and (i) follows.

Conclusion (ii) follows trivially from (i), and the definition of λ. Conclusion (iii) follows from (i) for if $\xi = \text{ord}_q(c)$, then $B^{q-1} \equiv 1$ modulo q^α if and only if $B^{c(q-1)} \equiv 1$ modulo $q^{\alpha+\xi}$. ∎

Corollary 10.1(ii,iii) can be used to eliminate prime factors of w. See [7] for results of a computer search for solutions to (ii) with small D. For example, the only primes less than 3×10^9 which satisfy $2^{q-1} \equiv 1$ modulo q^2 are $q = 1093$ and 3511.

When $q = 2$ and λ is even, the second congruence in the following corollary is a result of Ljunggren [17].

COROLLARY 10.2. *Let the notation be as given in Theorem 10 with $\lambda \geqslant 2$ if $q = 2$ or 3. Define γ by $D^{-s} = \pm(1 + \gamma q^\eta)$ and σ to be λ if $q > 3$, $\lambda - 1$ if $q = 3$, and $\lambda + 1$ if $q = 2$. Then*

$$sr \equiv -\gamma D q^\eta / a w^2 \quad \text{modulo } q^\sigma.$$

If $\eta = \lambda$ and $q = 2$ or 3, then

$$r \equiv \alpha\delta \quad \text{modulo } Q,$$

where $\delta = (D \pm 1)/q^\lambda$, $\alpha = a$ if λ is even, $\alpha = a/q$ if λ is odd, $Q = 3$ if $q = 3$, and $Q = 8$ if $q = 2$.

Proof. The first congruence follows from (23) when q is odd. Suppose $q = 2$. Note that $s = 1$ and $\eta = \lambda \geqslant 2$. Setting $-aw^2/2^\lambda D = c_0$ in (21) and dividing the equation by $2^\lambda t$ we obtain modulo $2^{\lambda+1}$

$$(2t + 1)c_0 + (2t + 1)(2t - 1)(t - 1)2^{\lambda-1}c_0^2/3 \equiv \gamma + (t - 1)\gamma^2 2^{\lambda-1},$$

which reduces to

$$(2t + 1)c_0 + 5(t - 1)2^{\lambda-1} \equiv \gamma + (t - 1)2^{\lambda-1}.$$

Since $r = 2t + 1$ and $\lambda \geqslant 2$, the first congruence in the corollary follows.

The second congruence follows from the first when $q = 2$ or 3 because $s = 1$, $q^\lambda/aw^2 \equiv \alpha$ modulo Q, and $D \equiv \pm(1 - \gamma q^\lambda + q^{2\lambda})$ modulo Qq^λ. ∎

Theorem 10 does not limit the number of solutions per family when $aw^2 = 1, 2, 3, 4, 6, 12$. We anticipate future results by remarking: (i) Theorem 14 tells us there are at most two solutions per family when D is even; (ii) Theorem 12 tells us there are at most $q - 1$ when D is divisible by an odd prime q provided $a \equiv D/3$ modulo 3 when $q = 3$. Hence we have relatively small bounds unless $D = 3$ and $aw^2 = 4$. The remaining case is $x^2 + 3 = 7^n$. Nagell [24] has shown that $x^2 + 3 = y^n$ has no solutions with $n > 1$ except $1^2 + 3 = 2^2$.

The half-integral case is simpler: only $aw^2 = \frac{1}{4}$ or $\frac{3}{4}$ can occur and Theorem 12 applies unless $aw^2 = \frac{1}{4}$ and $D = \frac{3}{4}$, but then $p = 1$.

THEOREM 11. *Suppose $u^2 + D = p^n$ has a solution at $n = 2k$, then $ax^2 + D = q^n$ has no solutions with $q^n \leqslant p^k$. This holds for half-integral solutions except when $q^n = p^k$, $a = 1$, and $x = \frac{1}{2}$.*

Proof. By assumption, $u^2 = p^{2k} - D < p^{2k}$ and so $u \leqslant p^k - 1$. Thus $D = p^{2k} - u^2 \leqslant 2p^k - 1$, and so $p^k \leqslant (D + 1)/2 \leqslant D$. If $ax^2 + D = q^n$, then $q^n > D \geqslant p^k$, a contradiction.

For half-integral solutions we get $u \leqslant p^k - \frac{1}{2}$ and so $p^k \leqslant D + \frac{1}{4}$. This gives the exceptional case. ∎

COROLLARY 11.1. *Suppose $a = 1$ and there is a solution in a family at an even value of n, then there are no other solutions in the family. This holds for half-integral solutions unless $p^m = D + \frac{1}{4}$.*

Proof. Let $n = m$ at the smallest solution in the family and let $n = n'$ at any solution in the family. By Theorem 9 and the hypothesis, m and n' are even and $m \mid n'$. By Theorem 11 with "q" $= p$, $m > n'/2$. Hence $m = n'$. ∎

COROLLARY 11.2. *Suppose $a = 1$. Let* (1) *have the solution (x, y, l). If $\mathrm{ord}_2(y) \neq \mathrm{ord}_2(f)$, then* (17) *has at most one solution in the family.*

Proof. By Theorem 2(iii), any n for which $\mathrm{ord}_2(y(n)) = \mathrm{ord}_2(f)$ must be an even multiple of l. Apply Corollary 11.1. ■

7. *The Case D Odd*

The next theorem does not require that D be odd; however, Theorem 14 contains a better result when D is even.

THEOREM 12. *Suppose D is divisible by an odd prime q. If $q = 3$, assume that $a \not\equiv D/3$ modulo 3. Let $(x, n) = (w, m)$ be the smallest solution in a family and let s be the least positive integer such that $p^{2ms} \equiv 1$ modulo q. Then there are at most $2s \leqslant q - 1$ solutions in the family.*

The above results hold for half-integral solutions except that there is one additional half-integral solution at $2m$ when $a = 1$ and $p^m = D + \frac{1}{4}$.

Proof. Let m be the smallest exponent for which a (half-integral) solution exists. By Theorem 9, all other (half-integral) solutions (x, n) have $n = rm$ where r is odd with the one exception noted in the half-integral case. Suppose (half-integral) solutions exist at the exponents n and k where $n - k$ is divisible by $2sm$. By Lemma 6(ii″)

$$\pm f \equiv \pm f + Aq^{\tau+\sigma} \text{ modulo } q^{\tau+\sigma+1},$$

where $\sigma = \mathrm{ord}_q(f)$ and $\tau = \mathrm{ord}_q((n-k)/m)$. Hence $\tau = 0$ and the signs cannot agree. By Lemma 6(ii) we therefore have

$$y(n)/y(k) \equiv -p^{(n-k)/2} \text{ modulo } q.$$

If there were another (half-integral) solution at the exponent

$$n' \equiv k \text{ modulo } 2sm,$$

we could divide two of these congruences to obtain $y(n')/y(n) \equiv +p^{(n'-n)/2}$ modulo q, a contradiction. Hence there are at most two (half-integral) solutions in each equivalence class modulo $2sm$. Since n is an odd multiple of m, only s of the equivalence classes can occur. It follows trivially from the definition of s that $s \mid (q-1)/2$. ■

COROLLARY 12.1. *Suppose (half-integral) solutions occur at the exponents rm and $r'm$ where $r \equiv r'$ modulo $2s$, then $\pm 1 = y(rm)/y(r'm) \equiv -p^{m(r-r')/2}$ modulo q.*

Proof. This follows from the proof of the theorem. ■

If D is odd and we are *not* looking at half-integral solutions, then aw^2 is even. If we divide (5) by $v = f = \pm y(rm)$, take the resulting equation modulo 8, and then consider cases, we obtain

$$\epsilon \equiv (-p^m)^{(r-1)/2} \text{ modulo } 8, \tag{24}$$

where ϵ is defined by Lemma 7. (When $4|aw^2$, this follows from Lemma 6(i) with $q = 2$.) This formula for ϵ can sometimes be used with (9) to prove that there is at most one solution per family.

We illustrate this. Suppose (17) has more than one solution in some family, D is a multiple of the odd prime q, $p^m \equiv \pm 1$ modulo q, and $a \not\equiv D/3$ modulo 3 if $q = 3$. By Corollary 12.1 with $r' = 1$ and $s = 1$,

$$\epsilon \equiv -p^{m(r-1)/2} \text{ modulo } q. \tag{25}$$

Comparing (9a) and (25), we see that $r \equiv -1$ modulo q.

Suppose D is odd. If $r \equiv 1$ modulo 4, then, by (25) and the assumption that $p^m \equiv \pm 1$ modulo q, $\epsilon \equiv -1$ modulo q, contradicting (24). Thus $r \equiv -1$ modulo 4. Combining (24), (25), and $p^m \equiv \pm 1$ modulo q, we obtain $\epsilon \equiv -p^m$ modulo $8q$. Thus necessarily $p^m \equiv \pm 1$ modulo $8q$.

Suppose further that $D = f^2$ or qf^2. Apply (9g). If D is a square, then $\epsilon = -1$; otherwise

$$\epsilon = (-q|r) = (-1)^{(q+1)/2}(r|q) = (-1)^{(q+1)/2}(-1|q) = -1.$$

Thus necessarily $p^m \equiv 1$ modulo $8q$.

We now apply these ideas with $q = 3$. Suppose $3|D$ and $a \not\equiv D/3$ modulo 3. Then $p^m \equiv a$ modulo 3, and if a is odd, $p^m \equiv D \equiv (-1)^{(D-1)/2}$ modulo 4. Thus we have a contradiction to $p^m \equiv -\epsilon$ modulo 12 if a is odd and $a \not\equiv (-1)^{(D-1)/2}$ modulo 3, or if $a \equiv 2$ modulo 3 and $D = f^2$ or $3f^2$.

Observation 7. Suppose that D is an odd multiple of 3, that $a \not\equiv D/3$ modulo 3, and that either (i) a is odd and $a \not\equiv (-1)^{(D-1)/2}$ modulo 3, or (ii) $a \equiv 2$ modulo 3 and D or $D/3$ is a square. Then (17) has at most one solution per family.

By combining Corollary 10.2 and (9), we can obtain results of Ljunggren [21]. For example, suppose a is odd, $D = 1 + 2^{2k}\delta$, and some family has two solutions. By (9b), $\epsilon \equiv (-1|r) \equiv r$ modulo 4. By Corollary 10.2 with $q = 2$, $r \equiv a\delta$ modulo 8. Note also that since $\lambda = \eta$, $au^2 - D \equiv -1$ modulo 8. By (9c), $\epsilon \equiv (-2|r)(2au^2)^{(r-1)/2}$ modulo $au^2 - D$. Suppose that $au^2 > D$. If $a\delta \equiv 5$ modulo 8, then -1 is a square modulo $au^2 - D$, a contradiction since there is a prime divisor q of $au^2 - D$ with $(-1|q) = -1$. If $a\delta \equiv 3$ modulo 8, then $-2a$ is a square modulo $au^2 - D$. When $a = 1$, this is a contradiction since there is a prime divisor q of $au^2 - D$ with $(-2|q) = -1$.

8. *The Case D Even*

In this section we present some results for even D. These include all the cases we are aware of in the literature except for $D = 2$ and 4, which are treated by Nagell. See Section 12 of [27]. Note that since D is even, there are no half-integral solutions to consider.

In this section we frequently use the integer v which we define as follows. When $4 \nmid D$, $v = (a - 1)/2$, and when $4 | D$, $v = (a + 1)/2$.

THEOREM 13. *Suppose D is even. Then each family has at most two solutions* $(x, n) = (w, m)$, (w', m') *where* $m < m'$. *If both these solutions exist, the following conditions must be satisfied*:

(i) $m' = rm$ *for some prime* $r \equiv 3$ modulo 4;
(ii) *the value of* ϵ *in Lemma 7 is* $(-1)^v$;
(iii) $(D | q) = (-1 | q)^{v+1}$ *for all primes* $q | aw$.

(*In the notation of* (19), w *can be replaced by* u *if* m *is an odd multiple of* l.)

Proof. We limit our attention to one family. Let (w, m) be the smallest solution in the family. By Theorem 9 any other solution (x, n) has $n = rm$ where r is odd. Suppose solutions exist for the exponents n and t where $n - t$ is divisible by $4m$. We can apply Lemma 6(ii′). Thus

$$\pm f \equiv \pm f + 2^{\tau+\sigma} \text{ modulo } 2^{\tau+\sigma+1},$$

a contradiction since $\sigma = \operatorname{ord}_2(f)$ and $\tau \geqslant 2$. Hence there is at most one solution in each of the residue classes m and $3m$ modulo $4m$. Thus there are at most two solutions m and rm where $r \equiv 3$ modulo 4. By Theorem 5, $n = sm$ is a solution if $s | r$. Hence r is prime.

Suppose we have two solutions (w, m) and (w', m'). By (i) and Lemma 6(ii),

$$y(m') \equiv p^{m(r-1)/2} y(m) + A2^{\sigma+1} \text{ modulo } 2^{\sigma+2}.$$

Thus

$$\begin{aligned} \epsilon = y(m')/y(m) &\equiv -p^{m(r-1)/2} \equiv -p^m \\ &\equiv -a - D \equiv (-1)^v \text{ modulo } 4. \end{aligned}$$

This proves (ii). From (9b) and the fact that $r \equiv 3$ modulo 4 we have, after taking Legendre symbols, $(-\epsilon | q) = (p^m | q)$. Since $p^m \equiv D$ modulo q, this proves (iii).

The last statement in the theorem follows from Theorem 2(v, iv). ∎

Theorem 13(iii) can sometimes be used to prove that (17) has at most one solution. For example,

Observation 8. If $a \equiv 1$ modulo 4 and d is not a square modulo a, then $ax^2 + 4d = p^n$ has at most one solution per family.

Just as Theorem 13(iii) was derived from (9b), we can obtain quadratic residue conditions from (9c), (9d), and (9f); however, (9c) must be combined with the following corollary, which determines r modulo 8.

COROLLARY 13.1. *Suppose a family has solutions at the exponents m and rm. Then*

$$r \equiv (-1)^{\nu}(a + D) \text{ modulo } 8.$$

Proof. Divide (5) by $y(m)$ and consider the result modulo 8. Using the facts from Theorem 13 that $r \equiv 3$ modulo 4 and that $\epsilon = (-1)^{\nu}$ we have

$$(-1)^{\nu} \equiv ra - (r-1)D/2 + (r-3)D^2/4 \text{ modulo } 8.$$

Hence

$$(-1)^{\nu} r \equiv a + (r - 1 + (1 - 3r)D/2)D/2 \text{ modulo } 8.$$

Using $r \equiv 3$ modulo 4 and, when $4 \nmid D$, also $(1-3r)D/2 \equiv r - 3$ modulo 8, the result follows. ■

The following corollary strengthens Theorem 13(iii) when $q \equiv 1$ modulo 8. For example, if $a = 17$, D is even, and $D \not\equiv \pm 1$ modulo 17, then there is at most one solution per family.

COROLLARY 13.2. *We use the notation of the theorem. Suppose both solutions exist and let q be a prime dividing aw^2. Then $D^{2s} \equiv 1$ modulo q for some odd s.*

Proof. Without loss of generality we can assume s is the least integer such that $D^{2s} \equiv 1$ modulo q. By Theorem 10(i), $r = 2st + 1$ for some t. By Theorem 13(i), $r \equiv 3$ modulo 4. Hence st is odd. ■

To illustrate how some of the above results can be applied, we derive a result of Ljunggren [18] as well as some related new results. Suppose $\text{ord}_2(D) = 2$ and (17) has more than one solution in a family. By Theorem 13(ii), $\epsilon = -1$ if $a \equiv 1$ modulo 4 and $\epsilon = +1$ otherwise. According to Corollary 13.1, $r \equiv 3$ modulo 8 if $a \equiv \pm 1$ modulo 8, and $r \equiv -1$ otherwise. Consequently $2\epsilon(-2|r)$ equals -2 if $a \equiv 1, 3$ modulo 8 and $+2$ otherwise. If a (or equivalently D) is a quadratic residue modulo $aw^2 - D$, (9c) implies that $2\epsilon(-2|r)$ is a square modulo $aw^2 - D$. If $a \equiv 1$ modulo 8, this implies that each prime factor of $aw^2 - D$ (and hence $|aw^2 - D|$ itself) is congruent to 1 or 3 modulo 8. Since $aw^2 - D \equiv 5$ modulo 8, this is a contradiction when $aw^2 > D$. In a similar fashion, one can derive a contradiction if either

$a \equiv 7$ modulo 8, or $aw^2 > D$ and $a \equiv 3$ modulo 8. When D is a square, (9g) leads to a contradiction for $a \equiv 3$ modulo 8 without the restriction that $aw^2 > D$. By specializing these results we obtain

Observation 9. Suppose the smallest solution in a family is at the exponent m, $\text{ord}_2(D) = 2$, $a \not\equiv 5$ modulo 8, $p^m > 2D$ when $a \equiv 1$ modulo 8, and that either $a = 1$ or D is a square. Then there is no other solution in the family.

In a strictly analogous way we can derive results when D is divisible by other powers of 2 by using (9) together with Theorem 13(ii) and Corollary 13.1. Specializing as in Observation 9 we have

Observation 10. Suppose the smallest solution in a family is at the exponent m. Suppose either (i) D is twice an odd square, and $a \equiv \pm 3$ modulo 8; (ii) D is 8 times a square and $a \equiv 5, 7$ modulo 8; (iii) D is 16 times a square and $a \equiv 3, 7$ modulo 8; or (iv) D is 16 times a square, $p^m > 2D$, and $a \equiv 5$ modulo 8. Then there is no other solution in the family.

We can use (9) in other ways. Suppose $3 \nmid D$. Then $aw^2 \equiv 0, -D$, or D modulo 3. Applying (9b, f, c), respectively, we obtain $\epsilon \equiv -D$ modulo 3 in the first two cases and $\epsilon \equiv -(-2|r)D$ in the last case. The last case cannot occur if $a \not\equiv D$ modulo 3. Thus when $\epsilon \equiv D$ modulo 3 and either $r \equiv 3$ modulo 8, or $a \not\equiv D$ modulo 3, there is no second solution. Applying Corollary 13.1 we have

Observation 11. Let $t = \text{ord}_2(D)$. Suppose that $(-1)^v \equiv D$ modulo 3 and that either

(i) $t = 1$ and $(a, D) \not\equiv \pm(1, 22), \pm(11, 14)$ modulo 24,
(ii) $t = 2$ and $a \not\equiv \pm 5$ modulo 24, or
(iii) $t \geqslant 3$ and $a \not\equiv \pm 7$ modulo 24.

Then there is at most one solution per family.

This extends a result of Nagell [26] for $t = 3$.

Another way of using two determinations of ϵ is to combine (25) with Theorem 13. For example suppose $6|D$ and that $a \not\equiv D/3$ modulo 3. Since $r \equiv 3$ modulo 4 and $p^m \equiv a$ modulo 3, (25) becomes $(-1)^v \equiv -a$ modulo 3. From the definition of v it is then easily deduced:

Observation 12 Suppose that $6|D$, that $a \not\equiv D/3$ modulo 3, and that either $4 \nmid D$ and $a \equiv \pm 1$ modulo 12 or $4|D$ and $a \equiv \pm 5$ modulo 12. Then there is at most one solution per family.

As a final example with D even, suppose that $5 \nmid pD$. Then $aw^2 \equiv 0, D$, $3D$ or $2D$ modulo 5. Apply (9b)–(9e) respectively together with the fact that

$(r-1)/2$ is odd to deduce that either $r = 3$ or $p^m \equiv \pm 1$ modulo 5. Combining this with Corollary 11.1 we have

Observation 13. Suppose $a = 1$ or E^* is odd, D is even, $5 \nmid D$, and $p \equiv \pm 2$ modulo 5. Then each family has at most one solution, except that if $p^m = (4D \pm 1)/3$, then one family has solutions at $n = m$ and $n = 3m$.

In deriving Observation 13 we needed D even only to conclude that $r \equiv 3$ modulo 4. Hence similar results can be obtained for some odd D. For example, in the discussion following (25) it was shown that if D is an odd multiple of 3 and $a \not\equiv D/3$ modulo 3, then $r \equiv -1$ modulo 4.

To illustrate some of the above results, we consider the equation

$$3Ax^2 + 8 = p^n$$

with A square-free and $A \equiv 1$ modulo 4, an equation of interest elsewhere [14]. Congruences modulo 3, 4, and 8 show that if there are any solutions, n must be odd, $p \equiv -1$ modulo 12, and $Ap \equiv 3$ modulo 8.

By setting $a = A/3$ if $3 | A$ and $a = 3A$ otherwise, the equation reduces to (17). Since n must be odd, it follows from Theorem 2(iii) that if there is a solution, then there must be one at the exponent l. Since $ab \equiv 2$ modulo 4, $l = L$.

Suppose some family has more than one solution. Write $3Au^2 + 8 = p^L$. By Theorems 9 and 13 and Corollary 13.1, there are exactly two solutions, one at the exponent L and one at the exponent rL, where $r \equiv 3A$ modulo 8. Combining Theorem 13(ii) with (9g) and the congruences above we see that $r \equiv 3$ modulo 8 and so, $A \equiv 1$ modulo 8 and $p \equiv 3$ modulo 8.

We can get a lower bound on those p that have a family with two solutions by applying Theorem 13(iii) and Corollary 10.1(iii) to conclude that if q is a prime dividing u (or $u/3$ if $3 | A$), then

$$q \equiv 1 \text{ or } 3 \text{ modulo } 8 \qquad \text{and} \qquad 2^q \equiv 2 \text{ modulo } q^2.$$

By [7], no primes q less than 3×10^9 satisfy these two congruences. Hence either

$$p^L = 3A + 8 \qquad \text{or} \qquad p^L > A \times 2.7 \times 10^{19}.$$

In the latter case, it follows from (18) that

$$r > 5.7 \times 10^9 \sqrt{A}.$$

We conclude this section by combining results from Sections 6 and 8 to show that there are usually at most two solutions per family.

THEOREM 14. *Suppose that $a \neq 2$ or 6 and that, when $a = 1$ or 3, $D \not\equiv \pm 3$ modulo 8. Then* (17) *has at most 2 solutions per family.*

Proof. We can apply Theorem 10 to those families for which the solution (w_i, m_i) with smallest exponent m_i satisfies $aw_i^2 \neq 1, 2, 3, 4, 6, 12$. We can apply Theorem 13 if $aw_i^2 = 1$ or 3 since D is even. By assumption, $aw_i^2 \neq 2$ or 6 for any family. The two remaining cases to be considered are $aw_i^2 = 4$ and 12. Since D is odd, we have, by the assumption on D,

$$p^{m_i} \equiv 4 \pm 1 \text{ modulo } 8.$$

Hence Observation 2 applies, and there is no other solution in these families. Thus no family contains more than two solutions.

9. $ax^2 + D = N^z$

In this section we study the equation

$$ax^2 + D = N^z \tag{26}$$

where the parameters a, D, and N are positive integers and $(N, 2D) = 1$. Except that we substitute N for "p" and z for "n" to emphasize the nature of the equation, the notation is as in Sections 2 and 3. *Let δ denote the number of families containing solutions of* (26).

LEMMA 10. *If M is a multiple of L but not l, $N^M > D/4$, and* (26) *has a solution with $z = 3M$, then $N^M = (D \pm 2)/3$ and* (26) *has no solution with $z = M$.*

Proof. Since $l \nmid M$ and $L \mid M$, there are odd positive integers x and y such that $a(x/2)^2 + b(y/2)^2 = N^M$ and the solution at $3M$ is in the same family as $(x/2, y/2)$. By (5) with "u" $= x/2$, "v" $= y/2$, and "r" $= 3$,

$$\pm f = (3N^M - by^2)y/2. \tag{27}$$

Since y is a divisor of f, either $y = f$ and $N^M = (D \pm 2)/3$ or $y \leqslant f/2$. In the former case, $N^M = aw^2 + D$ implies that $w^2 \leqslant 0$, a contradiction. Suppose $y \leqslant f/2$. Then, since $N^M > D/4$,

$$3N^M - by^2 > 3(D/4) - bf^2/4 = D/2.$$

Hence $\pm f > D/4 = (bf/4)f$ and so $bf < 4$ and the sign in (27) is "+." Since $f \geqslant 2y \geqslant 2$, we have $b = 1$, $y = 1$, and $f = 2$ or 3. By (27), $N^M = (2f + 1)/3$, which is not an integer when $f = 2$ or 3. ■

The following theorem generalizes a result of Lewis [15] on the number of families which contain solutions. Recall that E^* was defined in Section 2 to be either the exponent or half the exponent of the ideal class group and τ was defined in Section 3 to be the number of distinct prime factors of N.

THEOREM 15. *We have*

$$\delta \leqslant 2^{\tau-1} \leqslant N/3.$$

Suppose D is not an odd square. Let d(n) denote the number of divisors of n. Then

$$\delta \leqslant d(f)\,d(E^*)$$

and, if $N > D/4$, *then*

$$d \leqslant d(E^*).$$

Proof. We have already proved $\delta \leqslant 2^{\tau-1}$ in Theorem 4.

Suppose D is not an odd square. By Theorem 4 each family which has a solution belongs to $\mathfrak{p}_i$ for some i and thus is associated with an l_i and a $\beta_i = au_i + \omega v_i$ with norm aN^{l_i}. By Corollary 3.1 $v_i | f$, and by Theorem 1 $l_i | 3E^*$. Since $l_i = l_j$ and $v_i = v_j$ imply $\beta_i = \beta_j$, we have $\delta \leqslant d(f)\,d(3E^*)$. When $l_i = L_i$ for all i, we can replace $3E^*$ by E^* in this inequality. If $l_i \neq L_i$ for some i, then $\beta_i = ((ax/2) + (y/2)\omega)^3/a$ with x and y odd. Equating coefficients of ω and rearranging we obtain

$$N^{L_i} = (by^2 \pm 2v_i/y)/3, \tag{28}$$

where $y | v_i$. Even if this situation occurs, we could still associate the solution with the divisor L_i of E^* and the divisor v_i of f unless there is a family with $l_j = L_i$ and $v_j = v_i$. If this happens, $N^{L_i} > bv_i^2$. Combining this with (28) we have $3bv_i^2 < by^2 + 2v_i/y$, and so $2bv_i^2 < 2v_i/y$ since $y \leqslant v_i$. This is a contradiction.

Now suppose further that $N > D/4$. With each family containing a solution we associate a least exponent m as in Theorem 9. These m's are distinct since m and D determine x and hence the family. We want to show that the map $m \to L$ is one-to-one, for then there will be no more than $d(E^*)$ families.

By Theorem 8, either $m = l$ or $a = 1$, $N^l = 1 + D/4$, and $m = 2l$. Since $l = L$ or $3L$, we have $m = L$, $2L$, $3L$, or $6L$. We now show that if the m of some family is one of these four numbers then none of the other three can occur as an m for any family.

If $6L$ occurs, then kL cannot occur for any $k < 6$ by Theorem 11 with "q" = "p" = N.

Similarly, Theorem 11 implies that if $2L$ occurs then L cannot.

If $3L$ occurs, then by Lemma 10, $N^L = (D \pm 2)/3$. Hence L cannot occur. If $2L$ were to occur, then, by Theorem 8, $N^L = 1 + D/4$, and it would follow that $N^L = 2$ or 6, contradicting the assumption that N is odd. ∎

Theorem 15 can be combined with the results on the number of solutions per family obtained in Sections 6–8 to obtain bounds on the total number of solutions of (26). For example, Theorem 10 leads to

Observation 14. If $N > D + 12$ and D is not an odd square, then (26) has at most $2d(E^*)$ solutions.

To further illustrate this idea, we will give a table containing bounds on the number of solutions of (26) when $aD < 25$. The following explanations should make the table clear.

The class number equals χ, ϕ bounds the number of solutions per family, and $\sigma \leqslant \delta\phi$ bounds the total number of solutions. When we cannot do as well as known results, we have left E^* and ϕ blank. "Reasons" indicate bibliographic references, Theorems, Corollaries, or Observations, except for "Above," which will be discussed shortly, and "G," which means the class group is of type (2, 2). (See Wada [33] for a table of class groups.) "Exceptions" give values of N for which the value of σ may be incorrect (i.e., those for which $aw^2 = 2$, 4, 6, or 12 and so Theorem 10 is inapplicable) as well as improvements in σ for special cases. "Reasons" in parentheses relate to these improvements.

We have applied the following arguments without comment. The conditions $(N, 2D) = 1$ and $aD < 25$ together insure that either $N > D/4$ or $f = 1$, unless $(a, D, N) = (1, 20, 3)$ or $(1, 24, 5)$. Hence, we can apply Theorem 15 to get the bound $\delta \leqslant d(E^*)$, while for the two exceptional (a, D, N), $\delta = 1$ since N is a prime. When $a = 1$ and E^* is even, note that in the range of the table $ab \not\equiv 3$ modulo 8. Thus $l = L$. By Theorem 11 with "a" = 1, "q" = "p" = N, and "$2k$" equal to the largest value of l, at most one family has solutions. Hence $\delta = 1$ in these cases too. By inspection, $E^* = 1$. Thus $\delta = 1$ except possibly for $x^2 + 23$, which is discussed below. Theorems 10 and 14 are frequently used to obtain the bound $\phi = 2$. The reference to Theorem 10 in the reasons column refers to 10(ii).

A variety of arguments can be used to lower σ in special cases. We give three examples which are referred to as "Above" in the reasons column of the table.

If $D \equiv \pm 4$ modulo 10, $D \not\equiv \pm 1$ modulo 25, and $5 \nmid N$, then $x^2 + D = N^z$ has at most one solution per family. We prove this. Corollary 10.1(ii) applies with $q = 5$ when $5 | x$. Otherwise, $x^2 + D \not\equiv \pm 1$ modulo 5 and Observation 13 applies. This result applies to the cases $D = 6$ and 14 in the table. We can sometimes eliminate $5 | N$: applying (9f) gives $\epsilon \equiv D$ modulo 5 and Theorem 13(ii) also gives the value of ϵ. When $D = 14$, these values disagree.

Factors of 3 can be removed from E^* when estimating δ via the inequality $\delta \leqslant d(E^*)$ if the complete solution of $ax^2 + D = z^3$ is known. (See Coghlan and Stephens [8] and Hemer [13] for information on solutions with $a = 1$

aD	χ	a,D	E^*	ϕ	σ	Reasons	Exceptions
2	1	1,2			1	[27]	$\sigma = 2$ if $N = 3$
3	1	1,3			1	[24]	
5	2	1,5			1	[24]	
6	2	1,6	2	1	1	O12	
		2,3	1	1	1	O7	
		3,2	1	1	1	[19]; O10	
7	1	1,7	1	1	1	[18]; O3	
8	1	1,8	1	1	1	O11	
10	2	1,10	2	1	1	O11	
		2,5	1	2	2	(O4, C10.1)	$N = 7$; $\sigma = 1$ if $12 \nmid (N-7)$
		5,2	1	1	1	[27]; O10	
11	1	1,11	1	2	2	(O5)	$N = 15$; $\sigma = 1$ if $3 \nmid N$
12	1	1,12	1	1	1	O9, Above, [18]	
		3,4	1	1	1	[27]; O11	
13	2	1,13	1	1	1	[21]	$\sigma = 2$ if $N = 17$
14	4	1,14	4	1	1	Above	
		2,7	2	2	4	(O5)	$N = 3$; $\sigma = 2$ if $3 \nmid N$
		7,2			1	[27]	
15	2	1,15	2	1	1	O7	
		3,5	1	2	2	(T10)	$N = 17$; $\sigma = 1$ if $9 \nmid (N+1)$
		5,3	1	1	1	O7	
17	4	1,17	4	2	2	T14	
18	1	1,18	1	1	1	O12	
19	1	1,19	1	2	2		$N = 23$
20	2	1,20	2	1	1	O11	
		5,4			1	[27]	
21	4	1,21	2	1	1	[21]	
		3,7	1	1	1	G, O3	
		7,3	1	2	2	G	
22	2	1,22	2	2	2	(C10.1, O13)	$\sigma = 1$ if $15 \nmid (N-11)$
		2,11	1	2	2	(O4, C10.1)	$N = 13$; $\sigma = 1$ if $12 \nmid (N-1)$
		11,2	1	1	1	[27], O10	
23	3	1,23	3	1	1	Above, O3	
24	2	1,24	2	1	1	O9, C11.1($N = 5$)	
		3,8	1	2	2		

and $D \leqslant 100$.) For example, the only solution of $x^2 + 23 = z^3$ is $z = 3$, so the case $a = 1$, $D = 23$ in the table has $l_i = 3$ only for $N = 3$. Hence there is only one family unless $N = 3$. Since 3 is a prime, there is only one family in this case as well.

For the equation $x^2 + 12 = 13^z$, $z = 1$ is a solution and, by Corollary 13.1, $z \equiv 3$ modulo 8 for any other solutions. Since $z = 3$ is not a solution and, by Theorem 13, $\epsilon = -1$, (9e) modulo 3 yields $-1 = (-3|z)$ and so $z \equiv 2$ modulo 3. On the other hand, $z \equiv 7$ modulo 12 by (9f), a contradiction. Since x cannot equal 2 or 3, $\sigma = 1$ for the equation $x^2 + 12 = N^z$ for $N < 2D$.

When $\phi = 1$ note that $z|6L$ for all z at which the equations in the table have solutions. This follows from Theorem 9 if either $N > D/4$ or $f = 1$. If $(a, D, N) = (1, 20, 3)$, then equation (17) has no solutions modulo 8; while if $(a, D, N) = (1, 24, 5)$ then $z = 2$ is a solution of (17) and $\delta = 1$. For each pair (a, D), $\phi = 1$ with a finite number of exceptions: Those entries with $D = 4$ have been treated elsewhere [27] and Theorem 6(ii) applies to the remainder. On the other hand, Corollary 10.1(ii) provides a method for bounding N from below: See the discussion near the end of Section 8 regarding $3Ax^2 + 8 = p^n$.

10. $ax^2 + D = z^n$

We now discuss the equation

$$ax^2 + D = z^n \qquad (D = bf^2), \tag{29}$$

where we assume the parameters a, D, and n are positive, satisfy $(a, D) = 1$ and $D > 1$, and a is square-free. (For $D = 1$ and n and z odd see Nagell [27].)

The following lemma is useful when $(2D, z) \neq 1$.

LEMMA 11. *Suppose q is a prime and* $\text{ord}_q(aD) = t$. *If t is odd then*

(a) $\text{ord}_q(z^n) = 2j < t$ *or*
(b) $\text{ord}_q(z^n) = \text{ord}_q(D)$.

If t is even and q doesn't split in $\mathbb{Q}(\omega)$ then:

(c) $\text{ord}_q(z^n) = 2j \leqslant t$ or
(d) *if* $q = 2$, $\text{ord}_2(z^n) = t + \text{ord}_2(a + D2^{-t})$.

Proof. Suppose t is odd. Then $\text{ord}_q(ax^2) \neq \text{ord}_q(D)$ and so

$$\text{ord}_q(z^n) = \min(\text{ord}_q(ax^2), \text{ord}_q(D)).$$

By considering cases, we obtain (a) and (b).

Suppose t is even. Then $q \nmid a$ and the above argument gives (c) unless $\text{ord}_q(ax^2) = \text{ord}_q(D)$. Suppose this equality occurs. We may write $ax^2 + D = q^t M$ where $M = ax'^2 + Dq^{-t}$ and $q \nmid ax'D$. If $q \neq 2$, then, since it does not split in $\mathbb{Q}(\omega)$, $1 \nmid M$. If $q = 2$, then $M \equiv a + D2^{-t} \equiv a(1 + aD2^{-t})$ modulo 8. Since 2 does not split, $aD2^{-t} \not\equiv -1$ modulo 8. This leads to (*d*). ■

Suppose n does not divide E^*. If $n/(n, E^*)$ contains an odd factor we can use (5) to find the solutions of (29) with $(z, 2D) = 1$ by letting v run through all divisors of f, letting r run through all odd prime divisors of $n/(n, E^*)$, and solving the resulting equations for $au^2 = z^m - bv^2$. If $n/(n, E^*)$ is even and D is not an odd square, it follows from Theorem 3 with $m = n/2$ that $z^{n/2} = aD/4bv^2 + bv^2$ where $2v$ is a divisor of f. Hence we can find all solutions in this case too. Thus, *if D is not an odd square, we can determine all triples (a, x, z) such that $ax^2 + D = z^n$, $(z, 2D) = 1$, and $n \nmid E^*$ in a finite number of operations.*

When $n = 3$, $l = 1$, and $r = 3$, the method is well known.

Besides the straightforward calculation just described, we have basically three ways of dealing with (29) when $(z, 2D) = 1$. They generally require that $n \nmid E^*$. The "arithmetical" method is to use results from previous sections which assert that each family contains at most one solution, and then show that this solution occurs at $n = l$, usually by Theorem 9. When D is square-free this almost always leads to equations with no solutions; however, if D is not square-free, there may be solutions with $z^l < D/4$. Application of Observations 9 and 10 may lead to solutions when $z^l \leqslant 2D$.

The "gap" method is to apply Theorem 5 to obtain bounds on the largest possible z for which solutions may exist. This bound can be quite practical. Thus this method leads to equations which have a finite number of solutions.

The "algebraic" method is to argue directly concerning the roots of the polynomial equation (5). In this way one can bound the number of solutions of (29).

We begin by illustrating the arithmetical method. Suppose that $2^k \mid n$, $k \geqslant 1$, and D is square-free. (D square-free makes Lemma 11 easy to apply, and avoids solutions with $z \leqslant D/4$.) The case $a = 1$ is simple because it suffices to find all representations of D as a difference of two squares. If $a \neq 1$ and D is not a square, then by Theorem 1 there are no solutions with $(z, 2D) = 1$ when E is not a multiple of 2^{k+1}. Lemma 11 then gives numerous examples for which there are no solutions for any z.

Suppose again that n is even and also that D is divisible by a prime q not dividing z. By taking (29) modulo q (or 8 if $q = 2$), we obtain conditions for a solution:

(a) if q is odd, $(a \mid q) = 1$ and
(b) if $q = 2$, $D \equiv 1 - a$ modulo 8.

As another example note that by combining Observation 3, Lemma 11, and (28) with $y = v_i = f = 1$ we obtain a generalization of a result of Ljunggren [17]:

Observation 15. If a is odd, $aD \not\equiv -1$ modulo 8, D is square-free, $\operatorname{ord}_2(D^2 - 1)$ is even, and n does not divide the class number of $\mathbb{Q}(\omega)$, then (29) has no solutions, except possibly $z^n = (D \pm 2)^3/27$.

Stolt [32, Theorem 1] has given a long list of conditions, each implying that there are no solutions to (29). His results follow easily from (9g) and (24) when D is odd, and from (9g), Theorem 13(ii) and Corollary 13.1 when D is even.

There are a variety of other possible results, particularly when D is even. Suppose $D = bf^2 = d2^t$ where $t > 0$ and d is odd and square-free. Let $r2^k = n/(n, 3E^*)$. By combining Theorems 2(iii) and 13(i), Corollary 13.1, and Lemma 11, we see that (29) has no solutions with z odd unless $r = 1$ or r is a prime satisfying the congruence in Corollary 13.1. The factor of 3 multiplying E^* can be eliminated unless $l = 3L$ and $n/(n, E^*) = 3r'2^k$. When z is odd, (28) implies that b is odd (and so $b = d$ and t is even). By replacing L_i by L_i2^c, where $n/(L_i2^c)$ is odd, we can proceed as in the derivation of (28) and conclude, since $\operatorname{ord}_2(v_i) = t/2$ by Theorem 2(iii), $z^m = (b \pm 2f)/3$ for some $m \mid (n/3)$. When $k \neq 0$, it can be shown that $a = r = 1$ and $k \leqslant 2$. The solutions can be obtained explictly in these cases. When $k = 1$, $z^n = (1 + D/4)^2$ or $(b + f^2/4)^2$. When $k = 2$, $z^n = (2^{t-3} \pm 1)^2$ and b must equal $2^{t-4} \pm 1$. Thus we have the following generalization of a result of Ljunggren [18]:

Observation 16. Suppose that D is even and is a power of 2 times a square-free number and that $n/(n, E^*) = s$. If (29) has a solution with z odd then either (i) $s = 1, 2, 4$, or r where r is a prime satisfying Corollary 13.1 or (ii) $ab \equiv 3$ modulo 8, $s = 3, 6, 12$, or $3r$, and z is an odd power of $(b \pm 2f)^3/27$.

We now consider gap methods. By Theorem 6(ii), if $b \neq 1$, then (29) has no solutions with $n > 2$, $(z, 2D) = 1$, and $z > \Gamma(a, D)$. Sometimes we can use Lemma 11 to dispose of z for which $(z, 2D) \neq 1$. For example:

Observation 17. If $D > 1$ is square-free and $aD \not\equiv -1$ modulo 8, then there exists an effectively computable function $\Gamma(a, D)$ such that (17) has no solutions with $n \geqslant 3$ and $z > \Gamma(a, D)$.

When the following theorem applies, it provides a relatively small bound on the size of those z for which solutions of (29) exist and $(z, 2D) = 1$. Solutions with $(z, 2D) \neq 1$ can often be eliminated by Lemma 11.

THEOREM 16. *Let E^* be as defined in Section 2. Suppose that $n \nmid E^*$ and, if D is an odd square, $n/(n, E^*)$ has an odd factor. Suppose further that*

(i) $z > 4f^2 + 2\sqrt{D}$, *and*
(ii) $z > D(1 + ((2n + 1)/2\pi)^2)$.

Then (29) *has no solutions with* z *prime to* $2D$.

Suppose that $mm'(n, E^*)|n$, $1 < m \leqslant m'$, *and, if* D *is an odd square, that* m *and* m' *are odd. Suppose further that*

(i′) $z^n > (\Lambda f^2 + 2f\sqrt{b})^2$, *where* $\Lambda = 4$ *if* (1) *has an integral solution at* $n = L$ *and* $\Lambda = 16$ *otherwise, and*

(ii′) $z^n > D^m(1 + (2m' + 1)^2/4\pi^2)^m$.

Then (29) *has no solutions with* z *prime to* $2D$.

Proof. Suppose a solution exists. Then $l_i|n$ for some family. If D is an even square, then n/l_i is odd by Theorem 3(ii). Hence we can write $n = rK$ where $L_i|K$, $r > 1$, r is odd if $b = 1$, and $r = mm'$ in the last paragraph of the theorem. Thus $y(K)|y(n)$ by Theorems 2(iv) and 3(iv).

The result in the last paragraph of the theorem follows from Theorem 7.

We now consider the first paragraph in the theorem. The right-hand side of (28) achieves a maximum of $(D + 2)/3 < D$. It follows from (ii) that $l_i = L_i$ for all families. By (i) we may apply Theorem 5 with "k" = 2, "C" = f, "n" = "m" = K, and "n" = rK to deduce $r - 1 \geqslant \pi(z^m/D - 1)^{1/2} - \frac{3}{2}$, contradicting (ii). ■

Remarks. Somewhat surprisingly, the value of "a" influences the bound on z only indirectly by determining E^* and thus restricting n.

If D is fixed and n is sufficiently large ($n \geqslant 16(\log_2 D)^2$ suffices), then (i′) and (ii′) are of the form $z^n > (1 + \epsilon)^n$ with $\epsilon < 1$. To see this for (ii′), note that $mm' \leqslant n$ and so $m \leqslant \sqrt{n}$. Hence (29) has no solutions in this case.

We illustrate Theorem 16. Suppose $n = 5$, $aD < 25$, and D is not an odd square. If $(z, 2D) \neq 1$, Lemma 11 applies unless $aD = 7, 15, 18$, or 23. For the other 29 values of (a, D), Theorem 16 shows that all solutions have $z \leqslant 73$. By computer we find that there is only one solution, namely $1(22434)^2 + 19 = 55^5$.

We now sketch the algebraic method which Persson [29], Stolt [32], and Blass [5] have used to bound the number of solutions when either a or D is also considered to be a variable. Suppose n is odd and not a multiple of 3, D is not divisible by an nth power, $aD \not\equiv -1$ modulo 8, the class number of $\mathbb{Q}(\omega)$ is not divisible by n, and either $z > D/4$ or D is square-free. Without loss of generality we may assume that n is a prime. By Lemma 11, we may assume that $(z, 2D) = 1$. By Theorem 9, if there is a solution of (29) in some family, there is a solution at the exponent l in that family. By assumption $(n, L) = 1$, and so, since $n \neq 3$, we must have $l = 1$ to get a solution at n. (If $aD \not\equiv 3$ modulo 4 we can allow $n = 3$.)

We now regard D and n as fixed and z and a as variable. Set "m" = 1, "r" = n, "bv^2" = D, and "au^2" = $z - D$ in (5). Dividing by $ny(n)$, rearranging, and using the binomial identity

$$\sum_{j=0}^{k} \binom{2t+1}{2j+1}\binom{t-j}{t-k} = 2^{2k}\binom{t+k}{2k}(2t+1)\Big/(2k+1)$$

valid for $0 \leqslant k \leqslant t$, Stolt obtains

$$c_t + \sum_{k=0}^{t-1} (-1)^k c_k (4D)^k z^{t-k} = 0, \tag{30}$$

where $t = (n-1)/2$, $c_t = ((-4D)^t - \epsilon)/n$, and

$$c_k = \binom{t+k}{2k}\Big/(2k+1) \qquad \text{for} \quad k < t.$$

Since n is a prime, it can be shown that c_k is an integer for all k. See Stolt [32] for details. Since $c_0 = 1$, the roots of (30) are algebraic integers, say z_i, $1 \leqslant i \leqslant t$. Suppose z_i is an odd integer for $i \leqslant m$. By expressing the coefficients of (30) in terms of $z_1, \ldots, z_m$ and the symmetric functions in $z_{m+1}, \ldots, z_t$, Stolt and Persson obtain arithmetic conditions which allow them to bound m under appropriate conditions. For example, if $n \equiv 3$ modulo 4, then $m \leqslant 1$ and so (29) has at most one solution.

When $n = 5$, Blass [5] uses a different argument in which D is regarded as a variable. To insure that a solution occurs at l and that $5 \nmid l$, we require that $5 \nmid E$ and either D is square-free or $N > D/4$. Equation (5) can be rearranged to give

$$20a^2u^4 \pm 1 = (D - 5au^2)^2.$$

A congruence modulo 4 shows that the sign is plus. Hence we have an equation of the form $20a^2X^4 + 1 = Y^2$. Cohn [10, Theorem 7] has shown that there is at most one solution to this equation in positive integers X, Y. Thus, for fixed a, (29) has at most two solutions (x, D, z). In particular, if $a = 1$, then $X = 6$; while if $a = 2$, then $X = 1$. When $a = 1$, we obtain the result Blass proved for D square-free: the only solutions are

$$(x, D, z) = (22434, 19, 55) \qquad \text{and} \qquad (2859646, 341, 377).$$

When $a = 2$ we find the only solution is

$$(x, D, z) = (1429, 19, 21).$$

11. *Divisibility of Class Numbers*

Since, by Theorem 1, L divides the class number of $\mathbb{Q}(\omega)$, (1) may be used to create imaginary quadratic fields with large class numbers. This was done by Nagell [28] as follows. Set $a = 1$ and let p be any prime. The solutions D of $(-D|p) = +1$ lie in arithmetic progressions. For any such D

which is square-free (by Dirichlet's theorem, there are infinitely many D's which are primes), (1) has a possibly half-integral solution at $n = L$ by Theorem 1. Hence $p^L > D/4$. If we choose D large, then L must be large and so the class number of $\mathbb{Q}(\omega)$ is large.

A more interesting question is the construction of an infinite collection of fields whose class numbers are divisible by some specified integer h. We expand on some ideas of Nagell [24]. Suppose we simply set $n = 3h$, choose p and D, and then use (17) to determine a and x. We then consider the field $\mathbb{Q}(\omega)$. Two problems arise: first, there is no guarantee that $h = L$ and so no guarantee that h divides the class number; second, there is no guarantee that the values of ab (and hence the fields) are distinct.

Solving the first problem is a two-step process: (a) guarantee that (17) has a solution at the exponent l, and (b) infer results about l from the existence of a solution at the exponent n (which may not equal l). For (a), we can choose D square-free or (by Theorem 8) choose $p > D/4$ where D is not an odd square. We can handle (b) in two ways:

(i) Set $n = 3h$ and choose p and D so that each family contains at most one solution. (Hence $n = l$.)

(ii) Set $n = 3h^2$ and choose p and D so that $p > D + 12$.

In the second case, Theorem 10 implies that $n = rl$ where $r = 1$ or a prime, and so h divides L. (The factor of 3 in (i) and (ii) is only necessary when 3 divides h and $l = 3L$, and so it can frequently be removed.) Nagell [24] used the first idea with $D = 1$. (See Observation 6.)

Here are three ways to generate an infinite number of fields:

(iii) Fix $n > 2$ and D and let p range over an infinite set of odd numbers. By applying the Thue–Siegel–Roth theorem to $ax^2 + D = z^n$, we see that each field occurs at most a finite number of times.

(iv) Let $p = p(k) = \pi_k^k$, where π_k is the kth odd prime, and let $D = D(k) = \pi_1 \cdot \ldots \cdot \pi_{k-1}$. Since $p(i)$ is prime to $a(i)D(i)$ and not prime to $D(k)$ whenever $i < k$, the ith and kth fields are distinct.

(v) Fix $D > 1$ and let p be a fixed prime. Let $n = n(k) = 2^k n'$, where n' is chosen according to (i) or (ii). If the ith and kth fields were the same, there would be a family in which $\operatorname{ord}_2(n)$ was not a constant, contradicting Theorem 9.

The following observation illustrates the above ideas.

Observation 18. Let $p > D + 12$ be prime to $2D$ and suppose D is not an odd square. The fields $\mathbb{Q}(-Da(k))^{1/2}$, where $a(k) = p^{3 \cdot 2^k} m^2 - D$, are all distinct and the kth class number is divisible by $2^k m$.

III. Other Equations

12. $ax^2 + D\mathbf{q}^{\mathbf{z}} = p^n$

Boldface symbols such as **q** and **z** will denote vectors, and $\mathbf{q}^{\mathbf{z}}$ denotes the integer $\prod q_i^{z_i}$. We also use the notation of Section 2. In particular we assume that (1) has a (possibly half-integral) solution at the exponent m, that $x(m) = u$, $y(m) = v$, and that $ax(rm)^2 + by(rm)^2 = p^{rm}$.

We now turn our attention to the more general equation in the integer variables $x, \mathbf{z}, n$,

$$ax^2 + D\mathbf{q}^{\mathbf{z}} = p^n, \tag{31}$$

where we assume $\mathbf{z} \geqslant \mathbf{0}$, p is an odd positive integer prime to $D\prod q_i$, the q_i are primes not dividing a, and a and D are positive integers. When $D = 1$, we also assume that not all of the z_i are even.

As was the case in Part II, we focus our attention first on solutions of (1) in a single family. In particular, this means that we have specified the values for the parities of the z_i's, since changing the parity changes $\mathbb{Q}(\omega)$. The lemma below gives some additional information relating $\mathrm{ord}_q(by(m)^2)$ and $\mathrm{ord}_q(by(rm)^2)$ to r. (This complements Lemma 8, which applies when $\mathrm{ord}_q(by(m)^2) = 0$.) We then prove a theorem which limits the number of values of the exponent **z**.

Lemma 12. *Suppose that q divides $by(m)^2$ and that r is a positive integer (odd if $a \neq 1$). If $q = 2$, suppose that u and v are integral; if $q = 3$ and $3 \mid r$, suppose also that $3a \not\equiv by(m)^2$ modulo 9. Then*

$$\mathrm{ord}_q(y(rm)) = \mathrm{ord}_q(r) + \mathrm{ord}_q(y(m)). \tag{32}$$

Proof. Set $r = t2^s$ where t is odd. If $s \neq 0$, then $a = 1$ by the assumption on r. By squaring $x(n) + y(n)\omega$, we see that $y(2n) = 2x(n)y(n)$. Since $q \mid bv^2$ and $q \nmid x(n)$ we see by induction that (32) is true when r is a power of 2. By setting the pair "r," "m" in (32) first equal to $2^s, m$, and then equal to $t, 2^s m$, we see that it suffices to consider (32) when r is odd.

Suppose $q = 2$. By Theorem 2, $\mathrm{ord}_2(v) = \mathrm{ord}_2(y(rm))$. Hence (32) follows.

Suppose q is odd. Set $r = tq^s$ where $s = \mathrm{ord}_q(r)$. By (9a), $y(tm)/y(m) \not\equiv 0$ modulo q, and so it suffices to prove the lemma when r is a power of q. By induction it suffices to prove the lemma when $r = q$. (The hypothesis for $q = 3$ is automatically true when $9 \mid by(m)^2$.) Dividing (5) by v and taking the resulting equation modulo q^2 we find that

$$y(qm)/y(m) = \epsilon(q) \equiv (au^2)^{(q-3)/2} q(au^2 - bv^2/3).$$

We easily read off the following: (i) $\epsilon(q)$ is divisible by q; (ii) if $q \neq 3$, $\epsilon(q)$ is not divisible by q^2; (iii) if $q = 3$, $\epsilon(q)$ is divisible by q^2 if and only if $3au^2 \equiv bv^2$ modulo 9. The lemma follows. ■

Remarks. Lemma 8 (see Section 3) and Lemma 12 have some interesting applications to (17). Let $q \mid (D/by(l)^2)$. For simplicity assume that $a \neq 1$ and $b \neq 1$ so that (17) will have no solutions with n/l even. Suppose first that $q \nmid by(l)^2$. By applying Lemma 8 we see that there are no solutions if $q = 3$ and that n is a multiple of $3l$ if $q = 5$, 7, or 17.

Now suppose $q \mid by(l)^2$. Let m be the smallest exponent in a family of solutions and let $s(q) = \mathrm{ord}_q(D/by(l)^2)$. Then, by Lemma 12, $\mathrm{ord}_q(m/l) \leqslant s(q)$ with equality unless $q = 3$, $s(3) > 0$, and $3au^2 \equiv bv^2$ modulo 9. Suppose further that every prime dividing D also divides $by(l)^2$, and also, if 3 divides $f/y(l)$, that $3a \not\equiv by(l)^2$ modulo 9, then $m = lf/y(l)$.

LEMMA 13. *Suppose there are two solutions of* (31) *in the same family, one at the exponent* $n = m$, *and having* $\mathbf{z} = \mathbf{c}$, *the other at the exponent* $n = m'$, *and having* $\mathbf{z} = \mathbf{c}'$. *Then*:

(i) *There is a solution in the same family at the exponent* $n = (m, m')$ *with* $\mathbf{z} = \mathbf{w}$, *where* w_i *is the minimum of* c_i *and* c_i'.

(ii) *If* $m' = rm$, *then* $\mathbf{c} \leqslant \mathbf{c}'$.

(iii) *If* $m' = rm$, *and* $s \mid r$, *there is a solution in the same family at the exponent* $n = sm$ *with* $\mathbf{z}$ *satisfying* $\mathbf{c} \leqslant \mathbf{z} \leqslant \mathbf{c}'$.

Proof. First note that by assumption, when all the z_i are even, $D \neq 1$. Hence we can imitate the proof of Theorem 9 to show that when $b = 1$, $\mathrm{ord}_2(n) = \mathrm{ord}_2(m)$. Thus in all cases Theorems 2(iv) and 3(iv) imply that a solution also exists at $n = (m, m')$, with $\mathbf{z} = \mathbf{w}$.

To prove (ii), note that (i) implies that there is a solution with $n = m$ and $\mathbf{z} = \mathbf{w} = \min(\mathbf{c}, \mathbf{c}')$. Since the exponent m determines the value of $\mathbf{z}$ within each family, $\mathbf{w} = \mathbf{c}$.

To prove (iii), apply the same divisibility argument that was used in proving (i) to show that there is a solution of (31) at the exponent $n = sm$. The inequality $\mathbf{c} \leqslant \mathbf{z} \leqslant \mathbf{c}'$ also follows from this argument. ■

The next lemma tells us that if there are solutions with $\mathbf{z} = \mathbf{c}'$ and $\mathbf{z} = \mathbf{c} < \mathbf{c}'$, there are usually at least as many solutions with $\mathbf{z} = \mathbf{c}$ as there are with $\mathbf{z} = \mathbf{c}'$.

LEMMA 14. *Suppose a family contains at least two solutions. Suppose one solution occurs at the exponent* $n = m$, *and that* $\mathbf{c}$ *is the associated value of* $\mathbf{z}$. *Suppose another solution in the family occurs at the exponent* $n = tm$, *and has*

$\mathbf{z} = \mathbf{c}'$, *with* $\mathbf{c}' > \mathbf{c}$. *Finally suppose that there is also a solution with* $n = rtm$ *and* $\mathbf{z} = \mathbf{c}'$, *and that* $(r, t) = 1$. *Then there is also a solution with* $n = rm$ *and* $\mathbf{z} = \mathbf{c}$.

Proof. By Lemma 13(iii), there is a solution at the exponent $n = rm$ with $\mathbf{z} = \mathbf{w}$ satisfying $\mathbf{c} \leqslant \mathbf{w} \leqslant \mathbf{c}'$. Then by Lemma 13(i), since $(r, t) = 1$, there is a solution at the exponent $n = m$ with $z_i = \min(c_i', w_i)$. Since n determines $\mathbf{z}$ within a family, $c_i = \min(c_i', w_i)$, and so $\mathbf{w} = \mathbf{c}$. ∎

THEOREM 17. *Suppose at least two solutions of* (31) *exist in the same family, and have different* $\mathbf{z}$*'s. Let* $n = m$ *at the solution with the smallest exponent and let* $\mathbf{c}$ *be the associated value of* $\mathbf{z}$. *Suppose* $\mathbf{z} = \mathbf{w} \neq \mathbf{c}$ *at a second solution. Set* $\mathbf{w} - \mathbf{c} = \mathbf{e}$ *and suppose that* $q_i | D\mathbf{q}^{\mathbf{c}}$ *whenever* $e_i \neq 0$. *Then* $\mathbf{e}$ *is a vector consisting of zeros and a single positive nonzero component* e_i. *If there are solutions at the exponents n and rn when* $\mathbf{z} = \mathbf{w}$, *then there is also a solution at the exponent* $n = rm$ *with* $\mathbf{z} = \mathbf{c}$. *The value of* e_i *must be* 2 *except for*:

(i) $x^2 + 2^z = 3^n$ (*z odd*), *for which there are solutions with* $z = 1, 3$, *and* 5 *and no other* z;
(ii) $q_1 = 3$ *and* $3a \equiv D\mathbf{q}^{\mathbf{c}}$ modulo 9, *for which* $e_1 > 2$ *and there may be another solution with* $z_1 = w_1 + 2$.

In all cases, the smallest value of $e_i \neq 0$ *is associated with the exponent* $n = q_i m$.

Proof. By Lemma 13(i), we can replace D by $D\mathbf{q}^{\mathbf{c}}$, and so look at solutions with $\mathbf{z} \geqslant \mathbf{0}$ and even, and n a multiple of m.

Since there is now a solution at $\mathbf{z} = \mathbf{0}$,

$$D < p^m \tag{33}$$

and, since $q_i | D$ for all i,

$$\prod q_i \leqslant D. \tag{34}$$

Let another solution exist at $n = rm$ with $\mathbf{z} = \mathbf{e} \neq \mathbf{0}$.

Suppose that $\mathbf{e}$ is not a vector consisting of zeros and a single nonzero component $e_i = 2$. We will show that the following four cases lead either to a contradiction or to one of the listed exceptions:

(a) two distinct components of $\mathbf{e}$ are nonzero;
(b) only one $e_i \neq 0$ and $q_i \geqslant 5$;
(c) only one $e_i \neq 0$ and $q_i = 2$;
(d) only one $e_i \neq 0$ and $q_i = 3$.

By Lemma 12, $q_i | r$ whenever $e_i \neq 0$. By Lemma 13(iii), solutions exist at $n = sm$ whenever $s | r$. Thus, in case (a) we can assume $r = q_1 q_2$ and in the remaining cases we can assume $r = q_1^2$ since when $r = q_1$ Lemma 12 implies that $\mathbf{e}$ is of the form $(0, \ldots, 2, \ldots, 0)$ or we have exceptional case (ii). Note that in all four cases there is a solution of (31) at $n = q_1 m$ and, in case (a), also at $n = q_2 m$.

We wish to apply Corollary 8.1 with "m" $= q_i m$ and "n" $= rm$. If

$$D\mathbf{q}^{\mathbf{e}}/4 < p^{mq_i}, \tag{35}$$

then rm must equal $q_i m$ or $2q_i m$ and, in the latter case, $p^{mq_i} = 1 + q_i^2 D$. Since $r \neq q_i$, the latter case must apply. From (33) and (34) we have for $q_i \geqslant 3$

$$p^{mq_i} \geqslant p^{3m} > q_i^2(D + 1) > 1 + q_i^2 D;$$

a contradiction. Hence showing that (35) with $q_i \geqslant 3$ holds establishes a contradiction.

We now turn to case (a). Suppose $q_1 \neq 3$ and $q_2 \neq 3$. Lemma 12 applies and so $\mathbf{q}^{\mathbf{e}} = r^2 \leqslant D^2$ by (34). Hence (33) implies (35) with q_i equal to the maximum of q_1 and q_2.

Suppose $q_1 = 3$. Again, since $q_2 \neq 3$, Lemma 12 applies, and so we can assume $e_2 = 2$. The solution at $n = 3m$ yields the inequality $3^{e_1}D < p^{3m}$.

Suppose further that $q_2 \neq 2$. Then $q_2 \geqslant 5$. Set $i = 2$ in (35). We have by the above and (33)

$$p^{mq_2} \geqslant p^{5m} = (p^m)^2 p^{3m} > D^2 3^{e_1} D,$$

and so (35) follows from (34).

Suppose instead that $q_2 = 2$. Then $D \geqslant 6$ and so $p^{3m} > 3^{e_1} \cdot 2^2/4$, which is (35) with $i = 1$. This completes case (a).

Case (b) is easy: By (33) and (34), $Dq_1^4 \leqslant p^{5m}$, which proves (35) with $i = 1$, since $q_1 \geqslant 5$.

Next consider case (c). The solutions at $n = 2m$ and $n = 4m$ imply that $a = 1$. By Theorem 3(i), $x(m)$ and $x(2m)$ have absolute value 1. Since $x(2m) = x(m)^2 - D$, $D = 2$. This leads to exception (i) in the statement of the theorem. There are no other z's in this case since $x(4) = \pm 7$.

Finally consider case (d). Since p is prime to $2 \cdot 3$, $p \geqslant 5$. Lemma 12 applies when "m" $= 3m$ since "D" is then divisible by 9. Suppose Lemma 12 also applies at m. Then $e_1 = 4$, and (35) holds since $p \geqslant 5$ and (33) yield $p^{3m} > (9/2)^2 D$. If Lemma 12 does not apply at m, we must show that there are no solutions at $n = 3^t m$ with $t > 2$, to verify that the second exceptional case in the statement of the theorem holds. Let $\mathbf{z} = \mathbf{w}$ be the solution at $n = 3m$. Apply the above argument with m replaced by $3m$ and D by $3^{w_1}D$.

Suppose finally that there are solutions at the exponents n and rn when $\mathbf{z} = \mathbf{w}$. As we saw in the first paragraph of the proof, we may assume n is a multiple of m. Say $n = tm$. By Theorem 9, we may also assume that tm is the smallest exponent with $\mathbf{z} = \mathbf{w}$, for if $s|t$, and there is a solution at the exponent sm with $\mathbf{z} = \mathbf{w}$, then there must also be a solution at the exponent $n = rsm$ with $\mathbf{z} = \mathbf{w}$ since $rs \mid rt$. By Lemma 12, t is a power of that q_i, for which $e_i \neq 0$, and so $q_i \nmid r$. By Lemma 13, there is a solution at the exponent $n = rm$ with $\mathbf{z} = \mathbf{c}$. ■

Remark. We complement Theorem 17(i) by showing that the equation $x^2 + 2^z = 3^n$ has just the four solutions $(x, z, n) = (1, 1, 1)$, $(5, 1, 3)$, $(1, 3, 2)$, and $(7, 5, 4)$. Since 3 does not split in $\mathbb{Q}(i)$, z must be odd. In the above proof we saw that there are only three values for z. For each of these values of z we have to solve equation (17) with $D = 2^z$. When $z = 1$, there are two solutions $(x, n) = (1, 1)$ and $(5, 3)$. There are no others by Theorem 13. When $z = 3$, by Observation 11, there is only one solution to (17). Observation 11 also applies when $z = 5$.

COROLLARY 17.1. *Among all the solutions of* (31) *in a single family having $z_i > 0$ for all i, there is a minimal one. Let this minimal value of* $\mathbf{z}$ *be* $\mathbf{w}$, *and suppose the smallest exponent at which there is a solution with* $\mathbf{z} = \mathbf{w}$ *is m. Then for every solution with* $\mathbf{z} > \mathbf{w}$, *the vector* $\mathbf{z} - \mathbf{w} = \mathbf{e}$ *consists of zeros and a single non-zero component e_i, and the associated exponent n is a multiple of $q_i m$. The value of e_i must be* 2 *except for the two exceptions noted in Theorem* 17.

Proof. By Lemma 13(i), a solution exists with $n = m$ and $\mathbf{z} = \mathbf{w}$, where $\mathbf{w}$ is the minimum of all the values of $\mathbf{z}$ which satisfy $z_i > 0$ for all i, and m is the greatest common divisor of all the associated exponents n. That the exponents n are actually multiples of $q_i m$ when $\mathbf{z} > \mathbf{w}$ follows from Lemma 12. The last statement of the corollary follows by applying Theorem 17 with "D" $= D\mathbf{q}^{\mathbf{w}}$. ■

Assume now that $\mathbf{q}$ *and* $\mathbf{z}$ *have a single component.* We continue to restrict our attention to a single family of solutions.

If $q \nmid D$, we can replace D by qD and apply Theorem 17 to those solutions with $z > 0$. Thus we have

Observation 19. Each family of solutions of the equation $ax^2 + Dq^z = p^n$ contains at most 2 distinct values of z with the following exceptions for which there are at most 3 values of z:

(i) there is a solution at $z = 0$ and $q \nmid D$;
(ii) the equation is $x^2 + 2^z = 3^n$ and z is odd;
(iii) $q = 3$ and there is a solution at z with $3a \equiv q^z D$ modulo 9.

Remark. Note that by Theorem 9, if we have families of solutions with both odd and even z, then $(q \mid p) = +1$.

As an example, we show that the equation

$$x^2 + 3 \cdot 2^z = 7^n$$

has precisely the five solutions $(x, z, n) = (2, 0, 1), (1, 1, 1), (5, 3, 2), (1, 4, 2)$, and $(47, 6, 4)$. Since $p = 7$ is a prime, there is only one family with z even and one with z odd. By Observation 19, there are at most 5 values for z. By the table in Section 9, there are no other solutions with $z = 0$. By Observation 13, there cannot be two solutions with the same nonzero value of z.

The reader will find it easy to work out additional details when $q = 2$, even if $D = 1$ and z is even.

13. $ax^2 + D = 2p^n$ *with* $aD \equiv 1$ modulo 4

We now briefly turn our attention to some related equations which can be dealt with by minor modifications of the above methods. Specifically, we will investigate $ax^2 + D$ when a, x, and D are all odd. Since $ax^2 + D \equiv a(1 + aD)$ modulo 8, we have three cases:

(i) $aD \equiv 1$ modulo 4: In this case $ax^2 + D \equiv 2$ modulo 4 so we consider the equation $ax^2 + D = 2p^n$ with p odd.

(ii) $aD \equiv 3$ modulo 8: In this case $ax^2 + D \equiv 4$ modulo 8 so we consider the equation $ax^2 + D = 4p^n$ with p odd.

(iii) $aD \equiv 7$ modulo 8: In this case $ax^2 + D \equiv 0$ modulo 8 and we consider the equation $ax^2 + D = 4p^n$ with p even.

In cases (i) and (ii), the power of 2 on the right-hand side is determined by the congruence modulo 8. In case (iii), the power of 2 on the right-hand side is determined by the limitations of the theory. Case (i) was considered by Ljunggren [19]. When $p = 2$ and $a = 1$, case (iii) becomes the generalization of the Ramanujan–Nagell equation considered by Hasse [12].

In case (i), $(2) = \mathfrak{q}^2$ in $\mathbb{Q}(\omega)$ and so we handle this case by replacing "$\mathfrak{a}$" by "$\mathfrak{q}\mathfrak{a}$" in Part I. We shall make this comment more precise below. In cases (ii) and (iii) the factor of 4 can be handled by considering the half-integral solutions of (1). The parity of p is due to the fact that 2 remains prime in case (ii) and splits in case (iii). Because of this similarity, it is convenient to treat cases (ii) and (iii) together in the next section.

Consider the equation

$$ax^2 + D = 2p^n \qquad (D = bf^2), \tag{36}$$

where $D > 1$ and p are odd, $(D, p) = 1$, and $ab > 1$. We now indicate how the results in Sections 2–7, 9, and 10 must be modified to deal with (36). (Section 8, the case D even, does not apply.)

The definitions in Section 2 are unchanged except that $\mathfrak{a}$ is defined by $\mathfrak{a}^2 = (2a)$. Lemma 2 holds without the assumption $a > 1$; viz., $aD \equiv 1$ modulo 4 and $8a = w^2 + aby^2$ imply that w and y are even. Thus, by an obvious adaptation of the argument in Theorem 1 for the case "$a > 1$ and $b > 1$," *all solutions occur at odd multiples of l.* Note that since $ax^2 + D \equiv 2$ modulo 4, half-integer solutions do not occur. It is easy to show that Theorem 2 remains valid provided (4) is replaced by

$$ax(rm) + y(rm)\omega = \beta^r/(2a)^{(r-1)/2}$$

and (5) is replaced by

$$v \sum_{j=0}^{(r-1)/2} \binom{r}{2j} (-bv^2)^{(r-2j-1)/2}(au^2)^j = 2^{(r-1)/2} y(rm).$$

(Note that $\gamma = \beta^2/2a \equiv \omega$ modulo 2 and so $y(n)$ is odd.)

The theory of families developed in Section 3 carries over unchanged.

The results of Section 4 can be modified to cover this case but we will not go into this here except to note that (i) both sides of (9a)–(9e) (but not the moduli) should be multiplied by $2^{(r-1)/2}$, (ii) the factors of 4 in (9f) should be replaced by 2, and (iii) (9g) should be replaced by $\epsilon = (-2bv^2 \mid r)$.

Theorems 5, 7, and 8 are correct if "a" and "b" are replaced by $a/2$ and $b/2$ in the statements and proofs. Corollary 7.1 is valid and the proof is simpler since n/M and D are both odd.

The general structure theorem, Theorem 9, now becomes:

THEOREM 18. *Suppose that* (37) *has a solution. Then 2 ramifies and every prime divisor of p splits in $\mathbb{Q}(\sqrt{-ab})$. Let the smallest solution in a particular family occur at the exponent $n = m$. If another solution in the family occurs at $n = m'$, then m' is an odd multiple of m and solutions occur at all m'' such that $m \mid m'' \mid m'$.*

If $p^l \geq 5$, the gaps between values of n for which solutions exist in the family exceed

$$m(\pi(2p^m/D - 1)^{1/2} - 3/2).$$

For each family there are unique positive integers u and v prime to $2p$ such that

$$au^2 + bv^2 = 2p^l.$$

If $p^l > D/4$, $m = l$.

We always have m/l odd, $v \mid f$, uvf odd, and $(u, f) = 1$.

Theorem 10 is correct if we substitute $D/2$ for D. Since "aw^2" must be odd and the proof is essentially carried out over the q-adic numbers, there are no difficulties.

To obtain the analog of (30) for (37), replace "$4D$" by $2D$ in (30) and the definition of c_t. Congruence arguments like those used by Stolt limit the number of solutions of (37) that are also integer solutions to (36). For example: If $n \equiv 3$ modulo 4 is a prime not dividing the class number of $\mathbb{Q}(\sqrt{-aD})$, then $ax^2 + D = 2z^n$ has at most one solution with $(z, D) = 1$.

Ljunggren [19] has used another method of studying (36) which is similar to Theorem 10. Briefly, it is as follows. Suppose that $au^2 + D = 2p^m$; and that there is another solution to (36) at the exponent $n = rm$. Set $\gamma = w + uf\omega$, $t = (r-1)/2$, and $\epsilon = y(rm)/y(m) = \pm 1$. Since $y(rm)$ is the coefficient of ω in of $\beta\gamma^t$, we get

$$\pm 1 = \epsilon = \sum_{j=0}^{t/2} \binom{t}{2j} (w)^{t-2j} (au^2)^j + \binom{t}{2j+1} (w)^{t-2j-1} (au^2)^{j+1} (-D)^j.$$

Ljunggren substitutes $2w + D$ for au^2, expands, rearranges, transposes $(-D)^t$ to the right-hand side, and studies the 2-adic value of the two sides of the resulting equation. He also uses 3-adic and 5-adic analyses of this equation to obtain additional results when special assumptions are made about D.

14. $ax^2 + d = 4p^n$ *with* $ad \equiv 3$ modulo 4

We now consider the equation

$$ax^2 + d = 4p^n \qquad (D = d/4 = bf^2) \tag{38}$$

where $ad \equiv 3$ modulo 4 and $(p, d) = 1$. (Note that we use D for the quarter-integer $d/4$.)

Since $ad \equiv 3$ modulo 4, (1) has half-integral solutions for some values of p. The solutions of (38) correspond to those half-integral solutions of (1) with $y = \pm f/2$. Note that "x" in (38) is twice the half-integral "x" in (17).

As noted in the previous section, p is odd if and only if $ad \equiv 3$ modulo 8. In this case $L \neq l$ and, by Lemma 1, n/L is not divisible by 3.

If $ad \equiv 7$ modulo 8, then p must be even and the prime 2 splits. The theory developed previously covers (38) because all the primes dividing p must split; however, Lemma 1 is no longer valid: no power of a half-integral solution of (1) yields an integral solution.

In either case, the theory of half-integral solutions is at our disposal. Since x must be odd, Theorem 10 now has only two exceptional cases: $aw^2 = \frac{1}{4}$ or $\frac{3}{4}$. Section 8 is not relevant since d is odd. The ideas in Section 10 apply. In particular, Persson [29] and Stolt [31] used (30) with "$4D$" $= d$ to limit the number of solutions to (38) with fixed d and n.

When p is fixed in (38), study has been devoted primarily to the exceptional case $aw^2 = \frac{1}{4}$ of Theorem 10. For a survey, see Hasse's treatment [12] of the generalized Ramanujan–Nagell equation:

$$x^2 + d = 2^{n+2}.$$

As a supplement to the bibliography in [12] we mention the paper [22] where applications of the classical case $d = 7$ are mentioned.

We now discuss the two exceptional cases of Theorem 10. Suppose $a = 1$ or 3 and $d = 4p^m - a$. We will study the family of solutions of (38) containing $x = 1, n = m$. (By Theorem 10, all other families contain at most two solutions.) Suppose a solution occurs at the exponent $n = rm$. By Theorem 9, r is odd unless $a = 1$ and $r = 2$; hence we may suppose r is odd and $r \neq 1$.

Since $\beta = (a + f\omega)/2$, $\gamma = \beta - p^m$ and it is easy to prove by induction that for $t \geqslant 0$

$$\gamma^{t+1} \equiv a^t\beta - (a^t + 2ta^{t-1}\beta)p^m \text{ modulo } p^{2m}.$$

By looking at $\beta\gamma^{(r-1)/2}$ and remembering that $y(m)$ and $y(rm)$ are half-integers, we find that

$$(2y(rm)) \equiv a^{(r-1)/2}(2y(m))(1 - p^m(r-2)/a) \text{ modulo } p^{2m}.$$

Thus

$$\epsilon(r) \equiv a^{(r-1)/2}(1 - p^m(r-2)/a) \text{ modulo } p^{2m} \tag{39}$$

when $r > 2$.

Now suppose $a = 1$. By taking (39) modulo p^m we see that when $p^m \neq 2$, $\epsilon(r) = 1$. Hence $r - 2$ is a multiple of p^m. This is a contradiction if p is even (i.e., if $ad \equiv 7$ modulo 8). The exceptional case $p^m = 2$ yields the Ramanujan–Nagell equation: $x^2 + 7 = 2^{n+2}$. Since $p = 2$ is a prime, there is only one family. We now show that this family contains exactly five solutions. Corollary 12.1 yields $\epsilon(r)/\epsilon(r') \equiv -1$ modulo 7 whenever $r \equiv r'$ modulo 6 while (39) yields $\epsilon(r) = -1$ whenever $r > 2$. Hence there is at most one r in each of the congruence classes 3 and 5 modulo 6. By the proof of Theorem 12, there are at most two in the class 1 modulo 6. Since there are solutions at $n = 1, 2, 3, 5$, and 13, there are no others.

Now suppose $a = 3$. The results are less satisfactory than the case $a = 1$. As an example of what can be done, we show that the only solutions to the equation

$$3x^2 + 5 = 2^{n+2}$$

occur at the exponents $n = 1, 3$, and 7. By Theorem 12 (with "s" = 2), we need only show there is no solution at an exponent $n = r > 1$ with $r \equiv 1$ modulo 4. In this case $p^m = 2$ and so, by (39), $\epsilon(r) = -1$. Since $(-1|r) = 1$, (9b) leads to $\epsilon(r) = +1$, a contradiction. Hence there are exactly three solutions.

REFERENCES

1. R. ALTER AND K. K. KUBOTA, The diophantine equation $x^2 + D = p^n$, *Pacific J. Math.* **46** (1973), 11–16.
2. R. APÉRY, Sur une équation diophantienne, *C. R. Acad. Sci. Paris* **251** (1960), 1451–1452; MR **22** No. 10950.
3. A. BAKER, Linear forms in the logarithms of algebraic numbers (IV), *Mathematika* **15** (1968), 204–216; MR **41** No. 3402.
4. A. BAKER, Bounds for the solutions of the hyperelliptic equation, *Proc. Cambridge Philos Soc.* **65** (1969), 439–444; MR **38** No. 3226.
5. J. BLASS, On the diophantine equation $Y^2 + K = X^5$, *Bull. Amer. Math. Soc.* **80** (1974), 329.
6. Z. I. BOREVICH AND I. R. SHAFAREVICH "Number Theory, Pure and Applied Math" (N. Greenleaf, translator), Vol. 20, Academic Press, New York, 1966.
7. J. BRILLHARDT, J. TONASCIA, AND P. WEINBERGER, On the fermat quotient, *in* "Computers in Number Theory" (A. O. L. Atkin and B. J. Birch, eds), pp. 213–222, Academic Press, New York, 1971.
8. F. B. COGHLAN AND N. M. STEPHENS, The diophantine equation $x^3 - y^2 = k$, *in* "Computers in Number Theory" (A. O. L. Atkin and B. J. Birch, eds.), pp. 199–205, Academic Press, New York, 1971.
9. E. L. COHEN, Sur certaines équations diophantiennes quadratiques, *C. R. Acad. Sci. Paris Sér. A–B*, **274** (1972), A139–A140; MR **45** No. 169.
10. J. H. E. COHN, Squares in some recurrent sequences, *Pacific J. Math.* **41** (1972), 631–646.
11. L. J. GOLDSTEIN, "Analytic Number Theory," Prentice-Hall, Englewood Cliffs, New Jersey, 1971.
12. H. HASSE, Über eine diophantische Gleichung von Ramanujan–Nagell und ihre Verallgemeinerung, *Nagoya Math. J.* **27** (1966), 77–102; MR **34** No. 136.
13. O. HEMER, Notes on the diophantine equation $y^2 - k = x^3$, *Ark. Mat.* **3** (1954), 67–77; MR **15** No. 776.
14. N. P. HERZBERG, Integer solutions of $by^2 + p^n = x^3$, *J. Number Theory*, to appear.
15. D. J. LEWIS, Two classes of diophantine equations, *Pacific J. Math.* **11** (1961), 1063–1076; MR **25** No. 3005.
16. W. LJUNGGREN, On the diophantine equation $x^2 + p^2 = y^n$, *Norske Vid. Selsk. Forh.* (*Trondheim*) **16**, No. 8 (1943), 27–30; MR **8** No. 442.
17. W. LJUNGGREN, On a diophantine equation, *Norske Vid. Selsk. Forh.* (*Trondheim*) **18**, No. 32 (1945), 125–128; MR **8** No. 136.
18. W. LJUNGGREN, On the diophantine equation $Cx^2 + D = y^n$, *Pacific J. Math.* **14** (1964), 585–596; MR **28** No. 5035.
19. W. LJUNGGREN, On the diophantine equation $Cx^2 + D = 2y^n$, *Math. Scand.* **18** (1966), 69–86; MR **34** No. 4200.
20. W. LJUNGGREN, Über die Gleichungen $1 + Dx^2 = 2y^n$ und $1 + Dx^2 = 4y^n$, *Norske Vid. Selsk. Forh.* (*Trondheim*) **15**, No. 30 (1952), 115–118; MR **8** No. 422.
21. W. LJUNGGREN, New theorems concerning the diophantine equation $Cx^2 + D = y^n$, *Norske Vid. Selsk. Forh.* (*Trondheim*) **29**, No. 1 (1956), 1–4; MR **17** No. 1185.
22. D. G. MEAD, The equation of Ramanujan–Nagell and $[y^2]$, *Proc. Amer. Math. Soc.* **41**, No. 2 (1973), 333–342.
23. L. J. MORDELL, Diophantine equations, *Pure and Applied Math.* **30** (1969).
24. T. NAGELL, Sur l'impossibilité de quelques équations à deux indéterminées, *Norske Mat. Forenings Skr. Ser. 1*, No. 13 (1923), 65–82.
25. T. NAGELL, Verallgemeinerung eines Fermatschen Satzes, *Arch. Math.* (*Basel*) **5** (1954), 153–159; MR **15** No. 855.

26. T. Nagell, On the diophantine equation $x^2 + 8D = y^n$, *Ark. Mat.* **3** (1955), 103–112; MR **16** No. 903.
27. T. Nagell, Contributions to the theory of a category of diophantine equations of the second degree with two unknows, *Nova Acta Soc. Sci. Upsal.* **16**, No. 2 (1955); MR **17** No. 13.
28. T. Nagell, Über die Klassenzahl imaginärquadratischer Zahlkörper, *Abh. Math. Sem. Univ. Hamburg* **1** (1922), 140–150.
29. B. Persson, On a diophantine equation in two unknowns, *Ark. Mat.* **1** (1949), 45–57; MR **11** No. 328.
30. Th. Skolem, S. Chowla, and D. J. Lewis, The diophantine equation $3^{n+2} - 7 = x^2$ and related problems, *Proc. Amer. Math. Soc.* **10** (1959), 663–669; MR **22** No. 25.
31. B. Stolt, Die Anzahl von Lösungen gewisser diophantischer Gleichungen, *Arc. Math.* **8** (1957), 393–400; MR **21** No. 1952
32. B. Stolt, Über einen verallgemeinerten Fermatschen Satz, *Acta Arith.* **5**(1959), 267–276; MR **22** No. 1541.
33. H. Wada, A table of ideal class groups of imaginary quadratic fields, *Proc. Japan Acad.* **46** (1970), 401–403.

AMS(MOS) 1970 Subject Classifications: 1035, 10B25, 10B05.

STUDIES IN ALGEBRA AND NUMBER THEORY
ADVANCES IN MATHEMATICS SUPPLEMENTARY STUDIES, VOL. 6

The Left Regular Representation of a *p*-Adic Algebraic Group Is Type I

ELLIOT C. GOOTMAN†

Department of Mathematics, University of Georgia, Athens, Georgia

AND

ROBERT R. KALLMAN‡

Department of Mathematics, University of Florida, Gainesville, Florida

DEDICATED TO THE MEMORY OF NORMAN LEVINSON

We present further evidence that Type I locally compact groups with a countable basis for their topologies are stable under perturbations. We prove a general theorem which has as corollaries that CCR groups are stable under perturbations and that the left regular representation of a p-adic algebraic group is Type I. One may furthermore use this general theorem to give a new proof that the left regular representation of a connected real semisimple Lie group is Type I.

1. INTRODUCTION

In this paper we make the standard conventions that all topological groups G are locally compact and have a countable basis for their topologies, all multiplier representations of G are unitary and Borel, and all Hilbert spaces are separable.

The main result of this paper is the following theorem. It is motivated by, and verifies in a very important special case, a slight generalization of Conjecture I of Kallman [13].

THEOREM 1.1. *Let G be a locally compact group and let H be a closed subgroup of G such that G/H is compact. Let α be a multiplier for G, and*

† Supported in part by NSF Grant MCS 77-02134.

‡ Supported in part by NSF Grant GP-38023. Manuscript received by the editors in January 1974.

ISBN 0-12-599153-3

suppose that every α-representation of H is Type I. Then every unitary α-representation of G which is weakly contained in $[U^\pi \mid \pi$ is an α-representation of $H]$ is Type I, where U^π is the α-representation of G induced by π.

Theorem 1.1 is proved in Section 2. In Section 3 we show that the dual space of a certain C^*-algebra is homeomorphic to $\hat{H}^\alpha$, the α-dual space of H. We show that CCR groups are stable under perturbations in Section 4. Finally, in Section 5 we show that the left regular representation of a p-adic algebraic group is Type I.

See the books by Dixmier [5, 6], Auslander–Moore [1], and Mackey [15] for the basic results and notation in operator theory and group representations which we shall use.

2. Proof of Theorem 1.1

We prove Theorem 1.1 in this section. We first recall a few basic facts about multiplier representations of groups, their associated C^*-algebras, and the C^*-algebras associated with transformation groups.

Let G be a group. A multiplier for G is a Borel mapping α of $G \times G$ into the complex numbers of modulus one such that: (1) $\alpha(e, a) = \alpha(a, e) = 1$ for all a in G; (2) $\alpha(ab, c)\alpha(a, b) = \alpha(a, bc)\alpha(b, c)$ for all a, b, c in G. A Borel mapping π of G into the unitary operators on a separable Hilbert space is called an α-representation of G in case $\pi(ab) = \alpha(a, b)\pi(a)\pi(b)$ for all a, b in G. There exists a locally compact group with a countable basis, (G, α), such that certain ordinary unitary representations of (G, α) and the α-representations of G are in a natural one-to-one correspondence. As an abstract group (G, α) is the set $G \times T$, where T is the circle group. The multiplication $(g, t) \cdot (h, s) = (gh, ts/\alpha(g, h))$ makes $G \times T$ into a standard Borel group. There exists on $G \times T$ a unique locally compact topology with a countable basis which generates this given Borel structure. Note that T is central in (G, α) and we have an exact sequence of topological groups $1 \to T \to (G, \alpha) \to G \to e$. If π is an α-representation of G, then $\pi^0(g, t) = t\pi(g)$ is an ordinary unitary representation of (G, α). One checks that $\pi \to \pi^0$ is a one-to-one correspondence between α-representations of G and ordinary representations of (G, α) whose restriction to T is a multiple of $(e, t) \to t$. One may identify (H, α) with a closed subgroup of (G, α) in a natural manner such that $(G, \alpha)/(H, \alpha) = G/H$ (see Kallman [13, Proposition 3.1)]. Let $C^*((G, \alpha))$ be the C^*-algebra associated to the group (G, α). Let J_1 be the two-sided ideal in $C^*((G, \alpha))$ which is the intersection of all the kernels of representations of (G, α) whose restriction to T is a multiple of the character $(e, t) \to t$. There is a natural one-to-one correspondence between the α-representations of G and representations of the C^*-algebra $C^*((G,\alpha))/J_1$. We identify $\hat{G}^\alpha$, the unitary equivalence classes of α-representations of G, as a topological and Borel space with the dual space of $C^*((G,\alpha))/J_1$. For simplicity of notation, let $C^*(G,\alpha) = C^*((G,\alpha))/J_1$.

If X is any locally compact Hausdorff space, let $\mathscr{K}(X)$ be the space of complex-valued continuous functions on X with compact support. Let Z be a locally compact Hausdorff space with a countable basis for its topology, and let the group G act as a topological transformation group on Z. $\mathscr{K}(G \times Z)$ is a *-algebra with the multiplication $(f \cdot g)(t,z) = \int f(s,z)g(s^{-1}t, s^{-1}z)\,ds$ and the involution $f^*(s,z) = \bar{f}(s^{-1}, s^{-1}z)\,\Delta(s^{-1})$ (see Glimm [11] and Effros and Hahn [7, Sect. 3]). $\mathscr{K}(G \times Z)$ can be completed in a natural manner to be a separable C^*-algebra, denoted $C^*(G,Z)$.

Now suppose that Z is a homogeneous space G/H. Then every representation of $C^*(G,G/H)$ can be associated in a natural one-to-one manner with a system of imprimitivity for G on G/H (see Glimm [11, pp. 892–894] and the discussion in Effros and Hahn [7, pp. 35–39]). Recall very briefly how this natural association is made. If f is an element of $\mathscr{K}(G,G/H)$, h is an element of $\mathscr{K}(G/H)$, and s is an element of G, let $(sf)(t,z) = f(s^{-1}t, s^{-1}z)$ and $(hf)(t,z) = h(z)f(t,z)$. If L is a representation of $C^*(G,G/H)$, there is a uniquely determined unitary representation V of G and a norm-decreasing *-homomorphism M of $\mathscr{K}(G/H)$ into the bounded linear operators on $\mathscr{H}(L)$ such that $L(sf) = V(s)L(f)$ and $L(hf) = M(h)L(f)$ for all s in G, h in $\mathscr{K}(G/H)$, and f in $\mathscr{K}(G,G/H)$. Furthermore, V and M are related by $V(s)M(h)V(s)^{-1} = M(sh)$ for all s in G and h in $\mathscr{K}(G/H)$, where $(sh)(z) = h(s^{-1}z)$. Conversely, if we are given such a V and M, we obtain a *-representation of $\mathscr{K}(G \times G/H)$ by setting $L(f) = \int M(f(s,\cdot))V(s)\,ds$. Hence, every representation of $C^*(G,G/H)$ arises naturally by inducing a representation of H up to G. Mackey Theory shows that the corresponding representation of $C^*(G,G/H)$ is irreducible (Type I) if and only if the associated representation of H is irreducible (Type I).

Now let G, H, and α be as in Theorem 1.1. Consider $C^*((G,\alpha),(G,\alpha)/(H,\alpha))$. Let J_2 be the two-sided ideal in $C^*((G,\alpha),(G,\alpha)/(H,\alpha))$ which is the intersection of the kernels of representations associated with those representations of (H,α) which restrict to a multiple of the character $(e,t) \to t$ on T. For simplicity of notation, let $C^*(G,H,\alpha) = C^*((G,\alpha),(G,\alpha)/(H,\alpha))/J_2$.

LEMMA 2.1. *Every representation of $C^*(G,H,\alpha)$ is associated with a representation of (H,α) whose restriction to T is the character $(e,t) \to t$.*

Proof. It suffices to show that every irreducible representation of $C^*(G,H,\alpha)$ is associated with an irreducible representation of (H,α) whose restriction to T is the character $(e,t) \to t$. This statement is equivalent to the assertion that the set of elements of the dual space of $C^*((G,\alpha),(G,\alpha)/(H,\alpha))$

which are associated with an irreducible representation of (H, α) whose restriction to T is the character $(e, t) \to t$ is closed in the Fell topology. Denote this latter subset of the dual space of $C^*((G, \alpha), (G, \alpha)/(H, \alpha))$ by $\mathscr{S}$. Let L be an irreducible representation of $C^*((G, \alpha), (G, \alpha)/(H, \alpha))$ such that $\hat{L}$ is in the closure of $\mathscr{S}$. For any x in $H(L)$, $\langle L(\cdot)x, x\rangle$ can be approximated in the weak *-topology by some net ψ_γ of positive linear functionals associated to elements of $\mathscr{S}$. Since an easy computation shows that $\psi_\gamma((e, t)f) = t\psi_\gamma(f)$ for every f in $\mathscr{K}((G, \alpha) \times (G, \alpha)/(H, \alpha))$, we have that $\langle V((e, t))L(f)x, x\rangle = t\langle L(f)x, x\rangle$. Since this holds for every f and x, we have that $V((e, t)) = t$, and that L is an element of $\mathscr{S}$. Q.E.D.

Recall that $(G, \alpha)/(H, \alpha) = G/H$ is compact. For f in $\mathscr{K}((G, \alpha))$, let $\theta(f)$ be that element of $\mathscr{K}((G, \alpha) \times (G, \alpha)/(H, \alpha))$ such that $\theta(f)(s, z) = (f \otimes 1)(s, z) = f(s)$. One easily checks that $f \to \theta(f)$ is a *-homomorphism of the convolution algebra on $\mathscr{K}((G, \alpha))$ into the convolution algebra on $\mathscr{K}((G, \alpha) \times (G, \alpha)/(H, \alpha))$.

LEMMA 2.2. *θ extends to a *-homomorphism, also denoted by θ, of $C^*((G, \alpha))$ into $C^*((G, \alpha), (G, \alpha)/(H, \alpha))$, and $\theta(J_1)$ is contained in J_2.*

Proof. Let L be any representation of $C^*((G, \alpha), (G, \alpha)/(H, \alpha))$, and let π be the associated representation of (H, α). For f in $\mathscr{K}((G, \alpha))$, $L(\theta(f)) = U^\pi(f)$. Hence, $\|\theta(f)\| = \sup_L \|L(\theta(f))\| = \sup_\pi \|U^\pi(f)\| \leqslant \sup[\|\rho(f)\| \mid \rho$ any representation of $(G, \alpha)] = \|f\|$ in $C^*((G, \alpha))$. Thus θ is norm-decreasing and so extends to a *-homomorphism of $C^*((G, \alpha))$ into $C^*((G, \alpha), (G, \alpha)/(H, \alpha))$. Finally, for a in J_1, we have that $\rho(a) = 0$ for all α-representations of G. In particular, $U^\pi(a) = 0$ for all α-representations π of H. Let $\mathscr{S}$ be as in Lemma 2.1. If L is in $\mathscr{S}$, then the associated π is an α-representation of H, and $L(\theta(a)) = U^\pi(a) = 0$. Thus $\theta(a)$ is in J_2. Q.E.D.

Hence, from this lemma one easily sees that there is a natural *-homomorphism of $C^*(G, \alpha)$ into $C^*(G, H, \alpha)$, which we continue to denote by θ.

Proof of Theorem 1.1. If every α-representation of H is Type I, then $C^*(G, H, \alpha)$ is Type I by Lemma 2.1 and the discussion preceding it. Hence, $\theta(C^*(G, \alpha))$ is Type I, for every C^*-subalgebra of a Type I C^*-algebra is Type I (Dixmier [6, Proposition 4.3.5 and Théorème 9.1]). But if a is an element of $C^*(G, \alpha)$, $\theta(a) = 0$ if and only if $U^\pi(a) = 0$ for every π which is an α-representation of H. Let J be the two-sided ideal in $C^*(G, \alpha)$ which is the intersection of the kernels of the U^π. Hence, $C^*(G, \alpha)/J$ is Type I. Thus, if L is any representation of $C^*(G, \alpha)$ such that $\ker L$ contains J, then L is Type I. Hence, if L is any α-representation of G which is weakly contained in $[U^\pi \mid \pi$ is an α-representation of $H]$, then L is Type I. Q.E.D.

COROLLARY 2.3. *Let G, H, and α be as in Theorem* 1.1. *Then the left regular α-representation of G is Type I.*

Proof. The left regular α-representation of H induces the left regular α-representation of G. Now use Theorem 1.1. Q.E.D.

3. THE TOPOLOGY OF THE DUAL SPACE OF $C^*(G, H, \alpha)$

The purpose of this section is to show that the dual space of $C^*(G, H, \alpha)$ is homeomorphic to the dual space of $C^*(H, \alpha)$. We do this in fairly easy fashion by extending slightly some results of Glimm [11]. Recall that Fell [8] has defined a topology on the closed subgroups of a fixed locally compact topological group which has a countable basis for its topology.

LEMMA 3.1. *Let G be a locally compact group and H a closed subgroup of G. For each m in G/H, let G_m be the G-stability group at m. Then the mapping $m \to G_m$ is continuous from G/H into the closed subgroups of G with the Fell topology.*

Proof. Let $m_n = g_n \cdot H$, $m = g \cdot H$, and $m_n \to m$ in G/H. We may choose the g_n's so that $g_n \to g$, for the natural quotient mapping of $G \to G/H$ is open. Note that if $x_n = g_n g^{-1}$, then $x_n \to e$ and $G_{m_n} = x_n G_m x_n^{-1}$. Choose a compact set K in G which has empty intersection with G_m, and choose open sets $U_1, \ldots, U_p$ which have nonempty intersection with G_m. Then for large n, K has empty intersection with G_{m_n}. If not, one could choose a sequence of integers n_q such that K had an element, say k_{n_q}, in common with $G_{m_{n_q}}$. But since K is compact, we may assume that k_{n_q} converges to some k in K, and then $x_{n_q}^{-1} k_{n_q} x_{n_q}$, elements of G_m, would also converge to k. Hence, K would have nonempty intersection with G_m, a contradiction. Next, choose an element g_i which is in both G_m and U_i. Then $x_n g_i x_n^{-1}$ is in G_{m_n} and for large n is still an element of U_i, by the continuity of multiplication. Hence, for large n, G_{m_n} is in the neighborhood of G_m defined by $K, U_1, U_2, \ldots, U_p$. Q.E.D.

Lemma 3.1 has also been proved by Schochetman and Wu [16], but has been included for completeness.

One concludes from Lemma 3.1 that for every m there is a choice of Haar measure $d(m, \cdot)$ on G_m such that $m \to \int_{G_m} f(g)\, d(m, g)$ is continuous for all f in $\mathscr{K}(G)$. Let Y be the subset of $G/H \times G$ which is the union of the subsets $(m) \times G_m$. Y is a closed subset of $G/H \times G$.

Glimm [11] has pointed out how to make $\mathscr{K}(Y)$ into a *-algebra of interest in representation theory in a natural manner. For each m in G/H, let Δ_m be the modular function for G_m, so that $d(m, gg') = d(m, g)\,\Delta_m(g')$. The mapping $(m, g) \to \Delta_m(g)$ is continuous on Y. If f and g are elements of $\mathscr{K}(Y)$, set $(f \cdot g)(m, t) = \int_{G_m} f(m, s)g(m, s^{-1}t)\,d(m, s)$ and $f^*(m, t) = \bar{f}(m, s^{-1})\,\Delta_m(s^{-1})$. This *-algebra structure on $\mathscr{K}(Y)$ may be completed, after a few preliminaries, in a standard fashion to be a C^*-algebra, $C^*(Y)$. Every irreducible representation of $C^*(Y)$ has the form $\pi(f(m, \cdot))$ for some m in G/H and π an irreducible unitary representation of G_m.

Lemma 3.2. *The natural mapping $\hat{G}_m \to \widehat{C^*(Y)}$ is a homeomorphism onto its range. The range of this mapping is closed.*

Proof. The range of this natural mapping is closed, for if not, let $\hat{\pi}_\gamma$ be a net in $\hat{G}_m$ which converges in $\widehat{C^*(Y)}$ to an element $\hat{\pi}$ in $\hat{G}_{m'}$, $m \neq m'$. Choose an element f of $\mathscr{K}(Y)$ such that $f(m, \cdot) = 0$ and $\|\hat{\pi}(f)\| > 0$. Then $[\hat{\rho}\,|\,\|\hat{\rho}(f)\| > \|\hat{\pi}(f)\|/2]$ is an open neighborhood of $\hat{\pi}$ in $\widehat{C^*(Y)}$, and no $\hat{\pi}_\gamma$ is in this neighborhood, for $\|\hat{\pi}_\gamma(f)\| = 0$ for all γ. This is a contradiction. Hence, the range of this mapping is closed.

This mapping is continuous, for a basic open set in $\widehat{C^*(Y)}$ is of the form $[\hat{\rho}\,|\,\hat{\rho}$ in $\widehat{C^*(Y)}, \|\hat{\rho}(f)\| > 1]$ for some f in $\mathscr{K}(Y)$. But the intersection of such a set with $\hat{G}_m$ is of the form $[\hat{\pi}\,|\,\hat{\pi}$ in $\hat{G}_m$, $\|\hat{\pi}(f(m, \cdot))\| > 1]$, an open set, for $f(m, \cdot)$ is an element of $\mathscr{K}(G_m)$.

This mapping is open onto its range, for a basic open set in G_m is of the form $[\hat{\pi}\,|\,\hat{\pi}$ is in $\hat{G}_m, \|\hat{\pi}(f)\| > 1]$ for some f in $\mathscr{K}(G_m)$. We may view f as a continuous function on $(m) \times G_m$, a closed subset of Y. Since Y is locally compact and has a countable basis for its topology, there is a continuous function with compact support on Y, say g, whose restriction to $(m) \times G_m$ is f. But then the intersection of $[\hat{\rho}\,|\,\hat{\rho}$ is in $\widehat{C^*(Y)}$, $\|\hat{\rho}(g)\| > 1]$ with $\hat{G}_m$ is precisely $[\hat{\pi}\,|\,\hat{\pi}$ is in $\hat{G}_m$, $\|\hat{\pi}(f)\| > 1]$. Hence, this natural mapping is open onto its range. Q.E.D.

Next, there is a natural G-action on $\widehat{C^*(Y)}$. Let π be an irreducible unitary representation of G_m. If g is in G and f is an element of $\mathscr{K}(Y)$, let $(g \cdot \pi)(f) = \int_{G_{g \cdot m}} f(g \cdot m, s)\pi(g^{-1}sg)\,d(g \cdot m, s)$. Glimm [11] has shown that this G-action makes G into a topological transformation group on $\widehat{C^*(Y)}$. Furthermore, suppose that G/H is compact. If $\hat{\pi}$ is an element of $\hat{G}_m$, viewed as a subset of $\widehat{C^*(Y)}$, then $G/G_{\hat{\pi}}$ is compact, for $G_{\hat{\pi}}$ contains G_m, a conjugate of H.

Lemma 3.3. *$\widehat{C^*(Y)}/G$ is homeomorphic to $\hat{G}_m$, for any m in G/H.*

Proof. Fix m in G/H. One easily checks that $\hat{G}_m$ is a cross section for the G-orbits in $\widehat{C^*(Y)}$. As any quotient mapping is continuous, it is clear that the injection of $\hat{G}_m$ into $\widehat{C^*(Y)}/G$ is continuous. To prove the lemma it suffices to show that this natural injection is a closed mapping. Hence, if C is a closed subset of $\hat{G}_m$, we must show that $G \cdot C$ is a closed subset of $\widehat{C^*(Y)}$. But by the remarks preceding this lemma, there is a compact set K in G such that $G \cdot C = K \cdot C$. Suppose that $k_\gamma \cdot \hat{\pi}_\gamma \to \hat{\pi}$ in $\widehat{C^*(Y)}$. Since K is compact in G, we may suppose that $k_\gamma \to k$. But then $\hat{\pi}_\gamma \to k^{-1} \cdot \hat{\pi}$. Since C is closed in $\hat{G}_m$, and since $\hat{G}_m$ is closed in $\widehat{C^*(Y)}$, we have that $k^{-1} \cdot \hat{\pi}$ is an element of C. Hence, $\hat{\pi}$ is an element of $G \cdot C$, and $G \cdot C$ is closed in $\widehat{C^*(Y)}$.
Q.E.D.

PROPOSITION 3.4. *Let G be a group, H a closed subgroup with G/H compact, and α a multiplier for G. Then the dual space of $C^*(G, H, \alpha)$ is homeomorphic to the dual space of $C^*(H, \alpha)$.*

Proof. Recall that $(G, \alpha)/(H, \alpha) = G/H$ is compact. By Lemma 3.3 and Theorem 2.1 of Glimm [11], the dual space of $C^*((G, \alpha), (G, \alpha)/(H, \alpha))$ is homeomorphic to $\widehat{(H, \alpha)}$. Let J_2 be as in Section 2. By examining the explicit form of this isomorphism, one easily checks that an element of the dual space of $C^*((G, \alpha), (G, \alpha)/(H, \alpha))$ vanishes on J_2 if and only if the corresponding element of $\widehat{(H, \alpha)}$ restricts to the character $(e, t) \to t$ on T. Therefore, the dual space of $C^*(G, H, \alpha) =$ the dual space of $C^*((G, \alpha), (G, \alpha)/(H, \alpha))/J_2$ is homeomorphic to the subset of $\widehat{(H, \alpha)}$ consisting of those elements which restrict to the character $(e, t) \to t$ on T. But this latter set is homeomorphic to the dual space of $C^*(H, \alpha)$. Q.E.D.

4. CCR GROUPS ARE STABLE UNDER PERTURBATIONS

LEMMA 4.1. *Let G be a group and H a closed subgroup such that G/H has a finite invariant measure. Let α be a multiplier for G. Then every α-representation of G is weakly contained in some U^π, where π is an α-representation of H.*

Proof. The proof of this lemma is almost identical with that of Proposition 5.1 of Fell [10]. Note that there is a finite (G, α)-invariant measure on $(G, \alpha)/(H, \alpha)$. Let I be the one-dimensional identity representation of (H, α). Note that the one-dimensional identity representation of (G, α) is a direct summand of U^I, the regular representation of (G, α) on $(G, \alpha)/(H, \alpha)$. Let π

be an irreducible representation of (G, α) whose restriction to T is the character $(e, t) \to t$. Then $\pi|(H, \alpha)$ is a representation of (H, α) whose restriction to T is the character $(e, t) \to t$. But by Fell [9, Corollary 1, p. 260], $U^{\pi|(H,\alpha)}$ is unitarily equivalent to $U^I \otimes \pi$. But the identity representation of (G, α) a direct summand of U^I implies that $I \otimes \pi = \pi$ is weakly contained in $U^{\pi|(H,\alpha)}$. Q.E.D.

PROPOSITION 4.2. *Let G be a group, α a multiplier for G, and H a subgroup such that G/H is compact and has a finite invariant measure. Then $C^*(G, \alpha)$ is CCR if $C^*(H, \alpha)$ is CCR.*

Proof. A C^*-algebra is CCR if and only if its dual space is T_1. Hence, if $C^*(H, \alpha)$ is CCR, then $C^*(G, H, \alpha)$ is CCR by Proposition 3.4. But Lemma 4.1 implies that the natural mapping of $C^*(G, \alpha)$ into $C^*(G, H, \alpha)$ is one-to-one, by the proof of Theorem 1.1. Hence, $C^*(G, \alpha)$ is CCR, for any C^*-subalgebra of a CCR C^*-algebra is again a CCR C^*-algebra (Dixmier [6, Proposition 4.2.4, p. 86]). Q.E.D.

We remark that this proposition should still hold if one assumes only that there is a finite G-invariant measure on G/H.

The following conjecture seems quite reasonable.

CONJECTURE 4.3. *Any CCR group is unimodular.*

J. Dixmier has informed us by letter that he has verified this conjecture for connected groups, and that C. C. Moore has independently verified this conjecture for simply connected Lie groups.

COROLLARY 4.4. *Let G be a group and H a closed subgroup such that G/H is compact. Assume that H is CCR and Conjecture* 4.3 *holds for H. Then G is CCR.*

Proof. If Conjecture 4.3 holds for H, then H is unimodular. Since G/H is compact, Proposition 8 of Kallman [14] implies that G is unimodular. Hence, there is an invariant measure on G/H. Now apply Proposition 4.2. Q.E.D.

The following corollary is a rather weak version of Theorem 1 of Kallman [12] and of Proposition 2.2 of Kallman [13].

COROLLARY 4.5. *Let G be a group, α a multiplier for G, and H a subgroup of G such that G/H is compact and has a G-invariant measure. Then $C^*(G, \alpha)$ is Type* I *if $C^*(H, \alpha)$ is Type* I.

Proof. This corollary follows immediately from Lemma 4.1 and Theorem 1.1. Q.E.D.

5. The Left Regular Representation of a p-Adic Algebraic Group Is Type I

In this section k will be a field which is a finite algebraic extension of the p-adic numbers, Q_p. It is well known that such fields together with the reals and complexes are the only locally compact fields of characteristic zero. We will only consider linear algebraic groups which are defined over k, and when speaking of such a group we will always mean its k points. We start off this section by showing that solvable algebraic groups are always Type I. The ideas for the proof of this are essentially in Dixmier [4], where the analogous theorem for real algebraic groups is proved. The p-adic case differs in a few key respects, and we present most of the details for completeness. See Borel [2] for most of the basic results on algebraic groups that we use. If G is an algebraic group, we denote its Lie algebra by $L(G)$.

The Heisenberg group H is the locally compact group structure on k^{2n+1} such that $(x_0, x_1, \ldots, x_{2n}) \cdot (y_0, y_1, \ldots, y_{2n}) = (x_0 + y_0 - x_1 y_{n+1} - x_2 y_{n+2} - \cdots - x_n y_{2n}, x_1 + y_1, \ldots, x_{2n} + y_{2n})$. This group is nilpotent and its center is $[H, H]$ and coincides with all elements of the form $(x, 0, 0, \ldots, 0)$. H is topologically isomorphic to a unique algebraic group, which we also denote by H, whose Lie algebra, $L(H)$, has a basis $e_0, e_1, \ldots, e_{2n}$ which satisfies $[e_1, e_{n+1}] = e_0$, $[e_2, e_{n+2}] = e_0, \ldots, [e_n, e_{2n}] = e_0$ and all other undefined commutators zero.

Lemma 5.1. *Suppose $G = T \cdot H$, where H is the Heisenberg group and T is a k-split torus. If the center of H is central in G, then there is an abelian subgroup A in H,* $\dim A = 2$, *such that A is normal in G.*

Proof. T has a natural algebraic representation on $L(H)$. Since T is k-split, this action is diagonalizable. T leaves e_0 fixed, for the center of H is central in G. Choose any other nonzero element v in $L(H)$ such that T leaves the line generated by v invariant. Let A be the algebraic subgroup of H whose Lie algebra is generated by e_0 and v. Dim $A = 2$, and A is normal in H, for the center of H is contained in A. A is normal in G, for $L(A)$ is invariant under T. Q.E.D.

Lemma 5.2. *Let G be an algebraic group, A a normal connected abelian subgroup of the unipotent radical of G of dimension greater than one. Let χ be a character of A which is G-invariant—i.e., $\chi(gag^{-1}) = \chi(a)$ for all a in G.*

Then there is a normal connected algebraic subgroup A' of A, $\dim A' \geqslant 1$, *such that A' is contained in the kernel of χ.*

Proof. We may regard A as a finite-dimensional vector space in a natural manner such that $g \to g \cdot g^{-1}$ is an algebraic representation of G into the invertible matrices on A. There is a character χ_0 on k and f in the linear dual space of A such that $\chi(a) = \chi_0(f(a))$. Since A has dimension greater than one, $\ker f$ has dimension greater than or equal to one. Let A' be the smallest subspace of A which is G-invariant and contains $\ker f$. Then A' is a normal connected algebraic subgroup of A which is contained in $\ker \chi$. Q.E.D.

PROPOSITION 5.3. *Let G be a solvable algebraic group defined over k. Then G is Type* I.

Proof. By Proposition 9.3 of Borel and Tits [3], G has a trigonalizable algebraic subgroup H, defined over k, such that G/H is compact. It suffices to show that H is Type I by Corollary 2.3 of Kallman [13] plus Kakutani's fixed point theorem. Hence, it suffices to prove this proposition for trigonalizable groups which are defined over k. Any such group is split solvable since k has characteristic zero. We prove this last result by induction on the dimension of G, following Dixmier [4] very closely.

If the dimension of G is one, then G is abelian and therefore Type I. Let U be the unipotent radical of G. If A is an abelian algebraic subgroup of U which is normal in G, then $\hat{A}/G$ is countably separated, for the natural action of G on the linear dual space of A is algebraic and therefore countably separated, and $\hat{A}$ may be naturally identified with the linear dual of A.

Let π be a factor representation of G. For any element χ of $\hat{A}$, let G_χ be the G-stability group at χ. G_χ is an algebraic subgroup of G. The dimension of G_χ is less than the dimension of G if and only if the G-orbit which contains χ is nontrivial. Since $\hat{A}/G$ is countably separated, π restricts to a G-orbit. But if this G-orbit is nontrivial, the stability group at any point in this orbit will be an algebraic group of dimension less than the dimension of G. Hence, this stability group will be Type I by our induction assumption, and π will be Type I by Mackey theory. Otherwise for every abelian algebraic subgroup A of U which is normal in G, we have that π restricts to a character of A. If the dimension of this A is greater than one, then Lemma 5.2 implies that there exists an abelian algebraic subgroup A' of A which is normal in G and which is contained in the kernel of π. But G/A' is an algebraic group defined over k with dimension less than the dimension of G, and π in actuality is a representation of G/A'. Hence, π is Type I by our induction assumption. Finally, assume that G has no normal abelian algebraic subgroup of dimension greater than one contained in U. Then either U is the identity, in which case G is abelian and therefore Type I, or else U is a Heisenberg group (Dixmier [4, Lemme 10, p. 325]). But this latter case cannot occur by Lemma 5.1. Q.E.D.

THEOREM 5.4. *Let G be an algebraic group defined over k. Then the left regular representation of G is Type I.*

Proof. By Proposition 9.3 of Borel and Tits [3], G has a solvable algebraic subgroup H, defined over k, such that G/H is compact. H is Type I by Proposition 5.3. The left regular representation of G is Type I by Corollary 2.3. Q.E.D.

The following corollary is presented merely to illustrate some of the techniques of this paper. Of course it is known that connected semisimple Lie groups are Type I.

COROLLARY 5.5. *Let G be a connected real semisimple Lie group. Then the left regular representation of G is Type I.*

Proof. Let $G = KAN$ be the Iwasawa decomposition of G. AN is known to be a real algebraic group. Hence, $S = AN$ is Type I by Dixmier [4]. Let Z be the center of G. Z is a subgroup of K, and $Z \cdot S$ is a closed subgroup of G. $Z \cdot S$ is Type I, for S and Z are both Type I, and they commute. $G/Z \cdot S$ is compact. Hence, the left regular representation of G is Type I by Corollary 2.3 Q.E.D.

Note Added in Proof. This addendum is necessary because of the extended period of time from the date of submission of this paper (January 1974) to the present. There have been many overlapping results in other papers and several of the questions posed in this paper have been answered in whole or in part.

I. N. Bernshtein (All reductive p-adic groups are tame, *Functional Anal. Appl.* **8** (1974), 91–93) has shown that all reductive p-adic algebraic groups are in fact CCR. The result for which the present paper is named is much weaker than Bernshtein's theorem. However, the main result of this paper (Theorem 1.1) is of very general applicability and does not depend for its proof on the detailed structure of the group G, as does the proof of Bernshtein. M. A. Rieffel (Induced representations of C^*-algebras, *Advances in Math.* **13** (1974), 176–257) has proved theorems which generalize Proposition 3.4 to the case in which G/H is not necessarily compact. C. C. Moore and J. Rosenberg (Groups with T_1 primitive ideal spaces, *J. Functional Analysis* **22** (1976), 204–224) have proved Conjecture 4.3 for almost connected groups. Several people apparently thought of Conjecture 4.3 more or less simultaneously in 1971 or 1972, among others one of the authors (R. R. Kallman), C. C. Moore, and R. Lipsman. One of the authors (E. C. Gootman, Induced representations and finite volume homogeneous spaces, II, preprint) has generalized Proposition 4.2 to the case in which G/H has finite volume and α is trivial.

REFERENCES

1. L. AUSLANDER AND C. MOORE, Unitary representations of solvable Lie groups, *Mem. Amer. Math. Soc.* **62** (1966).
2. A. BOREL, "Linear Algebraic Groups," Benjamin, New York, 1969.
3. A. BOREL AND J. TITS, Groupes reductifs, Inst. Hautes Etudes Scientifiques, Publi. Math. No. 27, pp. 55–151.
4. J. DIXMIER, Sur les représentations des groupes des Lie algébriques, *Ann. Inst. Fourier* (*Grenoble*) **7** (1957), 315–328.
5. J. DIXMIER, "Les Algèbres d'opérateurs dans l'espace Hilbertien," Gauthier-Villars, Paris, 1957.
6. J. DIXMIER, "Les C^*-Algèbres et Leurs Representations," Gauthier-Villars, Paris, 1964.
7. E. G. EFFROS AND F. HAHN, Locally compact transformation groups and C^*-algebras, *Mem. Amer. Math. Soc.* **75** (1967).
8. J. M. G. FELL, A Hausdorff topology for the closed subsets of a locally compact non-Hausdorff space, *Proc. Amer. Math. Soc.* **13** (1962), 472–476.
9. J. M. G. FELL, Weak containment and induced representations of groups, *Can. J. Math.* **14** (1962), 237–268.
10. J. M. G. FELL, Weak containment and induced representations of groups, II, *Trans. Amer. Math. Soc.* **110** (1964), 424–447.
11. J. G. GLIMM, Families of induced representations, *Pacific J. Math.* **12** (1962), 885–911.
12. R. R. KALLMAN, Certain topological groups are Type I, *Bull. Amer. Math. Soc.* **76** (1970), 404–406.
13. R. R. KALLMAN, Certain topological groups are Type I. Part II, *Advances in Math.* **10** (1973), 221–255.
14. R. R. KALLMAN, The existence of invariant measures on certain quotient spaces, *Advances in Math.* **11** (1973), 387–391.
15. G. MACKEY, Unitary representations of group extensions I, *Acta Math.* **99** (1958), 265–311.
16. I. SCHOCHETMAN AND Y.-C. WU, Continuity of stability groups and conjugation, preprint.

AMS (MOS) 1970 subject classifications: 22D10, 22025, 22030.

STUDIES IN ALGEBRA AND NUMBER THEORY
ADVANCES IN MATHEMATICS SUPPLEMENTARY STUDIES, VOL. 6

Lattices in Semisimple Groups over Local Fields

GOPAL PRASAD

Tata Institute of Fundamental Research,
Bombay, India

TO MY FATHER ON HIS SIXTIETH BIRTHDAY

Contents

INTRODUCTION

A discrete subgroup Γ of a locally compact topological group G is said to be a *lattice* in G if the homogeneous space $\Gamma\backslash G$ carries a finite G-invariant Borel measure. A lattice Γ in G is *uniform* if $\Gamma\backslash G$ is compact, otherwise it is *nonuniform*. A lattice Γ is said to be *irreducible* (see Section 1) if no subgroup of Γ of finite index is a direct product of two infinite normal subgroups.

Let n be a positive integer and for $i \leqslant n$, let $\mathbf{H}_i$ be a connected adjoint semisimple algebraic group, defined, simple and isotropic over a nondiscrete locally compact field k_i of arbitrary characteristic. For $i \leqslant n$, let H_i be the group of k_i-rational points of $\mathbf{H}_i$ with the canonical locally compact topology induced from the topology on k_i. Let $H = \prod_{i=1}^{n} H_i$. In the sequel we call a

This is a revised version of an earlier preprint, with the title "Lattices in p-adic semi-simple groups," circulated in January 1974. The original version was written while I visited Yale University (New Haven) and the Institute for Advanced Study (Princeton). I want to thank these institutions for their generous hospitality and support.

ISBN 0-12-599153-3

group of this form a *non-archimedean semisimple group* if all the k_i are non-archimedean. The object of this paper is to study lattices in non-archimedean semisimple groups.

We shall now indicate the main results of this paper.

Let n and for $i \leqslant n$ let k_i, $\mathbf{H}_i$, and H_i be as before, and let $H = \prod_{i=1}^{n} H_i$. The polar rank of H is, by definition, $\sum_{i=1}^{n} k_i$-rank $\mathbf{H}_i$. Let Λ be a uniform lattice in H. We show, in Section 3, that any abelian subgroup of Λ is finitely generated and the polar rank of H equals the rank of any abelian subgroup of Λ of maximal rank. This result has been used later (in Section 4) to prove that the cohomological dimension of a torsion free lattice in a non-archimedean semisimple group equals the polar rank of the semisimple group. We also introduce a notion of rank of an abstract group (Subsection 3.4) and prove (see Theorem 3.7) that rank(Λ) = polar rank H. When H is real analytic (i.e., all the k_i are archimedean), then it was shown in Prasad–Raghunathan [26, Sect. 3] that for any (not necessarily uniform) lattice Δ, polar rank H = $\mathbf{R}$-rank H = rank(Δ).

In Section 4 we use the results of Section 3, together with the results of Margulis [22] and Raghunathan [28] on arithmeticity of nonuniform lattices in real analytic semisimple groups, to prove that an irreducible lattice Λ in a non-archimedean semisimple group H can be isomorphic to a lattice in a real analytic semisimple group only if Λ has a subgroup of finite index which is free and in that case polar rank $H = 1$. This answers a question of A. Borel. (Also see Remark 4.11.)

A lattice Λ in a non-archimedean semisimple group H is said to be *strongly rigid* if, given a lattice Λ' in a non-archimedean semisimple group H', any isomorphism $\theta: \Lambda \to \Lambda'$ extends to an isomorphism $H \to H'$ of topological groups. We prove that if polar rank $H \geqslant 2$, then any irreducible uniform lattice in H is strongly rigid (for precise statements see Theorems 8.6 and 8.7). The proof is modeled on Mostow's proof [23] of strong rigidity of uniform lattices in real analytic semisimple groups; our indebtedness to his work [23] would be obvious to the reader.

The main idea of the proof of strong rigidity is to use Bruhat–Tits buildings (Subsection 5.1) to show that an isomorphism $\theta: \Lambda \to \Lambda'$ induces a θ-equivariant isomorphism θ_* from the Tits building (Subsection 7.1) of H to the Tits building of H'. In case all the $\mathbf{H}_i$ are of k_i-rank $\geqslant 2$, this in view of the fundamental theorem of Tits geometry (Proposition 8.2; see Tits [35, Sect. 5] for the statement in complete generality and its proof) proves at once that θ extends to an isomorphism $H \to H'$ of topological groups. In case, for some i, k_i-rank $\mathbf{H}_i = 1$, then the fundamental theorem of Tits geometry does not suffice, and to prove strong rigidity we show, following an idea of Mostow [23], that the restriction of θ_* to the space of minimal parabolic subgroups of H is a homeomorphism onto the space of minimal parabolic subgroups of H'. This, along with an unpublished result of A. Borel (Proposition 8.5) and a simple consequence (Proposition 8.3) of the main theorem of Borel–Tits [8], is used to prove strong rigidity in case H has a rank 1 factor.

If polar rank $H = 1$, then it is known (Serre [32, Corollary on p. 121]) that any torsion free discrete subgroup of H is free. Hence, in particular, any torsion free lattice in H is (non-abelian and) free; therefore, its group of outer-automorphisms is infinite. This, it can be shown, implies that such a lattice is not strongly rigid. Thus, the restriction on the polar rank in the strong rigidity theorem is necessary.

Later in Section 8, we give an application of the theorems on strong rigidity to the group of units of integral quadratic forms. We also show that strong rigidity implies local (or deformation) rigidity (Subsection 8.9).

Finally, we note that a simple but curious property of Bruhat–Tits buildings (see Corollary 5.7) plays an important role in the proof of strong rigidity.

0. Notation and Conventions

0.1. As usual $\mathbf{Q}$, $\mathbf{R}$, and $\mathbf{C}$ will denote the fields of rational, real, and complex numbers, respectively. $\mathbf{Z}$ will denote the ring of rational integers. For a prime p, we let $\mathbf{Q}_p$ denote the field of p-adic numbers, i.e., the p-adic completion of $\mathbf{Q}$.

0.2. In the sequel, by a local field we mean a nondiscrete locally compact topological field.

0.3. Let $\mathbf{V}$ be a variety defined over a field k. We denote by $\mathbf{V}(k)$ the set of k-rational points of $\mathbf{V}$, and by $k[\mathbf{V}]$ the algebra of regular k-functions on $\mathbf{V}$.

In case k is a local field, there is a natural locally compact Hausdorff topology on $\mathbf{V}(k)$ (see A. Weil, "Foundations of Algebraic Geometry," Appendix III [39]).

0.4. Simple or almost simple algebraic groups are always assumed to be semisimple.

1. Preliminaries

1.1. Let $\mathscr{S}$ be a finite set. For $\alpha \in \mathscr{S}$, let k_α be a local field and let $\mathbf{G}_\alpha$ be a connected semisimple linear algebraic group defined over k_α. Let G_α be the group of k_α-rational points of $\mathbf{G}_\alpha$. Let $G = \prod_{\alpha \in \mathscr{S}} G_\alpha$, and for $\alpha \in \mathscr{S}$ let

π_α denote the natural projection of G on G_α. In the sequel, the topology on G, which is the Cartesian product of the Zariski topologies on G_α for $\alpha \in \mathscr{S}$, will be referred to as the *weak Zariski topology* on G. It is known that the product of finitely many noetherian spaces is noetherian and hence G, in the weak Zariski topology, is noetherian. Also, each G_α carries a natural locally compact Hausdorff topology (induced by the topology on k_α, cf. Subsection 0.3), and G_α with this topology is a topological group. Thus there is a natural locally compact Hausdorff topology on G and in this topology G is a topological group. Unless explicitly stated to the contrary, we shall view G as a topological group endowed with this topology.

The following two lemmas are essentially known. For the sake of expository completeness we include their proof here.

1.2. LEMMA. *Let Γ be a lattice in G. Assume that for every $\alpha \in \mathscr{S}$, k_α is non-archimedean, and either Γ is torsion free or all the k_α are of characteristic zero. Then Γ is uniform.*

Proof. There exists an open compact subgroup U of G such that for all $g \in G$, $g\Gamma g^{-1} \cap U$ is trivial. In fact if Γ is torsion free, then for all $g \in G$ and any open compact subgroup U of G, $g\Gamma g^{-1} \cap U$ is trivial. On the other hand, it is well known (see [31, LG4, Sect. 9, Theorem 5]) that if k is a non-archimedean local field of *characteristic zero*, then there is an open and compact subgroup of $GL(n, k)$ which is torsion free. From this it is evident that, when all the k_α are of characteristic zero, G has a compact-open torsion free subgroup U, and clearly, $g\Gamma g^{-1} \cap U$ is trivial for all $g \in G$.

Now for $g \in G$, the map $\lambda_g : u \mapsto \Gamma g u$ of U into $\Gamma\backslash G$ is injective. Let μ be a Haar measure on G. We denote the G-invariant measure on $\Gamma\backslash G$ by μ_Γ. Since λ_g is injective, $\mu_\Gamma(\Gamma g U) = \mu(gUg^{-1}) = \mu(U)$. Now since $\mu_\Gamma(\Gamma\backslash G) < \infty$, we conclude that $\Gamma\backslash G/U$ is finite and hence $\Gamma\backslash G$ is compact.

1.3. LEMMA. *Let G be as in Subsection* 1.1 *and let Γ be a uniform lattice in G. Let $\alpha \in \mathscr{S}$ be such that k_α is of characteristic zero. Then for all $\gamma \in \Gamma$, $\pi_\alpha(\gamma)$ is semisimple.*

Proof. Let γ be an element of Γ. The orbit $G(\gamma) = \{g\gamma g^{-1} \mid g \in G\}$ of γ is closed, for the orbit $\Gamma(\gamma) = \{\lambda\gamma\lambda^{-1} \mid \lambda \in \Gamma\}$ of γ under Γ is closed and, Γ being uniform, there exists a compact subset C of G such that $G = C \cdot \Gamma$ and then $G(\gamma) = \{c\mu c^{-1} \mid c \in C \text{ and } \mu \in \Gamma(\gamma)\}$. Let $\pi_\alpha(\gamma) = \pi_\alpha(\gamma)_s \cdot \pi_\alpha(\gamma)_u = \pi_\alpha(\gamma)_u \cdot \pi_\alpha(\gamma)_s$ be the Jordan decomposition of $\pi_\alpha(\gamma)$, with $\pi_\alpha(\gamma)_s$ (resp. $\pi_\alpha(\gamma)_u$) semisimple (resp. unipotent). It follows from an elementary application of the Jacobson–Morosov theorem that the closure of the orbit (under G_α) of $\pi_\alpha(\gamma)$ contains $\pi_\alpha(\gamma)_s$. From this it is obvious that the orbit $G(\gamma)$ (which is closed) has a point μ such that $\pi_\alpha(\mu) = \pi_\alpha(\gamma)_s$ and $\pi_\beta(\mu) = \pi_\beta(\gamma)$ for $\beta \neq \alpha$. Now if $\pi_\alpha(\gamma)$ is nonsemisimple, then for all $g \in G$, $\pi_\alpha(g\gamma g^{-1})$ is nonsemisimple, whereas, as we have seen, there is a point μ in $G(\gamma)$ such that $\pi_\alpha(\mu)$ $(=\pi_\alpha(\gamma)_s)$ is semisimple. Thus $\pi_\alpha(\gamma)$ is semisimple.

1.4. DEFINITION. A subgroup M of a locally compact topological group L is said to have property (S) in L if given any element $g \in L$ and a neighborhood Ω of the identity in L, there exists a positive integer n such that $\Omega g^n \Omega \cap M \neq \varnothing$.

It was first observed by A. Selberg that if $M \backslash L$ carries a finite L-invariant Borel measure, then M has property (S) in L. (See [27, Chap. V] for a proof.)

1.5. LEMMA. *Let $\mathscr{S}$; k_α, $\mathbf{G}_\alpha$ (for $\alpha \in \mathscr{S}$), and G be as in Subsection 1.1. We assume that for every $\alpha \in \mathscr{S}$, $\mathbf{G}_\alpha$ is an almost direct product of connected k_α-isotropic almost simple groups. Let Γ be a lattice in G. Then the centralizer $Z(\Gamma)$ of Γ in G is central in G. Moreover, the normalizer $N(\Gamma)$ of Γ is a discrete subgroup, and Γ is of finite index in $N(\Gamma)$.*

Proof. Let z be an element of $Z(\Gamma)$. Then for $\alpha \in \mathscr{S}$, $\pi_\alpha(z)$ commutes with $\pi_\alpha(\Gamma)$. Now since $\pi_\alpha(\Gamma)$ has property (S) in $\mathbf{G}_\alpha$, it is Zariski-dense in $\mathbf{G}_\alpha$ [36], it follows that $\pi_\alpha(z)$ is central in G_α for all $\alpha \in \mathscr{S}$, and therefore z is central in G. This proves the first assertion.

We shall now prove the second assertion. Since Γ is discrete, it is closed and countable, and $N(\Gamma)$ is a closed subgroup of G. It follows that for any $\gamma \in \Gamma$, the centralizer of γ in $N(\Gamma)$ is a closed subgroup of (at most) countable index in $N(\Gamma)$. Hence, by Baire's category theorem, the centralizer of γ in $N(\Gamma)$ is an open subgroup of $N(\Gamma)$.

Since the weak Zariski topology is noetherian, there is a finite subset Λ of Γ such that if an element of G commutes with every element of Λ, then it centralizes Γ and hence is central in G. Now, since Λ is finite, the centralizer Z of Λ in $N(\Gamma)$ is an open subgroup of $N(\Gamma)$. On the other hand, as Z is central, it is discrete. This implies that $N(\Gamma)$ is discrete.

Since $N(\Gamma) \supset \Gamma$ and $\Gamma \backslash G$ has finite measure, $\Gamma \backslash N(\Gamma)$ is finite.

1.6. DEFINITION. Let G be as in Subsection 1.1. A lattice Γ in G is said to be *irreducible* if no subgroup of Γ of finite index is a direct product of two infinite normal subgroups.

The following lemma is a consequence of Lemma 1.5. A well-known argument of A. Weil can be used to establish it, we therefore omit the details.

1.7. LEMMA. *Let G be as in Lemma 1.5. Let $\mathscr{S}_1$ be a subset of $\mathscr{S}$ and $\mathscr{S}_2$ be the complement of $\mathscr{S}_1$ in $\mathscr{S}$. Let $G_1 = \prod_{\alpha \in \mathscr{S}_1} G_\alpha$ and $G_2 = \prod_{\alpha \in \mathscr{S}_2} G_\alpha$, and π_1 (resp. π_2) be the natural projection of G on G_1 (resp. on G_2). Let Γ be a* *lattice in* G. *Assume that* $\pi_1(\Gamma)$ *is a discrete subgroup of* G_1. *Then* $\pi_2(\Gamma)$ *is discrete. Moreover,* $\pi_1(\Gamma)$ *and* $\Gamma_1 = \Gamma \cap G_1$ *(resp.* $\pi_2(\Gamma)$ *and* $\Gamma_2 = \Gamma \cap G_2$*) are lattices in* G_1 *(resp. in* G_2*) and* $\Gamma_1 \cdot \Gamma_2$ *is a subgroup of finite index in* Γ.

1.8. LEMMA. *Let* G *be as in Lemma* 1.5. *We assume that for all* $\alpha \in \mathscr{S}$, $\mathbf{G}_\alpha$ *is almost simple over* k_α. *Let* Γ *be a lattice in* G, *and let* $\Gamma = \Gamma_1 \cdot \Gamma_2$ *(direct product) where* Γ_1 *and* Γ_2 *are infinite normal subgroups of* Γ. *Then there is a disjoint partition* $\mathscr{S}_1 \cup \mathscr{S}_2 = \mathscr{S}$ *of* $\mathscr{S}$ *such that if* $G_1 = \prod_{\alpha \in \mathscr{S}_1} G_\alpha$ *(resp.* $G_2 = \prod_{\alpha \in \mathscr{S}_2} G_\alpha$*), then a subgroup of finite index of* Γ_1 *(resp.* Γ_2*) is contained in* G_1 *(resp.* G_2*) and* $G_1 \cap \Gamma$ *(resp.* $G_2 \cap \Gamma$*) is a lattice in* G_1 *(resp.* G_2*).*

Proof. For $\alpha \in \mathscr{S}$, let $\mathbf{G}_\alpha^1$ be the connected component of the identity in the Zariski closure of $\pi_\alpha(\Gamma_1)$ in $\mathbf{G}_\alpha$. Since Γ_1 is a normal subgroup of Γ and $\pi_\alpha(\Gamma)$ has property (S) in G_α, it follows that $\mathbf{G}_\alpha^1$ is a normal subgroup of $\mathbf{G}_\alpha$ and it is clearly defined over k_α. Now since $\mathbf{G}_\alpha$ is simple, $\mathbf{G}_\alpha^1$ is either trivial or else equals $\mathbf{G}_\alpha$. Let $\mathscr{S}_1 = \{\alpha \in \mathscr{S} \mid \mathbf{G}_\alpha^1 = \mathbf{G}_\alpha\}$ and let $G_1 = \prod_{\alpha \in \mathscr{S}_1} G_\alpha$. It is evident that a subgroup of Γ_1 of finite index is contained in G_1.

Let $\mathscr{S}_2$ be the subset of $\mathscr{S}$ obtained similarly with the help of Γ_2 and let $G_2 = \prod_{\alpha \in \mathscr{S}_2} G_\alpha$. We claim that $\mathscr{S}_2 = \mathscr{S} - \mathscr{S}_1$. In fact, since a subgroup of Γ_1 (resp. Γ_2) of finite index is contained in G_1 (resp. G_2), it follows that a subgroup of Γ of finite index is contained in $G_1 \cdot G_2$. Therefore, $G/G_1 \cdot G_2 \approx \prod_{\alpha \in \mathscr{S} - \mathscr{S}_1 \cup \mathscr{S}_2} G_\alpha$ carries a finite Haar measure. This implies that $\prod_{\alpha \in \mathscr{S} - \mathscr{S}_1 \cup \mathscr{S}_2} G_\alpha$ is compact and since for all $\alpha \in \mathscr{S}$, $\mathbf{G}_\alpha$ is isotropic and hence G_α is noncompact, it follows that $\mathscr{S} = \mathscr{S}_1 \cup \mathscr{S}_2$. On the other hand, since Γ_1 commutes with Γ_2, G_1 commutes with G_2 and from this it is obvious that $\mathscr{S}_1 \cap \mathscr{S}_2 = \varnothing$. Now all the assertions of the lemma are evident.

The following is a simple consequence of Lemmas 1.7 and 1.8.

1.9. COROLLARY. *Let* G *be as in the previous lemma and let* Γ *be a lattice in* G. *Then there is a disjoint partition* $\mathscr{S} = \bigcup_{i=1}^{r=r(\Gamma)} \mathscr{S}_i$ *of* $\mathscr{S}$ *such that if for* $i \leqslant r$, $G_i = \prod_{\alpha \in \mathscr{S}_i} G_\alpha$, *then* $\Gamma_i = G_i \cap \Gamma$ *is an irreducible lattice in* G_i *and (obviously then)* $\prod_{i=1}^{r} \Gamma_i$ *is a subgroup of* Γ *of finite index.*

In the following lemma (Lemma 1.10) we collect, for convenient reference, some known results on finite generation of lattices.

1.10. LEMMA. *Let* $\mathscr{C}$ *be a finite set. For* α *in* $\mathscr{C}$ *let* k_α *be either* $\mathbf{R}$ *or a non-archimedean local field, and let* $\mathbf{S}_\alpha$ *be a connected semisimple algebraic group defined over* k_α. *Let* S_α *be the group of* k_α*-rational points of* $\mathbf{S}_\alpha$ *with the natural locally compact Hausdorff topology. Let* $S = \prod_{\alpha \in \mathscr{C}} S_\alpha$, *and let* Λ *be a lattice* *in S. Assume that at least one of the following holds.*

(i) *For no* α, $\mathbf{S}_\alpha$ *has a factor of* k_α*-rank* 1.
(ii) *For all* $\alpha \in \mathscr{C}$, $k_\alpha = \mathbf{R}$.
(iii) Λ *is uniform.*

Then Λ *is finitely generated.*

Finite generation of lattices in S, when for no α, $\mathbf{S}_\alpha$ has a factor of k_α-rank 1, is due to Kazdan (see [14]). In case for all $\alpha \in \mathscr{C}, k_\alpha = \mathbf{R}$, finite generation follows from the decomposability of lattices in semisimple groups (Corollary 1.9), the result of Kazdan, and a result of Raghunathan (cf. [27, Corollary 13.15]) according to which any irreducible lattice in a real analytic semisimple group with an $\mathbf{R}$-rank 1 factor is finitely generated (in fact, finitely presentable). When Λ is uniform, its finite generation is a simple consequence of the fact that it acts properly, with compact quotient, on a connected locally compact space. For details see the proof of Theorem 2.7 in [15].

1.11. Lemma. *Let* $\mathscr{C}$ *and for* $\alpha \in \mathscr{C}$, k_α, $\mathbf{S}_\alpha$, *and* S_α *be as in the preceding lemma. Let* $p_\alpha : \mathbf{M}_\alpha \to \mathbf{S}_\alpha$ *be a universal covering of* $\mathbf{S}_\alpha$ *over* k_α. *Let* M_α *be the group of* k_α*-rational points of* $\mathbf{M}_\alpha$ *with the natural locally compact Hausdorff topology. Let* $S = \prod_{\alpha \in \mathscr{C}} S_\alpha$ *and* $M = \prod_{\alpha \in \mathscr{C}} M_\alpha$, *and let* $p : M \to S$ *be the natural projection. Let* Λ *be a finitely generated lattice in S. We assume that either all the* k_α *are of characteristic zero or* Λ *is uniform and all the* k_α *are non-archimedean. Then there is a torsion free lattice* Γ *in M such that* $p(\Gamma)$ *is a torsion free normal subgroup of* Λ *of finite index.*

This follows easily, in case all the k_α are of characteristic zero, from a result of Honda [17, Proposition 3] (cf. also [4, Sect. 6.4] and [8, Sects. 3.19–3.20]) which implies that for all α, $S_\alpha / p_\alpha(M_\alpha)$ is finite and a theorem of Selberg [30] according to which any finitely generated linear group (over a field of characteristic zero) has a torsion free subgroup of finite index. The assertion is proved in Borel [3, Lemme 3.4] in case all the k_α are non-archimedean.

1.12. Proposition. *Let* $\mathscr{S}$; k_α, $\mathbf{G}_\alpha$ (*for* $\alpha \in \mathscr{S}$), *and G be as in Subsection 1.1. We assume that for every* $\alpha \in \mathscr{S}$, $\mathbf{G}_\alpha$ *is an almost direct product of connected almost simple* k_α*-isotropic subgroups. Let* Γ *be a closed subgroup of G such that* $\Gamma \backslash G$ *carries a finite G-invariant Borel measure. Then* Γ *is dense in G in the weak Zariski topology.*

Proof. Let H be the closure of Γ in G in the weak Zariski topology. Since $\mathscr{S}$ is finite and for all $\alpha \in \mathscr{S}$, the Zariski topology on G_α is noetherian, $H \cdot G_\alpha$

is closed in the weak Zariski topology, and hence it is closed in the Hausdorff topology on G. We easily conclude from this (see, for example, [27, Theorem 1.13]) that $G_\alpha \cap H\backslash G_\alpha$ carries a finite invariant measure, hence $G_\alpha \cap H$ has property (S) in G_α. Now according to a theorem of Wang [36], $G_\alpha \cap H$ is Zariski dense in G_α. But since $G_\alpha \cap H$ is obviously closed in G_α in the Zariski topology on G_α, it follows that $G_\alpha \cap H = G_\alpha$. Hence, $H \supset G_\alpha$ for all $\alpha \in \mathcal{S}$. Therefore, $H = G$. This proves the proposition.

1.13. Let $\mathbf{H}$ be a connected algebraic group defined over a field k. Let $H = \mathbf{H}(k)$. In the sequel, we denote by H^+ the (normal) subgroup of H generated by the k-rational points of the unipotent radicals of the parabolic k-subgroups of $\mathbf{H}$.

1.14. Now let $\mathbf{H}$ be a connected algebraic group defined, almost simple, and isotropic over an infinite field k. Then it follows from the main theorem of Tits [34] that any proper normal subgroup of H^+ is central in H, and hence in particular, H^+ has no proper subgroup of finite index.

1.15. LEMMA. *Let $\mathbf{G}$ be a connected algebraic group defined, almost simple, and isotropic over a local field K of characteristic zero. Let G be the group of K-rational points of $\mathbf{G}$ with the Hausdorff topology, and H be a closed nondiscrete subgroup of G such that $H\backslash G$ carries a finite G-invariant measure. Then H is an open subgroup of G of finite index. Hence* (*by* 1.14), *$H \supset G^+$.*

Proof. It is known (cf. Weil [38, Theorem 5 in Chap. I]) that a local field of characteristic zero is a finite extension of either the field $\mathbf{R}$ of real numbers or the field $\mathbf{Q}_p$ of p-adic numbers (for a suitable prime p). So we can use the functor $R_{K/k}$ (see Weil [37, Chap. I]) to reduce the proof to the case when K is either $\mathbf{R}$ or $\mathbf{Q}_p$. So we assume that $\mathbf{Q}$ is dense in K. Then H, being a closed subgroup of G, is an analytic subgroup (cf. Serre [31, LG 5, Sect. 9]). Let $\mathfrak{g}$ be the Lie algebra of G and $\mathfrak{h}$ be the subalgebra corresponding to H. Since $\mathbf{G}$ is simple over K and H is nondiscrete, $\mathfrak{g}$ is simple and $\mathfrak{h} \neq 0$. Moreover since $H\backslash G$ carries a finite invariant measure, H has property (S) in G and hence (Wang [36]) it is Zariski dense in G. This implies that $\mathfrak{h}$ is an ideal of $\mathfrak{g}$ and therefore $\mathfrak{h} = \mathfrak{g}$. Thus H is an open subgroup of G. Now since the discrete space $H\backslash G$ carries a finite G-invariant measure, it is finite. This proves the lemma.

1.16. QUESTION. Let $\mathbf{G}$ be a connected algebraic group defined, isotropic, and almost simple over a local field K of positive characteristic. Let

$G = \mathbf{G}(K)$ and let H be a closed *nondiscrete* subgroup of G such that $H\backslash G$ carries a finite G-invariant measure. Then is it true that $H \supset G^+$? An affimative answer to this question will have important bearing on the strong rigidity theorems of Section 8.

The following well-known result will be frequently used in this paper.

1.17. LEMMA (R. Arens). *Let H be a locally compact σ-compact topological group and M be a Baire space on which H operates transitively. Then for every $m \in M$, the map $h \mapsto h \cdot m$ of H into M is open.*

(For a proof see [10, Chap. VII, Appendix 1, Lemme 2].)

1.18. DEFINITION. An algebraic group $\mathbf{H}$ defined over a field k is said to be *k-anisotropic* (or anisotropic over k) if there exists no nontrivial homomorphism of $\mathbf{GL}(1)$ into $\mathbf{H}$ defined over k, or, equivalently, if $\mathbf{H}$ contains no nontrivial k-split torus.

The following result has been announced by Bruhat and Tits in arbitrary characteristic.

1.19. THEOREM. *Let $\mathbf{M}$ be a reductive algebraic group defined and anisotropic over a local field k. Then the group $\mathbf{M}(k)$ is compact in the Hausdorff topology.*

We now use this result of Bruhat–Tits to prove the following lemma.

1.20. LEMMA. *Let k be a local field and let $\mathbf{H} = \mathbf{M} \cdot \mathbf{T}$ be an almost direct product of a reductive k-anisotropic group $\mathbf{M}$ and a k-split torus $\mathbf{T}$. Let $H = \mathbf{H}(k)$ with the natural Hausdorff topology. Then H has a unique maximum compact subgroup C, and $C \cdot \mathbf{T}(k)$ is a subgroup of finite index in H.*

Proof. The lemma is obvious in case $k = \mathbf{C}$ or $\mathbf{R}$, we therefore assume that k is non-archimedean. Let $X(\mathbf{H})$ (resp. $X(\mathbf{T})$) be the group of k-rational characters on $\mathbf{H}$ (resp. $\mathbf{T}$). Let $\mathfrak{o}$ be the ring of integers of k and $\mathfrak{o}^\times$ be the group of units of $\mathfrak{o}$. Then $\mathfrak{o}^\times$ is the (unique) maximum compact subgroup of $k^\times$. Let $C = \{h \in H \mid \chi(h) \in \mathfrak{o}^\times \text{ for all } \chi \in X(\mathbf{H})\}$. Evidently, any compact subgroup of H is contained in C and therefore, in particular, C is open. We shall show that C is compact (hence it is the unique maximum compact subgroup of H), and $C \cdot \mathbf{T}(k)$ is a subgroup of finite index in H.

The restriction map $X(\mathbf{H}) \to X(\mathbf{T})$ maps $X(\mathbf{H})$ onto a subgroup of $X(\mathbf{T})$ of finite index. From this it is obvious that $C \cap \mathbf{T}(k)$ is a compact (in fact the maximum compact) subgroup of $\mathbf{T}(k)$. Now let $\pi: \mathbf{H}(k) \to \mathbf{H}(k)/\mathbf{T}(k)$ be the natural projection. Since $\mathbf{H}/\mathbf{T}$ is k-anisotropic, by Theorem 1.19 $(\mathbf{H}/\mathbf{T})(k)$ is compact. But since $\mathbf{T}$ is k-split, according to [2, Corollary 15.7] the homomorphism $\mathbf{H}(k) \to (\mathbf{H}/\mathbf{T})(k)$ is surjective, and hence in view of Lemma 1.17 $\mathbf{H}(k)/\mathbf{T}(k)$ is isomorphic, as topological group, to $(\mathbf{H}/\mathbf{T})(k)$. This implies that $\mathbf{H}(k)/\mathbf{T}(k)$ is compact. Now C being open in $H(= \mathbf{H}(k))$, $\pi(C)$ is an open subgroup of $\mathbf{H}(k)/\mathbf{T}(k)$ and hence $\pi(C)$ is a compact subgroup of $\mathbf{H}(k)/\mathbf{T}(k)$ of finite index. This implies that $C \cdot \mathbf{T}(k)$ is of finite index in H. Moreover, since $\mathbf{T}(k) \cap C$ is compact, C is compact.

1.21. Lemma. *Let* $\mathbf{H}$ *be an algebraic group defined over a local field and let* $\mathbf{P}$ *be a parabolic* k *subgroup. Let* ρ *be a rational representation of* $\mathbf{H}$, *defined over* k, *on a vector space* $\mathbf{V}$. *Let* $H = \mathbf{H}(k)$, $P = \mathbf{P}(k)$ *and* $V = \mathbf{V}(k)$. *Let* v $(\in V)$ *be an eigenvector of* $\mathbf{P}$. *Then the orbit* $\rho(H) \cdot v$ *is locally compact.*

Proof. Let χ be any character on $\mathbf{P}$ defined over k. Then $\chi(\mathbf{P}(k))$ is a closed subgroup of $k^\times$. Since v $(\in V)$ is an eigenvector $\mathbf{P}$, this implies that $\rho(P) \cdot v$ is a closed subset of V-$\{0\}$.

Now, since $\mathbf{P}$ is a parabolic k-subgroup, H/P is compact. Therefore, there is a compact subset C of H such that $H = C \cdot P$. As $\rho(P) \cdot v$ is a closed subset of V-$\{0\}$ and $\rho(C)$ is compact, we conclude that the orbit $\rho(H) \cdot v = \rho(C) \cdot \rho(P) \cdot v$ is closed in V-$\{0\}$, and hence the orbit is locally compact.

We close this section with a well-known lemma which shall be used in Sections 2 and 8.

1.22. Lemma. *Let* $\mathbf{V}$ *be a variety and* $\mathbf{H}$ *be an algebraic group, both defined over a local field* k. *Let* V *(resp.* H*) be the set of* k*-rational points of* $\mathbf{V}$ *(resp.* $\mathbf{H}$*) with the natural locally compact Hausdorff topology. Let* $\boldsymbol{\sigma}$ *be an action of* $\mathbf{H}$ *on* $\mathbf{V}$ *defined over* k *and let* σ *be the induced action of* H *on* V. *Let* $v \in V$ *be a point such that the morphism* $h \mapsto \boldsymbol{\sigma}(h)v$ *of* $\mathbf{H}$ *into* $\mathbf{V}$ *is separable. Then the orbit* $\mathcal{O}_v = \sigma(H) \cdot v$ *of* v *is locally compact, and the map* $H \to \mathcal{O}_v$ *defined by* $h \mapsto \sigma(h) \cdot v$ *is an open map.*

Proof. Let $\boldsymbol{\mathcal{O}}_v$ be the orbit of v under $\mathbf{H}$. Then according to Proposition 6.7 of [2], $\boldsymbol{\mathcal{O}}_v$ is a smooth variety, defined over k, which is open in its closure in $\mathbf{V}$. It follows that the subspace $\boldsymbol{\mathcal{O}}_v(k)$ of V, consisting of the k-rational points of $\boldsymbol{\mathcal{O}}_v$, is a k-analytic manifold and it is locally closed in V. In particular, $\boldsymbol{\mathcal{O}}_v(k)$ is locally compact. Now since the morphism $h \mapsto \boldsymbol{\sigma}(h) \cdot v$ of $\mathbf{H}$ into $\mathbf{V}$ is separable, the map $h \mapsto \sigma(h) \cdot v$ of H into $\boldsymbol{\mathcal{O}}_v(k)$ is a k-analytic submersion and therefore, the orbit $\mathcal{O}_v$ of v is open in $\boldsymbol{\mathcal{O}}_v(k)$. Hence the orbit is locally compact, and the orbit map is an open map.

2. k-Regular Elements and Lattices

Let k be a local field (i.e., a nondiscrete locally compact field) with absolute value $| \ |$. Let $\mathbf{k}$ be a fixed algebraic closure. We let $| \ |$ also denote the unique absolute value on $\mathbf{k}$ which restricts to the given absolute value on k (cf. Jacobson [20, Chap. V, Theorem 17]). Let $\mathbf{G}$ be a connected semisimple linear algebraic group defined over k. Let $G = \mathbf{G}(k)$. Let $\mathfrak{g}$ (resp. $\mathbf{\mathfrak{g}}$) be the Lie algebra of G (resp. $\mathbf{G}$). As usual, we denote the adjoint representation by Ad. For $g \in G$, let $m(g)$ denote the number of eigenvalues (counted with multiplicity) of $\operatorname{Ad} g$ of absolute value 1. We introduce a:

2.1. Definition. An element $g \in G$ is said to be *k-regular* if $m(g) \leqslant m(x)$ for all $x \in G$. It is evident that an element of G is k-regular if and only if any nonzero power of it is k-regular.

Notation. For an algebraic group $\mathbf{H}$ over K and $x \in H$ $(=\mathbf{H}(K))$, we denote the centralizer of x in H (resp. $\mathbf{H}$) by H_x (resp. $\mathbf{H}_x$). We let $\mathbf{H}^0$ denote the connected component of the identity, and $L(\mathbf{H})$ the Lie algebra of $\mathbf{H}$.

2.2. Lemma. *Let $g \in G$ be a k-regular element. Then*

(i) *In case k is of characteristic* 0, *g is semisimple.*

(ii) *Let m be the smallest positive integer such that g^m is semisimple. Then $\mathbf{G}_{g^m}$ has a unique maximum k-split torus $\mathbf{T}$ of $\mathbf{G}$, and $\mathbf{G}^0_{g^m}$ is an almost direct product of $\mathbf{T}$ and a k-anisotropic reductive subgroup $\mathbf{B}$.*

Moreover, the group of k-rational points of any maximal k-split torus contains a k-regular element.

Proof. In case $\operatorname{char}(k) = 0$ let $g = u \cdot v = v \cdot u$ be the Jordan decomposition of g with u (resp. v) unipotent (resp. semisimple); otherwise, let m be the smallest positive integer such that g^m is semisimple, and let $v = g^m$. Let $\mathbf{S}$ $(\subset \mathbf{G})$ be a maximal torus defined over k, which contains v (cf. Borel–Tits [7, Proposition 10.3 and Théorème 2.14], note that according to Proposition 18.1 of [1] v is contained in a maximal torus of $\mathbf{G}$). Let $S = \mathbf{S}(k)$. Let $\mathbf{S}^d$ (resp. $\mathbf{S}^a$) be the maximum k-split (resp. k-anisotropic) subtorus of $\mathbf{S}$, and let S^d be the group of k-rational points of $\mathbf{S}^d$. Then $\mathbf{S} = \mathbf{S}^a \cdot \mathbf{S}^d$ (almost direct product) and it follows from Lemma 1.20 that S has a unique maximum compact subgroup C. Moreover, $C \cdot S^d$ is of finite index in S. Thus there is a positive integer s such that $v^s \in C \cdot S^d$. Let $v^s = x \cdot y$ with $x \in C$ and $y \in S^d$. Since C is compact, all the eigenvalues of $\operatorname{Ad} x$ have absolute value 1, and since x, y commute, $m(y) = m(v^s)$. On the other hand, it is obvious that

$m(y) = m(v^s) = m(v) = m(g)$. Thus y is k-regular. Now let $\mathbf{T}$ be a maximal k-split torus of $\mathbf{G}$ containing $\mathbf{S}^d$. Since $y \in \mathbf{T}(k)$, $\mathbf{T}(k)$ has a k-regular element. As any two maximal k-split tori are conjugate to each other over k [7, Théorème 4.21], it follows that every maximal k-split torus contains a k-regular element.

Let $X(\mathbf{T})$ be the set of rational characters on $\mathbf{T}$. In the sequel we view $X(\mathbf{T})$ as an additive group. For $\alpha \in X(\mathbf{T})$, let

$$\mathfrak{g}^\alpha = \{X \in \mathfrak{g} \mid \operatorname{Ad} tX = \alpha(t)X \text{ for } t \in \mathbf{T}\}.$$

Let

$$\Phi = \{\alpha \mid \alpha \in X(\mathbf{T}), \alpha \neq 0 \text{ and } \mathfrak{g}^\alpha \neq \{0\}\}$$

and

$$\mathfrak{z} = \mathfrak{g}^0 = \{X \in \mathfrak{g} \mid \operatorname{Ad} tX = X \text{ for all } t \in \mathbf{T}\}.$$

Then

$$\mathfrak{g} = \mathfrak{z} \oplus \coprod_{\alpha \in \Phi} \mathfrak{g}^\alpha.$$

From this decomposition of $\mathfrak{g}$, which is known as the root space decomposition of $\mathfrak{g}$ with respect to $\mathbf{T}$, it is now evident (since $\exists t \in \mathbf{T}(k)$ such that $|\alpha(t)| \neq 1\ \forall \alpha \in \Phi$) that an element $t \in \mathbf{T}(k)$ is k-regular if and only if $|\alpha(t)| \neq 1$ $\forall \alpha \in \Phi$. This implies in particular that if $t \in \mathbf{T}(k)$ is k-regular, then

$$\begin{aligned}\text{the Lie algebra of } \mathbf{G}_t &= \{X \in \mathfrak{g} \mid \operatorname{Ad} tX = X\} = \mathfrak{z} \\ &= \text{the Lie algebra of the centralizer } Z(\mathbf{T}) \text{ of } \mathbf{T} \text{ in } \mathbf{G}.\end{aligned}$$

Now since $Z(\mathbf{T}) \subset \mathbf{G}_t$ and since the centralizer of a torus is connected (cf. Borel [1, Proposition 18.4]), it follows that $Z(\mathbf{T}) = \mathbf{G}_t^0$. Therefore, since $y \in S^d$ $(\subset \mathbf{T}(k))$ is k-regular, $Z(\mathbf{T}) = \mathbf{G}_y^0$, and (since $y \in \mathbf{S}$) $\mathbf{S} \subset \mathbf{G}_y^0 = Z(\mathbf{T})$. Since $\mathbf{S}$ is a maximal torus of $\mathbf{G}$ and $\mathbf{T}$ is central in $Z(\mathbf{T})$, it follows that $\mathbf{T} \subset \mathbf{S}$ and hence $\mathbf{T} = \mathbf{S}^d$.

From well-known results on algebraic groups it follows that $Z(\mathbf{T})$ is a reductive group defined over k and $Z(\mathbf{T}) = \mathbf{M} \cdot \mathbf{T}$ (almost direct product), where $\mathbf{M}$ is a reductive group defined over k. Since $\mathbf{T}$ is a maximal k-split torus in $\mathbf{G}$, $\mathbf{M}$ is k-anisotropic.

As x commutes with y and all the eigenvalues of $\operatorname{Ad} x$ are of absolute value 1 and for every $\alpha \in \Phi$, $|\alpha(y)| \neq 1$, $\operatorname{Ad} xy$ fixes an element X of $\mathfrak{g}$ if and only if $\operatorname{Ad} xX = X = \operatorname{Ad} yX$. Thus $\mathfrak{g}_{xy} = \mathfrak{g}_x \cap \mathfrak{g}_y$, where for $z \in G$ $\mathfrak{g}_z = \{X \in \mathfrak{g} \mid \operatorname{Ad} zX = X\}$. Now since x, y are commuting semisimple elements, it follows from [7, Lemme 10.1] that $L(\mathbf{G}_x) = \mathfrak{g}_x$, $L(\mathbf{G}_{xy}) = \mathfrak{g}_{xy}$, and moreover since $\mathbf{G}_x \cap \mathbf{G}_y = (\mathbf{G}_x)_y$, $L(\mathbf{G}_x \cap \mathbf{G}_y) = \mathfrak{g}_x \cap \mathfrak{g}_y = \mathfrak{g}_{xy}$. Thus $L(\mathbf{G}_x \cap \mathbf{G}_y) =$

$L(\mathbf{G}_{xy})$. Since clearly $\mathbf{G}_x \cap \mathbf{G}_y \subset \mathbf{G}_{xy}$, we conclude that $\mathbf{G}_{xy}^0 = (\mathbf{G}_x \cap \mathbf{G}_y)^0$. Now since $v \in \mathbf{S}$, $(\mathbf{T} \subset)\mathbf{S} \subset \mathbf{G}_v^0$ and we have

$$\mathbf{T} \subset \mathbf{G}_v^0 \subset \mathbf{G}_{vs}^0 = \mathbf{G}_{xy}^0 = (\mathbf{G}_x \cap \mathbf{G}_y)^0 \subset \mathbf{G}_y^0 = Z(\mathbf{T}) = \mathbf{M} \cdot \mathbf{T}.$$

Hence

$$\mathbf{G}_v^0 = \mathbf{B} \cdot \mathbf{T} \text{ (almost direct product)}, \qquad \text{where} \quad \mathbf{B} = \mathbf{M} \cap \mathbf{G}_v^0.$$

Since $\mathbf{M}$ is k-anisotropic, $\mathbf{B}$ $(\subset \mathbf{M})$ is k-anisotropic (and hence by Theorem 1.19, $\mathbf{B}(k)$ is compact). Also, since $\mathbf{T}$ is central in $\mathbf{G}_v^0$ $(\subset Z(\mathbf{T}))$, it is the unique maximal k-split torus of $\mathbf{G}_v$.

In case $\text{char}(k) = 0$, v is the semisimple Jordan component of g. Hence, it follows from [2, Corollary 11.12] that $g \in \mathbf{G}_v^0(k)$. Now since $\mathbf{B}$ is k-anisotropic, in view of [7, Corollaire 8.5], $\mathbf{G}_v^0(k)$ consists entirely of semisimple elements and hence g is semisimple. Now, clearly $g = v$ and hence, in case k is of characteristic zero,

$$\mathbf{G}_g^0 = \mathbf{G}_v^0 = \mathbf{B} \cdot \mathbf{T} \quad \text{(almost direct product)}.$$

This proves the lemma.

2.3. Definition. The dimension of a maximal k-split torus of $\mathbf{G}$ is called the *k-rank* of $\mathbf{G}$.

We shall now give a characterization of k-regular elements which turns out to be very useful.

2.4. Let $\mathbf{T}$ be a maximal k-split torus in $\mathbf{G}$. Let $\mathfrak{g} = \mathfrak{z} \oplus \coprod_{\alpha \in \Phi} \mathfrak{g}^\alpha$ be the root space decomposition of $\mathfrak{g}$ with respect to $\mathbf{T}(k)$, $\mathfrak{z}$ being the centralizer of $\mathbf{T}(k)$. We fix an ordering on the set Φ of roots. Let $\Delta = \{\alpha_i\}_{i=1}^r$ (resp. Φ^+) be the set of simple roots (resp. positive roots) and let $\mathfrak{u}^+ = \coprod_{\alpha \in \Phi^+} \mathfrak{g}^\alpha$. Then $\mathfrak{u}^+$ is a nilpotent subalgebra of $\mathfrak{g}$. Let $n = \dim \mathfrak{u}^+$ and $\tau = \sum_{\alpha \in \Phi^+} (\dim \mathfrak{g}^\alpha) \cdot \alpha$. Let ρ denote the representation of G on $\bigwedge^n \mathfrak{g}$ induced by Ad. It is evident that the weights of ρ with respect to $\mathbf{T}(k)$ are of the form $\tau - \sum m_i \alpha_i (m_i \geqslant 0)$. The multiplicity of the weight τ is 1 and, obviously, for $i \leqslant r$, $\tau - \alpha_i$ is a weight.

2.5. Lemma. *The following two conditions on an element $g \in G$ $(= \mathbf{G}(k))$ are equivalent.*

(1) *g is k-regular.*

(2) *$\rho(g)$ has a unique eigenvalue of maximum absolute value which occurs in addition with multiplicity* 1.

Proof. In case $\operatorname{char}(k) = 0$ let $g = u \cdot v = v \cdot u$ be the Jordan decomposition of g with u (resp. v) unipotent (resp. semisimple), otherwise let $n = n(g)$ be the smallest positive integer such that g^n is semisimple, and let $v = g^n$. Then g satisfies (1) (resp. (2)) if and only if v satisfies (1) (resp. (2)). Hence we may (and we shall) assume that $g = v$, i.e., g is semisimple. Let $\mathbf{S}$ be a maximal k-torus of $\mathbf{G}$ containing g and let $\mathbf{S}^d$ be the maximum k-split subtorus of $\mathbf{S}$. Let S^d (resp. S) be the group of k-rational points of $\mathbf{S}^d$ (resp. $\mathbf{S}$) with the natural Hausdorff topology. Then (cf. the proof of Lemma 2.2) S has a unique maximum compact subgroup C and $C \cdot S^d$ is of finite index in S. Therefore there exists a positive integer s such that $g^s \in C \cdot S^d$. It is obvious that g satisfies (1) (resp. (2)) if and only if g^s satisfies (1) (resp. (2)), hence replacing g by g^s, we can assume that $g \in C \cdot S^d$. Let $g = x \cdot y$ with $x \in C$ and $y \in S^d$. Since C is compact, the eigenvalues of $\rho(x)$ are of absolute value 1. Thus g satisfies (1) (resp. (2)) if and only if y satisfies (1) (resp. (2)). Thus we may further assume that $g = y$, and then $g \in S^d$. Since any two maximal k-split tori are conjugate to each other by an element of $G\,(=\mathbf{G}(k))$ we can, after replacing g by a conjugate, assume that $\mathbf{S}^d \subset \mathbf{T}$. In view of Bourbaki [9, Chap. VI, Sect. 1, Proposition 22], there exists an ordering on the set Φ of roots such that the subset $\{\varphi \mid \varphi \in \Phi, |\varphi(g)| > 1\}$ is contained in the set of roots positive with respect to this ordering. Now since any two orderings on the set Φ of roots are conjugate to each other by an element of the Weyl group $N(\mathbf{T})(k)/Z(\mathbf{T})(k)$ ($N(\mathbf{T})$ denotes the normalizer of $\mathbf{T}$ in $\mathbf{G}$ and $Z(\mathbf{T})$ the centralizer of $\mathbf{T}$), after replacing g by a suitable conjugate under the group $N(\mathbf{T})(k)$, we can assume that $\forall \varphi \in \Phi^+$, $|\varphi(g)| \geqslant 1$. Recall that Φ^+ is the set of positive roots in the ordering on Φ fixed above (2.4).

It has been observed in the course of the proof of Lemma 2.2 that an element $t \in \mathbf{T}(k)$ is k-regular if and only if

$$|\varphi(t)| \neq 1 \qquad \text{for all} \quad \varphi \in \Phi.$$

Now since any root φ is a nonnegative or nonpositive integral linear combination of the simple roots $\{\alpha_i\}_{i=1}^r$ and since $|\alpha_i(g)| \geqslant 1$ for all i, it follows that there exists a $\varphi \in \Phi$ such that $|\varphi(g)| = 1$ if and only if for some $m \leqslant r$, $|\alpha_m(g)| = 1$. Thus g is k-regular if and only if for all $i \leqslant r$, $|\alpha_i(g)| \neq 1$. It has been noted in Subsection 2.4 that the weights of ρ are of the form $\tau - \sum_{i=1}^r m_i\alpha_i$ with $m_i \geqslant 0$. Also,

$$|\tau(g)| \geqslant \left|\left(\tau - \sum_i m_i\alpha_i\right)(g)\right| = |\tau(g)| \cdot \prod_i |\alpha_i(g)|^{-m_i}$$

and equality holds if and only if for every i such that $|\alpha_i(g)| \neq 1$, $m_i = 0$. Now if g is k-regular, then $|\alpha_i(g)| \neq 1$ for all $i \leqslant r$ and therefore, $\rho(g)$ has a unique eigenvalue $(=\tau(g))$ of maximum absolute value and the multiplicity of this eigenvalue is 1 since in the representation ρ the multiplicity of weight

τ is 1. Conversely, if g is not k-regular, then for some $i \leqslant r$, $|\alpha_i(g)| = 1$ and hence $|\tau(g)| = |\tau(g)|\,|\alpha_i(g)|^{-1} = |(\tau - \alpha_i)(g)|$. Since $\tau - \alpha_i$ is a weight, there are at least two eigenvalues (counted with multiplicity) of $\rho(g)$ of maximum absolute value. This proves the lemma.

2.6. PROPOSITION. *Let $g \in G$ be a k-regular semisimple element. Then there is a Zariski open neighborhood $\mathcal{U}_g$ of the identity in G such that for $h \in \mathcal{U}_g$, hg^m (and hence $g^m h = g^m(hg^m)g^{-m}$) is k-regular for large positive integers m.*

Proof. Let $W = \bigwedge^n \mathfrak{g} \otimes_k \mathbf{k}$ and $W = \coprod W^\lambda$ be the decomposition of W into eigenspaces for $\rho(g)$. Let μ be the unique eigenvalue of $\rho(g)$ of maximum absolute value (cf. Lemma 2.5). Let $\{w_i\}_{i=1}^l$ be a basis of W such that $w_1 = W^\mu \cap \{w_i\}_{i=1}^l$ and suitable subsets of it span different eigenspaces. Let $\{w_i^*\}$ be the dual basis of $W^* = \mathrm{Hom}_{\mathbf{k}}(W, \mathbf{k})$. Let

$$\mathcal{U}_g = \{h \in G \,|\, w_1^*(\rho(h)w_1) \neq 0\}.$$

Clearly $\mathcal{U}_g$ is a Zariski open neighborhood of the identity. We shall prove that for $h \in \mathcal{U}_g$, hg^m is k-regular for large positive integers m.

For $s \leqslant l$, let ξ_s be the set of s-tuples $\{I = (i_1, \ldots, i_s) \,|\, i_1 < i_2 < \cdots < i_s \leqslant l\}$. Let $\xi = \bigcup_{s \leqslant l} \xi_s$. For $I \in \xi_s$ we set $w_I = w_{i_1} \wedge \cdots \wedge w_{i_s} \in \bigwedge^s W$ and let $\{w_I^*\}_{I \in -\xi}$ be the dual basis of $(\bigwedge W)^*$. We define μ_i by the equation

$$\rho(g)e_i = \mu_i e_i \qquad (\text{thus } \mu_1 = \mu).$$

Also for $I = (i_1, \ldots, i_s)$, let $\mu_I = \prod_{j=1}^s \mu_{i_j}$. In the sequel for an $x \in G$ we let $\bigwedge^s x$ denote the automorphism of $\bigwedge^s W$ induced by $\rho(x)$. The characteristic polynomial $P_m(X)$ ($\in k[X]$) of $\rho(hg^m)$ is then given by:

$$P_m(X) = \sum_{s=0}^{l} (-1)^s X^{l-s} \operatorname{tr}(\textstyle\bigwedge^s hg^m).$$

(For an endomorphism A, $\operatorname{tr}(A)$ denotes the trace of A.)

In view of the preceding lemma, to prove the proposition it suffices to show that for large positive integers m, $P_m(X)$ has a unique root of maximum absolute value and it has multiplicity 1. We shall now investigate the behavior of the roots of $P_m(X)$ as $m \to \infty$. Let

$$F_m(X) = \sum_{s=0}^{l} (-1)^s X^{l-s} \frac{\operatorname{tr}(\bigwedge^s hg^m)}{\mu^{sm}}.$$

Then the roots of $F_m(X)$ are of the form λ/μ^m, where λ is a root of $P_m(X)$. Since $F_m(X)$ is monic, we see that to prove the proposition it is enough to show that $F_m(X)$ tends to the polynomial $X^l + w_1^*(h(w_1))X^{l-1}$ (note that, by definition, $w_1^*(h(w_1)) \neq 0$ for $h \in \mathscr{U}_g$). Now

$$c_s(m) = \frac{\operatorname{tr}(\bigwedge^s hg^m)}{\mu^{sm}} = \frac{1}{\mu^{sm}} \sum_{I \in \xi_s} w_I^*(hg^m(w_I))$$
$$= \sum \left(\frac{\mu_I}{\mu^s}\right)^m w_I^*(h(w_I)).$$

For $s > 1$, $|\mu_I| < |\mu_1|^s = |\mu|^s$, so that $c_s(m) \to 0$ as $m \to \infty$; whereas for $s = 1$, $|\mu_i| < |\mu|$ if $i > 1$ so that $c_1(m)$ tends to $w_1^*(h(w_1))$. This proves the proposition.

2.7. *Remark.* It is obvious from the proof of Proposition 2.6 that if C is a compact (in the Hausdorff topology on G) subset of $\mathscr{U}_g$, then there exists a positive integer $m_0 = m_0(C)$ such that hg^m and $g^m h$ are k-regular for all $h \in C$ and $m \geqslant m_0$.

2.8. Let $\mathscr{S}$ be a finite set and for every $\alpha \in \mathscr{S}$, let k_α be a local field of characteristic zero. For $\alpha \in \mathscr{S}$, let $\mathbf{G}_\alpha$ be a connected semisimple algebraic group defined over k_α and G_α be the group of k_α-rational points of $\mathbf{G}_\alpha$ with the canonical Hausdorff topology. We assume that for no $\alpha \in \mathscr{S}$, $\mathbf{G}_\alpha$ has a nontrivial connected k_α-anisotropic factor. Let $G = \prod_{\alpha \in \mathscr{S}} G_\alpha$ and, for $\alpha \in \mathscr{S}$, let $\pi_\alpha : G \to G_\alpha$ be the canonical projection of G on G_α.

2.9. DEFINITIONS. An element $g \in G$ is said to be *polar regular* (resp. *polar regular* semisimple) if $\pi_\alpha(g)$ is k_α-regular (resp. k_α-regular and semisimple) for all $\alpha \in \mathscr{S}$.

The *polar rank* $r(G)$ of G is defined as follows:

$$r(G) = \sum_{\alpha \in \mathscr{S}} (k_\alpha\text{-rank of } \mathbf{G}_\alpha).$$

We shall now give two applications of Proposition 2.6. Let G be as in Subsection 2.8, and let Γ be a lattice in G.

2.10. LEMMA. *Given a polar regular s.s. element $g \in G$ and a neighborhood (in the Hausdorff topology) Ω of the identity in G, there is a positive integer a such that $\Omega g^a \Omega \cap \Gamma$ has a polar regular element. In particular, any lattice in G has a polar regular element.*

Proof. For $\alpha \in \mathscr{S}$, let $\mathscr{U}_g^\alpha$ be a neighborhood of the identity in G_α such that for $h_\alpha \in \mathscr{U}_g^\alpha$, $\pi_\alpha(g)^m h_\alpha$ is k_α-regular for large positive integers m (cf. Proposition 2.6). Let $\mathscr{U}_g = \prod_{\alpha \in \mathscr{S}} \mathscr{U}_g^\alpha$. Let ω be a compact neighborhood of the identity in G such that $\omega^2 \subset \Omega \cap \mathscr{U}_g$. In view of Remark 2.7, it is obvious that there is a positive integer m_0 such that $g^m h$ is polar regular for $m \geqslant m_0$ and all $h \in \omega^2$. Now since Γ has property (S), there exists a positive integer n such that

$$\omega(g^{m_0})^n \omega \cap \Gamma \neq \varnothing$$

Let $\gamma = h_1 g^{m_0 n} h_2 \in \Gamma$ with $h_1, h_2 \in \omega$. We claim that γ is polar regular. In fact $h_1^{-1}\gamma h_1 = g^{m_0 n} h_2 h_1$, and since $h_2 h_1 \in \omega^2$ and $m_0 n \geqslant m_0$, it follows that $h_1^{-1}\gamma h_1$ is polar regular and hence γ is polar regular. Let $a = m_0 n$. Then since $\omega \subset \Omega, \gamma \in \Omega g^a \Omega \cap \Gamma$ and we have proved the lemma.

The following lemma will be used in the next section.

2.11. Lemma. *Given finitely many elements $\{\gamma_i\}_{1 \leqslant i \leqslant t}$ and a polar regular element γ of the lattice Γ, there exists a $\delta \in \Gamma$ such that for infinitely many positive integers m, $\gamma^m \delta \gamma_i^{-1}$ is polar regular s.s. for $1 \leqslant i \leqslant t$.*

Proof. Let n be a positive integer such that $\pi_\alpha(\gamma^n)$ is semisimple for all $\alpha \in \mathscr{S}$ (Lemma 2.2). Now for $\alpha \in \mathscr{S}$, let $\mathscr{U}_\gamma^\alpha$ be a Zariski open neighborhood of the identity in G_α such that for $h_\alpha \in \mathscr{U}_\gamma^\alpha$, $\pi_\alpha(\gamma^{ns}) \cdot h_\alpha$ is k_α-regular for large positive integers s (Proposition 2.6), and let $\mathscr{R}^\alpha$ be the set of regular elements of G_α. Then $\mathscr{R}^\alpha$ is a Zariski open subset of G_α consisting entirely of semisimple elements. Let $\mathscr{U}_\gamma = \prod_{\alpha \in \mathscr{S}} \mathscr{U}_\gamma^\alpha$ and $\mathscr{R} = \prod_{\alpha \in \mathscr{S}} \mathscr{R}^\alpha$. Now $(\bigcap_{i \leqslant t} \mathscr{U}_\gamma^\alpha \cdot \pi_\alpha(\gamma_i)) \cap (\bigcap_{i \leqslant t} \mathscr{R}^\alpha \cdot \pi_\alpha(\gamma_i))$, being intersection of nonempty Zariski open subsets of G_α, is nonempty. This implies that $(\bigcap_{i \leqslant t} \mathscr{U}_\gamma \cdot \gamma_i) \cap (\bigcap_{i \leqslant t} \mathscr{R} \cdot \gamma_i)$ is a nonempty open (in the weak Zariski topology) subset of G. Since Γ is dense in G (Proposition 1.18), there is a $\delta \in \Gamma \cap ((\bigcap_{i \leqslant t} \mathscr{U}_\gamma \cdot \gamma_i) \cap (\bigcap_{i \leqslant t} \mathscr{R} \cdot \gamma_i))$. Now since for $i \leqslant t$, $\delta\gamma_i^{-1} \in \mathscr{U}_\gamma$, it follows that for all large s, $(\gamma^n)^s \delta \gamma_i^{-1}$ is polar regular for all $i \leqslant t$. Also, since for all i, $\delta\gamma_i^{-1} \in \mathscr{R}$, $\mathscr{V} = \bigcap_{i \leqslant t} \mathscr{R}(\delta\gamma_i^{-1})^{-1}$ is a set (containing the identity), open in the weak Zariski topology on G. Now since the weak Zariski topology is noetherian, there is a positive integer p such that $\bigcup_{s=1}^\infty \gamma^{ns} \cdot \mathscr{V} = \bigcup_{s \leqslant p} \gamma^{ns} \cdot \mathscr{V}$. This implies that for infinitely many s, $\gamma^{ns} \in \mathscr{V}$. Thus for infinitely many s, $\gamma^{ns} \delta \gamma_i^{-1} \in \mathscr{R}$. Since $\pi_\alpha(\mathscr{R}) = \mathscr{R}^\alpha$ consists entirely of semisimple elements, the assertion of the lemma follows.

Notation. For subsets X, Y of a group G we shall let $X[Y]$ denote the set $\{xyx^{-1} | x \in X, y \in Y\}$. For $x, y \in G, x[y]$ will denote the element xyx^{-1}.

The following proposition is needed to prove Proposition 2.14, which in turn is needed in Section 6 to show that given a simply connected nonarchimedean semisimple group (cf. Section 4) and a lattice Γ in it, the set of

Γ-adopted apartments, in the Bruhat–Tits building $\mathscr{X}$ associated with the semisimple group, is dense in the set of all apartments in $\mathscr{X}$.

2.12. PROPOSITION. *Let* $\mathbf{G}$ *be a connected semisimple algebraic group defined over a non-archimedean local field* k, *and let* $G = \mathbf{G}(k)$. *Let* Ω *be a neighborhood of the identity in* G. *Let* $\mathbf{T}$ *be a maximal* k-*split torus in* $\mathbf{G}$. *Let* $T = \mathbf{T}(k)$ *and* $Z(T)$ *be the centralizer of* T *in* G. *Let* T_r *be the set of* k-*regular elements of* T. *Then there exists a neighborhood* V *of the identity in* G *such that* $V \cdot T_r \cdot V$ *consists entirely of* k-*regular elements and*

$$V \cdot T_r \cdot V \subset \Omega[Z(T)].$$

Proof. Let $\mathfrak{g}$ be the Lie algebra of G and let $\mathfrak{g} = \mathfrak{z} \oplus \prod_{\alpha \in \Phi} \mathfrak{g}^\alpha$ be the root space decomposition of $\mathfrak{g}$ with respect to T; here $\mathfrak{z}$ denotes the Lie subalgebra corresponding to $Z(T)$ and Φ is a subset of $\mathrm{Hom}(T, k^\times)$. We fix a Weyl chamber; this gives rise to an ordering on the set Φ of roots. Let Φ^+ (resp. Φ^-) be the set of positive (resp. negative) roots under this ordering and let C be the subset of T_r given by

$$C = \{t \in T_r \,|\, |\alpha(t)| > 1, \ \forall \alpha \in \Phi^+\}.$$

Since the set of Weyl chambers is finite and, given any $t \in T_r$, there is a Weyl chamber such that t lies in the "corresponding" C; to prove the proposition it suffices to show that there is a neighborhood V of the identity such that every element of $V \cdot C \cdot V$ is k-regular, and

$$V \cdot C \cdot V \subset \Omega[Z(T)].$$

Let $\mathfrak{u}^+$ (resp. $\mathfrak{u}^-$) be the Lie subalgebra $\sum_{\alpha \in \Phi^+} \mathfrak{g}^\alpha$ (resp. $\sum_{\alpha \in \Phi^-} \mathfrak{g}^\alpha$). Let $n = \dim \mathfrak{u}^+$, and let e_+ (resp. e_-) be a nonzero vector of the one-dimensional subspace $\bigwedge^n \mathfrak{u}^+$ (resp. $\bigwedge^n \mathfrak{u}^-$) of $\bigwedge^n \mathfrak{g}$. Let $\mathbf{U}^+$ (resp. $\mathbf{U}^-$) be the connected unipotent algebraic subgroup of $\mathbf{G}$, normalized by $Z(\mathbf{T})$, corresponding to the subalgebra $\mathfrak{u}^+$ (resp. $\mathfrak{u}^-$). Let $U^+ = \mathbf{U}^+(k)$ and $U^- = \mathbf{U}^-(k)$. Evidently $Z(T)$ normalizes both U^+ and U^-. Let ρ be the natural representation of G on $\bigwedge^n \mathfrak{g}$. Since $Z(T) \cdot U^+$ (resp. $U^- \cdot Z(T)$) is the normalizer of $\mathfrak{u}^+$ (resp. $\mathfrak{u}^-$), it follows that

$$Z(T) \cdot U^+ = \{x \in G \,|\, \rho(x) \textstyle\bigwedge^n \mathfrak{u}^+ = \bigwedge^n \mathfrak{u}^+\} \tag{1}$$

and

$$U^- \cdot Z(T) = \{x \in G \,|\, \rho(x) \textstyle\bigwedge^n \mathfrak{u}^- = \bigwedge^n \mathfrak{u}^-\}. \tag{2}$$

Let $\{e_1, \ldots, e_l\}$ be a basis of $\bigwedge^n \mathfrak{g}$ such that: (i) $e_1 = e_+$, $e_l = e_-$; and (ii) for all $i \leqslant l$ there is a character $\lambda_i: T \to k^\times$ such that

$$\rho(t)e_i = \lambda_i(t)e_i \qquad \text{for} \quad t \in T.$$

Let $\{e_i^*\}$ be the dual basis of $\mathrm{Hom}_k(\bigwedge^n \mathfrak{g}, k)$ and let e_+^* (resp. e_-^*) denote the element e_1^* (resp. e_l^*). We use the map $\sum x_i e_i \mapsto (x_1, \ldots, x_l)$ to identify $\bigwedge^n \mathfrak{g}$ with k^l and introduce a norm $\| \ \|$ on $\bigwedge^n \mathfrak{g}$ by setting $\|x\| = \max_{1 \leqslant i \leqslant l} |x_i|$ for $x = (x_1, \ldots, x_l)$. For $m > 0$, on the ring $M(m, k)$ of $m \times m$ matrices with entries in k, we introduce a norm $\| \ \|$ by setting

$$\|A\| = \max_{1 \leqslant i,\, j \leqslant m} |a_{ij}| \qquad \text{for} \quad A = (a_{ij}).$$

Using the basis $\{e_i\}_{i=1}^l$ we can identify $\mathrm{End}_k(\bigwedge^n \mathfrak{g})$ with $M(l, k)$, thus the norm on $M(l, k)$ induces a norm on $\mathrm{End}_k(\bigwedge^n \mathfrak{g})$ which we shall denote again by $\| \ \|$. It is obvious that for any permutation σ of $\{1, \ldots, l\}$, the basis $\{e_{\sigma(i)}\}_{i=1}^l$ induces the same norm.

Let $\epsilon < 1$ be a positive number and let

$$V^\epsilon = \{g \in G \mid \|\rho(g) - \mathbf{I}\| < \epsilon^l\}.$$

($\mathbf{I}$ denotes the identity transformation in End $\bigwedge^n \mathfrak{g}$.)

Clearly, V^ϵ is an open neighborhood of the identity in G. Let $v \in V^\epsilon$ and $c \in C$. We claim[2] that

(i) Given an eigenvalue ν of $\rho(vc)$, there is an integer $a \leqslant l$ such that $|\nu - \lambda_a(c)| \leqslant \epsilon |\lambda_a(c)|$.

(ii) $\rho(vc)$ has a unique eigenvalue of maximum absolute value and this eigenvalue has multiplicity 1. (Thus according to Lemma 2.5, vc is k-regular.)

(iii) We can choose an eigenvector f_+ of $\rho(vc)$, corresponding to the eigenvalue of maximum absolute value, such that $e_+^*(f_+) = 1$ and $\|e_+ - f_+\| < \epsilon$.

We shall now establish the claim. Let ν be an eigenvalue of $\rho(vc)$. We shall first show that $|\nu| = |\lambda_i(c)|$ for some $i \leqslant l$.

After reindexing $\{e_2, \ldots, e_{l-1}\}$, we assume that for all $i \leqslant j$, $|\lambda_i(c)| \geqslant |\lambda_j(c)|$. Now if for no i, $|\nu| = |\lambda_i(c)|$, then there exists an integer $a \leqslant l$ such that

$$|\lambda_a(c)| > |\nu| > |\lambda_{a+1}(c)|.$$

(If $a = 0$ or l, only the meaningful inequality is assumed to hold.)

Since $\det(\nu\mathbf{I} - \rho(vc)) = 0$, we have

$$\prod_{i=1}^{l} (\nu - v_{ii}\lambda_i(c)) = -\sum_{\sigma \neq 1} \epsilon_\sigma \beta_{\sigma(1),1} \cdots \beta_{\sigma(l),l}, \tag{3}$$

[2] This was suggested by Proposition 1 of Dani and Dani [13] and the argument used here to establish the claim is an adaptation of their proof of that proposition.

where $(v_{ij}) = \rho(v)$, $\beta_{ij} = v\,\delta_{ij} - v_{ij}\lambda_j(c)$ (δ_{ij} the Kronecker delta), $\epsilon_\sigma = \pm 1$, and σ runs over all permutations of $\{1, \ldots, l\}$ except the identity. The right-hand side of (3) can be majorized by $(\epsilon^l)^2|\lambda_1(c)| \cdots |\lambda_a(c)| \cdot |v|^{l-a}$, and thus

$$\left|\prod_{i=1}^{l} (v - v_{ii}\lambda_i(c))\right| = |\lambda_1(c)| \cdots |\lambda_a(c)|\,|v|^{l-a} \leqslant \epsilon^{2l}|\lambda_1(c)| \cdots |\lambda_a(c)|\,|v|^{l-a},$$

which is a contradiction (since $\epsilon < 1$). Hence $|v| = |\lambda_i(c)|$ for some $i \leqslant l$. Now let s be the smallest integer such that $|v| = |\lambda_s(c)|$ and let t be the largest integer such that $|v| = |\lambda_t(c)|$; then from (3) we get

$$|\lambda_1(c)| \cdots |\lambda_{s-1}(c)|\,|v - v_{ss}\lambda_s(c)| \cdots |v - v_{tt}\lambda_t(c)|\,|v|^{l-t} \leqslant \epsilon^{2l}|\lambda_1(c)| \cdots |\lambda_t(c)|\,|v|^{l-t}.$$

Thus

$$|v - v_{ss}\lambda_s(c)| \cdots |v - v_{tt}\lambda_t(c)| \leqslant \epsilon^{2l}|\lambda_s(c)| \cdots |\lambda_t(c)|,$$

and therefore, for an a ($s \leqslant a \leqslant t$),

$$|v - v_{aa}\lambda_a(c)| \leqslant \epsilon^{2l/(t-s+1)}|\lambda_a(c)| < \epsilon|\lambda_a(c)|$$

(since $t - s + 1 \leqslant l$ and $\epsilon < 1$). Now since

$$|v_{aa} - 1| < \epsilon^l \leqslant \epsilon,$$

$$|v - \lambda_a(c)| = |(v - v_{aa}\lambda_a(c)) + (v_{aa} - 1)\lambda_a(c)| < \epsilon|\lambda_a(c)|,$$

i.e.,

$$|v - \lambda_a(c)| < \epsilon|\lambda_a(c)|, \tag{4}$$

and of course

$$|v| = |\lambda_a(c)|.$$

Let $X^l + a_1X^{l-1} + \cdots + a_l$ (resp. $X^l + b_1X^{l-1} + \cdots + b_l$) be the characteristic polynomial of $\rho(c)$ (resp. $\rho(vc)$). Since $c \in C$, $|\alpha(t)| > 1$ for all $\alpha \in \Phi^+$. From this it is evident that

$$|\lambda_1(c)| > |\lambda_i(c)| \qquad \text{for} \quad 1 < i \leqslant l.$$

Now it can be seen, by direct computation, that

$$|b_1| = |a_1| = |\lambda_1(c)|, \tag{5}$$

and for all $r \leqslant l$

$$|a_r| \leqslant |\lambda_1(c)| \cdots |\lambda_r(c)|, \qquad |b_r| \leqslant |\lambda_1(c)| \cdots |\lambda_r(c)|. \tag{6}$$

Let $\mu_1, \ldots, \mu_l$ be the eigenvalues of $\rho(vc)$ indexed so that

$$|\mu_1| \geqslant \cdots \geqslant |\mu_l|,$$

and let s be the largest integer such that $|\mu_s| = |\mu_1|$, then

$$|b_s| = |\mu_1| \cdots |\mu_s| = |\mu_1|^s \tag{7}$$

and for all $r \leqslant l$

$$|b_r| \leqslant |\mu_1| \cdots |\mu_r|. \tag{8}$$

Now since for every μ_i there is a λ_j such that $|\mu_i| = |\lambda_j(c)|$, it follows that for all i, $|\mu_i| \leqslant |\lambda_1(c)|$. Thus from (5) and (8) we get

$$|\mu_1| = |\lambda_1(c)|.$$

Now if $s > 1$, then by (6)

$$|b_s| \leqslant |\lambda_1(c)| \cdots |\lambda_s(c)| < |\lambda_1(c)|^s = |\mu_1|^s,$$

i.e.,

$$|b_s| < |\mu_1|^s.$$

This contradicts (7). Thus $s = 1$, i.e., $\rho(vc)$ has a unique eigenvalue of maximum absolute value and it has multiplicity 1. This, in view of Lemma 2.5, implies that vc is k-regular.

Let $f_+ = \sum y_i e_i$, with $y_1 = 1$ and

$$\begin{bmatrix} y_2 \\ \vdots \\ y_l \end{bmatrix} = \lambda_1(c)\, \Delta^{-1} \begin{bmatrix} v_{21} \\ \vdots \\ v_{l1} \end{bmatrix}$$

where Δ is the $(l-1)$-rowed minor of $(\mu_1 \mathbf{I} - \rho(vc))$ obtained by omitting the first row and column. It is not hard to see that Δ is nonsingular, $\|\Delta^{-1}\| \leqslant |\mu_1|^{-1}$ and $\rho(vc)f_+ = \mu_1 f_+$. Clearly, $e_+^*(f_+) = y_1 = 1$, and

$$\begin{aligned} \|f_+ - e_+\| &= \max_{i \neq 1} |y_i| \leqslant |\lambda_1(c)|\, |\mu_1|^{-1} \max_{2 \leqslant j \leqslant l} |v_{j1}| \\ &= \max_{2 \leqslant j \leqslant l} |v_{j1}| < \epsilon^l \leqslant \epsilon. \end{aligned}$$

Thus we have completely proved our claim.

We shall now show that $f_+ \in \rho(G)e_+$. Let $Z(\mathbf{T})$ be the centralizer of $\mathbf{T}$ in $\mathbf{G}$. Then $Z(\mathbf{T})$ is reductive and it is an almost direct product of $\mathbf{T}$ with a k-anisotropic reductive subgroup. So, by Lemma 1.20, $Z(T)\,(=Z(\mathbf{T})(k))$ has a unique maximum compact subgroup M, and $M \cdot T$ is a subgroup of $Z(T)$ of finite index. Now, since vc is k-regular, there is a positive integer m such that a conjugate of $(vc)^m$ is contained in $Z(T)$ (cf. Lemma 2.2) and hence there exists a positive integer s such that a conjugate of $(vc)^s$ is contained in $M \cdot T$. Thus there is a $g \in G$ such that $g^{-1}(vc)^s g = xy$, with $x \in M$ and

$y \in T$. Note that $xy = g^{-1}(vc)^s g$ is k-regular, $\rho(g^{-1})f_+$ is an eigenvector of $\rho(xy)$ corresponding to its unique eigenvalue of maximum absolute value, and x commutes with y. Since $x \in M$, the eigenvalues of $\rho(x)$ are of absolute value 1 and therefore y is k-regular. Now let $N(T)$ be the normalizer of T in G, then since the Weyl group $N(T)/Z(T)$ acts transitively on the Weyl chambers, we can assume that g is so chosen that $y \in C$. As $k\rho(g^{-1})f_+$ is the eigenspace of $\rho(xy)$ corresponding to the unique eigenvalue of maximum absolute value, it is the eigenspace of $\rho(y)$ corresponding to its unique eigenvalue of maximum absolute value. But, as can be easily seen, for any $y \in C$, ke_+ is the eigenspace of $\rho(y)$ corresponding to the unique eigenvalue of maximum absolute value. Thus $\rho(g^{-1})f_+ = \lambda e_+$, i.e., $f_+ = \lambda\rho(g)e_+$, for a $\lambda \in k^\times$. We shall now use the Bruhat decomposition to prove that if for a $g \in G$ and $\lambda \in k^\times$, $f_+ = \lambda\rho(g)e_+$ is such that $e_+^*(f_+) = 1$, then $f_+ \in \rho(G)e_+$. We first note that $Z(T)$ acts by a character $\chi^+ : Z(T) \to k^\times$ (resp. $\chi^- : Z(T) \to k^\times$) on $\bigwedge^n \mathfrak{u}^+ = ke_+$ (resp. $\bigwedge^n \mathfrak{u}^- = ke_-$). Note also that the action of the unipotent group U^+ (resp. U^-) on ke_+ (resp. ke_-) is trivial. Now let $g = u_- z u_+$ be a Bruhat decomposition of g, with $u_- \in U^-$, $z \in N(T)$, and $u_+ \in U^+$. Then $f_+ = \lambda\rho(g)e_+ = \lambda\rho(u_- z)e_+$. Now since $e_+^*(f_+) = 1$, $\lambda e_+^*(\rho(u_-)\rho(z)e_+) = \lambda\rho^*(u_-^{-1})e_+^*(\rho(z)e_+) = 1$, where ρ^* is the contragradient of ρ. But since for every $\alpha \in U^-$, $\rho^*(\alpha)e_+^* = e_+^*$, we get

$$\lambda e_+^*(\rho(z)e_+) = 1. \tag{9}$$

If $z \notin Z(T)$, then $\operatorname{Ad} z$ takes the root space of at least one positive root to the root space of a negative root; from this we conclude at once that if $z \notin Z(T)$, then $\rho(z)e_+$ is contained in the subspace spanned by $\{e_i\}_{i \geq 2}$, and then $e_+^*(\rho(z)e_+) = 0$. This proves that $z \in Z(T)$. Then $\rho(z)e_+ = \chi^+(z)e_+$, and hence (by (9)) $\lambda\chi^+(z) = 1$. This implies that $\lambda = \chi^+(z^{-1})$ and therefore $\lambda e_+ = \rho(z^{-1})e_+$. Now $f_+ = \rho(g)\lambda e_+ = \rho(gz^{-1})e_+$, and we have proved that $f_+ \in \rho(G)e_+$. We note that a similar argument proves that if for a $g \in G$ and $\lambda \in k^\times$, $f_- = \lambda\rho(g)e_-$ is such that $e_-^*(f_-) = 1$, then $f_- \in \rho(G)e_-$. This will be used later in the proof of this proposition.

Since for $g \in G$ the multiplicity of an eigenvalue of $\operatorname{Ad} g$ is equal to the multiplicity of its inverse, it is evident that the multiplicity of an eigenvalue ν of $\rho(g)$ is equal to the multiplicity of the eigenvalue ν^{-1}. It follows, in particular, that for $v \in V^\epsilon$ and $c \in C$, there is a unique eigenvalue of $\rho(vc)$ of minimum absolute value and it has multiplicity 1. In fact, this eigenvalue is the inverse of the eigenvalue μ_1 of maximum absolute value. Let now $f_- = \sum z_i e_i$ with $z_l = 1$, and

$$\begin{bmatrix} z_1 \\ \vdots \\ z_{l-1} \end{bmatrix} = \lambda_l(c)\, \nabla^{-1} \begin{bmatrix} v_{1l} \\ \vdots \\ v_{l-1\,l} \end{bmatrix},$$

where ∇ is the $(l-1)$-rowed minor of $(\mu_1^{-1}\mathbf{I} - \rho(vc))$ obtained by omitting the last row and column. It can be shown that ∇ is nonsingular, $\|\nabla^{-1}\| \leqslant |\lambda_l(c)|^{-1}$ and $\rho(vc)f_- = \mu_1^{-1}f_-$, i.e., f_- is an eigenvector of $\rho(vc)$ corresponding to the eigenvalue μ_1^{-1}. Also, it is obvious that $e^*(f_-) = z_l = 1$, and

$$\begin{aligned}\|f_- - e_-\| = \max_{i\neq l} |z_i| &\leqslant |\lambda_l(c)|\,\|\nabla^{-1}\| \max_{1\leqslant j\leqslant l-1} |v_{jl}| \\ &\leqslant |\lambda_l(c)|\,|\lambda_l(c)|^{-1} \max_{1\leqslant j\leqslant l-1} |v_{jl}| \\ &< \epsilon^l \leqslant \epsilon.\end{aligned}$$

Moreover, arguing as in the preceding paragraph, we can conclude that $f_- \in \rho(G)e_-$.

Now let ρ also denote the natural representation of G on $\bigwedge^n \mathfrak{g} \oplus \bigwedge^n \mathfrak{g}$ and let $\mathcal{O} = \rho(G)(e_+, e_-)$ be the orbit of (e_+, e_-). Let $\mathcal{O}_+$ (resp. $\mathcal{O}_-$) be the orbit of e_+ (resp. e_-). Let π be the map $g \mapsto \rho(g)(e_+, e_-)$ of G onto $\mathcal{O}$, and let π_+ (resp. π_-) be the map $g \mapsto \rho(g)e_+$ (resp. $g \mapsto \rho(g)e_-$) of G onto $\mathcal{O}_+$ (resp. $\mathcal{O}_-$). Then, by Lemma 1.22, in characteristic zero all the three maps π, π_+, and π_- are open. In case k is of positive characteristic, Lemma 1.21 combined with Lemma 1.17 implies that π_+ and π_- are open maps. One can now use this to conclude that π is also open. Let Ω_0 be a neighborhood of the identity such that $\Omega_0^2 \subset \Omega$. Then there is a positive δ such that the given $(h_+, h_-) \in \mathcal{O}$ with $\|h_+ - e_+\| < \delta$ and $\|h_- - e_-\| < \delta$, there is a $g \in \Omega_0$ such that $\rho(g)(e_+, e_-) = (h_+, h_-)$.

Now since $U^+ \cdot Z(T) \cdot U^-$ is an open subset of G and $\pi_- : G \to \mathcal{O}_-$ is open, it follows that there is a $\xi > 0$ such that any $h_- \in \mathcal{O}_-$ with $\|h_- - e_-\| < \xi$ belongs to $\rho(U^+ \cdot Z(T))e_- = \rho(U^+ \cdot Z(T) \cdot U^-)e_-$. Also for $u \in U^+$ and $z \in Z(T)$, $e_-^*(\rho(uz)e_-) = \chi^-(z)e_-^*(\rho(u)e_-) = \chi^-(z)\rho^*(u^{-1})e_-^*(e_-) = \chi^-(z)$. Thus every $h_- \in \mathcal{O}_-$, with $\|h_- - e_-\| < \xi$ and $e_-^*(h_-) = 1$, is actually contained in $\rho(U^+)e_-$. Let N be an open neighborhood of the identity in U^- and η be a positive constant such that for all $n \in N$ and $x \in \bigwedge^n \mathfrak{g}$ with $\|x - e_-\| < \eta$

$$e_-^*(\rho(n^{-1})x) \neq 0,$$

$$\left\| \frac{1}{e_-^*(\rho(n^{-1})x)} x - e_- \right\| < \delta, \qquad \text{and} \qquad \left\| \frac{\rho(n^{-1})x}{e_-^*(\rho(n^{-1})x)} - e_- \right\| < \xi.$$

Now since $N \cdot Z(T) \cdot U^+$ is an open neighborhood of the identity in G and π_+ is an open map, there exists a $\kappa > 0$ such that every $h_+ \in \mathcal{O}_+$, with $\|h_+ - e_+\| < \kappa$, is contained in $\rho(N \cdot Z(T))e_+ = \rho(N \cdot Z(T) \cdot U^+)e_+$, and since for $n \in N$ and $z \in Z(T)$, $e_+^*(\rho(nz)e_+) = \chi^+(z)e_+^*(\rho(n)e_+) = \chi^+(z)$, it follows that if further $e_+^*(h_+) = 1$, then h_+ is contained in $\rho(N)e_+$.

Let ϵ be a positive number less than $\min(1, \delta, \xi, \eta, \kappa)$. Let $v \in V^\epsilon$ and c be an element of C. Then, as we have shown above, vc is k-regular. Moreover, there exist eigenvectors f_+ $(\in\mathcal{O}_+)$ and f_- $(\in\mathcal{O}_-)$ of $\rho(vc)$, corresponding to its unique eigenvalues of maximum and minimum absolute values, respectively, such that $e_+^*(f_+) = 1 = e_-^*(f_-)$, $\|f_+ - e_+\| < \epsilon$ and $\|f_- - e_-\| < \epsilon$. Since $\epsilon < \kappa$, there is an $n \in N$ such that $f_+ = \rho(n)e_+$. Moreover, as

$$\|f_- - e_-\| < \epsilon < \eta, \qquad e_-^*(\rho(n^{-1})f_-) \neq 0,$$

$$\|e_-^*(\rho(n^{-1})f_-)^{-1}f_- - e_-\| < \delta, \qquad \text{and} \qquad \|h_- - e_-\| < \xi,$$

where $h_- = \rho(n^{-1})f_-/e_-^*(\rho(n^{-1})f_-)$. Now

$$(f_+, e_-^*(\rho(n^{-1})f_-)^{-1}f_-) = \rho(n)(e_+, h_-),$$

and since $f_- \in \mathcal{O}_-$, $\rho(n^{-1})f_-$ and (therefore) h_- $(= e_-^*(\rho(n^{-1})f_-)\rho(n^{-1})f_-)$ are elements of $\mathcal{O}_-$. But since $\|h_- - e_-\| < \xi$ and, obviously, $e_-^*(h_-) = 1$, there is a $u \in U^+$ such that $\rho(u)e_- = h_-$. Therefore,

$$(f_+, e_-^*(\rho(n^{-1})f_-)^{-1}f_-) = \rho(n)(e_+, \rho(u)e_-) = \rho(nu)(e_+, e_-).$$

This shows that $(f_+, e_-^*(\rho(n^{-1})f_-)^{-1}f_-) \in \mathcal{O}$. Now since $\|f_+ - e_+\| < \epsilon < \delta$ and $\|e_-^*(\rho(n^{-1})f_-)^{-1}f_- - e_-\| < \delta$, there is a $g \in \Omega_0$ such that $\rho(g)(e_+, e_-) = (f_+, e_-^*(\rho(n^{-1})f_-)^{-1}f_-)$. As f_+ and $e_-^*(\rho(n^{-1})f_-)^{-1}f_-$ are eigenvectors of $\rho(vc)$, it follows that e_+ and e_- are eigenvectors of $\rho(g^{-1}vcg)$. This implies, in particular (cf. (1) and (2)), that

$$g^{-1}vcg \in Z(T) \cdot U^+ \cap U^- \cdot Z(T) \subset Z(T).$$

Thus $vc \in g[Z(T)] \subset \Omega_0[Z(T)]$. Now let $V_0 = \Omega_0 \cap V^\epsilon$, and V be a symmetric neighborhood of the identity such that $V^2 \subset V_0$, then

$$V \cdot C \cdot V \subset V[V^2 \cdot C] \subset \Omega_0[V^\epsilon \cdot C] \subset \Omega_0^2[Z(T)] \subset \Omega[Z(T)],$$

and since every element of $V^\epsilon \cdot C$ is k-regular and $V \cdot C \cdot V \subset \Omega_0[V^\epsilon \cdot C]$, $V \cdot C \cdot V$ consists entirely of k-regular elements. This proves the proposition.

2.13. *Remark.* Proposition 2.12 is a non-archimedean analogue of a result of Mostow [23, Lemma 8.2]. (A different proof of Mostow's result can be found in Prasad and Raghunathan [26, Sect. 2].) Both the proof of Mostow and the proof given in [26] make essential use of the geometry of euclidean spaces.

The following proposition is a simple consequence of Proposition 2.12. We record it here for later use.

2.14. PROPOSITION. *Let $\mathscr{C}$ be a finite set and for every $\alpha \in \mathscr{C}$, let $\mathbf{G}_\alpha$ be a connected semisimple linear algebraic group defined over a non-archimedean local field k_α, and let G_α be the group of k_α-rational points of $\mathbf{G}_\alpha$ with the natural*

Hausdorff topology. For $\alpha \in \mathscr{C}$, let $\mathbf{T}_\alpha$ be a maximal k_α-split torus in $\mathbf{G}_\alpha$, and let T_α be the group of k_α-rational points of $\mathbf{T}_\alpha$. Let $G = \prod_{\alpha \in \mathscr{C}} G_\alpha$, $T = \prod_{\alpha \in \mathscr{C}} T_\alpha$, and $Z(T)$ be the centralizer of T in G. Let Γ be a lattice in G and Ω be a neighborhood of the identity in G. Then $\Omega[Z(T)] \cap \Gamma$ contains a polar regular element.

Proof. Let T_r be the set of polar regular elements in T, and let V be a neighborhood of the identity in G such that $V \cdot T_r \cdot V$ consists entirely of polar regular elements and $V \cdot T_r \cdot V \subset \Omega[Z(T)]$. Existence of such a neighborhood V is guaranteed by Proposition 2.12. Now let $t \in T$ be a polar regular element. Then for every nonzero integer m, t^m is polar regular. Hence

$$\bigcup_{m \neq 0} V \cdot t^m \cdot V \subset \Omega[Z(T)]$$

and elements of $\bigcup_{m \neq 0} V \cdot t^m \cdot V$ are polar regular. On the other hand, since Γ has property (S) in G, $(\bigcup_{m \neq 0} V \cdot t^m \cdot V) \cap \Gamma \neq \varnothing$ and thus $\Omega[Z(T)] \cap \Gamma$ has a polar regular element. This proves the proposition.

3. Determination of Polar Rank from the Group Theoretic Structure of a Lattice—Two Methods

3.1. Let $\mathscr{S}$ be a finite set and for every $\alpha \in \mathscr{S}$, let $\mathbf{G}_\alpha$ be a connected semisimple algebraic group defined over a local field k_α and G_α be the group of k_α-rational points of $\mathbf{G}_\alpha$ with the natural locally compact Hausdorff topology. We assume that for no $\alpha \in \mathscr{S}$, $\mathbf{G}_\alpha$ has a nontrivial connected k_α-anisotropic factor. Let $G = \prod_{\alpha \in \mathscr{S}} G_\alpha$ and let Γ be a *uniform* lattice in G. The object of this section is to give two methods of computing polar rank of G in terms of the group theoretic structure of Γ. We begin with a

3.2. Lemma. *Let G and Γ be as above. Let $r =$ polar rank G. Then any abelian subgroup of Γ is finitely generated and is of rank $\leqslant r$. Moreover, if $\rho \in \Gamma$ is a polar regular s.s. element, then the centralizer Γ_ρ of ρ in Γ has an abelian subgroup of finite index which is of rank r.*

Proof. To prove the first assertion of the lemma let Λ be an abelian subgroup of Γ. After replacing Λ by a subgroup of finite index we assume (cf. Lemma 1.11) that Λ is torsion free. Now for $\alpha \in \mathscr{S}$ let $\pi_\alpha : G \to G_\alpha$ be the natural projection. By Lemma 1.3, for $\alpha \in \mathscr{S}$ such that k_α is of characteristic zero, $\pi_\alpha(\gamma)$ is semisimple for all $\gamma \in \Gamma$. On the other hand, for any $\alpha \in \mathscr{S}$ such that k_α is of positive characteristic, there exists a positive integer n_α such that for all $g_\alpha \in G_\alpha$, $g_\alpha^{n_\alpha}$ is semisimple. Therefore we can choose a positive integer n such that for all $\gamma \in \Gamma$ and $\alpha \in \mathscr{S}$, $\pi_\alpha(\gamma^n)$ is semisimple. Since Λ is torsion free, the map $\lambda \mapsto \lambda^n$ is injective. Thus replacing Λ by a subgroup *isomorphic* to Λ we may (and we shall) assume that for all $\lambda \in \Lambda$ and $\alpha \in \mathscr{S}$, $\pi_\alpha(\lambda)$ is semisimple.

Now let $\mathbf{T}_\alpha$ be the connected component of the identity in the Zariski closure of $\pi_\alpha(\Lambda)$ in $\mathbf{G}_\alpha$. Since $\pi_\alpha(\Lambda)$ is abelian and it consists entirely of semisimple elements, $\mathbf{T}_\alpha$ is a torus. Let $\mathbf{T}_\alpha^d$ be the maximum k_α-split subtorus of $\mathbf{T}_\alpha$ and let T_α (resp. T_α^d) be the group of k_α-rational points of $\mathbf{T}_\alpha$ (resp. $\mathbf{T}_\alpha^d$) with the natural Hausdorff topology. Then (cf. the proof of Lemma 2.2) T_α has a unique maximum compact subgroup C_α and $C_\alpha \cdot T_\alpha^d$ is a subgroup of finite index in T_α. Let $C = \prod_{\alpha \in \mathscr{S}} C_\alpha$ and $T^d = \prod_{\alpha \in \mathscr{S}} T_\alpha^d$. Once again after replacing Λ by a subgroup of finite index we assume that $\Lambda \subset C \cdot T^d$.

Since T_α^d is the group of k_α-rational points of a k_α-split torus in $\mathbf{G}_\alpha$, $T_\alpha^d \approx (k_\alpha^\times)^{m_\alpha}$, where $m_\alpha = \dim \mathbf{T}_\alpha^d \leqslant k_\alpha$-rank $\mathbf{G}_\alpha$. Now if $k_\alpha = \mathbf{C}$ or $\mathbf{R}$, then $k_\alpha^\times \approx S \times \mathbf{R}$ where S is the (compact) subgroup of k_α consisting of elements of absolute value 1; whereas if k_α is non-archimedean, then $k_\alpha \approx \mathfrak{o}_\alpha^\times \times \mathbf{Z}$, where $\mathfrak{o}_\alpha^\times$ is the subgroup of k_α consisting of elements of absolute value 1. Thus it follows that (note that $\mathfrak{o}_\alpha^\times$ is a compact subgroup of k_α) $T^d = \prod_\alpha T_\alpha^d$ is isomorphic to the direct product of a compact group and $\mathbf{R}^m \times \mathbf{Z}^n$ with $m + n \leqslant r$. Since Λ is a discrete subgroup of $C \cdot T^d$ and C is compact, it readily follows that Λ is finitely generated and is of rank $\leqslant m + n \leqslant r$. This proves the first assertion of the lemma. We shall now prove the second assertion.

For an element $\gamma \in \Gamma$, let G_γ (resp. Γ_γ) denote the centralizer of γ in G (resp. in Γ). Then it can be shown easily (see for example [27, Lemma 1.14]) that $G_\gamma \cdot \Gamma$ is closed and hence $G_\gamma \cdot \Gamma/\Gamma$, being a closed subset of the compact space G/Γ, is compact. Now by a simple application of Lemma 1.17, it follows that the natural map $G_\gamma/\Gamma_\gamma \to G \cdot \Gamma_\gamma/\Gamma$ is a homeomorphism and hence for all $\gamma \in \Gamma$, G_γ/Γ_γ is compact. Thus, in particular, G_ρ/Γ_ρ is compact. Now let $G_\rho^0 = \prod_\alpha H_\alpha$, where H_α is the group of k_α-rational points of the connected component of the identity in the centralizer of $\rho_\alpha = \pi_\alpha(\rho)$ in $\mathbf{G}_\alpha$. Then, of course, G_ρ/G_ρ^0 is finite and it can be easily seen, using Lemma 2.2 and Lemma 1.20, that G_ρ^0 has a unique maximum compact subgroup and the quotient of G_ρ^0 by this subgroup is isomorphic to $\mathbf{R}^a \times \mathbf{Z}^b$, with $a + b = r$. Now since G_ρ/Γ_ρ (and hence $G_\rho^0/\Gamma_\rho \cap G_\rho^0$) are compact, it follows easily that ${}^0\Gamma_\rho = \Gamma_\rho \cap G_\rho^0$ has an abelian subgroup, of finite index, of rank $a + b$ $(=r)$. Since $\Gamma_\rho/{}^0\Gamma_\rho$ is finite, this proves the second assertion of the lemma.

The following proposition provides us with a method of determining polar rank of G.

3.3. PROPOSITION. *Let G and Γ be as in 3.1. Then the polar rank of G equals the rank of an abelian subgroup of Γ of maximum rank.*

Proof. Let $r =$ polar rank G. Then, by the preceding lemma, the rank of any abelian subgroup of Γ is at most r. Also, according to Lemma 2.10, Γ

contains a polar regular s.s. element ρ and by Lemma 3.2, the centralizer Γ_ρ of ρ (in Γ) has an abelian subgroup of rank r. This proves the proposition.

We shall now define rank of an abstract group and shall prove that if Γ is a *uniform* lattice in G, then $\operatorname{rank}(\Gamma) = \text{polar rank } G$. This turns out to be crucial for proving that in general a lattice in a real analytic semisimple group cannot be isomorphic to a lattice in a semisimple group over a non-archimedean local field (for precise statements see Section 4).

3.4. DEFINITION. Let Λ be an abstract group. For a nonnegative integer n, let $\Lambda(n)$ be the subset consisting of $\lambda \in \Lambda$ such that the centralizer Λ_λ of λ in Λ has a finitely generated abelian subgroup, of rank $\leqslant n$, as a subgroup of finite index. Now if for no n, Λ is a union of finitely many translates of $\Lambda(n)$, then the *rank* of Λ is ∞. Otherwise, the *rank* of Λ (denoted $\operatorname{rank}(\Lambda)$) is, by definition, the smallest nonnegative integer n such that there is a finite subset Σ of Λ with $\Lambda = \Lambda(n) \cdot \Sigma$.

3.5. *Remark.* Let $\{e\} \to Z \to \Gamma \xrightarrow{\pi} \Lambda \to \{e\}$ be a central extension of Λ, with Z finite. Then it is not too difficult to prove that $\operatorname{rank}(\Gamma) = \operatorname{rank}(\Lambda)$. In fact, it can be shown that, for all n, $\Gamma(n) = \pi^{-1}(\Lambda(n))$, and this immediately implies that $\operatorname{rank}(\Gamma) = \operatorname{rank}(\Lambda)$.

3.6. DEFINITION. Let $G = \prod_{\alpha \in \mathscr{S}} G_\alpha$ be as before, and let $\pi_\alpha : G \to G_\alpha$ be the natural projections. An element $g \in G$ is said to be *regular* if $\pi_\alpha(g)$ is regular in G_α for all $\alpha \in \mathscr{S}$. Since for $\alpha \in \mathscr{S}$, the set of regular elements of G_α is a Zariski open subset of G_α, the set $\mathscr{R}$ of regular elements of G is open in the weak Zariski topology.

3.7. THEOREM. *Let G and Γ be as in Subsection* 3.1. *Then*

$$\operatorname{rank}(\Gamma) = \text{polar rank } G.$$

Proof. By Proposition 1.12, Γ is dense in G in the weak Zariski topology. Now since the set $\mathscr{R}$ of regular elements of G is open, $\Gamma \cap \mathscr{R} \neq \varnothing$, i.e., Γ contains regular elements. Let $\gamma \in \Gamma$ be a regular element. Then the centralizer Γ_γ of γ in Γ has an abelian subgroup ${}^0\Gamma_\gamma$ of finite index; in fact, since for every $\alpha \in \mathscr{S}$, $\pi_\alpha(\gamma)$ is regular, the connected component of the identity in the centralizer of $\pi_\alpha(\gamma)$ in $\mathbf{G}_\alpha$, being a (maximal) torus, is abelian and thus the centralizer of γ in G has an abelian subgroup of finite index. Now according to Lemma 3.2, rank ${}^0\Gamma_\gamma \leqslant r = \text{polar rank } G$, and hence $\gamma \in \Gamma(r)$ (for notation, see Definition 3.4). This shows that $\Gamma \cap \mathscr{R} \subset \Gamma(r)$. Since the weak Zariski topology is noetherian, we can easily conclude that there is a finite subset Σ of Γ such that $(\Gamma \cap \mathscr{R}) \cdot \Sigma = \Gamma$. Now since $\Gamma \cap \mathscr{R} \subset \Gamma(r)$, this shows that $\operatorname{rank}(\Gamma) \leqslant r$. Thus to complete the proof of the theorem we have only to show that if $q < r$, then for no finite subset Σ of Γ, $\Gamma(q) \cdot \Sigma = \Gamma$. We first observe that, in view of Lemma 3.2, no element of $\Gamma(q)$ is polar regular s.s. Now if possible let $\Gamma = \bigcup_{i=1}^{t} \Gamma(q) \cdot \gamma_i$ with $\gamma_i \in \Gamma$. Let $\gamma \in \Gamma$ be a polar regular element (existence of polar regular elements in Γ is guaranteed by Lemma 2.10). Let $\delta \in \Gamma$ and s be a positive integer such that $\gamma^s \delta \gamma_i^{-1}$ is polar regular s.s. for $1 \leqslant i \leqslant t$ (see Lemma 2.11). Now since $\Gamma = \bigcup_{i=1}^{t} \Gamma(q)\gamma_i$, $\gamma^s \delta \gamma_j^{-1} \in \Gamma(q)$ for some $j \leqslant t$. This contradicts the fact that $\Gamma(q)$ has no polar regular s.s. elements. Hence, $\operatorname{rank}(\Gamma) = r$.

For real analytic semisimple groups the following stronger result holds.

3.8. THEOREM (Prasad–Raghunathan [26, Sect. 3.9]). *Let H be a real analytic linear semisimple group with no compact factors. Let Δ be a (not necessarily uniform) lattice in H. Then*

$$\operatorname{rank}(\Delta) = \mathbf{R}\text{-rank } H.$$

4. COMPARISON OF LATTICES IN REAL AND NON-ARCHIMEDEAN SEMISIMPLE GROUPS

We begin with some definitions:

4.1. DEFINITION. (Serre [32, Sect. 1.8]). An abstract group Λ is said to be *virtually torsion free* (resp. *virtually free*) if there is a torsion free (resp. free) subgroup of Λ of finite index.

Let Λ be a virtually torsion free group. By definition, the *virtual cohomological dimension* of Λ (denoted $\operatorname{vcd}(\Lambda)$) is the cohomological dimension of a torsion free subgroup of Λ of finite index. In view of Serre [32, Théorème 1], $\operatorname{vcd}(\Lambda)$ is well defined.

4.2. DEFINITION. (see also [25, Remark 1.6]). Let H be a real analytic linear semisimple group with trivial center and no compact factors. We call an irreducible nonuniform lattice Δ in H a $\mathbf{Q}$-*rank* 1 *lattice* if it has the following property:

(R1) Given a nontrivial unipotent element of Δ, there exists a *unique* maximal unipotent subgroup of Δ which contains it.

4.3. DEFINITION. Let L be a real Lie group with finitely many connected components. We define the *characteristic index* $\chi(L)$ (or $d(L)$ in Serre's notation, see [32, Sect. 2.1]) of L by

$$\chi(L) = \dim L - (\text{dimension of a maximal compact subgroup of } L).$$

The following is known (cf. Goto and Wang [16, Sect. 1]).

4.4. Let M be a closed subgroup of a real Lie group L. Assume that both L and M have finitely many connected components. Then $\chi(M) \leqslant \chi(L)$, and $\chi(M) = \chi(L)$ if and only if L/M is compact. Moreover, if M is normal, then $\chi(L) = \chi(M) + \chi(L/M)$.

The following lemma is a consequence of the recent results, due to Margulis and Raghunathan, on the arithmeticity of nonuniform lattices in real analytic linear semisimple groups and a theorem of Borel and Serre on the cohomological dimension of arithmetic groups.

4.5. Lemma. *Let H be a real analytic linear semisimple group with trivial center and no compact factors and let Δ be an irreducible lattice in H. Let* $r = \mathbf{R}\text{-rank } H$ $(=\text{rank}(\Delta)$; *see Theorem* 3.8). *Then*

$$\text{vcd}(\Delta) \geqslant \chi(H) - r.$$

Moreover, if Δ is uniform, then $\text{vcd}(\Delta) = \chi(H)$, *and if Δ is a* $\mathbf{Q}$*-rank* 1 *lattice, then* $\text{vcd}(\Delta) = \chi(H) - 1$.

Proof. When Δ is a uniform lattice in H, according to Serre [32, Proposition 18(a)] $\text{vcd}(\Delta) = \chi(H)$ $(> \chi(H) - r)$. Thus it only remains to prove the lemma in the case Δ is a nonuniform lattice. Let us first consider the case when Δ is a $\mathbf{Q}$-rank 1 lattice. Let Φ be a maximal unipotent subgroup of Δ and let U be the minimum unipotent analytic subgroup of H containing Φ. Let $N(U)$ be the normalizer of U in H and let

$$N^0(U) = \{h \in N(U) \mid \text{Int } h|_U \text{ preserves a Haar measure on } U\}.$$

Then by [25, Lemma 1.2 and its proof] $N(U)$ is a parabolic subgroup of H and $\Lambda = N^0(U) \cap \Delta$ is a uniform lattice in $N^0(U)$. Now since $N(U)$ is parabolic, $H/N(U)$ is compact and hence (cf. 4.4)

$$\chi(N^0(U)) = \chi(N(U)) - 1 = \chi(H) - 1.$$

Now (since Λ is a uniform lattice in $N^0(U)$), by Serre [32, Proposition 18(a)],

$$\text{vcd}(\Lambda) = \chi(H) - 1.$$

Also, according to [32, Proposition 5], $\text{vcd}(\Delta) \geqslant \text{vcd}(\Lambda)$. Thus $\text{vcd}(\Delta) \geqslant \chi(H) - 1$. On the other hand, since Δ is a nonuniform lattice, in view of [32, Proposition 18(b)], $\text{vcd}(\Delta) \leqslant \chi(H) - 1$ and thus $\text{vcd}(\Delta) = \chi(H) - 1$.

Since H is a nontrivial noncompact semisimple group, $r = \mathbf{R}\text{-rank } H \geqslant 1$ and so

$$\text{vcd}(\Delta) = \chi(H) - 1 \geqslant \chi(H) - r.$$

Let us now assume that Δ is not a $\mathbf{Q}$-rank 1 lattice. Then according to Margulis [22] and Raghunathan [28], Δ is an arithmetic lattice, i.e., there exists a connected linear algebraic simple group $\mathbf{H}$ defined over $\mathbf{Q}$ and an analytic isomorphism $\lambda: H \to \mathbf{H}(\mathbf{R})^0$ ($\mathbf{H}(\mathbf{R})^0$ is the identity component of the group of $\mathbf{R}$-rational points of $\mathbf{H}$) such that $\lambda(\Delta)$ is arithmetic. Since $\mathbf{H}(\mathbf{R})^0$ is isomorphic with H, $\mathbf{R}\text{-rank } \mathbf{H} = r$. Let $s = \mathbf{Q}\text{-rank } \mathbf{H}$, since any $\mathbf{Q}$-split torus is $\mathbf{R}$-split, $r \geqslant s$. According to Borel–Serre [5, Sect. 11],[3] $\text{cd}(\lambda(\Delta)) = \chi(H) - s$. Thus

$$\text{vcd}(\Delta) = \text{vcd}(\lambda(\Delta)) = \chi(H) - s \geqslant \chi(H) - r.$$

This completes the proof of the lemma.

4.6. Lemma. *Let H be a real analytic linear semisimple group with no compact factors and let Δ be an irreducible lattice in H. Then*

$$\text{vcd}(\Delta) \geqslant \text{rank}(\Delta).$$

Moreover, if $\text{vcd}(\Delta) = \text{rank}(\Delta)$, *then Δ is nonuniform and virtually free, and H is locally isomorphic to $SL(2, \mathbf{R})$.*

Proof. Let r be the $\mathbf{R}$-rank of H. Then according to Theorem 3.8 $\text{rank}(\Delta) = r$. After replacing Δ by a subgroup of finite index we can assume that Δ is torsion free and then dividing out the (finite) center of H we can assume that H has trivial center. Now by Lemma 4.5, $\text{vcd}(\Delta) \geqslant \chi(H) - r$. On the other hand, from Iwasawa decomposition $H = KAN$ of H it follows that $\chi(H) - r = \chi(H) - \dim A = \dim N$. Hence $\text{vcd}(\Delta) \geqslant \dim N$. But, as is well known (and follows for example by considering the root space decomposition of the Lie algebra of H with respect to $\text{Ad}\, A$), $\dim N \geqslant r = \dim A$ and $\dim N = r$ if and only if H is locally isomorphic to the direct product of r copies of $SL(2, \mathbf{R})$. Thus $\text{vcd}(\Delta) \geqslant r$ and if $\text{vcd}(\Delta) = r$, then H is locally isomorphic to the direct product of r copies of $SL(2, \mathbf{R})$.

To prove the second assertion of the lemma we assume that $\text{vcd}(\Delta) = r$. Then H is locally isomorphic to the direct product of r copies of $SL(2, \mathbf{R})$. Now since, by Lemma 4.5, virtual cohomological dimension of a uniform lattice in H is $\chi(H) = 2r$, Δ is nonuniform. But then according to [25, Lemma 1.1] Δ is a $\mathbf{Q}$-rank 1, lattice and so by Lemma 4.5, $\text{vcd}(\Delta) = \chi(H) - 1 - 2r - 1$. Since by hypothesis $\text{vcd}(\Delta) = r$, we get $r = 1$, i.e., H is locally iso-

[3] In fact, we need only the following: Let Γ be a torsion free arithmetic subgroup of $\mathbf{H}$, then $\text{cd}(\Gamma) \geqslant \chi(H) - s$. This statement can be proved directly.

morphic to $SL(2, \mathbf{R})$. It is well known that nonuniform lattices in (linear) groups locally isomorphic to $SL(2, \mathbf{R})$ are virtually free.

4.7. Let $\mathscr{C}$ be a finite set. For $\alpha \in \mathscr{C}$, let k_α be a non-archimedean local field and let $\mathbf{G}_\alpha$ be a connected semisimple algebraic group defined over k_α. We assume that for no $\alpha \in \mathscr{C}$, $\mathbf{G}_\alpha$ has a nontrivial connected normal k_α-anisotropic subgroup. Let G_α be the group of k_α-rational points of $\mathbf{G}_\alpha$. We endow $\prod_{\alpha \in \mathscr{C}} G_\alpha$ with the natural locally compact Hausdorff topology. In the sequel we call a group of the form $\prod_{\alpha \in \mathscr{C}} G_\alpha$ a *non-archimedean semisimple group.*

For uniform lattices in nonarchimedean semisimple groups we have the following result.

4.8. PROPOSITION. *Let G be a non-archimedean semisimple group and let Γ be a uniform lattice in G. Let $r =$ polar rank G. Then*

$$\operatorname{vcd}(\Gamma) = r = \operatorname{rank}(\Gamma).$$

Proof. By Théorème 2 of Serre [32] $\operatorname{vcd}(\Gamma) \leqslant r$. (This follows easily from the fact that any discrete torsion free subgroup of G acts simplicially and properly discontinuously on a contractible simplicial complex, namely the Bruhat–Tits building associated with G, of dimension r.)

On the other hand, by Proposition 3.3, Γ has a free abelian subgroup Λ of rank r. It is well known that $\operatorname{cd}(\Lambda) = r$, and according to Serre [32, Proposition 5], $\operatorname{vcd}(\Gamma) \geqslant \operatorname{cd}(\Lambda) = r$. Therefore $\operatorname{vcd}(\Gamma) = r$. Also, by Theorem 3.7, $\operatorname{rank}(\Gamma) = r$ and the proof of the proposition is complete.

From Lemma 4.6 and Proposition 4.8 we can easily deduce the following:

4.9. THEOREM. *Let G be a non-archimedean semisimple group and let H be a real linear analytic semisimple group. Let Γ (resp. Δ) be an irreducible lattice in G (resp. H). Assume that Γ is isomorphic to Δ. Then Δ is virtually free.*

(We note that since Δ is virtually torsion free (see Lemmas 1.10–1.11), Γ is virtually torsion free and hence, by Lemma 1.2, it is uniform.)

4.10. *Remark.* Since a subgroup of a free group is free, the centralizer of any nontrivial element in a free group F is infinite cyclic and thus $\operatorname{rank}(F) = 1$. This, in view of Theorem 3.7, implies that if a lattice in a non-archimedean semisimple group G is virtually free, then polar rank $G = 1$. Conversely, any torsion free discrete subgroup of a non-archimedean semisimple group of polar rank 1 is free (cf. Serre [32, Corollary on p. 121]). We note here that Ihara [19] has constructed uniform lattices in $PGL(2, \mathbf{Q}_p)$.

4.11. *Remark.* We sketch an alternative proof of Theorem 4.9 which does not use the arithmeticity result. We observe first that since Γ is isomorphic to Δ, it follows from Theorems 3.7 and 3.8 that polar rank $G = \mathbf{R}$-rank H. Now if possible assume that Δ is uniform; then $\mathrm{vcd}(\Delta) = \chi(H) > \mathbf{R}$-rank H. On the other hand, $\mathrm{vcd}(\Delta) = \mathrm{vcd}(\Gamma) =$ polar rank G, a contradiction. Therefore Δ is nonuniform. Now if $\mathbf{R}$-rank $H > 1$, then it can be shown that every irreducible nonuniform lattice in H has a solvable subgroup which is not virtually abelian, i.e., no subgroup, of the solvable subgroup, of finite index is abelian. But one can use Lemmas 1.2, 1.3 and the Lie–Kolchin theorem to show that every solvable subgroup of Γ is virtually abelian. Therefore, we conclude that $\mathbf{R}$-rank $H = 1$. This implies that polar rank $G = 1$ and (cf. the preceding remark) Γ and hence Δ is virtually free.

5. Geometry of Bruhat–Tits Buildings and Parabolic Subgroups

5.1. Let $\mathbf{G}$ be a connected, simply connected algebraic group defined, isotropic, and almost simple over a non-archimedean local field k. We fix a maximal k-split torus $\mathbf{T}$ of $\mathbf{G}$ and a minimal parabolic k-subgroup $\mathbf{P}$ containing $\mathbf{T}$. Let $N(\mathbf{T})$ (resp. $Z(\mathbf{T})$) be the normalizer (resp. centralizer) of $\mathbf{T}$ in $\mathbf{G}$. Note that $Z(\mathbf{T}) \subset \mathbf{P}$. Let G, P, T, $N(T)$, and $Z(T)$ be the groups of k-rational points of $\mathbf{G}$, $\mathbf{P}$, $\mathbf{T}$, $N(\mathbf{T})$, and $Z(\mathbf{T})$, respectively, with the natural Hausdorff topology. Let ${}_0W$ be the Weyl group $N(T)/Z(T)$, and let ${}_0S$ ($\subset {}_0W$) be the set of reflections corresponding to the elements in the basis of k-roots of $\mathbf{G}$ with respect to $\mathbf{T}$ associated with $\mathbf{P}$.

Let (G, B, N, S) be the standard *saturated* Tits system [12] of G ($=\mathbf{G}(k)$), whose "parabolic" subgroups are the parahoric subgroups of G, which is compatible with $\mathbf{T}$ and $\mathbf{P}$ fixed above. In particular, $N = N(T)$, B is an open compact subgroup, and $H = B \cap N$ is the maximum compact subgroup of $Z(T)$. The Weyl group $W = N/H$ of this Tits system is an extension of ${}_0W$ by the group $Z(T)/H$ which is free abelian of rank $s = k$-rank $\mathbf{G}$ ($=\dim \mathbf{T}$). Moreover, there is a canonical identification of ${}_0S$ with a subset of S.

Let $\mathscr{I}$ be the Tits complex [12, Sect. 2] associated with the Tits system (G, B, N, S). We recall that $\mathscr{I}$ is a locally finite simplicial complex of dimension s and the group G operates simplicially on it. Its geometric realization, known as the *Bruhat–Tits building of* $\mathbf{G}$ *over* k, is a contractible space and it carries a natural G-invariant metric d. In the sequel we shall let $\mathscr{I}$ also denote the Bruhat–Tits building with the G-invariant Bruhat–Tits metric d.

The group G operates transitively on the set of s-simplexes (chambers) of $\mathscr{I}$, and there is a unique s-simplex C, known as the *standard chamber*, fixed by the subgroup B.

5.2. *Notation.* For a simplex Δ of $\mathscr{I}$, $\bar{\Delta}$ denotes the closed simplex; it is the (compact) subspace of the Bruhat–Tits building corresponding to the subcomplex consisting of Δ and all its faces. Given a simplex Δ, the interior Δ^0 of Δ is the subset of $\bar{\Delta}$ consisting of points which do not lie on a proper face of $\bar{\Delta}$.

Now let C be the standard chamber. Then (cf. [12, Corollaire 2.1.6]) $G \cdot \bar{C} = \mathscr{I}$.

The subset $\mathsf{A} = \bigcup_{n \in N(T)} n \cdot \bar{C}$ is the *standard apartment* and its transforms by elements of G are the *apartments* of $\mathscr{I}$. The subgroup H fixes the standard apartment pointwise. It is well known that the apartments are isometric to the euclidean space $\mathbf{R}^s$.

Given two points $x, y \in \mathscr{I}$, there is a unique *geodesic*, denoted $[xy]$ of length $d(x, y)$ joining x to y, and it is contained in every apartment containing both x and y [12, Proposition 2.5.4]. A subset E of $\mathscr{I}$ is said to be *convex* if given any two points x, y of E, the unique geodesic $[xy]$ joining x to y is contained in E.

In the sequel, for a subset E of a metric space (M, d) and a nonnegative v, we let $T_v(E)$ denote the subset $\{x \in M \mid d(x, E) \leqslant v\}$.

5.3. LEMMA. *Any closed subset of $\mathscr{I}$ of finite diameter is compact.*

Proof. For $\delta > 0$ and $x \in \mathscr{I}$, let $B_\delta(x)$ denote the closed ball, of radius δ, centered at x. Since $\mathscr{I}$ is locally compact, for every $x \in \mathscr{I}$ there is an $\epsilon(x) > 0$ such that $B_{\epsilon(x)}(x)$ is compact. Now since $\bar{C}$ is compact, there exists a finite subset $\{c_i\}_{i=1}^m$ of $\bar{C}$ such that $\bigcup_{i=1}^m B_{\frac{1}{2}\epsilon(c_i)}(c_i) \supset \bar{C}$. Let $\epsilon = \frac{1}{4} \min_{1 \leqslant i \leqslant m} \epsilon(c_i)$. Given $c \in \bar{C}$, there is an $i \leqslant m$ such that $c \in B_{\frac{1}{2}\epsilon(c_i)}(c_i)$ and then clearly $B_{2\epsilon}(c) \subset B_{\frac{1}{2}\epsilon(c_i)}(c) \subset B_{\epsilon(c_i)}(c_i)$. Thus for all $c \in \bar{C}$, $B_{2\epsilon}(c)$ is compact. Since $G \cdot \bar{C} = \mathscr{I}$, for all $x \in \mathscr{I}$, $B_{2\epsilon}(x)$ is compact.

Given a compact subset E of $\mathscr{I}$, there exist finitely many elements $\{e_i\}_{i=1}^n$ of E such that $\bigcup_{i=1}^n B_\epsilon(e_i) \supset E$ and then for all $e \in E$, $B_\epsilon(e) \subset \bigcup_{i=1}^n B_{2\epsilon}(e_i)$. Now since $T_\epsilon(E) = \bigcup_{e \in E} B_\epsilon(e) \subset \bigcup_{i=1}^n B_{2\epsilon}(e_i)$, it follows that $T_\epsilon(E)$ is compact. We shall now prove, by induction, that for every positive integer r, $T_{r\epsilon}(E)$ is compact. So let us assume the statement proved for integers $\leqslant r - 1$; then $T_{(r-1)\epsilon}(E)$ and hence $T_\epsilon(T_{(r-1)\epsilon}(E))$ are compact. But since for any x, $y \in \mathscr{I}$ there is a geodesic of length $d(x, y)$ joining x to y, $T_\epsilon(T_{(r-1)\epsilon}(E)) = T_{r\epsilon}(E)$. This proves that $T_{r\epsilon}(E)$ is compact.

Now since for any $x \in \mathscr{I}$, $B_{r\epsilon}(x) = T_{r\epsilon}(\{x\})$, any closed ball (in $\mathscr{I}$) of finite radius is compact. As any set of finite diameter is contained in some closed ball, it follows that any closed subset of $\mathscr{I}$ that is of finite diameter is compact.

5.4. Let E be a *closed convex* subset of $\mathscr{I}$. We claim that given a point x $(\in \mathscr{I})$, there exists a unique point of E (we shall denote it by $\pi_E(x)$) such that

$$d(x, \pi_E(x)) \leqslant d(x, e) \qquad \text{for all} \quad e \in E. \tag{1}$$

Existence of a point $y \in E$, such that $d(x, y) \leqslant d(x, e)$ for all $e \in E$, follows at once from Lemma 5.3. Now if possible let y and z be two distinct points of E such that for all $e \in E$,

$$d(x, y) \leqslant d(x, e) \tag{2}$$

and

$$d(x, z) \leqslant d(x, e). \tag{3}$$

Then $d(x, y) \leqslant d(x, z) \leqslant d(x, y)$, hence $d(x, y) = d(x, z)$. Let m be the middle point of the geodesic $[yz]$. Since $[yz]$ lies in E, $m \in E$. According to Bruhat–Tits [12, Lemme 3.2.1]

$$d(x, y)^2 + d(x, z)^2 \geqslant 2d(x, m)^2 + \tfrac{1}{2}d(y, z)^2.$$

Since $d(x, y) = d(x, z)$, we get

$$d(x, y)^2 \geqslant d(x, m)^2 + \tfrac{1}{4}d(y, z)^2,$$

i.e.,

$$d(x, m)^2 \leqslant d(x, y)^2 - \tfrac{1}{4}d(y, z)^2 < d(x, y)^2.$$

This contradicts (2). Thus we have established the claim. Now let π_E denote the mapping from $\mathscr{I}$ to E defined by

$$x \mapsto \pi_E(x) \qquad \text{for} \quad x \in \mathscr{I}.$$

It is obvious that for $e \in E$, $\pi_E(e) = e$. We call π_E *the projection* of $\mathscr{I}$ *on* E. We shall now show that π_E is continuous.

Let p, q be two points of $\mathscr{I}$ and let m be the middle point of the geodesic $[\pi_E(p)\pi_E(q)]$ joining $\pi_E(p)$ and $\pi_E(q)$. Then

$$\begin{aligned} d(p, \pi_E(q)) &\leqslant d(p, q) + d(q, \pi_E(q)) \\ &\leqslant d(p, q) + d(q, \pi_E(p)) \end{aligned}$$

(since $d(q, \pi_E(q) \leqslant d(q, \pi_E(p))$ by definition of π_E)

$$\leqslant d(p, q) + d(q, p) + d(p, \pi_E(p));$$

thus

$$d(p, \pi_E(q)) \leqslant 2d(p, q) + d(p, \pi_E(p)). \tag{4}$$

Also, by Bruhat–Tits [12, Lemme 3.2.1]:

$$d(p, \pi_E(p))^2 + d(p, \pi_E(q))^2 \geqslant \tfrac{1}{2}d(\pi_E(p), \pi_E(q))^2 + 2d(p, m)^2.$$

Thus

$$\begin{aligned} &d(\pi_E(p), \pi_E(q))^2 \\ &\quad \leqslant 2\{d(p, \pi_E(p))^2 + d(p, \pi_E(q))^2 - 2d(p, m)^2\} \\ &\quad \leqslant 2\{d(p, \pi_E(p))^2 + (2d(p, q) + d(p, \pi_E(p)))^2 - 2d(p, m)^2\} \quad (by (4)) \\ &\quad \leqslant 2\{2d(p, \pi_E(p))^2 - 2d(p, m)^2 + 4d(p, q)(d(p, q) + d(p, \pi_E(p)))\} \\ &\quad \leqslant 8d(p, q)\{d(p, q) + d(p, \pi_E(p))\} \end{aligned}$$

(since by (1) $d(p, \pi_E(p)) \leqslant d(p, m)$). This shows that if $d(p, q)$ is small, then $d(\pi_E(p), \pi_E(q))$ is small, proving the continuity of π_E.

5.5. *Remark.* It can in fact be shown that for all $p, q \in \mathscr{I}$, $d(\pi_E(p), \pi_E(q)) \leqslant d(p, q)$. Since this will not be used, we do not prove it here.

We shall now prove a simple lemma which turns out to be very useful.

5.6. Lemma. *Let A be an apartment of $\mathscr{I}$, and let π_A be the projection of $\mathscr{I}$ on A. Let Δ be a chamber (i.e., an s-simplex) of A and a be a point in the interior Δ^0 of Δ. Then for $x \in \mathscr{I}$, $\pi_A(x) = a$ if and only if $x = a$.*

Proof. Let $x \in \mathscr{I}$ be such that $\pi_A(x) = a$. If $x \in A$, then since $\pi_A|_A =$ Identity, $x = a$. So if possible assume that $x \notin A$ and let A_0 be an apartment which contains x and $\bar{\Delta}$ (cf. Bruhat–Tits [12, Proposition 2.3.1]). From the definition of π_A it follows that $d(x, y) \geqslant d(x, a)$ for all $y \in A$ and hence, in particular, for all $y \in \bar{\Delta}$. Consider now the geodesic $[xa]$ joining x and a; since both x and a are in A_0, it lies in A_0. Now it is obvious that, since $a \in \Delta^\circ$ whereas $x \notin \bar{\Delta}$, the geodesic $[xa]$ intersects the boundary ($= \bar{\Delta} - \Delta^0$) of Δ at a point b, and clearly, $d(x, b) < d(x, a)$, which contradicts the fact that for all $y \in \bar{\Delta}$, $d(x, y) \geqslant d(x, a)$. This proves that $x \in A$, and then of course, $x = a = \pi_A(x)$.

5.7. Corollary. *Let A be an apartment of $\mathscr{I}$ and E be a closed subset of $\mathscr{I}$ such that $\pi_A(E) \supset A$. Then $E \supset A$.*

Proof. Since $\pi_A(E) \supset A$, it follows from the previous lemma that E contains the union U of the interiors of all the s-simplexes contained in A. Now since U is dense in A, it follows that $E \supset A$.

Before we give an application of Corollary 5.7, we prove a simple geometric result.

5.8. Lemma. *Let d denote the usual metric on the n-dimensional euclidean space $\mathbf{R}^n$, and let $f: \mathbf{R}^n \to \mathbf{R}^n$ be a continuous map. Assume that there exists a constant m such that*

$$d(\xi, f(\xi)) \leqslant m \qquad \textit{for all} \quad \xi \in \mathbf{R}^n.$$

Then f is surjective.

Proof. Let E denote the closed ball in $\mathbf{R}^n$ of radius m and with center at the origin. Let $y \in \mathbf{R}^n$ and let $\varphi_y: E \to \mathbf{R}^n$ be the map

$$x \mapsto (x + y) - f(x + y) \qquad \text{for} \quad x \in E.$$

Then since for all $\xi \in \mathbf{R}^n, d(\xi, f(\xi)) \leqslant m$, it follows that $\varphi_y(E) \subset E$. Hence by Brouwer's fixed point theorem [33, Theorem 4.7.5] φ_y has a fixed point, i.e., there is an $e \in E$ such that

$$\varphi_y(e) = e + y - f(e + y) = e$$

i.e.,

$$f(e + y) = y.$$

This proves that y is in the image of f and thus f is surjective.

5.9. *Remark.* I believe that Lemma 5.8 is known but was unable to trace a reference for it. The proof given above is simpler than my original proof and it was suggested by John J. Millson.

5.10. LEMMA. *Let A be an apartment of $\mathscr{I}$ and let v be a nonnegative real number. Then A is the only apartment of $\mathscr{I}$ contained in $T_v(A)$.*

Proof. Let E be an apartment contained in $T_v(A)$. Let π_A (resp. π_E) be the projection of $\mathscr{I}$ on A (resp. on E). Then, as can be seen easily,

$$d(e, \pi_E \pi_A(e)) \leqslant 2v \qquad \text{for all} \quad e \in E.$$

According to Lemma 5.8, therefore, $\pi_E \pi_A: E \to E$ is surjective. Now since $E \supset \pi_E(A) \supset \pi_E(\pi_A(E)) = E$, $\pi_E(A) = E$. According to Bruhat–Tits [12, Théorème 2.5.12] apartments (being complete) are closed and hence by Corollary 5.7, $A \supset E$. Now the theorem on the invariance of domain [33, Theorem 4.7.16] implies that E is open in A. Since A is connected, $E = A$ and this proves the lemma.

5.11. Let a, b, c be three points of $\mathscr{I}$ and let $[ab]$ (resp. $[ac]$) be the geodesic joining a and b (resp. a and c). Let $f: [ab] \to \mathbf{R}$ be the map given by

$$f(p) = \frac{d(a, c)^2 + d(a, p)^2 - d(p, c)^2}{2d(a, p) \cdot d(a, c)} \qquad \text{for} \quad p \in [ab].$$

The *angle* θ between $[ab]$ and $[ac]$ is defined by

$$0 \leqslant \theta \leqslant \pi, \qquad \cos\theta = \lim_{p \to a} f(p).$$

We assert that θ is well defined. To prove this we have to show that $\lim_{p \to a} f(p)$ exists and is of modulus $\leqslant 1$. Let Δ_1, Δ_2 be s-simplexes such that $\bar{\Delta}_1$ contains a and a nontrivial segment, beginning at a, of the geodesic $[ab]$ and $\bar{\Delta}_2$ contains c. Let A be an apartment which contains both Δ_1 and Δ_2. Then since $a, c \in A$, the geodesic $[ac]$ lies in A whereas, since a nontrivial segment of $[ab]$ is contained in $\bar{\Delta}_1$ and hence in A, for $p \in [ab]$ so close to a that the geodesic $[ap]$ lies in A

$$0 \leqslant \varphi \leqslant \pi, \qquad \cos\varphi = f(p)$$

defines the euclidean angle between the lines $[ap]$ and $[ac]$ in the euclidean space A. This shows that on a nontrivial segment of $[ab]$, beginning at a, the function f is constant and $|f(p)| \leqslant 1$. This proves that θ is well defined and in fact it coincides with the euclidean angle between the lines $[ap]$ and $[ac]$ in A.

We shall now show that Bruhat–Tits buildings share a well-known property of negative curvature spaces. More precisely, we shall prove the following:

5.12. Lemma. *Let a, b, c be three points of $\mathscr{I}$ and let θ, φ, ψ be the interior angles of the geodesic triangle abc at the vertices a, b, and c, respectively. Then*

(1) $d(b,c)^2 \geqslant d(a,b)^2 + d(a,c)^2 - 2d(a,b)d(a,c)\cos\theta$,
(2) $\theta + \varphi + \psi \leqslant \pi$.

Proof. Let A be an apartment which contains the geodesic $[ac]$ and a nontrivial segment, beginning at a, of the geodesic $[ab]$ (cf. Subsection 5.11 above). Let Δ be an s-simplex (i.e., a chamber) of A such that $a \in \bar{\Delta}$ and let $\rho = \rho_{A;\Delta}$ be the retraction of $\mathscr{I}$ on A with center Δ (cf. [12, Théorème 2.3.4 and Définition 2.3.5]). Since $\rho|_A =$ Identity, it is clear that θ is the interior angle, at the vertex $a = \rho(a)$, of the euclidean triangle with vertices at $a = \rho(a)$, $\rho(b)$, and $c = \rho(c)$ and therefore,

$$d(\rho(b), \rho(c))^2 = d(\rho(b), c)^2 = d(a,c)^2 + d(a, \rho(b))^2 - 2d(a, \rho(b))d(a,c)\cos\theta.$$

But according to [12, Proposition 2.5.3]:

$$d(b,c) \geqslant d(\rho(b), \rho(c)) \quad \text{and} \quad d(a, \rho(b)) = d(a,b).$$

Thus we get

$$d(b,c)^2 \geqslant d(a,c)^2 + d(a,b)^2 - 2d(a,b)d(a,c)\cos\theta. \tag{5}$$

This proves the first assertion of the lemma. To prove the second assertion we consider a euclidean triangle $\mathfrak{abc}$ (lying in $\mathbf{R}^2$) with sides $\mathfrak{ab}$, $\mathfrak{bc}$, and $\mathfrak{ca}$ of lengths $d(a,b)$, $d(b,c)$, and $d(c,a)$, respectively. We let α, β, γ denote the interior angles of the triangle $\mathfrak{abc}$ at vertices $\mathfrak{a}, \mathfrak{b}$, and $\mathfrak{c}$, respectively. Then

$$\cos \alpha = \frac{d(a,b)^2 + d(a,c)^2 - d(b,c)^2}{2d(a,b)d(a,c)}.$$

So by (5)

$$\cos \alpha \leqslant \cos \theta \quad \text{and hence} \quad \theta \leqslant \alpha.$$

We can similarly prove that $\varphi \leqslant \beta$ and $\psi \leqslant \gamma$. Therefore

$$\theta + \varphi + \psi \leqslant \alpha + \beta + \gamma = \pi.$$

We shall need the following simple lemma for the proof of our next result.

5.13. Lemma. *Let a, b, c, d be four distinct points of the Bruhat–Tits building $\mathscr{I}$. Let φ (resp. ψ) be the angle between the geodesics $[ab]$ and $[ac]$ (resp. between $[ac]$ and $[ad]$) and θ be the angle between $[ab]$ and $[ad]$. Then*

$$\theta \leqslant \varphi + \psi.$$

Proof. Let A be an apartment containing the geodesic $[ad]$ and a nontrivial segment (beginning at a) of $[ab]$. Let Δ be a chamber (i.e., an s-simplex) of A such that $a \in \bar{\Delta}$ and $\rho = \rho_{A;\Delta}$ be the retraction of $\mathscr{I}$ on A with center Δ. Note that $\rho(a) = a$ and $\rho(d) = d$.

It is clear, from the definition of angles, that $\theta =$ angle between $[a\rho(b)]$ and $[ad]$. Let $\bar{\varphi}$ (resp. $\bar{\psi}$) be the angle between $[a\rho(b)]$ and $[a\rho(c)]$ (resp. between $[a\rho(c)]$ and $[a\rho(d)]$). Then, since A is a euclidean space, it follows that

$$\theta \leqslant \bar{\varphi} + \bar{\psi}.$$

Thus to establish the lemma it suffices to show that $\bar{\varphi} \leqslant \varphi$ and $\bar{\psi} \leqslant \psi$. Now let p be an arbitrary point on the geodesic $[ab]$, then

$$\begin{aligned} d(\rho(p), \rho(c))^2 &= d(a, \rho(p))^2 + d(a, \rho(c))^2 - 2d(a, \rho(p))d(a, \rho(c)) \cos \bar{\varphi} \\ &= d(a,p)^2 + d(a,c)^2 - 2d(a,p)d(a,c) \cos \bar{\varphi} \end{aligned}$$

(cf. [12, Proposition 2.5.3]), but according to Proposition 2.5.3 of [12], $d(\rho(p), \rho(c)) \leqslant d(p,c)$. Thus

$$\cos \bar{\varphi} \geqslant \frac{d(a,p)^2 + d(a,c)^2 - d(p,c)^2}{2d(a,p)d(a,c)}$$

whereas, as we have seen in 5.11, if p is close enough to a, then

$$\cos \varphi = \frac{d(a,p)^2 + d(a,c)^2 - d(p,c)^2}{2d(a,p)d(a,c)},$$

and hence,

$$\cos \bar{\varphi} \geqslant \cos \varphi,$$

which implies that $\bar{\varphi} \leqslant \varphi$. It can similarly be shown that $\bar{\psi} \leqslant \psi$. This completes the proof of the lemma.

5.14. LEMMA. *Let E be a convex subset of $\mathscr{I}$. Then $T_v(E)$ is convex for all $v \geqslant 0$.*

Proof. In view of Lemmas 5.12 and 5.13, the proof of Lemma 3.5 of Mostow [23] can be used verbatim to give a proof of the present lemma.

5.15. (We continue with the notation introduced in Subsection 5.1.) In the sequel we call a subgroup of G *parabolic* if it is the group of k-rational points of a parabolic k-subgroup of $\mathbf{G}$. It is well known that every parabolic subgroup of G is equal to its own normalizer.

For a subset I of ${}_0S$, let ${}_0W_I$ be the subgroup of ${}_0W$ generated by I and let $\mathbf{P}_I = \mathbf{P} \cdot {}_0W_I \cdot \mathbf{P}, P_I = P \cdot {}_0W_I \cdot P$. It is well known [7, Sect. 4] that the groups $\mathbf{P}_I$ are parabolic k-subgroups of $\mathbf{G}$ and any parabolic k-subgroup of $\mathbf{G}$ is conjugate, by an element of G, to a unique $\mathbf{P}_I$. The parabolic subgroups $\mathbf{P}_I$ $(I \subset {}_0S)$ are called the *standard parabolic k-subgroups*; any subgroup of $\mathbf{G}$ containing $\mathbf{P}$ is a standard parabolic subgroup. It is known that $P_I = \mathbf{P}_I(k)$ and hence the subgroups P_I are parabolic. The map $I \mapsto P_I$ from the power set of ${}_0S$ into the set of subgroups of G is one-to-one and onto the set of subgroups of G containing P. It is obvious that any parabolic subgroup of G is conjugate to a (unique) P_I.

For a subset I of S we denote by W_I the subgroup of W generated by I. The subgroup B is called the *standard Iwahori subgroup*, and the *standard parahoric subgroups* B_I $(I \subset S)$ are the subgroups $B_I = B \cdot W_I \cdot B$. If $I \neq S$, then W_I is finite and B_I is compact. In particular the group $B_{{}_0S}$ is a compact subgroup; in the sequel we shall let K denote this subgroup. It is known [12] that $G = K \cdot P$ ("Iwasawa decomposition").

Let $\mathbf{o}$ be the vertex, of the standard chamber C, fixed by K. We identify ${}_0W$ with the subgroup of W fixing $\mathbf{o}$ and ${}_0S$ with the subset of S consisting of elements which fix $\mathbf{o}$. Let D be the smallest cone in the standard apartment A, with vertex at $\mathbf{o}$, which contains $\bar{C}$. For a subset I of ${}_0S$, let D_I be the subset of D fixed by every element of I. The transforms of D by elements of G shall be called the *closed quarters* of $\mathscr{I}$ and the transforms of D_I $(I \subset {}_0S)$ *the closed quarter-walls*. (Since $D = D_\varnothing$, a closed quarter is a closed quarter-wall.)

Let $p, q \in \mathscr{I}$. Then q is said to be *conjugate* to p if there is a $g \in G$ such that $q = g \cdot p$. We note that the vertex of any closed quarter-wall is conjugate to $\mathbf{o}$.

A *ray* in $\mathscr{I}$, by definition, is a ray (i.e., a half-line), lying in an apartment of $\mathscr{I}$, which originates at a point conjugate to $\mathbf{o}$. Evidently any ray is the transform (by a suitable element of G) of a ray, contained in D, which originates at $\mathbf{o}$.

5.16. DEFINITION. Let (M, d) be a metric space. For subsets E and F of M, the *Hausdorff distance* between E and F, denoted $\mathrm{hd}(E, F)$, is defined by

$$\mathrm{hd}(E, F) = \inf\{v \mid E \subset T_v(F), F \subset T_v(E)\}.$$

(By convention $\mathrm{hd}(E, \varnothing) = 0$ if $E = \varnothing$, otherwise $\mathrm{hd}(E, \varnothing) = \infty$.)

Two subsets E and F of M are said to be *equivalent* (we write $E \sim F$) if $\mathrm{hd}(E, F) < \infty$. It is obvious that $\sim$ is an equivalence relation and for any $E\ (\subset M)$ and $v < \infty$, $T_v(E) \sim E$.

5.17. DEFINITION. Two subsets of the Bruhat–Tits building $\mathscr{I}$ are said to be *equipollent* if there is an apartment, containing both of them, in which they are translates of each other.

5.18. LEMMA. *Let I be a subset of ${}_0S$. Then*

(i) $\{g \in G \mid gD_I \sim D_I\} = P_I$.

Moreover if L is a ray in D, originating at $\mathbf{o}$, *such that D_I is the minimum wall of D containing L, then*

(ii) $P_I = \{g \in G \mid gL \sim L\}$.

Proof. For convenience we denote $\{g \in G \mid gD_I \sim D_I\}$ by $P(D_I)$. Since $\sim$ is an equivalence relation and the metric on $\mathscr{I}$ is G-invariant, $P(D_I)$ is a subgroup. To prove the first assertion of the lemma we shall show first that $P \subset P(D_I)$. In fact, since P is generated by $Z(T)$ and $B \cap P$, and the action of elements of $Z(T)$ on the standard apartment A is by affine translations, to prove that $P(D_I) \subset P$ we have only to show that $P(D_I)$ contains $B \cap P$. It is known (cf. Bruhat–Tits [12, Sects. 5.2.34 and 7.4.5], see also a forthcoming paper [6] of Borel and Serre), that $B \cap P$ in fact fixes D (and hence D_I for any $I \subset {}_0S$) pointwise. Thus $B \cap P \subset P(D_I)$ and therefore P is contained in $P(D_I)$. Now since P_I is generated by P and I, and every element of I fixes D_I pointwise, $P_I \subset P(D_I)$. On the other hand, since the action of any element of ${}_0S$ on A is by reflection in the corresponding root hyperplane, it is obvious that for no $\alpha \in {}_0S - I$, $\alpha D_I \sim D_I$. As any subgroup of G containing P equals P_J for a $J \subset {}_0S$, we conclude that $P(D_I) = P_I$. This proves the first assertion; the proof of the second assertion is similar.

5.19. *Remark.* For a closed quarter-wall E, let $P(E) = \{g \in G \mid gE \sim E\}$. Then, since any closed quarter-wall is transform of a D_I, $P(E)$ is a parabolic subgroup of G. Conversely, given any parabolic subgroup Q of G, it is obvious that there is a closed quarter-wall E such that $P(E) = Q$.

5.20. *Remark.* Let L be a ray in $\mathscr{I}$. We claim that there is a unique minimal closed quarter-wall which contains L. Evidently it suffices to establish the claim in the case L is a ray originating at $\mathbf{o}$ and lying in (the standard closed quarter) D. So let L be a ray in D with origin at $\mathbf{o}$ and let D_I be the minimal wall of D containing L. Let A_0 be any other apartment which contains L. In view of [12, Proposition 2.5.8], there is a $g \in G$ such that $A_0 = g \cdot \mathsf{A}$ and $g \cdot x = x$ for all $x \in L$. In particular, $g \cdot \mathbf{o} = \mathbf{o}$ and hence $g \in K$. Since $g \cdot L = L$ and D_I is the minimal wall of D containing L, by Lemma 5.18(ii) $g \in P_I$. Thus $g \in K \cap P_I$. But it is known (cf. the forthcoming paper of Borel–Serre quoted above) that the subgroup $K \cap P_I$ fixes D_I pointwise. Hence the closed quarter-wall D_I is contained in A_0 and therefore it is the minimal closed quarter-wall in A_0 containing L. Since any closed quarter-wall is contained in an apartment, this proves that D_I is the unique minimal closed quarter-wall containing L.

Let now L be any ray in $\mathscr{I}$ and let E be the minimal closed quarter-wall which contains L. Let $P(L) = \{g \in G \mid gL \sim L\}$ and, as before, let $P(E) = \{g \in G \mid gE \sim E\}$. Then it follows from Lemma 5.18 that $P(L) = P(E)$.

5.21. Lemma. *Let E, F be closed quarter-walls and let $P(F) \supset P(E)$. Then exists a $g \in P(F)$ such that $g \cdot F$ is a wall of E.*

Proof. It is enough to prove the lemma in the case $E = D_I$ for an $I \subset {}_0S$. In this case, by Lemma 5.18, $P(E) = P_I$ and since $P(F) \supset P(E)$, there is a subset J of ${}_0S$ containing I such that $P(F) = P_J$. In view of Lemma 5.18, it is obvious now that there is a $g \in G$ such that $g \cdot F = D_J$, and since $I \subset J$, $g \cdot F\ (= D_J)$ is a wall of $E\ (= D_I)$. Moreover, since

$$P_J = P(F) = P(g^{-1} \cdot D_J) = g^{-1}P(D_J)g = g^{-1}P_J g, \qquad g \in P_J\ (= P(F)).$$

This proves the lemma.

5.22. Corollary. *Let E, F be closed quarter-walls such that $P(E) = P(F)$. Then $E \sim F$.*

Proof. According to the previous lemma there exist $g, h \in P(E) = P(F)$ such that gF is a wall of E and hE is a wall of F. By dimension consideration then $E = g \cdot F$ and $F = h \cdot E$. Since $g \in P(F)$, $E = g \cdot F \sim F$.

5.23. Lemma. (i) *Let L and M be equivalent rays. Then $M = p \cdot L$ for a $p \in P(L) = \{g \in G \mid gL \sim L\}$.*

(ii) *Let E and F be equivalent closed quarter-walls. Then $F = p \cdot E$ for a $p \in P(E) = \{g \in G \mid gE \sim E\}$.*

Proof. We shall prove the second assertion first. Since $E \sim F$, $P(E) = P(F)$. Now it immediately follows from Lemma 5.21 that there is a $p \in P(E)$ such that $F = p \cdot E$.

To prove the first assertion, let E (resp. F) be the minimum closed quarter-wall containing L (resp. M). Then $P(E) = P(L)$, $P(F) = P(M)$ (5.20). But since $L \sim M$, $P(L) = P(M)$. Therefore, $P(E) = P(F)$. Hence, there is a $p \in P(E) = P(F)$ such that $p \cdot E = F$. Now note that since $p \in P(E) = P(L)$, $p \cdot L \sim L$. So $p \cdot L \sim M$. But since the rays $p \cdot L$ and M have the same origin and are contained in a euclidean space, they can be at finite Hausdorff distance from each other only if they are identical. So $M = p \cdot L$. This completes the proof of the lemma.

5.24. Corollary. *Let E be a closed quarter-wall. Then there is a unique closed quarter-wall equivalent to E and with vertex at* $\mathbf{o}$.

Proof. Let $E = g \cdot F$, where F is a wall of the standard closed quarter D and $g \in G$. As $G = K \cdot P$, $g = \kappa \cdot p$ with $\kappa \in K$ and $p \in P$. Now since $P \subset P(F)$, $p \cdot F \sim F$ and hence $E = \kappa p \cdot F \sim \kappa \cdot F$. Since κ fixes $\mathbf{o}$, the vertex of the closed quarter-wall $\kappa \cdot F$ is at $\mathbf{o}$.

To prove uniqueness, it suffices to show that if E_1 and E_2 are two equivalent closed quarter-walls with vertices at $\mathbf{o}$, then $E_1 = E_2$. It is obvious that we can assume one of E_1, E_2 (say E_1) to be a wall of D. Since $E_1 \sim E_2$, it follows from the preceding lemma that there is a $p \in P(E_1)$ such that $p \cdot E_1 = E_2$. As $\mathbf{o}$ is the vertex of both E_1 and E_2, p fixes $\mathbf{o}$ and hence $p \in K \cap P(E_1)$. But every element of $K \cap P(E_1)$ fixes E_1 pointwise, in particular $p \cdot E_1 = E_1$. This proves that $E_1 = E_2$.

5.25. Given two apartments A_0 and A and a point $a \in A$ conjugate to $\mathbf{o}$, we let $A_0 \wedge_a A$ denote the union of all rays L in A originating at a and contained in $T_r(A_0)$ for some $r \geqslant 0$. If E is a closed quarter-wall in A with vertex at a, then we denote by $A_0 \wedge_a E$ the set $(A_0 \wedge_a A) \cap E$. Following Mostow [23], we call a subset of the form $A_0 \wedge_a A$ a *splice* in $\mathscr{I}$. A splice $\mathfrak{C}$ is said to be *irreducible* if, given splices $\mathfrak{C}_1, \ldots, \mathfrak{C}_n$ with $\mathfrak{C} \sim \mathfrak{C}_1 \cup \cdots \cdots \cup \mathfrak{C}_n$, for some $i \leqslant n$, $\mathfrak{C} \sim \mathfrak{C}_i$.

5.26. Lemma. (i) *Let A_0 and A be two apartments and let $a \in A$ be a point conjugate to* $\mathbf{o}$. *Then $A_0 \wedge_a A$ is convex and is union of certain closed quarter-walls in A with vertex at a. Moreover for a $v \geqslant 0$, $A_0 \wedge_a A \subset T_v(A_0) \cap A$.*

(ii) *Let A be an apartment and $g \in G$. Then a closed quarter-wall E with vertex at a is contained in $g \cdot A \wedge_a A$ if and only if $g \in P(E) \cdot N(A)$, where $N(A)$ is the stabilizer of A.*

(iii) *Given a closed quarter-wall* E $(\subset A)$ *with vertex at* a, *there exists a* $g \in G$ *such that* $gA \wedge_a A = E$.

Proof. Let $g \in G$ be an element of G such that $g \cdot A = A_0$. Let $L \subset g \cdot A \wedge_a A$ be a ray which originates at a and let E be the minimal closed quarter-wall containing L. We shall first show that $E \subset g \cdot A \wedge_a A$ and $g \in P(E) \cdot N(A)$; this will in particular prove that $A_0 \wedge_a A$ is a union of certain closed quarter-walls in A with vertex at a. As $L \subset g \cdot A \wedge_a A$, there exists a positive constant u such that $L \subset T_u(g \cdot A) \cap A$. Therefore, $L \subset T_u(T_u(L) \cap g \cdot A)$, and hence $T_u(L) \cap g \cdot A$ is unbounded. Since (cf. Lemma 5.14) $T_u(L)$ is convex, $T_u(L) \cap g \cdot A$ is convex and hence it contains a half-line. So, for a $v \geqslant u$, $T_v(L) \cap g \cdot A$ contains a ray M (equipollent to a half-line in $T_u(L) \cap g \cdot A$).

We shall now show that $L \sim M$. For this it will suffice to show that for some $t \geqslant 0, L \subset T_t(M)$. We shall in fact show that $L - T_v(M)$ is relatively compact. Let π_L denote the projection of $\mathscr{I}$ on L defined in Subsection 5.4. Then since $M \subset T_v(L)$, for all $y \in M, d(y, \pi_L(y)) \leqslant v$. Hence, $M \subset T_v(\pi_L(M))$ and $\pi_L(M) \subset T_v(M)$. From this it follows, since M is connected and unbounded, that $\pi_L(M)$ is a connected unbounded subset of the half-line L and hence it is a half-line. Therefore, $L - \pi_L(M)$ is relatively compact. This implies that $L - T_v(M)$ $(\subset L - \pi_L(M))$ is relatively compact; which proves that $L \sim M$. Now according to Lemma 5.23(i), there exists a $p \in P(L)$ $(=\{g \in G \mid gL \sim L\})$ such that $M = p \cdot L$. Since E is the minimal closed quarter-wall containing L, $E \subset A$ and $P(L) = P(E)$ (cf. Remark 5.20). It is obvious that $p \cdot E$ is the minimal closed quarter-wall containing $M = p \cdot L$ and hence, $p \cdot E \subset A_0 = g \cdot A$. Since $p \in P(E)$, $E \sim p \cdot E$ and therefore, for a suitable $c \geqslant 0$, $E \subset T_c(p \cdot E) \subset T_c(A_0)$. This shows that $E \subset A_0 \wedge_a A$. Since there are only finitely many closed quarter-walls in A with vertex at a, there is a $v \geqslant 0$ such that $A_0 \wedge_a A \subset T_v(A_0) \cap A$. Also, as $E \subset p^{-1}g \cdot A$, there exists (cf. [12, Proposition 2.5.8]) an $h \in G$ fixing E pointwise, such that $h \cdot A = p^{-1}g \cdot A$. Since $h \cdot E = E$, $h \in P(E)$. Now as $ph \cdot A = g \cdot A$, $g \in P(E) \cdot N(A)$. To prove the converse let E be a closed quarter-wall in A with vertex at a such that $g \in P(E) \cdot N(A)$. Then $g = pn$ with $p \in P(E), n \in N(A)$ and $gn^{-1} \cdot E = p \cdot E \sim E$. But since $gn^{-1} \cdot E$ is a closed quarter-wall in $g \cdot A$, it follows that $E \subset g \cdot A \wedge_a A$. Thus we have proved that a closed quarter-wall E in A, with vertex at a, is contained in $g \cdot A \wedge_a A$ if and only if $g \in P(E) \cdot N(A)$. By Lemma 5.14, for all $r \geqslant 0$, $T_r(A_0)$ is convex, and hence the union of rays originating at a and contained in $T_r(A_0) \cap A$ is also convex. This proves that $A_0 \wedge_a A$ is convex.

It clearly suffices to prove the last part of the lemma in the case A is the standard apartment and E is a wall of the standard closed quarter D. So assume that $E = D_I$ for an $I \subset {}_0S$, and A is the standard apartment. Any closed quarter-wall in A with vertex at $\mathbf{o}$ is of the form $\omega \cdot D_J$ for an $\omega \in {}_0W$

and $J \subset {}_0S$. Now let $g \in P_I$ be such that $g \notin P(\omega D_J) \cdot N$ $(=\omega P_J \omega^{-1} \cdot N)$ unless $J \supset I$ and $\omega \in P_J$. (Such a choice of g is possible because for $\omega \in {}_0W$, $\omega P_J \omega^{-1} \cdot N = \bigcup_{w \in {}_0W} \omega P_J w$ and for any $\omega, w \in {}_0W$, $\mathbf{P}_I \cap \omega \mathbf{P}_J w$ is a subvariety of $\mathbf{P}_I$ of strictly smaller dimension unless $\omega \mathbf{P}_J w \supset \mathbf{P}_I$. But it follows from the Bruhat decomposition that $\omega \mathbf{P}_J w \supset \mathbf{P}_I$ if and only if $J \supset I$ and $\omega, w \in {}_0W_J$.) Since by Lemma 5.18(i) $P(E) = P(D_I) = P_I$, it follows from the second part of this lemma that the splice $g \cdot A \wedge_a A = D_I = E$.

5.27. COROLLARY. *The irreducible splices are precisely the closed quarter-walls.*

Proof. According to the preceding lemma any splice is a finite union of closed quarter-walls and any closed quarter-wall is a splice. The assertion of the corollary is obvious from this.

We shall now prove a proposition (Proposition 5.29) which is crucial for the proof of strong rigidity of lattices in non-archimedean semisimple groups (see Sections 7–8). We need a lemma first.

5.28. LEMMA. *Let A and A_0 be two apartments and let E be a closed quarter-wall, with vertex at a, contained in A. Then $F = A_0 \wedge_a E$ is a closed quarter-wall. Furthermore, for all v sufficiently large,*

$$F \subset T_v(A_0) \cap E \qquad \textit{and} \qquad T_v(A_0) \cap E \sim F.$$

Proof. That $F = A_0 \wedge_a E = (A_0 \wedge_a A) \cap E$ is a closed quarter-wall, and for v large, $F \subset T_v(A_0) \cap E$ follow at once from Lemma 5.26(i).

It clearly suffices to prove the last assertion of the lemma in case A is the standard apartment, $a = \mathbf{o}$, and E is a wall of the standard closed quarter D. So assume that A is the standard apartment and E is a face of D. Let $g \in G$ be such that $A_0 = g \cdot A$. Then $A = \bigcup_{w \in {}_0W} w \cdot D$ and $A_0 = \bigcup_{w \in {}_0W} gw \cdot D$. Obviously, for any $v \geqslant 0$, $T_v(A_0) = \bigcup_{w \in {}_0W} T_v(gw \cdot D)$. Let ω be a fixed element of ${}_0W$. We shall first investigate the intersection $T_v(g\omega \cdot D) \cap E$. It follows from [12, Corollary 2.9.6] that there is a $\mathfrak{D}$ $(\subset D)$ equipollent to D such that both $\mathfrak{D}$ and $g\omega \cdot \mathfrak{D}$ are contained in an apartment Σ. Evidently $\mathfrak{D}$ is a closed cone. Let $\mathfrak{E}$ be the wall of $\mathfrak{D}$ parallel to (=equipollent to) the wall E of D. Now let $\mathscr{D}$ be the cone in Σ equipollent to $g\omega \cdot \mathfrak{D}$ and with the same vertex as $\mathfrak{D}$. Clearly, $\mathscr{D} \cap \mathfrak{D}$ is a wall of $\mathfrak{D}$. Let $\mathscr{D} \cap \mathfrak{E} = \mathfrak{F}_\omega$ and F_ω be the wall of E parallel to $\mathfrak{F}_\omega$. Then it is obvious that for any $r \geqslant 0$

$$T_r(\mathscr{D}) \cap \mathfrak{E} \sim \mathscr{D} \cap \mathfrak{E} = \mathfrak{F}_\omega \sim F_\omega. \tag{$*$}$$

Also since $\mathfrak{D}$ $(\subset D)$, $\mathfrak{F}_\omega$, $\mathfrak{E}$, and $\mathscr{D}$ are equipollent to D, F_ω, E, and $g\omega \cdot D$, respectively, there is an $r \geqslant 0$ such that $T_r(\mathfrak{D}) \supset D$; $T_r(\mathfrak{E}) \supset E$, $T_r(E) \supset \mathfrak{E}$; $T_r(\mathfrak{F}_\omega) \supset F_\omega$, $T_r(F_\omega) \supset \mathfrak{F}_\omega$; and $T_r(\mathscr{D}) \supset g\omega \cdot D$, $T_r(g\omega \cdot D) \supset \mathscr{D}$. Therefore, for $v \geqslant 2r$,

$$\begin{aligned} F_\omega &\subset T_r(\mathfrak{F}_\omega) \cap E = T_r(\mathscr{D} \cap \mathfrak{E}) \cap E \subset T_r(\mathscr{D}) \cap E \\ &\subset T_{2r}(g\omega \cdot D) \cap E \subset T_v(g\omega \cdot D) \cap E \subset T_r(T_{v+r}(g\omega \cdot D) \cap \mathfrak{E}) \\ &\subset T_r(T_{v+2r}(\mathscr{D}) \cap \mathfrak{E}) \sim F_\omega \qquad \text{(cf. (*))}. \end{aligned} \tag{**}$$

We readily conclude from (**) that

$$F_\omega \sim T_v(g\omega \cdot D) \cap E.$$

We also observe that $F_\omega \subset F$. In fact, since for $v \geqslant 2r$

$$F_\omega \subset T_v(g\omega \cdot D) \cap E, \qquad F_\omega \subset A_0 \wedge_a E = F.$$

Now we complete the proof of the lemma. For every $w \in {}_0W$, let F_w be the face of E such that for large v,

$$F_w \subset T_v(gw \cdot D) \cap E \sim F_w.$$

Then for v large enough:

$$\bigcup_{w \in {}_0W} F_w \subset F \subset T_v(g \cdot A) \cap E = \left(\bigcup_{w \in {}_0W} T_v(gw \cdot D) \right) \cap E \sim \bigcup_{w \in {}_0W} F_w.$$

Hence, for v sufficiently large,

$$\bigcup_{w \in {}_0W} F_w \sim F \sim T_v(g \cdot A) \cap E.$$

This completes the proof of the lemma.

5.29. PROPOSITION. *Let A be an apartment and let $a \in A$ be a point conjugate to* **o**. *Let A_0 be any other apartment. Then for v sufficiently large*

$$A_0 \wedge_a A \sim T_v(A_0) \cap A.$$

Proof. Let E_i $(1 \leqslant i \leqslant r)$ be the closed quarters in A with vertex at a. Then, by Lemma 5.28, for large v and for all $i \leqslant r$, $T_v(A_0) \cap E_i \sim A_0 \wedge_a E_i$. Hence:

$$T_v(A_0) \cap A = T_v(A_0) \cap \left(\bigcup_{i=1}^{s} E_i \right) \sim \bigcup_{i=1}^{r} A_0 \cap_a E_i = A_0 \wedge_a A.$$

This proves the proposition.

5.30. COROLLARY. *Let A be an apartment and let $a \in A$ be a point conjugate to* **o**. *Let A_0 be an apartment such that $E = A_0 \wedge_a A$ is a closed quarter. Then for $v \geqslant d(a, A_0)$,*

$$E \subset T_v(A_0) \cap A \sim E.$$

Proof. Since according to Proposition 5.29, for all v large, $E \subset T_v(A_0) \cap A \sim E$, to prove the corollary we have only to show that $E \subset T_v(A_0) \cap A$ for $v \geqslant d(a, A_0)$. Let $g \in G$ be such that $A_0 = g \cdot A$. Then according to Lemma 5.26(ii), $g \in P(E) \cdot N(A)$. Let $g = p \cdot n$ with $p \in P(E)$ and $n \in N(A)$. Then $E_0 = p \cdot E\ (= gn^{-1} \cdot E)$ is closed quarter of $A_0\ (= g \cdot A)$. According to [12, Corollary 2.9.6] there is an apartment Σ which contains cones $\mathfrak{C}\ (\subset E)$, $\mathfrak{C}_0\ (\subset E_0)$ equipollent to E and E_0, respectively. Now (since $p \in P(E)$) $E \sim E_0$ and hence $\mathfrak{C} \sim \mathfrak{C}_0$. But then clearly $\mathfrak{C} \cap \mathfrak{C}_0\ (\subset E \cap E_0)$ contains a cone F equipollent to both $\mathfrak{C}$ and $\mathfrak{C}_0$ (and hence F is also equipollent to both E and E_0). It is obvious that E is the closure of the convex hull of the set $F \cup \{a\}$ and therefore, for $v \geqslant d(a, A_0)$, $T_v(A_0)$, which is a closed convex subset containing a and F, contains E. Thus for $v \geqslant d(a, A_0)$, $T_v(A_0) \cap A \supset E$ and we have proven the corollary.

We close this section with the following:

5.31. Lemma. *Let E_1 and E_2 be two equivalent closed quarters with vertices at a_1 and a_2, respectively. Let $c = d(a_1, a_2)$. Then*

$$\mathrm{hd}(E_1, E_2) \leqslant c.$$

Proof. According to [12, Corollary 2.9.6] there is an apartment Σ which contains cones $\mathfrak{C}_1\ (\subset E_1)$ and $\mathfrak{C}_2\ (\subset E_2)$ equipollent to E_1 and E_2, respectively. Since $E_1 \sim E_2$, $\mathfrak{C}_1 \sim \mathfrak{C}_2$ and hence $\mathfrak{C}_1 \cap \mathfrak{C}_2$ contains a cone F equipollent to both $\mathfrak{C}_1$ and $\mathfrak{C}_2$. Clearly F is equipollent to E_1 and E_2, and hence E_1 (resp. E_2) is the closure of the convex hull of $F \cup \{a_1\}$ (resp. $F \cup \{a_2\}$). Therefore $T_c(E_2)$ (resp. $T_c(E_1)$), which is a closed and convex (see Lemma 5.14) subset containing $F \cup \{a_1\}$ (resp. $F \cup \{a_2\}$), contains E_1 (resp. E_2). This proves that $\mathrm{hd}(E_1, E_2) \leqslant c$.

6. Pseudo-Isometries and the Equivariant Map between the Sets of Apartments

The rest of this paper is devoted to proving strong rigidity of lattices in non-archimedean semisimple groups (see Theorems 8.6 and 8.7 for precise statements). As has been noted in the introduction, our proof of strong rigidity is modeled on Mostow [23] and, thanks to Corollary 5.7, it is simpler than the proof of strong rigidity of uniform lattices in real analytic semisimple groups.

Let n, n' be two positive integers. For $i \leqslant n$ (resp. $j \leqslant n'$) let $\mathbf{M}_i$ (resp. $\mathbf{M}'_j$) be a connected, simply connected, algebraic group defined, isotropic, and almost simple over a non-archimedean local field K_i (resp. K'_j). Let M_i (resp. M'_j) be the group of K_i (resp. K'_j) rational points of $\mathbf{M}_i$ (resp. $\mathbf{M}'_j$) with the canonical locally compact Hausdorff topology. For $i \leqslant n$ (resp. $j \leqslant n'$), let $\mathscr{X}_i$ (resp. $\mathscr{X}'_j$) be the Bruhat–Tits building of $\mathbf{M}_i$ over K_i (resp. of $\mathbf{M}'_j$ over K'_j) with the M_i- (resp. M'_j-) invariant Bruhat–Tits metric. (See Subsection 5.1.) Let

$$M = \prod_{i=1}^{n} M_i, \quad M' = \prod_{j=1}^{n'} M'_j \quad \text{and} \quad \mathscr{X} = \prod_{i=1}^{n} \mathscr{X}_i, \quad \mathscr{X}' = \prod_{j=1}^{n'} \mathscr{X}'_j.^{4}$$

The natural action of M (resp. M') on $\mathscr{X}$ (resp. $\mathscr{X}'$) is proper [15, Lemma 2.6].

A subspace of $\mathscr{X}$ (resp. $\mathscr{X}'$) is said to be an *apartment* if it is of the form $\prod_{i=1}^{n} A_i$ (resp. $\prod_{j=1}^{n'} A'_j$), where A_i (resp. A'_j) is an apartment of $\mathscr{X}_i$ (resp. $\mathscr{X}'_j$). We can similarly define closed quarters and closed quarter-walls of $\mathscr{X}$ and $\mathscr{X}'$.

Given an apartment A of $\mathscr{X}$ (resp. A' of $\mathscr{X}'$) the projection π_A of $\mathscr{X}$ on A (resp. the projection $\pi_{A'}$ of $\mathscr{X}'$ on A') is defined by associating to $x \in \mathscr{X}$ (resp. $x' \in \mathscr{X}'$) the point of A (resp. A') which is nearest to x (resp. x'). It can be checked (cf. Subsection 5.4) that both π_A and $\pi_{A'}$ are well-defined continuous maps; in fact, if $A = \prod A_i$ (resp. $A' = \prod A_j$), then for $(x_1, \ldots, x_n) \in \mathscr{X}$, $\pi_A(x_1, \ldots, x_n) = (\pi_{A_1}(x_1), \ldots, \pi_{A_n}(x_n))$ and for $(x'_1, \ldots, x'_{n'}) \in \mathscr{X}'$, $\pi_{A'}(x'_1, \ldots, x'_{n'}) = (\pi_{A'_1}(x'_1), \ldots, \pi_{A'_{n'}}(x'_{n'}))$.

Let Γ and Γ' be uniform lattices in M and M', respectively. In the sequel *we assume that both Γ and Γ' are torsion free.* Then Γ (resp. Γ') acts freely and properly on $\mathscr{X}$ (resp. $\mathscr{X}'$) [15, Lemma 2.6]. Let $\theta: \Gamma \to \Gamma'$ be an isomorphism. In this section we shall prove that θ induces a natural bijection from the set $\mathscr{A}$ of apartments of $\mathscr{X}$ to the set $\mathscr{A}'$ of apartments of $\mathscr{X}'$.

We begin with some definitions.

6.1. Definition (Mostow [23, Sect. 9]). Let Y, Y' be metric spaces with metrics d and d', respectively. Let $k \geqslant 1$ and b be a nonnegative real number. A map $\varphi: Y \to Y'$ is a (k, b) *pseudo-isometry* if

$$d'(\varphi(y_1), \varphi(y_2)) \leqslant k d(y_1, y_2) \qquad \text{for all} \quad y_1, y_2 \in Y$$

and

$$d'(\varphi(y_1), \varphi(y_2)) \geqslant k^{-1} d(y_1, y_2) \qquad \text{whenever} \quad d(y_1, y_2) \geqslant b.$$

A map $\varphi: Y \to Y'$ is said to be a *pseudo-isometry* if it is a (k, b) pseudo-isometry for some k and b.

6.2. Definition. Let H and H' be two groups and $\lambda: H \to H'$ be an isomorphism. A map $\varphi: Y \to Y'$ of an H-space Y to an H'-space Y' is said to be *λ-equivariant* if for all $h \in H$ and $y \in Y$, $\varphi(h \cdot y) = \lambda(h) \cdot \varphi(y)$.

[4] In the sequel we shall let d and d' denote the product metrics on $\mathscr{X}$ and $\mathscr{X}'$, respectively.

6.3. LEMMA. *Any continuous θ-equivariant map $\varphi: \mathscr{X} \to \mathscr{X}'$ is proper.*

Proof. Let C' be a compact subset of $\mathscr{X}'$ and assume, if possible, that $\varphi^{-1}(C')$ is noncompact. Then there is a sequence $\{x_i\} \subset \varphi^{-1}(C')$ which has no convergent subsequence. Since $\Gamma \backslash \mathscr{X}$ is compact, there is a subsequence of $\{x_i\}$ which is convergent modulo Γ. After replacing $\{x_i\}$ by a subsequence, we assume that $\{x_i\}$ itself is convergent modulo Γ. Then there is a sequence $\{\gamma_i\}$, contained in Γ, such that $\{\gamma_i \cdot x_i\}$ is convergent. But then $\{\varphi(\gamma_i \cdot x_i)\} = \{\theta(\gamma_i) \cdot \varphi(x_i)\}$ is convergent. Let $E' = C' \cup \{\theta(\gamma_i)\varphi(x_i)\}$. Then E' is a relatively compact subset of $\mathscr{X}'$ and since $\varphi(x_i) \in C' \subset E'$, $\theta(\gamma_i)E' \cap E' \neq \varnothing$ for all i. Therefore, since the action of M' on $\mathscr{X}'$ is proper, $\{\theta(\gamma_i)\}$ is a finite set and hence a subsequence of $\{\gamma_i\}$ is a constant sequence. This, in view of the fact that $\{\gamma_i \cdot x_i\}$ is convergent, implies that a subsequence of $\{x_i\}$ is convergent; a contradiction which proves the lemma.

6.4. LEMMA. *There exists a θ-equivariant pseudo-isometry $\varphi: \mathscr{X} \to \mathscr{X}'$.*

Proof. $\mathscr{X} = \prod \mathscr{X}_i$ and $\mathscr{X}' = \prod \mathscr{X}'_j$, being products of simplicial complexes, are cell complexes. The first barycentric subdivisions of these cell complexes are simplicial complexes; in the sequel we consider $\mathscr{X}$ and $\mathscr{X}'$ endowed with these simplicial complex structures. It is obvious that the action of M (resp. M') on $\mathscr{X}$ (resp. $\mathscr{X}'$) is simplicial.

Let π (resp. π') denote the natural projection $\mathscr{X} \to \Gamma \backslash \mathscr{X}$ (resp. $\mathscr{X}' \to \Gamma' \backslash \mathscr{X}'$). Since Γ and Γ' act freely and properly on $\mathscr{X}$ and $\mathscr{X}'$, respectively, π and π' are covering projections. We now note that simplicial action of a group on a simplicial complex becomes *regular* upon passage to the second barycentric subdivision (see for example, Bredon [11; Chap. III, Sect. 1]). Thus the second barycentric subdivisions of $\mathscr{X}$ and $\mathscr{X}'$ induce simplicial complex structures on $\Gamma \backslash \mathscr{X}$ and $\Gamma' \backslash \mathscr{X}'$, respectively. Moreover since both $\Gamma \backslash \mathscr{X}$ and $\Gamma' \backslash \mathscr{X}'$ are compact, these simplicial complexes are finite.

We identify Γ' with Γ with the help of the isomorphism θ. Since $\mathscr{X}$ and $\mathscr{X}'$ are contractible, both $\xi = (\mathscr{X}, \pi, \Gamma \backslash \mathscr{X})$ and $\xi' = (\mathscr{X}', \pi', \Gamma \backslash \mathscr{X}')$ are universal Γ bundles (cf. [18]) and hence there is a homotopy equivalence $\bar{\psi}: \Gamma \backslash \mathscr{X} \to \Gamma \backslash \mathscr{X}'$ such that $\xi \approx \bar{\psi}^*(\xi')$. Also the bundle induced from ξ' by any map homotopic to $\bar{\psi}$ is isomorphic to ξ. By the simplicial approximation theorem (cf. Spanier [33, Theorem 3.4.8]) there exists a map $\bar{\varphi}$, homotopic to $\bar{\psi}$, which is simplicial after an iterated barycentric subdivision of $\Gamma \backslash \mathscr{X}$. Now since $\bar{\varphi}^*(\xi') \approx \xi$, it is clear that there is a $\varphi: \mathscr{X} \to \mathscr{X}'$ which induces $\bar{\varphi}$. Evidently φ is simplicial after iterated barycentric subdivisions of $\mathscr{X}$ and $\mathscr{X}'$, and from this we can easily conclude that there is a constant k_1 such that

$$d'(\varphi(x_1), \varphi(x_2)) \leqslant k_1 d(x_1, x_2) \qquad \text{for all} \quad x_1, x_2 \in \mathscr{X}.$$

On the other hand, since π' is a covering map, there exists a positive constant α and a compact subset C' of $\mathscr{X}'$ such that any closed ball in $\mathscr{X}'$ of radius $\leqslant \alpha$ can be brought inside C' by a suitable element of Γ'. According to Lemma 6.3, $\varphi^{-1}(C')$ is compact; let $b = 2 \operatorname{diam.} \varphi^{-1}(C')$. It is obvious that if B' ($\subset \mathscr{X}'$) is a closed ball of radius $\leqslant \alpha$, then $\operatorname{diam.} \varphi^{-1}(B') \leqslant \frac{1}{2}b$. Thus for $x_1, x_2 \in \mathscr{X}$ with $d'(\varphi(x_1), \varphi(x_2)) \leqslant n\alpha$, there is a path in $\mathscr{X}$, of length less than $\frac{1}{2}nb$, joining x_1 to x_2. Given now $x_1, x_2 \in \mathscr{X}$ with $d(x_1, x_2) \geqslant b$, we can choose $n \geqslant 1$ so that

$$n\alpha \leqslant d'(\varphi(x_1), \varphi(x_2)) < (n+1)\alpha.$$

Then

$$d(x_1, x_2) < \frac{1}{2}(n+1)b = \frac{n+1}{2n} \cdot \frac{b}{\alpha} \cdot n\alpha \leqslant \frac{b}{\alpha} d'(\varphi(x_1), \varphi(x_2)).$$

Set $k = \max(k_1, b/\alpha)$, then

$$d(x_1, x_2) \leqslant kd'(\varphi(x_1), \varphi(x_2) \qquad \text{for all} \quad x_1, x_2 \in \mathscr{X} \quad \text{with} \quad d(x_1, x_2) \geqslant b$$

and

$$d'(\varphi(x_1), \varphi(x_2) \leqslant kd(x_1, x_2) \qquad \text{for all} \quad x_1, x_2 \in \mathscr{X}.$$

Thus φ is a (k, b) pseudo-isometry. This proves the lemma.

6.5. Let $\mathscr{A}$ (resp. $\mathscr{A}'$) be the set of apartments of $\mathscr{X}$ (resp. $\mathscr{X}'$). Since the group M (resp. M') acts transitively on $\mathscr{A}$ (resp. $\mathscr{A}'$), there is a natural locally compact Hausdorff topology on $\mathscr{A}$ (resp. $\mathscr{A}'$) induced by the topology on M (resp. M').

In the sequel we always assume $\mathscr{A}$ and $\mathscr{A}'$ endowed with these topologies.

Given a compact subset C (resp. C') of $\mathscr{X}$ (resp. $\mathscr{X}'$), the subgroup of M (resp. M') which fixes C (resp. C') pointwise is open. Hence, for any apartment A_0 of $\mathscr{X}$ (resp. A_0' of $\mathscr{X}'$), the subset $\mathscr{U}_{A_0}(C) = \{A \in \mathscr{A} \mid A \cap C = A_0 \cap C\}$ (resp. $\mathscr{U}_{A_0'}(C') = \{A' \in \mathscr{A}' \mid A' \cap C' = A_0' \cap C'\}$) is a neighborhood of A_0 (resp. A_0') in $\mathscr{A}$ (resp. $\mathscr{A}'$).

6.6. Let $A = \prod_{i=1}^n A_i$ be an apartment of $\mathscr{X}$, where for $i \leqslant n$, A_i is an apartment of $\mathscr{X}_i$. Let $N(A_i)$ be the stabilizer of A_i in M_i. Then $N(A_i)$ is the normalizer in M_i of the group T_{A_i} of the K_i-rational points of a maximal K_i-split torus $\mathbf{T}_{A_i}$ of $\mathbf{M}_i$. Let $Z(T_{A_i})$ be the centralizer of T_{A_i} in M_i. Then $Z(T_{A_i})\backslash N(T_{A_i})$ is finite and $T_{A_i}\backslash Z(T_{A_i})$ is compact. Let $T_A = \prod T_{A_i}$, $Z(T_A) = \prod Z(T_{A_i})$ and $N(A) = \prod N(A_i)$. Then $Z(T_A)\backslash N(A)$ is finite, $T_A\backslash Z(T_A)$ is compact. For $A' \in \mathscr{A}'$ we define $T_{A'}$, $Z(T_{A'})$, and $N(A')$ similarly.

DEFINITION. An apartment A (resp. A') of $\mathscr{X}$ (resp. $\mathscr{X}'$) is said to be *Γ-adopted* (resp. *Γ'-adopted*) if $Z(T_A) \cap \Gamma$ (resp. $Z(T_{A'}) \cap \Gamma'$) contains an abelian subgroup, of finite index, of rank $r =$ polar rank M ($=$polar rank M', see Proposition 3.3).

6.7. We shall now prove that the set $\mathscr{A}_\Gamma$ (resp. $\mathscr{A}'_{\Gamma'}$) of Γ-adopted (resp. Γ'-adopted) apartments of $\mathscr{X}$ (resp. $\mathscr{X}'$) is dense in $\mathscr{A}$ (resp. $\mathscr{A}'$).

It follows from Proposition 2.14 that given $A \in \mathscr{A}$ and a neighborhood Ω of the identity in M, $\Omega[Z(T_A)] \cap \Gamma$ contains a polar regular s.s. element ρ. Let $\rho = \omega z \omega^{-1}$ with $\omega \in \Omega$ and $z \in Z(T_A)$. Let M_ρ (resp. Γ_ρ) be the centralizer of ρ in M (resp. Γ). Then obviously, $\omega T_A \omega^{-1} = T_{\omega \cdot A} \subset M_\rho$, and it is an immediate consequence of Lemma 2.2 that a subgroup M_ρ^0 of M_ρ of finite index centralizes $T_{\omega \cdot A}$. Thus $M_\rho^0 \subset Z(T_{\omega \cdot A})$. According to Lemma 3.2, Γ_ρ has an abelian subgroup of finite index of rank r. Hence, $\omega \cdot A$ is Γ-adopted (i.e., $\omega \cdot A \in \mathscr{A}_\Gamma$). Since Ω and A were arbitrary, it follows that $\mathscr{A}_\Gamma$ is dense in $\mathscr{A}$. That $\mathscr{A}'_{\Gamma'}$ is dense in $\mathscr{A}'$ is proved similarly.

6.8. Let $r =$ polar rank M ($=$polar rank M', cf. Proposition 3.3). We shall now show that *given an abelian subgroup Δ of Γ of rank r, there is a unique apartment in $\mathscr{X}$ which is stabilized by Δ. Moreover, this apartment is Γ-adopted.*

Let Δ be an abelian subgroup of Γ of rank r. Then, arguing as in the proof of Lemma 3.2, we can show that for $i \leqslant n$ there is a maximal K_i-split torus $\mathbf{T}_i$ of $\mathbf{M}_i$ such that the centralizer of $\prod_{i=1}^n \mathbf{T}_i(K_i)$ contains Δ. Let A_i be the apartment in $\mathscr{X}_i$ which is stabilized by $\mathbf{T}_i(K_i)$, and let $A = \prod A_i$. Then $T_A = \prod \mathbf{T}_i(K_i)$, $\Delta \subset Z(T_A)$, and $Z(T_A)$ stabilizes A. Now since $\Delta \subset Z(T_A) \cap \Gamma$, $T_A \backslash Z(T_A)$ is compact; T_A is the direct product of a compact subgroup and a subgroup isomorphic to $\mathbf{Z}^r$ (cf. the proof of Lemma 3.2), and Δ has rank r, Δ is a subgroup of finite index in $Z(T_A) \cap \Gamma$.

This implies that A is Γ-adopted. The uniqueness assertion follows at once from the observation that any apartment which is stabilized by Δ is stabilized by $\prod \mathbf{T}_i(K_i)$.

It can similarly be proved that *given an abelian subgroup Δ' of Γ' of rank r, there is a unique apartment of $\mathscr{X}'$ which is stabilized by Δ'. Moreover, this apartment is Γ'-adopted.*

6.9. We shall now construct a θ-equivariant bijection $\mathscr{A}_\Gamma \to \mathscr{A}'_{\Gamma'}$. Let $A \in \mathscr{A}_\Gamma$ and let Δ_A be an abelian subgroup of $Z(T_A) \cap \Gamma$ of finite index. Then the rank of Δ_A is r. Let $\Delta'_A = \theta(\Delta_A)$. Then Δ'_A is an abelian subgroup of Γ' of rank r and hence there is a unique $A' \in \mathscr{A}'_{\Gamma'}$ which is stabilized by Δ'_A. We associate A' to A. It is obvious that this association is well defined and hence defines a map $\overline{\theta}_0 : \mathscr{A}_\Gamma \to \mathscr{A}'_{\Gamma'}$. Clearly $\overline{\theta}_0$ is θ-equivariant, i.e., for

$A \in \mathscr{A}_\Gamma$ and $\gamma \in \Gamma$, $\overline{\theta}_0(\gamma \cdot A) = \theta(\gamma) \cdot \overline{\theta}_0(A)$. We shall prove that $\overline{\theta}_0$ has a θ-equivariant extension $\overline{\theta}: \mathscr{A} \to \mathscr{A}'$ which is a homeomorphism.

We fix (cf. Lemma 6.4) once and for all a θ-equivariant pseudo-isometry $\varphi: \mathscr{X} \to \mathscr{X}'$ and a θ^{-1}-equivariant pseudo-isometry $\varphi': \mathscr{X}' \to \mathscr{X}$. We also fix constants $k \geqslant 1$ and $b > 0$ such that both φ and φ' are (k, b) pseudo-isometries. We now prove the following lemma.

6.10. LEMMA. *For all* $A \in \mathscr{A}_\Gamma$,

$$\overline{\theta}_0(A) \subset \varphi(A) \subset T_{kbr}(\overline{\theta}_0(A)).$$

Proof. Let $A \in \mathscr{A}_\Gamma$ and $A' = \overline{\theta}_0(A)$. Let $N(A)$ be the stabilizer of A in M and Δ be an abelian subgroup of $N(A) \cap \Gamma$ of finite index. Let $\Delta' = \theta(\Delta)$ and $N(A')$ be the stabilizer of A' in M'. Since A' is stabilized by Δ', $\Delta' \subset N(A')$. Let π_A (resp. $\pi_{A'}$) be the projection of $\mathscr{X}$ on A (resp. $\mathscr{X}'$ on A') introduced in Section 5. It follows from the construction that, since $N(A)$ (resp. $N(A')$) stabilizes A (resp. A'), π_A (resp. $\pi_{A'}$) is $N(A)$- (resp. $N(A')$-) equivariant; i.e., $\pi_A(\eta \cdot x) = \eta \cdot \pi(x)$ for all $\eta \in N(A)$ and $x \in \mathscr{X}$ (resp. $\pi_{A'}(\eta' \cdot x') = \eta' \cdot \pi_{A'}(x')$ for all $\eta' \in N(A')$ and $x' \in \mathscr{X}'$). This implies in particular that $\pi_{A'} \cdot \varphi|_A: A \to A'$ is $\theta|_\Delta$-equivariant. Since Δ (resp. Δ') acts properly discontinuously on A (resp. A'), A (resp. A') is the total space of a principal Δ- (resp. Δ'-) bundle ξ_A (resp. $\xi'_{A'}$) and since A (resp. A') is Γ- (resp. Γ'-) adopted, the base $\Delta\backslash A$ (resp. $\Delta'\backslash A'$) of this bundle is a compact manifold. Moreover, as A and A' are contractible (in fact A, A' are Euclidean spaces), the bundles ξ_A, $\xi'_{A'}$ are universal bundles. We identify Δ' with Δ with the help of the isomorphism θ. Then evidently the map $\pi_{A'} \cdot \varphi|_A$ induces a Δ-bundle map $\xi_A \to \xi'_{A'}$, and since ξ_A and $\xi'_{A'}$ are universal Δ-bundles, the induced map $\Delta\backslash A \to \Delta\backslash A'$ is a homotopy equivalence. It is well known that any homotopy equivalence between two compact manifolds is surjective. Therefore, the induced map $\Delta\backslash A \to \Delta\backslash A'$ is surjective and hence $\pi_{A'} \cdot \varphi(A) = A'$. Now since φ is proper (cf. Lemma 6.3), it is a closed map; so $\varphi(A)$ is a closed subset of $\mathscr{X}'$, and we can use Lemma 5.6 to conclude that $\varphi(A) \supset A' = \overline{\theta}_0(A)$.

To complete the proof of the lemma we have now only to show that $T_{kbr}(A') \supset \varphi(A)$. Since φ is a (k, b) pseudo-isometry, it would be enough to show that $A \subset T_{br}(\varphi^{-1}(A'))$. We shall in fact show that $A \subset T_{br}(\varphi^{-1}(A') \cap A)$. Let $E = \varphi^{-1}(A') \cap A$ and $F = T_{br}(E) \cap A$. Then since $\varphi(A) \supset A'$, $\varphi(E) = A'$. Evidently both E and F are stable under the action by Δ. The map φ induces a map $\varphi_\Delta: \Delta\backslash\mathscr{X} \to \Delta'\backslash\mathscr{X}'$, and the projection $\pi_{A'}: \mathscr{X}' \to A'$, being Δ'-equivariant, induces a map $\Delta'\backslash\mathscr{X}' \to \Delta'\backslash A'$ which we shall denote by $\overline{\pi}_{A'}$. The metrics on $\mathscr{X}$ and $\mathscr{X}'$ induce metrics $\overline{d}, \overline{d}'$ on $\Delta\backslash\mathscr{X}$ and $\Delta'\backslash\mathscr{X}'$, respectively. Since φ is a (k, b) pseudo-isometry, φ_Δ is a (k, b) pseudo-isometry. Now let λ (resp. λ') be the natural projection $A \to \Delta/A$ (resp. $A' \to \Delta'\backslash A'$). According to a classical theorem of Bieberbach (see [27, Chap. VIII] for a modern proof) any discrete subgroup Λ of the group of isometries (rigid motions!) of the r-dimensional euclidean space $\mathbf{R}^r$, such that $\Lambda\backslash\mathbf{R}^r$ is compact, has a subgroup of finite index consisting of translations, and it is a simple consequence of this result that if moreover Λ is abelian, then it consists entirely of translations. Thus the elements of Δ and Δ' act as translations on A and A', respectively. It is now obvious that after replacing Δ and Δ' by suitable subgroups of finite index, we may (and we shall) assume that the maps λ and λ' are injective on balls of radius $2br$ and $8kbr$, respectively. Then any two points of $\Delta\backslash A$ (resp. $\Delta'\backslash A'$) at a distance at most br (resp. $4kbr$) can be joined by a unique geodesic. Now let Σ' be a triangulation of $\Delta'\backslash A'$ having mesh less than b/k. Let Σ'_0 be a subset of $\Delta'\backslash A'$, which, for every simplex $\sigma' \in \Sigma'$, contains exactly one point $p'_{\sigma'}$ from the interior of σ'. For each $p' \in \Sigma'_0$ we select a $\psi(p')$ in $\lambda(E) \cap \varphi_\Delta^{-1}(p')$. We extend ψ to a continuous map (which we denote again by ψ) of $\Delta'\backslash A'$ into $\Delta\backslash A$ as follows: assume ψ to be so defined on the s-skeleton of Σ' that for every s-simplex σ', and p' in the closed simplex $\bar{\sigma}'$, $\bar{d}(\psi(p'), \psi(p'_{\sigma'})) \leqslant bs$. Let τ' be an $s+1$ simplex and σ' be an s-face of τ', then since $\bar{d}(p'_{\tau'}, p'_{\sigma'}) \leqslant b/k$ and φ_Δ is a (k, b) pseudo-isometry, $\bar{d}(\psi(p'_{\tau'}), \psi(p'_{\sigma'})) \leqslant b$, and hence $\bar{d}(\psi(p'_{\tau'}), \psi(p')) \leqslant b(s+1)\,(\leqslant br)$. We now map the unique geodesic segment $[p'_{\tau'}, p']$ to the unique geodesic segment $[\psi(p'_{\tau'}), \psi(p')]$, where p' varies over the faces of τ'. It is obvious from the definition that for all $x' \in \bar{\tau}'$, $\bar{d}(\psi(x'), \psi(p'_{\tau'})) \leqslant b(s+1)\,(\leqslant br)$. Thus

$$\psi(\Delta'\backslash A') \subset T_{br}(\psi(\Sigma'_0)) \cap \Delta\backslash A \subset T_{br}(\lambda(E)) \cap \Delta\backslash A \subset \lambda(F).$$

Also since $\varphi_\Delta\psi(p'_{\tau'}) = p'_{\tau'}$ and $\bar{d}(\psi(x'), \psi(p'_{\tau'})) \leqslant br$, $\bar{d}(\varphi_\Delta\psi(x'), p'_{\tau'}) \leqslant kbr$. As $\bar{d}'(x', p'_{\tau'}) \leqslant b/k\ (\leqslant kbr)$, we have $\bar{d}'(x', \varphi_\Delta\psi(x')) \leqslant 2kbr$ and hence

$$\bar{d}'(x', \bar{\pi}_{A'}\varphi_\Delta\psi(x')) \leqslant 4kbr.$$

Thus $\bar{\pi}_{A'}\varphi_\Delta\psi$ can be deformed, along the geodesic segments, into the identity map of $\Delta'\backslash A'$. This implies that ψ induces an isomorphism: $H_r(\Delta'\backslash A', \mathbf{Z}) \to H_r(\Delta\backslash A, \mathbf{Z})$, and hence ψ is surjective. Therefore, $\Delta\backslash A = \lambda(F)$. But since F is Δ-stable, $F \supset A$. This completes the proof of the lemma.

We shall now prove the main result of this section.

6.11. Proposition. *There is a unique homeomorphism $\bar{\theta}\colon \mathscr{A} \to \mathscr{A}'$ which extends $\bar{\theta}_0\colon \mathscr{A}_\Gamma \to \mathscr{A}'_{\Gamma'}$. Moreover, $\bar{\theta}$ has the following properties.*

(1) *For all $A \in \mathscr{A}$:*

$$\bar{\theta}(A) \subset \varphi(A) \subset T_{kbr}(\bar{\theta}(A)).$$

(2) *$\bar{\theta}$ is θ-equivariant, i.e., for all $A \in \mathscr{A}$ and $\gamma \in \Gamma$, $\bar{\theta}(\gamma \cdot A) = \theta(\gamma) \cdot \bar{\theta}(A)$.*

Proof. Since $\mathscr{A}_\Gamma$ is dense in $\mathscr{A}$ (Subsection 6.7), in order to prove that $\overline{\theta}_0$ extends to a continuous map of $\mathscr{A}$ into $\mathscr{A}'$ it suffices to show that given $A \in \mathscr{A}$ and a sequence $\{A_n\}_{n \geqslant 1}$ of elements in $\mathscr{A}_\Gamma$ converging to A, the sequence $\{\overline{\theta}_0(A_n)\}$ is convergent in $\mathscr{A}'$. So let A be an apartment in $\mathscr{X}$ and $\{A_n\}_{n \geqslant 1} \subset \mathscr{A}_\Gamma$ be a sequence converging to A. Let $a \in A$, and for any $s > 0$, let B_s and B'_s denote the closed balls of radius s and centers a and $\varphi(a)$ in $\mathscr{X}$ and $\mathscr{X}'$, respectively. Since the sequence $\{A_n\}$ converges to A, and $\mathscr{U}_A(B_{ks})$ is a neighborhood of A (Subsection 6.5), there is an $n_0(s)$ such that for $n \geqslant n_0(s)$

$$A_n \cap B_{ks} = A \cap B_{ks}.$$

Since φ is a (k, b) pseudo-isometry, for $s \geqslant b$, $\varphi^{-1}(B'_s) \subset B_{ks}$. It now follows easily that for $s \geqslant b$ and $n \geqslant n_0(s)$

$$\varphi(A_n) \cap B'_s = \varphi(A) \cap B'_s. \tag{1}$$

Also recall (Lemma 6.10) that for all n, $\overline{\theta}_0(A_n) \subset \varphi(A_n)$ and $T_v(\overline{\theta}_0(A_n)) \supset \varphi(A_n)$, where $v = kbr$. Thus for all $s > 0$ and all $n \geqslant 1$,

$$\overline{\theta}_0(A_n) \cap B'_s \subset \varphi(A_n) \cap B'_s \tag{2}$$

and

$$\varphi(A_n) \cap B'_s \subset T_v(\overline{\theta}_0(A_n) \cap B'_{s+v}). \tag{3}$$

It follows from (1) and (3) that for $n \geqslant n_0(s)$, $\varphi(a) \in T_v(\overline{\theta}_0(A_n))$ and hence $B'_v \cap \overline{\theta}_0(A_n) \neq \varnothing$ for $n \geqslant n_0(s)$. Since the set of apartments in $\mathscr{A}'$ intersecting a given compact subset of $\mathscr{X}'$ is compact; the sequence $\{\overline{\theta}_0(A_n)\}_{n \geqslant 1}$ is contained in a compact subset and hence in particular it has cluster points. Let A' be a cluster point (in $\mathscr{A}'$) of this sequence. Then since a subsequence of $\{\overline{\theta}_0(A_n)\}$ converges to A' and $\mathscr{U}_{A'}(B'_s)$ is a neighborhood of A', for suitable $n \geqslant n_0(s)$ we get from (2) and (1):

$$A' \cap B'_s = \overline{\theta}_0(A_n) \cap B'_s \subset \varphi(A_n) \cap B'_s = \varphi(A) \cap B'_s.$$

Letting $s \to \infty$ we get

$$A' \subset \varphi(A).$$

On the other hand, from (1) and (3) we get (for suitable $n \geqslant n_0(s)$)

$$\begin{aligned}\varphi(A) \cap B'_s = \varphi(A_n) \cap B'_s &\subset T_v(\overline{\theta}_0(A_n) \cap B'_{s+v}) \\ &= T_v(A' \cap B'_{s+v}).\end{aligned}$$

(For the last equality we have used the fact that $\mathcal{U}_{A'}(B'_{s+v})$ is a neighborhood of A'.)

Therefore, for all $s \geqslant b$

$$\varphi(A) \cap B'_s \subset T_v(A').$$

Again letting $s \to \infty$, we have

$$\varphi(A) \subset T_v(A'). \tag{5}$$

Let A'_1, A'_2 be two cluster points of the sequence $\{\overline{\theta}_0(A_n)\}_{n \geqslant 1}$. Then by (4) and (5) we have

$$A'_1 \subset \varphi(A) \subset T_v(A'_2).$$

So, by Lemma 5.10, $A'_1 = A'_2$. This shows that $\{\overline{\theta}_0(A_n)\}_{n \geqslant 1}$ has a unique cluster point, therefore this sequence is convergent. Moreover, if A' is the limit, then by (4) and (5):

$$A' \subset \varphi(A) \subset T_{kbr}(A').$$

This proves that the map $\overline{\theta}_0 : \mathscr{A}_\Gamma \to \mathscr{A}'_{\Gamma'}$ has a continuous extension $\overline{\theta} : \mathscr{A} \to \mathscr{A}'$, and for all $A \in \mathscr{A}$

$$\overline{\theta}(A) \subset \varphi(A) \subset T_{kbr}(\overline{\theta}(A)).$$

The uniqueness and θ-equivariance follow from the fact that $\mathscr{A}_\Gamma$ is dense in $\mathscr{A}$ and $\overline{\theta}_0$ is θ-equivariant. The map $\theta^{-1} : \Gamma' \to \Gamma$ similarly induces a continuous map $\mathscr{A}' \to \mathscr{A}$, which restricted to $\mathscr{A}'_{\Gamma'}$ is the inverse of $\overline{\theta}_0$ and thus this map is the inverse of $\overline{\theta}$. This proves that $\overline{\theta}$ is a homeomorphism.

6.12. COROLLARY. *The map $\varphi : \mathscr{X} \to \mathscr{X}'$ is surjective.*

Proof. It follows at once from the previous proposition that $\varphi(\mathscr{X})$ contains all the apartments of $\mathscr{X}'$. But since $\mathscr{X}'$ is union of its apartments, $\varphi(\mathscr{X}) = \mathscr{X}'$.

7. THE INDUCED ISOMORPHISM BETWEEN TITS BUILDINGS

We begin this section with some definitions.

7.1. Let **H** be a connected reductive algebraic group defined over a field K and let H be the group of K-rational points of **H**. As before, we call a subgroup P of H parabolic if $P = \mathbf{P}(K)$, for a parabolic K-subgroup **P** of **H**.

The *Tits building* $\Delta(\mathbf{H}, K)$ of $\mathbf{H}$ over K is the set of all parabolic subgroups of H ordered by the opposite of the inclusion relation.[5] It is known (cf. Tits [35, Sect. 5.14]) that $\Delta(\mathbf{H}, K)$ is a building of spherical type.

Now let a be a positive integer and for $i \leqslant a$, let $\mathbf{H}_i$ be a connected reductive algebraic group defined over a field K_i. Let H_i be the group of K_i-rational points of $\mathbf{H}_i$ and let $H = \prod_{i=1}^{a} H_i$. A subgroup of H is said to be parabolic if it is of the form $\prod P_i$, where for all $i \leqslant a$, P_i is a parabolic subgroup of H_i. The Tits building of H is, by definition, the Cartesian product $\prod_{i=1}^{a} \Delta(\mathbf{H}_i, K_i)$ of the Tits buildings $\Delta(\mathbf{H}_i, K_i)$; it is the set of all parabolic subgroups of $\prod_{i=1}^{a} H_i$ ordered by the opposite of the inclusion relation.

In this section we shall continue to use the notations introduced in Section 6. We define the equivalence class $[S]$ (resp. $[S']$) of a splice[6] S (resp. S') in $\mathscr{X}$ (resp. $\mathscr{X}'$) to be the set of all splices C (resp. C') in $\mathscr{X}$ (resp. $\mathscr{X}'$) such that $S \sim C$ (resp. $S' \sim C'$). It is evident from definitions that a splice S (resp. S') in $\mathscr{X}$ (resp. $\mathscr{X}'$) is irreducible if and only if every splice in $[S]$ (resp. $[S']$) is irreducible. Let $\mathscr{C}$ (resp. $\mathscr{C}'$) denote the set of equivalence classes of splices in $\mathscr{X}$ (resp. $\mathscr{X}'$) and let $\mathscr{T}$ (resp. $\mathscr{T}'$) be the set of equivalence classes of irreducible splices in $\mathscr{X}$ (resp. $\mathscr{X}'$). Obviously, there is a natural action of M (resp. M') on $\mathscr{C}$ (resp. $\mathscr{C}'$) and $\mathscr{T}$ (resp. $\mathscr{T}'$) is stable under this action.

For a closed quarter-wall E (resp. E') in $\mathscr{X}$ (resp. $\mathscr{X}'$) let

$$P(E) = \{m \in M \mid mE \sim E\}$$

(resp. $P(E') = \{m' \in M' \mid m'E' \sim E'\}$). Then it follows at once from Lemma 5.18 (also see Remark 5.19) that $P(E)$ (resp. $P(E')$) is a parabolic subgroup of M (resp. M') and if F (resp. F') is a wall of E (resp. E'), then $P(E) \subset P(F)$ (resp. $P(E') \subset P(F')$). Let now E_1, E_2 be two equivalent closed quarter-walls in $\mathscr{X}$. Then, by Lemma 5.23(ii), there exists an $m \in P(E_1)$ such that $E_2 = m \cdot E_1$ and hence $P(E_2) = mP(E_1)m^{-1} = P(E_1)$. Since, in view of Corollary 5.27, any irreducible splice is a closed quarter-wall, this shows that there is a well-defined M-map $\lambda : \mathscr{T} \to \mathfrak{P}$; where $\mathfrak{P}$ is the set of parabolic subgroups of M. *This map is bijective.* Its surjectivity follows readily from Lemma 5.26(iii) and injectivity from Corollary 5.22. There is a similarly defined bijective M'-map $\lambda' : \mathscr{T}' \to \mathfrak{P}'$, where $\mathfrak{P}'$ is the set of parabolic subgroups of M'.

We now introduce a partial order $<$ on $\mathscr{T}$. For $\xi, \eta \in \mathscr{T}$, $\eta < \xi$ if and only if there are quarter-walls $E \in \xi$ and $F \in \eta$ such that F is a wall of E. It is a simple consequence of Lemmas 5.18(i) and 5.21 that for $\xi, \eta \in \mathscr{T}$, $\eta < \xi$ if and only if $\lambda(\xi) \subset \lambda(\eta)$. Thus λ is an (order-preserving) isomorphism

[5] Since a parabolic subgroup of H is the group of K-rational points of a unique parabolic K-subgroup of $\mathbf{H}$ (see [7, Corollary 4.20]), our definition of the Tits building is equivalent to the definition given in Tits [35, Sect. 5.3] of the building of $\mathbf{H}$ over K.

[6] For definition of splice see Subsection 5.25.

of $\mathcal{T}$ onto $\prod_{i=1}^{n} \Delta(\mathbf{M}_i, K_i)$. We can introduce similarly a partial order, denoted again by $<$, on $\mathcal{T}'$ and it has the property that for $\xi', \eta' \in \mathcal{T}'$, $\eta' < \xi'$ if and only if $\lambda'(\xi') \subset \lambda'(\eta')$, and thus with this partial order on $\mathcal{T}'$, λ' is an order-preserving isomorphism of $\mathcal{T}'$ onto $\prod_{j=1}^{n'} \Delta(\mathbf{M}'_j, K'_j)$.

The object of this section is to prove that there is a θ-equivariant bijection $\mathcal{C} \to \mathcal{C}'$ which maps $\mathcal{T}$ onto $\mathcal{T}'$ and restricted to $\mathcal{T}$ it preserves the partial order and hence induces a θ-equivariant isomorphism

$$\theta_*: \prod_{i=1}^{n} \Delta(\mathbf{M}_i, K_i) \to \prod_{j=1}^{n'} \Delta(\mathbf{M}'_j, K'_j).$$

We fix a point $\mathbf{o}$ (resp. $\mathbf{o}'$) of $\mathcal{X}$ (resp. $\mathcal{X}'$) which is vertex of a closed quarter-wall of $\mathcal{X}$ (resp. $\mathcal{X}'$). We say that a point of $\mathcal{X}$ (resp. $\mathcal{X}'$) is conjugate to $\mathbf{o}$ (resp. $\mathbf{o}'$) if it is transform, by an element of M (resp. M'), of $\mathbf{o}$ (resp. $\mathbf{o}'$). We now begin by proving a lemma.

7.2. Lemma. *Let A_0 and A be two apartments of $\mathcal{X}$ and let a (resp. a') be a point of the apartment A (resp. $\bar{\theta}(A)$) conjugate to $\mathbf{o}$ (resp. $\mathbf{o}'$). Then*

$$\bar{\theta}(A_0) \wedge_{a'} \bar{\theta}(A) \sim \varphi(A_0 \wedge_a A).$$

Proof. We first note that by Proposition 6.11, for all $\mathbf{A} \in \mathcal{A}$

$$\bar{\theta}(\mathbf{A}) \subset \varphi(\mathbf{A}) \subset T_{kbr}(\bar{\theta}(\mathbf{A})).$$

Now, in view of Proposition 5.29, for v large we have

$$\bar{\theta}(A_0) \wedge_{a'} \bar{\theta}(A) \sim T_v(\bar{\theta}(A_0)) \cap \bar{\theta}(A).$$

But since

$$\begin{aligned} T_v(\bar{\theta}(A_0)) \cap \bar{\theta}(A) &\subset T_v(\varphi(A_0)) \cap \varphi(A) \\ &\subset T_{v+kbr}(\bar{\theta}(A_0)) \cap T_{kbr}(\bar{\theta}(A)) \\ &\subset T_{kbr}(T_{v+2kbr}(\bar{\theta}(A_0)) \cap \bar{\theta}(A)), \end{aligned}$$

and

$$\begin{aligned} T_{kbr}(T_{v+2kbr}(\bar{\theta}(A_0)) \cap \bar{\theta}(A)) &\sim T_{v+2kbr}(\bar{\theta}(A_0)) \cap \bar{\theta}(A) \\ &\sim \bar{\theta}(A_0) \wedge_{a'} \bar{\theta}(A), \end{aligned}$$

it follows that for all v sufficiently large,

$$\bar{\theta}(A_0) \wedge_{a'} \bar{\theta}(A) \sim T_v(\varphi(A_0)) \cap \varphi(A). \tag{1}$$

As φ is a (k, b) pseudo-isometry, for all $s \geqslant b$

$$T_{s/k}(\varphi(A_0)) \cap \varphi(\mathcal{X}) \subset \varphi(T_s(A_0)) \subset T_{ks}(\varphi(A_0)),$$

and hence

$$T_{s/k}(\varphi(A_0)) \cap \varphi(A) \subset \varphi(T_s(A_0)) \cap \varphi(A) \subset T_{ks}(\varphi(A_0)) \cap \varphi(A). \quad (2)$$

For v large, (2) together with (1) implies that

$$\bar{\theta}(A_0) \wedge_{a'} \bar{\theta}(A) \sim \varphi(T_v(A_0)) \cap \varphi(A). \quad (3)$$

But again since φ is a (k, b) pseudo-isometry, if for $x, y \in \mathscr{X}$, $\varphi(x) = \varphi(y)$, then $d(x, y) \leqslant b$. Therefore,

$$\varphi(T_v(A_0) \cap A) \subset \varphi(T_v(A_0)) \cap \varphi(A) \subset \varphi(T_{b+v}(A_0) \cap A) \quad (4)$$

On the other hand, it follows from Proposition 5.29 that, for v large

$$A_0 \wedge_a A \sim T_v(A_0) \cap A.$$

Thus for v large,

$$\varphi(T_v(A_0) \cap A) \sim \varphi(A_0 \wedge_a A) \sim \varphi(T_{b+v}(A_0) \cap A).$$

Now we can conclude from (4) that

$$\varphi(T_v(A_0)) \cap \varphi(A) \sim \varphi(A_0 \wedge_a A),$$

and then (3) implies that

$$\bar{\theta}(A_0) \wedge_{a'} \bar{\theta}(A) \sim \varphi(A_0 \wedge_a A).$$

7.3. Given a splice $A_0 \wedge_a A$ we associate to $A_0 \wedge_a A$ the equivalence class $[\bar{\theta}(A_0) \wedge_{a'} \bar{\theta}(A)] \in \mathscr{C}'$, where a' is any element of $\bar{\theta}(A)$ conjugate to $\mathbf{o}'$. It follows immediately from the preceding lemma that this association is well defined and gives rise to a map $\theta_*: \mathscr{C} \to \mathscr{C}'$. It is obvious that θ_* is surjective, and its injectivity can easily be deduced from Lemma 7.2 using the fact that, since φ is a pseudo-isometry, two subsets E and F of $\mathscr{X}$ are equivalent if and only if $\varphi(E) \sim \varphi(F)$. Also since $\bar{\theta}$ is θ-equivariant, θ_* is θ-equivariant, i.e., for all $\xi \in \mathscr{C}$ and $\gamma \in \Gamma$, $\theta_*(\gamma \cdot \xi) = \theta(\gamma) \cdot \theta_*(\xi)$. We shall now show that given $\xi \in \mathscr{C}$, $\xi \in \mathscr{T}$ if and only if $\theta_*(\xi) \in \mathscr{T}'$. To do this we first observe that, in view of Lemma 7.2, θ_* has the following property: given any splice C, every element of $\theta_*([C])$ is equivalent to $\varphi(C)$. Similarly, its inverse θ_*^{-1}, which is nothing but the map $\mathscr{C}' \to \mathscr{C}$ induced by $\bar{\theta}^{-1}$, has the property that for any splice C' in $\mathscr{X}'$, every element of $\theta_*^{-1}([C'])$ is equivalent to $\varphi'(C')$. Now let C be irreducible, and let C' be a splice representing the equivalence class $\theta_*([C])$. Assume if possible that C' is not irreducible and let $C' \sim C_1' \cup \cdots \cup C_m'$, where each C_i' is a splice inequivalent to C'. For $i \leqslant m$, choose a splice C_i representing $\theta_*^{-1}([C_i'])$. Then $C_i \sim \varphi'(C_i')$. On the other hand, $C \in \theta_*^{-1}([C']) = [C]$ and hence $C \sim \varphi'(C') \sim \varphi'(C_1') \cup \cdots \cup \varphi'(C_m') \sim C_1 \cup \cdots \cup C_m$, which leads to a contradiction. This proves that if $\xi \in \mathscr{T}$, then $\theta_*(\xi) \in \mathscr{T}'$. The converse can be proved similarly.

Now let E, F be two closed quarter-walls with vertex at a and contained in an apartment A of $\mathscr{X}$. Then it follows from Lemma 5.26(iii) that there exist apartments A_1 and A_2 such that $E = A_1 \wedge_a A$ and $F = A_2 \wedge_a A$. Let a' be a point of $A' = \bar{\theta}(A)$ conjugate to $\mathbf{o}'$ and for $i = 1, 2$ let $A_i' = \bar{\theta}(A_i)$. Then $\theta_*([E]) = [A_1' \wedge_{a'} A']$, $\theta_*([F]) = [A_2' \wedge_{a'} A']$, and both $A_1' \wedge_{a'} A'$ and $A_2' \wedge_{a'} A'$ are irreducible and hence are closed quarter-walls in $\mathscr{X}'$ (cf. Corollary 5.27). Moreover by Lemma 7.2, $A_1' \wedge_{a'} A' \sim \varphi(E)$ and $A_2' \wedge_{a'} A' \sim \varphi(F)$. Now assume that F is a wall of E. Then $F \subset E$ and hence $\varphi(F) \subset \varphi(E)$. Therefore $A_2' \wedge_{a'} A'$ is contained in a tubular neighborhood of $A_1' \wedge_{a'} A'$. But since both $A_1' \wedge_{a'} A'$ and $A_2' \wedge_{a'} A'$ are closed quarter-walls, in the apartment A', with common vertex a', it follows at once that $A_2' \wedge_{a'} A'$ is a wall of $A_1' \wedge_{a'} A'$. Thus $\theta_*([F]) < \theta_*([E])$. This proves that for $\xi, \eta \in \mathscr{T}$ if $\xi < \eta$, then $\theta_*(\xi) < \theta_*(\eta)$. The converse (that is, if $\theta_*(\xi) < \theta_*(\eta)$, then $\xi < \eta$) can be proved by showing that θ_*^{-1} is also order preserving.

The following proposition is obvious now.

7.4. Proposition. *θ induces a θ-equivariant isomorphism*

$$\theta_*: \prod_{i=1}^{n} \Delta(\mathbf{M}_i, K_i) \to \prod_{j=1}^{n'} \Delta(\mathbf{M}_j', K_j').$$

Let ${}^0\mathscr{T}$ (resp. ${}^0\mathscr{T}'$) be the set of equivalence classes of closed quarters in $\mathscr{X}$ and let $\mathscr{P}$ (resp. $\mathscr{P}'$) be the set of minimal parabolic subgroups of M (resp. M'). Then (cf. Lemma 5.18(i)) $\lambda|_{{}^0\mathscr{T}}: {}^0\mathscr{T} \to \mathscr{P}$ (resp. $\lambda'|_{{}^0\mathscr{T}'}: {}^0\mathscr{T}' \to \mathscr{P}'$) is a bijection. Also (cf. Subsection 7.3) $\theta_*|_{{}^0\mathscr{T}}: {}^0\mathscr{T} \to {}^0\mathscr{T}'$ and $\theta_*|_{\mathscr{P}}: \mathscr{P} \to \mathscr{P}'$ are θ-equivariant bijections. Now let P be a minimal parabolic subgroup of M. Then the natural map $M/P \to \mathscr{P}$ is bijective. The quotient topology on M/P thus induces compact Hausdorff topologies on $\mathscr{P}$ and ${}^0\mathscr{T}$. We can similarly introduce compact Hausdorff topologies on ${}^0\mathscr{T}'$ and $\mathscr{P}'$.

In the sequel we shall let θ_* denote also the restriction of $\theta_*: \mathscr{T} \to \mathscr{T}'$ to ${}^0\mathscr{T}$ and of $\theta_*: \mathfrak{P} \to \mathfrak{P}'$ to $\mathscr{P}$. We shall now use an argument due to G. D. Mostow to prove that $\theta_*: {}^0\mathscr{T} \to {}^0\mathscr{T}'$ is a homeomorphism. We need the following:

7.5. Lemma. *There exist constants c and v satisfying: for each closed quarter E of $\mathscr{X}$ with vertex a there is a closed quarter $E' \in \theta_*([E])$ with vertex a' such that*

$$\mathrm{hd}(\varphi(E), E') < c \qquad \textit{and} \qquad d'(\varphi(a), a') \leqslant kbr + v.$$

Proof. Let E be a closed quarter contained in an apartment A of $\mathscr{X}$. Let a be the vertex of E. Let

$$\mathscr{A}(E, A) = \{A_0 \in \mathscr{A} \mid A_0 \wedge_a A = E\}.$$

Let a' be any point of $A' = \bar{\theta}(A)$ conjugate to $\mathbf{o}'$. Since for any A_1, A_2 in $\mathscr{A}(E, A)$, $A_1 \wedge_a A = A_2 \wedge_a A = E$, and E is a closed quarter, it follows from Subsection 7.3 and Lemma 7.2 that for all $A_1', A_2' \in \bar{\theta}(\mathscr{A}(E, A))$

$$A_1' \wedge_{a'} A' = A_2' \wedge_{a'} A'.$$

Set $E' = A_0' \wedge_{a'} A'$, where $A_0' \in \bar{\theta}(\mathscr{A}(E, A))$. Then E' is a closed quarter. Also, it is obvious that

$$\bar{\theta}(\mathscr{A}(E, A)) = \mathscr{A}'(E', A') = \{A_0' \in \mathscr{A}' \,|\, A_0' \wedge_{a'} A' = E'\}.$$

In view of Corollary 5.30,

$$E \subset \bigcap_{A_0 \in \mathscr{A}(E,A)} T_{d(a,A_0)}(A_0) \cap A.$$

Therefore,

$$\begin{aligned} \varphi(E) &\subset \varphi\left(\bigcap_{A_0 \in \mathscr{A}(E,A)} T_{d(a,A_0)}(A_0) \cap A\right) \\ &\subset \bigcap_{A_0 \in \mathscr{A}(E,A)} \varphi(T_{d(a,A_0)}(A_0)) \cap \varphi(A) \\ &\subset \bigcap_{A_0 \in \mathscr{A}(E,A)} T_{kd(a,A_0)}(\varphi(A_0)) \cap \varphi(A), \end{aligned}$$

since φ is a (k, b) pseudo-isometry. By Proposition 6.11

$$T_{kbr}(\bar{\theta}(A_0)) \supset \varphi(A_0) \supset \bar{\theta}(A_0) \tag{1}$$

and hence

$$\varphi(E) \subset \bigcap_{A_0 \in \mathscr{A}(E,A)} T_{k(d(a,A_0)+br)}(\bar{\theta}(A_0)) \cap \varphi(A). \tag{2}$$

Again as φ is a (k, b) pseudo-isometry,

$$d'(\varphi(a), \varphi(A_0)) \leqslant kd(a, A_0),$$

and if $d(a, A_0) \geqslant b$, then

$$k^{-1}d(a, A_0) \leqslant d'(\varphi(a), \varphi(A_0)).$$

Now, in view of (1), we have

$$d'(\varphi(a), \varphi(A_0)) \leqslant d'(\varphi(a), \bar{\theta}(A_0)) \leqslant d'(\varphi(a), \varphi(A_0)) + kbr \leqslant k(d(a, A_0) + br). \tag{3}$$

Restricting the choice of A_0 so that $d(a, A_0) \geqslant br$, we get

$$\frac{k(d(a, A_0) + br)}{d'(\varphi(a), \bar{\theta}(A_0))} \leqslant \frac{2kd(a, A_0)}{d'(\varphi(a), \varphi(A_0))} \leqslant 2k^2.$$

Now since $\varphi(E) \subset \varphi(A) \subset T_{kbr}(A')$, we get from (2)

$$\varphi(E) \subset \bigcap T_{2k^2 d'(\varphi(a),\bar{\theta}(A_0))}(\bar{\theta}(A_0)) \cap T_{kbr}(A'), \tag{4}$$

where A_0 ranges over all the elements of $\mathscr{A}(E, A)$ such that $d(a, A_0) \geqslant br$. Now if $d'(\varphi(a), \bar{\theta}(A_0)) \geqslant 2kbr$, then by (3), $d(a, A_0) \geqslant br$. Hence by (4)

$$\varphi(E) \subset \bigcap T_{2k^2 d'(\varphi(a), A_0')}(A_0') \cap T_{kbr}(A'), \tag{5}$$

where A_0' ranges over all the elements of $\mathscr{A}'(E', A')$ such that $d'(\varphi(a), A_0') \geqslant 2kbr$.

It is obvious that there is a constant v such that given any point x' in an apartment A', there is a point $y' \in A'$ conjugate to $\mathbf{o}'$ such that $d'(x', y') \leqslant v$. Thus there is a point $a' \in A'$ conjugate to $\mathbf{o}'$ such that

$$d'(\varphi(a), a') \leqslant kbr + v. \tag{6}$$

From (5) now we get

$$\varphi(E) \subset \bigcap T_{kbr}(T_{4k^2 d'(a', A_0') + kbr}(A_0') \cap A'), \tag{7}$$

where A_0' ranges over all the elements of $\mathscr{A}'(E', A')$ with $d'(a', A_0') \geqslant 3kbr + v$. In fact, if $d'(a', A_0') \geqslant 3kbr + v$, then by (6)

$$2d'(a', A_0') \geqslant d'(\varphi(a), A_0') \geqslant 2kbr.$$

In view of Proposition 5.29, there is a constant c_1 such that for any apartment A' and closed quarter E' in A', the right-hand side of (7) is a subset of $T_{c_1}(E')$. Actually, since M' acts transitively on the set of pairs (A', E') consisting of an apartment A' and a closed quarter E' in A', any constant c_1 which is valid for a single (A', E') is valid for all (A', E'). Consequently for any closed quarter E in X, there is a quarter E' in X' such that

$$\varphi(E) \subset T_{c_1}(E').$$

We shall now show that there is a constant c_2 such that $E' \subset T_{c_2}(\varphi(E))$. In the following for any $A \in \mathscr{A}$, we let A' denote the apartment $\bar{\theta}(A)$. Since φ is a (k, b) pseudo-isometry and it is surjective (Corollary 6.12), for all $s \geqslant 0$ and $A_0 \in \mathscr{A}$,

$$T_s(A_0') \subset T_s(\varphi(A_0)) \subset \varphi(T_{ks+b}(A_0)).$$

Hence

$$\begin{aligned} A' \cap \bigcap T_{d'(a', A_0')}(A_0') &\subset \varphi(A) \cap \bigcap \varphi(T_{kd'(a', A_0')+b}(A_0)) \\ &\subset \varphi(A \cap \bigcap T_{kd'(a', A_0')+2b}(A_0)), \end{aligned}$$

where A_0 can vary over any subset of $\mathscr{A}$. In particular

$$E' \subset \varphi(A \cap \bigcap T_{kd'(a', A_0')+2b}(A_0)),$$

where A_0 varies over all elements of $\mathscr{A}(E, A)$ with $d(a, A_0) \geqslant 2k^2br + kv + 2b$. For A_0 in this subset $\mathscr{S}$ of $\mathscr{A}(E, A)$ we get

$$\begin{aligned}\frac{kd'(a', A'_0) + 2b}{d(a, A_0)} &\leqslant \frac{k(d'(\varphi(a), A'_0) + d'(a', \varphi(a))) + 2b}{d(a, A_0)} \\ &\leqslant \frac{k(d'(\varphi(a), A'_0) + kbr + v) + 2b}{d(a, A_0)} \qquad \text{(by (6))} \\ &\leqslant \frac{k(kd(a, A_0) + 2kbr + v) + 2b}{d(a, A_0)} \qquad \text{(by (3))} \\ &\leqslant k^2 + 1.\end{aligned}$$

Therefore:

$$E' \subset \varphi\left(A \cap \bigcap_{A_0 \in \mathscr{S}} T_{(k^2+1)d(a, A_0)}(A_0)\right).$$

Now select a constant c_2 so that

$$A \cap \bigcap_{A_0 \in \mathscr{S}} T_{(k^2+1)d(a, A_0)}(A_0) \subset T_{c_2/k}(E). \qquad (8)$$

Then

$$E' \subset \varphi(T_{c_2/k}(E)) \subset T_{c_2}(\varphi(E))$$

is valid for all chambers E, since M acts transitively on the set of pairs (A, E), where A is an apartment of $\mathscr{X}$ and E is a closed quarter in A. Set $c = \max\{c_1, c_2\}$. Then for all chambers E in $\mathscr{X}$,

$$\operatorname{hd}(\varphi(E), E') < c,$$

where a' is a point of A' ($= \bar{\theta}(A)$), conjugate to $\mathbf{o}'$, such that $d(a', \varphi(a)) \leqslant kbr + v$ and $E' = A'_0 \wedge_{a'} A'$. This proves the lemma.

7.6. For a positive real number s, let B_s (resp. B'_s) be the closed ball of radius s in $\mathscr{X}$ (resp. $\mathscr{X}'$) with center at $\mathbf{o}$ (resp. $\mathbf{o}'$). For $\xi \in {}^0\mathscr{T}$, let $\mathscr{U}(\xi, s)$ be the set of $\eta \in {}^0\mathscr{T}$ such that for the closed quarter $E \in \xi$, $F \in \eta$ with vertex at $\mathbf{o}$ (cf. Corollary 5.24) $E \cap B_s = F \cap B_s$. $\mathscr{U}(\xi, s)$ is a neighborhood of ξ (cf. Subsection 6.5). For $\xi' \in {}^0\mathscr{T}$ we define $\mathscr{U}(\xi', s)$ similarly.

7.7. We can now verify that the map $\theta_*: {}^0\mathscr{T} \to {}^0\mathscr{T}'$ is continuous. Let $\{\xi_n\}$ be any convergent sequence in ${}^0\mathscr{T}$ and suppose $\xi_\infty = \lim_{n\to\infty} \xi_n$. To prove that θ_* is continuous, we have to show that $\theta_*(\xi_n)$ is a convergent sequence converging to $\theta_*(\xi_\infty)$. But since ${}^0\mathscr{T}'$ is compact, to prove that $\{\theta_*(\xi_n)\}$ is a convergent sequence converging to $\theta_*(\xi_\infty)$ we have only to show that any subsequence of $\{\theta_*(\xi_n)\}$ converges to $\theta_*(\xi_\infty)$. After replacing $\{\theta_*(\xi_n)\}$ by a convergent subsequence we assume now that $\{\theta_*(\xi_n)\}$ is convergent and let ξ' be its limit. For $n \leqslant \infty$, let $E_n \in \xi_n$ be the closed quarter with vertex at $\mathbf{o}$ (cf. Corollary 5.24). For each $n \leqslant \infty$ choose (see Lemma 7.5) $E'_n \in \theta_*(\xi_n)$ with vertex at a point p'_n, such that $d(\varphi(\mathbf{o}), p'_n) \leqslant kbr + v$ and $\mathrm{hd}(\varphi(E_n), E') < c$. Let E''_n be the closed quarter equivalent to E'_n and with vertex at $\mathbf{o}'$. Let $E'' \in \xi'$ be the closed quarter with vertex at $\mathbf{o}'$. For $s \geqslant 0$, let B_s and B'_s denote the closed balls of radius s, with center $\mathbf{o}$ and $\mathbf{o}'$, in $\mathscr{X}$ and $\mathscr{X}'$, respectively.

Set $u = d(\varphi(\mathbf{o}), \mathbf{o}')$. Then $d(p'_n, \mathbf{o}') \leqslant kbr + u + v$, and then in view of Lemma 5.31, for all $n \leqslant \infty$

$$\mathrm{hd}(E'_n, E''_n) < kbr + u + v.$$

But, since for all $n \leqslant \infty$, $\mathrm{hd}(\varphi(E_n), E'_n) \leqslant c$,

$$\mathrm{hd}(\varphi(E_n), E''_n) \leqslant kbr + c + u + v. \tag{1}$$

As $\{\xi_n\}$ converges to ξ_∞ and $\{\theta_*(\xi_n)\}$ to ξ', given $s \geqslant 0$, there is an $n_0(s)$ such that for $n \geqslant n_0(s)$

$$E_n \cap B_{ks} = E_\infty \cap B_{ks}, \tag{2}$$

and

$$E''_n \cap B'_s = E'' \cap B'_s \tag{3}$$

We can use (1), (2), and (3) to conclude (cf. the proof of Proposition 6.11) that

$$\mathrm{hd}(E'', E''_\infty) \leqslant 2(kbr + c + u + v).$$

Thus $E'' \sim E''_\infty$ and hence $\xi' = [E''] = [E''_\infty] = \theta_*(\xi)$. This proves that θ_* is continuous. Now since both ${}^0\mathscr{T}$ and ${}^0\mathscr{T}'$ are compact Hausdorff spaces and θ_* is bijective, it is a homeomorphism. As a consequence we have the following proposition.

7.8. PROPOSITION. *The map $\theta_*: \mathscr{P} \to \mathscr{P}'$ is a θ-equivariant homeomorphism.*

8. STRONG RIGIDITY OF LATTICES AND ITS APPLICATIONS

In this section we shall prove strong rigidity of lattices in non-archimedean semisimple groups (see Theorems 8.6 and 8.7). We need the following three results (Propositions 8.2, 8.3, and 8.5). The first is a simple consequence of a theorem of J. Tits, the second follows from a recent result of A. Borel and J. Tits and the third is an unpublished result of A. Borel.

We begin with some preliminaries.

8.1. Let K, K' be two fields and $\alpha: K \to K'$ be a homomorphism. Given a variety $\mathbf{V}$ defined over K, we denote by ${}^\alpha\mathbf{V}$ the K'-variety obtained from $\mathbf{V}$ by "change of base" with the help of α. The canonical injection $\mathbf{V}(K) \to {}^\alpha\mathbf{V}(K')$ shall be denoted by α^0; it is obviously bijective if α is an isomorphism. We have $K'[{}^\alpha\mathbf{V}] = K[\mathbf{V}] \otimes_K K'$, where K' has the K-algebra structure given by α, the map α^0 is compatible with the homomorphism $f \mapsto f \otimes 1$ of $K[\mathbf{V}]$ in $K'[{}^\alpha\mathbf{V}]$, and this homomorphism defines α^0 when $\mathbf{V}$ is affine. If $\lambda: \mathbf{V} \to \mathbf{W}$ is a K-morphism of K-varieties, then there exists a canonical K'-morphism ${}^\alpha\lambda: {}^\alpha\mathbf{V} \to {}^\alpha\mathbf{W}$ and we have ${}^\alpha\lambda \cdot \alpha^0 = \alpha^0 \cdot \lambda$ on $\mathbf{V}(K)$. From this it follows that if $\mathbf{V}$ is a K-group, then ${}^\alpha\mathbf{V}$ is a K'-group and α^0 is a homomorphism of groups. If $\mathbf{V}$ is a subgroup of $\mathbf{GL}_n$, then ${}^\alpha\mathbf{V}$ is a subgroup of the group $\mathbf{GL}_n$ considered as a K'-group, and for $(g_{ij}) \in \mathbf{V}(K)$, $\alpha^0(g_{ij}) = (\alpha(g_{ij}))$. Moreover if α is an isomorphism, then

$${}^\alpha\mathbf{V}(K') = \{(\alpha(g_{ij})) \mid (g_{ij}) \in \mathbf{V}(K)\}.$$

In case $\alpha: K \to K'$ is an isomorphism and $\mathbf{V}$ is a connected K-group, it is also clear that the buildings $\Delta(\mathbf{V}, K)$ and $\Delta({}^\alpha\mathbf{V}, K')$ are canonically isomorphic. Also since any isomorphism between two non-archimedean local fields is a topological isomorphism (for a proof see for example [8, Sect. 2.3]), it follows that if K, K' are nonarchimedean local fields and $\alpha: K \to K'$ is an isomorphism, then the map $\alpha^0: \mathbf{V}(K) \to {}^\alpha\mathbf{V}(K')$ is an isomorphism of topological groups. In particular if Λ is a lattice in $\mathbf{V}(K)$, then $\alpha^0(\Lambda)$ is a lattice in ${}^\alpha\mathbf{V}(K')$.

8.2. Proposition (cf. Tits [35, Sect. 5.8]). *Let* $\mathbf{S}$ (*resp.* $\mathbf{S}'$) *be a connected adjoint absolutely simple algebraic group defined over a field* K (*resp.* K'). *We assume that* K-*rank* $\mathbf{S} \geqslant 2$, *and if either of the fields* K, K' *is of characteristic* 2 *or* 3, *then it is not perfect. Let* $\Delta(\mathbf{S}, K)$ (*resp.* $\Delta(\mathbf{S}', K')$) *be the Tits building of* $\mathbf{S}$ *over* K (*resp. of* $\mathbf{S}'$ *over* K'). *Let* $\psi: \Delta(\mathbf{S}, K) \to \Delta(\mathbf{S}', K')$ *be an isomorphism. Then there exists a unique isomorphism* $\alpha: K \to K'$ *and a unique* K'-*isomorphism* $\beta: {}^\alpha\mathbf{S} \to \mathbf{S}'$ *such that* β *induces* ψ.

8.3. Proposition (cf. Borel–Tits [8, Théorème 8.11]). *Let* $\mathbf{S}$ (*resp.* $\mathbf{S}'$) *be an adjoint absolutely simple connected algebraic group defined and isotropic over an infinite field* K (*resp.* K'). *Assume that if either of the fields* K, K' *is of characteristic* 2 *or* 3, *then it is not perfect. Let* S (*resp.* S') *be the group of* K (*resp.* K') *rational points of* $\mathbf{S}$ (*resp.* $\mathbf{S}'$). *Let* H (*resp.* H') *be a subgroup of* S (*resp.* S') *containing* S^+ (*resp.* S'^+). *Let* $\lambda: H \to H'$ *be an isomorphism. Then there exists a unique isomorphism* $\alpha: K \to K'$ *and a unique* K'-*isomorphism* $\mu: {}^\alpha\mathbf{S} \to \mathbf{S}'$ *such that* $\lambda = \mu \cdot \alpha^0|_H$.

8.4. For a locally compact Hausdorff topological space Y, we let $\mathrm{Top}(Y)$ denote the Hausdorff space (with the compact-open topology) of all self-homeomorphisms of Y. It is easy to see that the composition map $\mathrm{Top}(Y) \times \mathrm{Top}(Y) \to \mathrm{Top}(Y)$ defined by

$$(f, g) \mapsto f \cdot g, \qquad \text{for} \quad f, g \in \mathrm{Top}(Y),$$

is continuous; and if H is a topological group acting on Y, then the induced map $H \to \mathrm{Top}(Y)$ is continuous. In general $\mathrm{Top}(Y)$ is not a topological group but if Y is compact, then it is a topological group.

8.5. PROPOSITION (A. Borel). *Let* $\mathbf{V}$ *be a variety and* $\mathbf{H}$ *be an algebraic group, both defined over a local field K. Let V (resp. H) be the set of K-rational points of* $\mathbf{V}$ *(resp.* $\mathbf{H}$*) with the natural locally compact Hausdorff topology. Let* $\boldsymbol{\sigma}$ *be an action of* $\mathbf{H}$ *on* $\mathbf{V}$ *defined over K, and let* σ *be the induced action of H on V. Let W be a locally compact H-subspace of V. Let I be the kernel of the action of H on W. Assume that there exist finitely many points* $w_1, \ldots, w_m \in W$ *such that* $\bigcap_{i=1}^m H_{w_i} = I$*, where* H_{w_i} *denotes the isotropy group of* w_i*; and the morphism* $h \mapsto (\boldsymbol{\sigma}(h)w_1, \ldots, \boldsymbol{\sigma}(h)w_m)$ *of* $\mathbf{H}$ *into* $\mathbf{V}^m$ *is separable.*[7] *Then the induced homomorphism* $\rho: H/I \to \mathrm{Top}(W)$ *is a homeomorphism onto its closed image.*

Proof. Obviously, ρ is injective and since H/I is a topological group acting on W, $\rho: H/I \to \mathrm{Top}(W)$ is continuous. Thus to prove that ρ is a homeomorphism onto $\rho(H/I)$ we have only to show that the topology induced on H/I by ρ is finer than the topology on H/I. But since H/I is a topological group, and the composition map $\mathrm{Top}(W) \times \mathrm{Top}(W) \to \mathrm{Top}(W)$ is continuous, it is enough to prove that any neighborhood of the identity in H/I contains a neighborhood of the identity in the topology induced by the map ρ.

For a compact subset C, and an open subset U of W containing C, let $\langle C, U\rangle$ denote the set $\{h \in H/I \mid h \cdot C \subset U\}$. Then the finite intersections of the sets of the form $\langle C, U\rangle$ forms a fundamental system of neighborhoods of the identity in the topology induced from $\mathrm{Top}(W)$ on H/I. We shall now show that given any neighborhood Ω of the identity in H/I, there exist open neighborhoods U_i of w_i in W such that $\bigcap \langle\{w_i\}, U_i\rangle \subset \Omega$. This will prove that the homomorphism $\rho: H/I \to \mathrm{Top}(W)$ is a homeomorphism onto $\rho(H/I)$, and then it will also follow that $\rho(H/I)$, being a locally compact (and hence locally closed) subgroup of $\mathrm{Top}(W)$, is closed.

Consider the variety $\mathbf{V}^m$ with the natural $\mathbf{H}$ action. Let $\mathfrak{w} = (w_1, \ldots, w_m) \in W^m$, and let $\mathcal{O}_{\mathfrak{w}}$ $(\subset W^m)$ be the orbit of $\mathfrak{w}$ under H. According to Lemma 1.22, the map $h \mapsto h \cdot \mathfrak{w}$ of H onto $\mathcal{O}_{\mathfrak{w}}$ is an open map, and hence, the natural

[7] Note that, since the Zariski topology is noetherian, existence of a finite set of points $w_1, \ldots, w_m \in W$ such that $\bigcap_{i=1}^m H_{w_i} = I$ is obvious. Hence if $\mathrm{char}(K) = 0$, the hypothesis is always satisfied.

map $H/I \to \mathcal{O}_{\mathfrak{w}}$ is open. Now let Ω be a neighborhood of the identity in H/I. Then the set $\Omega \cdot \mathfrak{w} = \{\omega \cdot \mathfrak{w} | \omega \in \Omega\}$ is a neighborhood of $\mathfrak{w}$ in $\mathcal{O}_{\mathfrak{w}}$. Therefore, there is an open neighborhood of $\mathfrak{w} = (w_1, \ldots, w_m)$ in W^m of the form $U_1 \times \cdots \times U_m$ (where for $i \leqslant m$, U_i is an open neighborhood of w_i in W) such that $U_1 \times \cdots \times U_m \cap \mathcal{O}_{\mathfrak{w}} \subset \Omega \cdot \mathfrak{w}$. It is obvious that $\bigcap_{i=1}^m \langle \{w_i\}, U_i \rangle \subset \Omega$, and this proves the proposition.

We are in a position now to prove the main theorems of this section.

8.6. Theorem. *Let n, n' be positive integers. For $i \leqslant n$ (resp. $j \leqslant n'$), let $\mathbf{S}_i$ (resp. $\mathbf{S}'_j$) be an adjoint, absolutely simple, connected algebraic group, defined and isotropic over a non-archimedean local field K_i (resp. K'_j), and let S_i (resp. S'_j) be the group of K_i (resp. K'_j) rational points of $\mathbf{S}_i$ (resp. $\mathbf{S}'_j$) with the natural locally compact Hausdorff topology. Let $S = \prod_{i=1}^n S_i$, $S' = \prod_{j=1}^{n'} S'_j$; and let π_i (resp. π'_j) be the natural projection of S on S_i (resp. of S' on S'_j). Let Λ (resp. Λ') be a uniform lattice in S (resp. S'), and let $\theta: \Lambda \to \Lambda'$ be an isomorphism. Then $n = n'$, and there is an i such that K_i-rank $\mathbf{S}_i = 1$ if and only if for some j, K'_j-rank $\mathbf{S}'_j = 1$. Now assume that for all i (resp. j) $\leqslant n$ such that K_i-rank $\mathbf{S}_i = 1$ (resp. K'_j-rank $\mathbf{S}'_j = 1$), the closure of $\pi_i(\Lambda)$ (resp. $\pi'_j(\Lambda')$) contains S_i^+ (resp. $S_j'^+$). Then, if necessary after reindexing, there exist isomorphisms $\alpha_i: K_i \to K'_i$ and K'_i-isomorphisms $\theta_i: {}^{\alpha_i}\mathbf{S}_i \to \mathbf{S}'_i$ such that the isomorphism $\boldsymbol{\theta} = (\theta_1 \cdot \alpha_1^0, \ldots, \theta_n \cdot \alpha_n^0) : \prod_{i=1}^n S_i \to \prod_{i=1}^n S'_i$ is an extension of θ.*

Proof. For $i \leqslant n$ and $j \leqslant n'$, let $p_i: \mathbf{M}_i \to \mathbf{S}_i$ and $p'_j: \mathbf{M}'_j \to \mathbf{S}'_j$ be the universal coverings of $\mathbf{S}_i$ and $\mathbf{S}'_j$ over K_i and K'_j, respectively, and let M_i (resp. M'_j) be the group of K_i (resp. K'_j) rational points of $\mathbf{M}_i$ (resp. $\mathbf{M}'_j$) with the natural Hausdorff topology. Let $M = \prod_{i=1}^n M_i$ and $M' = \prod_{j=1}^{n'} M'_j$. Let p, p' be the natural projections: $M \to S$ and $M' \to S'$, respectively. Let Γ (resp. Γ') be a torsion free lattice in M (resp. M') such that $\Phi = p(\Gamma)$ (resp. $\Phi' = p'(\Gamma')$) is a normal subgroup of Λ (resp. Λ') of finite index (cf. Lemma 1.11). After replacing Γ and Γ' by suitable subgroups of finite index, we assume that $\theta(\Phi) = \Phi'$. Let ${}^0\theta$ denote the restriction of θ to Φ. Since the kernel of p (resp. p') is finite and Γ (resp. Γ') is torsion free, $p|_\Gamma$ (resp. $p'|_{\Gamma'}$) is an isomorphism onto Φ (resp. Φ') and hence the isomorphism ${}^0\theta: \Phi \to \Phi'$ induces an isomorphism $\Gamma \to \Gamma'$ which we denote again by θ.

For $i \leqslant n$ (resp. $j \leqslant n'$) let $\Delta(\mathbf{M}_i, K_i)$ (resp. $\Delta(\mathbf{M}'_j, K'_j)$) be the Tits building of $\mathbf{M}_i$ over K_i (resp. of $\mathbf{M}'_j$ over K'_j) and $\Delta(\mathbf{S}_i, K_i)$ (resp. $\Delta(\mathbf{S}'_j, K'_j)$) be the Tits building of $\mathbf{S}_i$ over K_i (resp. of $\mathbf{S}'_j$ over K'_j). Then (cf. Tits [35, Sect. 5.4]) p_i (resp. p'_j) induces a natural isomorphism $\Delta(\mathbf{M}_i, K_i) \to \Delta(\mathbf{S}_i, K_i)$ (resp. $\Delta(\mathbf{M}'_j, K'_j) \to \Delta(\mathbf{S}'_j, K'_j)$). According to Proposition 7.4, there is a θ-equivariant isomorphism $\prod_{i=1}^n \Delta(\mathbf{M}_i, K_i) \to \prod_{j=1}^{n'} \Delta(\mathbf{M}'_j, K'_j)$, and hence there is an isomorphism $\theta_*: \prod_{i=1}^n \Delta(\mathbf{S}_i, K_i) \to \prod_{j=1}^n \Delta(\mathbf{S}'_j, K'_j)$ which is ${}^0\theta$-equivariant. It is known [35] that decomposition of a building of spherical type into a product of irreducible factors is unique upto order. Therefore, we conclude that $n = n'$ and for $i \leqslant n$, the isomorphism θ_* takes $\Delta(\mathbf{S}_i, K_i)$ isomorphically onto $\Delta(\mathbf{S}'_j, K'_j)$ for a suitable $j \leqslant n$. Thus, if necessary after reindexing, we may (and we shall) assume that for all $i \leqslant n$, θ_* induces an isomorphism of $\Delta(\mathbf{S}_i, K_i)$ onto $\Delta(\mathbf{S}'_i, K'_i)$. Then it follows from Proposition 8.2 that for all $i\,(\leqslant n)$ such that K_i-rank $\mathbf{S}_i \geqslant 2$, there exists a unique isomorphism $\alpha_i\colon K_i \to K'_i$ and a unique K'_i-isomorphism $\theta_i\colon {}^{\alpha_i}\mathbf{S}_i \to \mathbf{S}'_i$ which induces the isomorphism between the Tits buildings $\Delta(\mathbf{S}_i, K_i)$ and $\Delta(\mathbf{S}'_i, K'_i)$. (Note that no local field of positive characteristic is perfect.) It is obvious from the uniqueness of θ_i and ${}^0\theta$-equivariance of θ_* that for $s \in {}^{\alpha_i}\mathbf{S}_i$ and $\varphi \in \Phi$, $\theta_i(\alpha_i^0(\pi_i(\varphi))s\alpha_i^0(\pi_i(\varphi))^{-1}) = \pi'_i(\theta(\varphi))\theta_i(s)(\pi'_i(\theta(\varphi)))^{-1}$. In particular, for all $\varphi, \psi \in \Phi$, $\theta_i(\alpha_i^0\pi_i(\varphi\psi\varphi^{-1})) = \pi'_i(\theta(\varphi))\theta_i(\alpha_i^0\pi_i(\psi))(\pi'_i(\theta(\varphi)))^{-1}$. Now since

$$\theta_i(\alpha_i^0\pi_i(\varphi\psi\varphi^{-1})) = \theta_i(\alpha_i^0\pi_i(\varphi))\theta_i(\alpha_i^0\pi_i(\psi))(\theta_i(\alpha_i^0\pi_i(\varphi))^{-1},$$

it follows that $(\theta_i\alpha_i^0(\pi_i(\varphi)))^{-1}\pi'_i(\theta(\varphi))$ centralizes $\theta_i(\alpha_i^0\pi_i(\Phi))$. As Φ is a lattice in S, $\alpha_i^0\pi_i(\Phi)$ is Zariski dense in ${}^{\alpha_i}\mathbf{S}_i$ and hence $\theta_i(\alpha_i^0\pi_i(\Phi))$ is Zariski dense in $\mathbf{S}'_i$. Therefore $(\theta_i(\alpha_i^0(\pi_i(\varphi))))^{-1}\pi'_i(\theta(\varphi))$ is central. But since $\mathbf{S}'_i$ has trivial center, $\theta_i(\alpha_i^0(\pi_i(\varphi))) = \pi'_i(\theta(\varphi))$ for all $\varphi \in \Phi$.

Now let i $(\leqslant n)$ be such that K_i-*rank* $\mathbf{S}_i = 1$. Then since $\Delta(\mathbf{S}_i, K_i)$ is isomorphic to $\Delta(\mathbf{S}'_i, K'_i)$, K'_i-*rank* $\mathbf{S}'_i = 1$. We now fix a minimal parabolic subgroup $\mathbf{P}_i$ (resp. $\mathbf{P}'_i$) of $\mathbf{S}_i$ (resp. $\mathbf{S}'_i$) defined over K_i (resp. K'_i), and let P_i (resp. P'_i) be the group of K_i (resp. K'_i) rational points of $\mathbf{P}_i$ (resp. $\mathbf{P}'_i$) with the natural Hausdorff topology. Let $\mathscr{P}_i$ (resp. $\mathscr{P}'_i$) be the set of minimal parabolic subgroups of $\mathbf{S}_i$ (resp. $\mathbf{S}'_i$) defined over K_i (resp. K'_i). Then there are canonical bijections of S_i/P_i onto $\mathscr{P}_i$, and of S'_i/P'_i onto $\mathscr{P}'_i$. We use these to get compact Hausdorff topology on $\mathscr{P}_i$ and $\mathscr{P}'_i$.

There is a natural action of $\mathbf{S}_i$ on the projective variety $\mathbf{S}_i/\mathbf{P}_i$, and since $\mathbf{S}_i$ is adjoint, simple, and isotropic, the induced action of S_i on S_i/P_i is effective. We shall now show that the intersection $\mathfrak{v}_i$ of the subalgebras of the Lie algebra of $\mathbf{S}_i$ corresponding to the minimal K_i-parabolic subgroups is trivial. In fact, given a maximal K_i-split torus of $\mathbf{S}_i$, considering the minimal K_i-parabolic subgroups containing it, we see that $\mathfrak{v}_i$ is centralized by the torus. It follows that $\mathfrak{v}_i$ is centralized by every K_i-split torus of $\mathbf{S}_i$, and hence $\mathfrak{v}_i$ is central in the Lie algebra of $\mathbf{S}_i$. But since $\mathbf{S}_i$ is of adjoint type, its Lie algebra has a trivial center [7, Complements: 2.23(a)] and therefore $\mathfrak{v}_i$ is trivial. Now it is obvious that we can apply Proposition 8.5 to conclude that the induced homomorphism $S_i \to \mathrm{Top}(S_i/P_i)$ is an isomorphism (of topological groups) onto its image, which is closed. (Note that S_i/P_i is compact, and hence $\mathrm{Top}(S_i/P_i)$ is a topological group.) Thus the natural homomorphism $S_i \to \mathrm{Top}(\mathscr{P}_i)$ is an isomorphism (of topological groups) onto its closed image. We use this isomorphism to identify S_i with a closed subgroup of $\mathrm{Top}(\mathscr{P}_i)$. We identify S'_i with a closed subgroup of $\mathrm{Top}(\mathscr{P}'_i)$ in a similar way. Now since θ_*

induces an isomorphism of $\Delta(\mathbf{S}_i, K_i)$ onto $\Delta(\mathbf{S}'_i, K'_i)$, it induces a bijection $\theta^i_*: \mathscr{P} \to \mathscr{P}'_i$, and it follows from Proposition 7.8 that θ^i_* is a ${}^0\theta$-equivariant homeomorphism. We identify $\mathscr{P}'_i$ with $\mathscr{P}_i$ with the help of this homeomorphism.

Let $\Phi_i = \pi_i(\Phi)$ and $\Phi'_i = \pi'_i(\Phi')$. When we identify $\mathscr{P}'_i$ with $\mathscr{P}_i$ with the help of the homeomorphism θ^i_*, Φ'_i, considered as a subgroup of $\mathrm{Top}(\mathscr{P}_i)$, gets identified with Φ_i. Hence the closure H'_i of Φ'_i gets identified with the closure H_i of Φ_i. This shows that there exists an isomorphism $\lambda_i: H_i \to H'_i$ of topological groups. It follows from the hypothesis (see also Subsection 1.14) that $H_i \supset S_i^+$ and $H'_i \supset S_i'^+$. Hence, by Proposition 8.3, there is an isomorphism $\alpha_i: K_i \to K'_i$ and an isomorphism $\theta_i: {}^{\alpha_i}\mathbf{S}_i \to \mathbf{S}'_i$ defined over K'_i such that $\lambda_i = \theta_i \cdot \alpha_i^0|_{H_i}$.

Now let $\boldsymbol{\theta}: \prod_{i=1}^n S_i \to \prod_{i=1}^n S'_i$ be the isomorphism $(\theta_1 \cdot \alpha_1^0, \ldots, \theta_n \cdot \alpha_n^0)$. Then, clearly, $\boldsymbol{\theta}(\varphi) = \theta(\varphi)$ for $\varphi \in \Phi$. We shall prove that in fact $\boldsymbol{\theta}(\lambda) = \theta(\lambda)$ for all $\lambda \in \Lambda$. Let $\lambda \in \Lambda$, then for all $\varphi \in \Phi$, $\lambda\varphi\lambda^{-1} \in \Phi$ and hence $\theta(\lambda\varphi\lambda^{-1}) = \boldsymbol{\theta}(\lambda\varphi\lambda^{-1}) = \boldsymbol{\theta}(\lambda)\boldsymbol{\theta}(\varphi)\boldsymbol{\theta}(\lambda^{-1}) = \boldsymbol{\theta}(\lambda)\theta(\varphi)\boldsymbol{\theta}(\varphi^{-1})$. On the other hand, $\theta(\lambda\varphi\lambda^{-1}) = \theta(\lambda)\theta(\varphi)\theta(\varphi^{-1})$. Hence for all $\varphi \in \Phi$, $\boldsymbol{\theta}(\lambda)^{-1}\theta(\lambda)$ commutes with $\theta(\varphi)$. Thus for all $\lambda \in \Lambda$, $\boldsymbol{\theta}(\lambda)^{-1}\theta(\lambda)$ centralizes $\Phi' = \theta(\Phi)$. We conclude, in view of the Zariski density of Φ', that $\boldsymbol{\theta}(\lambda)^{-1}\theta(\lambda)$ is central and hence $\boldsymbol{\theta}(\lambda) = \theta(\lambda)$ for all $\lambda \in \Lambda$. This completes the proof of the theorem.

Remark. In case $n > 1$ and Λ (and hence Λ' also) is irreducible, it follows from Lemmas 1.7 and 1.15 that the closure of $\pi_i(\Lambda)$ (resp. $\pi'_j(\Lambda')$) contains S_i^+ (resp. S_j^+) for all i (resp. j) $\leqslant n$ such that $\mathrm{char}(K_i) = 0$ (resp. $\mathrm{char}(K'_j) = 0$).

The following theorem is a corollary of Theorem 8.6.

8.7. THEOREM. *Let n, n' be positive integers. For $i \leqslant n$ (resp. $j \leqslant n'$) let $\mathbf{H}_i$ (resp. $\mathbf{H}'_j$) be a connected, adjoint, semisimple algebraic group, defined, simple, and isotropic over a local field k_i (resp. k'_j), and let H_i (resp. H'_j) be the group of k_i (resp. k'_j) rational points of $\mathbf{H}_i$ (resp. $\mathbf{H}'_j$) with the natural locally compact Hausdorff topology. Let $H = \prod_{i=1}^n H_i$, $H' = \prod_{j=1}^{n'} H'_j$, and let π_i (resp. π'_j) be the natural projection of H on H_i (resp. of H' on H'_j). Let Λ (resp. Λ') be a uniform lattice in H (resp. H') and let $\theta: \Lambda \to \Lambda'$ be an isomorphism. Then $n = n'$, and there is an i such that k_i-rank $\mathbf{H}_i = 1$ if and only if for some j, k'_j-rank $\mathbf{H}'_j = 1$. Now assume that for i (resp. j) $\leqslant n$ such that k_i-rank $\mathbf{H}_i = 1$ (resp. k'_j-rank $\mathbf{H}'_j = 1$), the closure of $\pi_i(\Lambda)$ (resp. $\pi'_j(\Lambda')$) contains H_i^+ (resp. $H_j'^+$). Then, if necessary after reindexing, there exist isomorphisms $\theta_i: H_i \to H'_i$ of topological groups such that θ is the restriction to Λ of $(\theta_1, \ldots, \theta_n): H \to H'$.*

Proof. For each $i \leqslant n$ (resp. $j \leqslant n'$), there is a finite separable extension K_i (resp. K'_j) and an adjoint absolutely simple connected algebraic group $\mathbf{S}_i$ (resp. $\mathbf{S}'_j$) such that $R_{K_i/k_i}\mathbf{S}_i = \mathbf{H}_i$ (resp. $R_{K'_j/k'_j}\mathbf{S}'_j = \mathbf{H}'_j$) [7, Sect. 6.21(ii)]. Therefore, in particular, K_i-rank $\mathbf{S}_i = k_i$-rank $\mathbf{H}_i$ and K'_j-rank $\mathbf{S}'_j = k'_j$-rank $\mathbf{H}'_j$.

Let S_i (resp. S'_j) be the group of K_i (resp. K'_j) rational points of $\mathbf{S}_i$ (resp. $\mathbf{S}'_j$) with the natural locally compact Hausdorff topology, and let $S = \prod_{i=1}^{n} S_i$ (resp. $S' = \prod_{j=1}^{n'} S'_j$). Then S (resp. S') is canonically isomorphic to H (resp. H') as topological groups. Now the theorem immediately follows from Theorem 8.6.

Some Applications of the Strong Rigidity Theorem

8.8. Groups of Units of Quadratic Forms. It is well known that the special orthogonal group of a nondegenerate integral quadratic form of rank $n \geqslant 3$ is a connected semisimple algebraic group defined over $\mathbf{Q}$, and the group of units is a lattice in the group of $\mathbf{R}$-rational points of the orthogonal group of the quadratic form. The latter group is compact (resp. locally isomorphic to $SL(2, \mathbf{R})$) if and only if the quadratic form is definite (resp. indefinite and of rank 3). Moreover the special orthogonal group is absolutely almost simple if $n \neq 4$; and if $n = 4$ and f is indefinite, then the group of its $\mathbf{R}$-rational points is locally isomorphic to $SL(2, \mathbf{C})$ or $SL(2, \mathbf{R}) \times SL(2, \mathbf{R})$, depending on whether the form is of Witt index 1 or 2 over $\mathbf{R}$.

Let f, f' be two nondegenerate integral quadratic forms of rank $\geqslant 5$ and assume that f is indefinite. Let Γ (resp. Γ') be the group of units of f (resp. f'). One can use strong rigidity of lattices in real analytic semisimple groups [25, Theorem B] to conclude that if a subgroup of Γ of finite index is isomorphic to a subgroup of Γ' of finite index, then the projective special orthogonal group $PSO(f)$ of f is isomorphic to the corresponding group $PSO(f')$ of f', and the isomorphism is defined over $\mathbf{Q}$. It can be deduced from this that then *f' is equivalent (over the field of rational numbers) to a rational multiple of f.*

It is not in general true that a nondegenerate indefinite integral quadratic form of rank 2, 3, or 4 is determined (upto a rational multiple) by its group of units. However, we can prove the following: *Let f and f' be nondegenerate indefinite integral quadratic forms of rank n and n', respectively. Assume that $n = 2$, 3, or 4 and a subgroup of finite index of the group of units of f is isomorphic to a subgroup of finite index of the group of units of f'. Then $n = n'$ and f represents 0 over $\mathbf{Q}$ if and only if f' does. Moreover, if f is isotropic (i.e., it represents 0) over $\mathbf{Q}$ and is of rank 4, then f' is equivalent (over $\mathbf{Q}$) to a rational multiple of f.* We note here, in this connection, that the group of units of a nondegenerate indefinite integral quadratic form of rank 2 is finite if the form is isotropic over $\mathbf{Q}$, otherwise it has a subgroup of finite index which is infinite cyclic. Also, according to the well-known Godement criterion, a nondegenerate integral quadratic form of rank ≥ 3 is isotropic over $\mathbf{Q}$ if and only if its group of units is a nonuniform lattice in the group of $\mathbf{R}$-rational points of its orthogonal group; further it is known that a uniform lattice in a real-analytic group cannot be isomorphic to a nonuniform lattice in any real-analytic group (this is proved in [24]). These observations, together with the strong rigidity theorem [25, Theorem B], take care of our first assertion. Now assume that f is isotropic over $\mathbf{Q}$ and is of rank 4. Let d (resp. d') be the discriminant of f (resp. f') and let $k = \mathbf{Q}(\sqrt{d})$ (resp. $k' = \mathbf{Q}(\sqrt{d'})$). Then in case the Witt index of f (resp. f') is 1, the projective special orthogonal group of f (resp. f') is $\mathbf{Q}$-simple and it is isomorphic, over $\mathbf{Q}$, to $R_{k/\mathbf{Q}}(PSL(2))$ (resp. $R_{k'/\mathbf{Q}}(PSL(2))$), and if the Witt index is 2 then it is isomorphic, over $\mathbf{Q}$, to the direct product of two copies of $PSL(2)$ and hence the group of units is irreducible if and only if the Witt index is 1. From this we conclude that the Witt index (over $\mathbf{Q}$) of f' is equal to that of f. Now since any two forms of rank 4 and Witt index 2 are equivalent, we have only to consider the case when f (and therefore f' also) is of Witt index 1 over $\mathbf{Q}$. It can be shown easily that, in this case, f' is equivalent (over $\mathbf{Q}$) to a rational multiple of f if and only if $d \equiv d' (\text{mod } \mathbf{Q}^{\times 2})$ or equivalently, if and only if $k \approx k'$. But, since the projective special orthogonal group of f (resp. f') is isomorphic to $R_{k/\mathbf{Q}}(PSL(2))$ (resp. $R_{k'/\mathbf{Q}}(PSL(2))$), the strong rigidity theorem can once again be applied to conclude that $R_{k/\mathbf{Q}}(PSL(2))$ is isomorphic to $R_{k'/\mathbf{Q}}(PSL(2))$, and then it follows (see for example Proposition 8.3) that $k \approx k'$. This proves our second assertion.

We now consider *definite* quadratic forms. For a nondegenerate integral quadratic form and a rational prime p, by the *group of p-units* of the quadratic form we mean the subgroup of the orthogonal group consisting of matrices with coefficients in $\mathbf{Z}[1/p]$. Now let f, f' be *definite* integral quadratic forms of rank ≥ 5 and let $p, p' \in \mathbf{Z}$ be primes. Let Γ_p (resp. $\Gamma'_{p'}$) be the group of p-units of f (resp. p'-units of f'). Then Γ_p (resp. $\Gamma'_{p'}$) is a lattice in $O(f)(\mathbf{Q}_p)$ (resp. $O(f')(\mathbf{Q}_{p'})$). We now assume that the Witt index of f over $\mathbf{Q}_p$ ($=$ the $\mathbf{Q}_p$-rank of $O(f)$) is at least 2. Then it follows from Theorem 8.6 that if a subgroup of Γ_p of finite index is isomorphic to a subgroup of $\Gamma'_{p'}$ of finite index, then $p = p'$ and $PSO(f)$ is isomorphic to $PSO(f')$ over $\mathbf{Q}$. As before, one can deduce from this that then the quadratic form f' is equivalent (over $\mathbf{Q}$) to a rational multiple of f.

It is obvious how to extend the above results on quadratic forms to (nondegenerate) skew-symmetric, hermitian, and skew-hermitian forms. We leave the details to the reader.

8.9. STRONG RIGIDITY IMPLIES LOCAL RIGIDITY. Let Γ be a finitely generated group. Let $R(\Gamma, G)$ be the space of all homomorphisms of Γ into a topological group G, with the topology of pointwise convergence. There is a natural action of G on $R(\Gamma, G)$. A homomorphism $u \in R(\Gamma, G)$ is said to be *locally rigid* if the orbit under G of u in $R(\Gamma, G)$ is open in $R(\Gamma, G)$.

Let now $H = \prod_{i=1}^{n} H_i$ be as in Theorem 8.7 and Λ be an irreducible discrete uniform subgroup of H. We assume that for all $i \leqslant n$, *the characteristic* *of the field k_i is zero.* Let $\iota: \Lambda \to H$ be the natural inclusion. We shall prove that Theorem 8.7 implies that ι is locally rigid. In fact it is a simple consequence of a theorem of Koszul (see [21, Chap. IV, Theorem 2] that there is a neighborhood W of i in $R(\Lambda, H)$ such that for all $w \in W$, w is an isomorphism of Λ onto a discrete uniform subgroup of H. One can now apply Theorem 8.7 to conclude that there is a neighborhood Ω ($\subset W$) of ι such that given $\omega \in \Omega$, there is an $h \in H$ so that $\omega = \operatorname{Int} h \cdot \iota$. This shows that the orbit of ι under H contains the neighborhood Ω of ι. The homogeneity of the orbit immediately gives that the orbit of ι is in fact open in $R(\Lambda, H)$. This proves that $\iota: \Lambda \to H$ is locally rigid.

According to a well-known result of A. Weil, $u \in R(\Lambda, H)$ is locally rigid if $H^1(\Lambda, \mathrm{Ad} \cdot u) = 0$. In a recent paper [15, Sect. 9] H. Garland has proved that for certain arithmetic $\Lambda, H^1(\Lambda, \mathrm{Ad} \cdot \iota) = 0$; where, as before, $\iota: \Lambda \to H$ is the natural inclusion.

Note added in proof: The following developments related to this paper have taken place since the paper was submitted:

(1) G. A. Margulis has announced very beautiful and far-reaching results on arithmeticity and finite dimensional representations of lattices in semisimple groups over local fields of characteristic zero at the International Congress of Mathematicians (Vancouver, 1974).

(2) Question 1.16 has been settled in the affirmative. This also leads to an affirmative solution of the problem of strong approximation in simply connected semisimple algebraic groups over global fields of arbitrary characteristics. See the author's paper in *Ann. of Math.* **105** (1977), and also a paper of Margulis in *Functional Anal. Appl.* **11** (1977).

(3) The extension of θ to an isomorphism of H onto H' in Theorem 8.7 has been shown to be unique, see the author's article in *Math. Ann.* **218** (1975).

(4) In positive characteristics, strong rigidity holds (Theorem 8.7) whereas local rigidity may fail; see the author's note in *Bull. Soc. Math. France* **105** (1977).

Remarks: (i) Let M, M' and $\mathfrak{X}, \mathfrak{X}'$ be as in Section 6. Let M_∞ (resp. M'_∞) be a real analytic semisimple group with trivial center and no compact factors, and let X (resp. X') be the symmetric Riemannian space associated with M_∞ (resp. M'_∞). Let Γ, Γ' be torsion-free discrete *uniform* subgroups of $M_\infty \times M$ and $M'_\infty \times M'$, respectively. Let $\theta: \Gamma \to \Gamma'$ be an isomorphism. Then we can apply a recent result of F. E. A. Johnson (see his forthcoming paper "On the triangulation of stratified sets and singular varieties") to prove that $\Gamma \backslash X \times \mathfrak{X}$ and $\Gamma' \backslash X' \times \mathfrak{X}'$ have *nice* triangulations, and then use this to construct a θ-equivariant pseudo-isometry from $X \times \mathfrak{X}$ to $X' \times \mathfrak{X}'$ and a θ^{-1}-equivariant pseudo-isometry from $X' \times \mathfrak{X}'$ to $X \times \mathfrak{X}$. Now one can combine the results of Mostow [23] with our results to extend Theorems 8.6 and 8.7 to the case where some of the local fields involved are archimedean.

(ii) We can use Weil's restriction of scalars $R_{k/\mathbf{Q}}$ to extend the results of Section 8.8 to forms over number fields. For related results see a paper of A. J. Hahn in *Crelle's Journal* **273** (1975).

(iii) In a preliminary version of this paper, Lemma 1.5 was proved under the assumption that Γ is finitely generated, consequently we had to have the same restriction on Γ in Lemmas 1.7, 1.8 and Corollary 1.9. Margulis in his paper in *Functional Anal. Appl.* **11** (1977) has also given a proof of the Lemma for all Γ; his proof is exactly the same as given here.

Acknowledgments

I express my deep gratitude to Professor A. Borel for several valuable suggestions, to Professor J. Tits for a very helpful communication and to Professors G. D. Mostow and M. S. Raghunathan for fruitful conversations.

References

1. A. Borel, Groupes algébriques linéaires, *Ann. of Math.* **64** (1956), 20–82.
2. A. Borel, "Linear Algebraic Groups," Benjamin, New York, 1969.
3. A. Borel, Cohomologie de certains groupes discrets et Laplacien p-adique, *Séminaire Bourbaki*, No. 437 (1973/74).
4. A. Borel and J-P. Serre, Théorèmes de finitude en cohomologie Galoisienne, *Comment. Math. Helv.* **39** (1964), 111–164.
5. A. Borel and J-P. Serre, Corners and arithmetic groups, *Comment. Math. Helv.* **48** (1973), 436–491.
6. A. Borel and J-P. Serre, Cohomologie d'immeubles et de groupes S-arithmetiques, *Topology* **15** (1976), 211–232.
7. A. Borel and J. Tits, Groupes réductifs, *Publ. Math. I.H.E.S.* **27** (1965), 55–150; Complements, *ibid.* **41** (1972), 253–276.
8. A. Borel and J. Tits, Homomorphismes "abstraits" de groupes algébriques simples, *Ann. of Math.* **97** (1973), 499–571.
9. N. Bourbaki, "Groupes de Lie et Algèbres de Lie," Chapter IV–VI, Hermann, Paris, 1968.
10. N. Bourbaki, "Intégration," Chapter VII and VIII, Hermann, Paris, 1963.
11. G. E. Bredon, "Introduction to Compact Transformation Groups," Academic Press, New York, 1972.
12. F. Bruhat and J. Tits, Groupes réductifs sur un corps local, *Publ. Math. I.H.E.S.* **41** (1972), 5–252.
13. J. S. Dani and S. G. Dani, Discrete groups with dense orbits, *J. Indian Math. Soc.* **37** (1973), 183–195.
14. C. Delaroche and A. Kirillov, Sur les relations entre l'espace dual d'un groupe et la structure de ses sours-groupes fermés, *Séminaire Bourbaki* No. 343 (1967/68).
15. H. Garland, p-adic curvature and the cohomology of discrete subgroups of p-adic groups, *Ann. of Math.* **97** (1973), 375–423.
16. M. Goto and H.-C. Wang, Non-discrete uniform subgroups of semi-simple Lie groups, *Math. Ann.* **198** (1972), 259–286.
17. T. Honda, Isogenies, rational points and section points of group varieties, *Japan J. Math.* **30** (1960), 84–101.
18. D. Husemoller, "Fiber Bundles," McGraw-Hill, New York, 1966.
19. Y. Ihara, On discrete subgroups of two by two projective linear groups over p-adic fields, *J. Math. Soc. Japan* **18** (1966), 219–235.
20. N. Jacobson, "Lectures on Abstract Algebra," Vol. III, Van Nostrand-Reinhold, Princeton, New Jersey, 1964.
21. J. L. Koszul, *in* "Lectures on Groups of Transformations," Chap. IV, Tata Inst. Fund. Res., Bombay, 1965.
22. G. A. Margulis, Arithmetic properties of discrete subgroups (in Russian), *Uspekhi Mat. Nauk* **29** (1974), 49–98.
23. G. D. Mostow, Strong rigidity of locally symmetric spaces, *Ann. of Math. Study No. 78*, Princeton Univ. Press, Princeton, N.J., 1973.

24. G. Prasad, Unipotent elements and isomorphism of lattices in semi-simple groups, *J. Indian Math. Soc.* **37** (1973), 103–124.
25. G. Prasad, Strong rigidity of **Q**-rank 1 lattices, *Invent. Math.* **21** (1973), 255–286.
26. G. Prasad and M. S. Raghunathan, Cartan subgroups and lattices in semi-simple groups, *Ann. of Math.* **96** (1972), 296–317.
27. M. S. Raghunathan, "Discrete subgroups of Lie groups," Springer-Verlag, Berlin and New York, 1972.
28. M. S. Raghunathan, Discrete groups and **Q**-structures on semi-simple Lie groups, to appear.
29. M. Rosenlicht, Some rationality questions on algebraic groups, *Annali di Mat.* (IV) **43** (1957), 25–50.
30. A. Selberg, On discontinuous groups in higher dimensional symmetric spaces, "Contributions to Function Theory," Tata Inst. Fund. Res., Bombay, 1960.
31. J-P. Serre, "Lie Algebras and Lie Groups," Benjamin, New York, 1965.
32. J-P. Serre, Cohomologie des groupes discrets, "Prospects in Mathematics," Ann. of Math. Study No. 70, Princeton Univ. Press, Princeton, New Jersey, 1971.
33. E. H. Spanier, "Algebraic Topology," McGraw-Hill, New York, 1966.
34. J. Tits, Algebraic and abstract simple groups, *Ann. of Math.* **80** (1964), 313–329.
35. J. Tits, Buildings of spherical type and finite B–N pairs, "Lecture Notes in Mathematics," Springer-Verlag, Berlin and New York, 1974.
36. S. P. Wang, On density properties of *S*-subgroups of locally compact groups, *Ann. of Math.* **94** (1971), 325–329.
37. A. Weil, "Adeles and Algebraic Groups," Notes by M. Demazure and T. Ono, The Institute for Advanced Study, Princeton, New Jersey, 1961.
38. A. Weil, "Basic Number Theory," Springer-Verlag, Berlin and New York, 1967.
39. A. Weil, "Foundations of Algebraic Geometry," Amer. Math. Soc., Providence, Rhode Island, reprinted 1962.

STUDIES IN ALGEBRA AND NUMBER THEORY
ADVANCES IN MATHEMATICS SUPPLEMENTARY STUDIES, VOL. 6

Commutative *R*-Subalgebras of *R*-Infinite *R*-Algebras and the Schmidt Problem for *R*-Algebras

THOMAS J. LAFFEY

Department of Mathematics, University College, Belfield, Dublin, Ireland

1. INTRODUCTION

Let R be a commutative associative ring with identity. A nonassociative (=not necessarily associative) ring A is called an R-algebra if

(i) A is a left unital R-module,
(ii) $r(ab) = (ra)b = a(rb)$ for all $a, b \in A$, $r \in R$.

Let M be a (left unital) R-module. We say that M is R-finite if M is finitely generated (f.g.) as an R-module, otherwise we say M is R-infinite.

A nonassociative ring B is called *alternative* if $(a, b, b) = (a, a, b) = 0$ for all $a, b \in B$ where, as usual, (x, y, z) denotes the associator $(xy)z - x(yz)$. An R-algebra A is called alternative if A is alternative as a ring.

We say that an R-algebra A *involves* an R-algebra B if there exist R-subalgebras D, E of A with D an ideal of E such that B is isomorphic, as an R-algebra, to E/D.

We now state our main result.

MAIN THEOREM. *Let R be a commutative associative ring with identity and let A be an R-infinite alternative R-algebra. Then A has an R-infinite commutative associative R-subalgebra.*

In particular if R is a field, the Main Theorem implies that every infinite-dimensional alternative algebra over a field has an infinite-dimensional commutative associative R-algebra. This generalizes the main result of [7].

Taking R to be $\mathbb{Z}$, the ring of integers, the Main Theorem implies that every infinite alternative ring has an infinite commutative associative subring. For associative rings, this result was first proved by the author in [6] and the result was extended to alternative rings by Bell [1] (cf. [10]).

By a result of Witt [12] all the nonzero commutative subalgebras of the free Lie algebra on two generators are one-dimensional. Thus one cannot

ISBN 0-12-599153-3

extend the Main Theorem to flexible algebras, i.e., algebras satisfying the identity $(x, y, x) = 0$.

In Section 4 of the paper we consider the structure of an R-infinite alternative R-algebra A all of whose proper R-subalgebras are R-finite. In particular either $A^2 = \{0\}$ or A is a field. For the case $R = \mathbb{Z}$, the structure of an infinite associative ring S with no proper infinite subrings was determined in [8], namely S is either the zero-ring on the p-quasicyclic group $C(p^\infty)$, for some prime p, or S is the union of the finite fields $\bigcup_{n=1}^{\infty} GF(p^{q^n})$ for some primes p, q. The R-algebras which arise in Section 4 are analogous to those.

The proof of the Main Theorem is by contradiction. Assuming A is a counterexample to the theorem, it is shown that A involves a counterexample D which can be embedded as an order in a finite-dimensional division algebra.

When an earlier draft of this paper was prepared we were not in fact able to eliminate the possibility of the existence of such a counterexample D but the reduction of the problem to this case is sufficient to obtain the results on the Schmidt problem in Section 4. A recent result of Formanek [2] can now be applied to eliminate the possibility of the existence of D in the case where A is associative, and for A alternative but not associative, D can be eliminated by using the multiplication table for Cayley–Dickson division algebras.

2. Notation and Preliminary Results

We begin by stating some well-known results on alternative rings. The proofs of these statements are given, for instance, in Chapter III of Schafer's book [11].

Let A be an alternative ring and let $x, y, z \in A$. Then

(1) (Artin) The subring generated by x, y is associative.

(2) If (x, y, z) denotes the associator $(xy)z - x(yz)$, then

- (a) $(x, y, z) + (x, z, y) = 0$,
- (b) $(x, y, z) + (y, x, z) = 0$,
- (c) $3(x, y, z) = (xy, z) - x(y, z) - (x, y)z$ where $(x, y) = xy - yx$,
- (d) $(x^2, y, z) = x(x, y, z) + (x, y, z)x$,
- (e) $(x, xy, z) = (x, y, z)x$,
- (f) $(x, yx, z) = x(x, y, z)$.

Throughout the paper, R will denote a commutative associative ring with identity. By an R-module is meant a left unital R-module.

Let A be an R-algebra and let $a \in A$. We write

$$\operatorname{Ann}(a) = \{b \in A \mid ab = ba = 0\}.$$

An easy consequence of the identities above is:

LEMMA 1. *Let A be an alternative R-algebra and let $a \in A$. Then* $\operatorname{Ann}(a)$ *is an R-subalgebra of A.*

Proof. The only nontrivial point is to check that if $x, y \in \operatorname{Ann}(a)$, so does xy. But $(xy)a = (x, y, a) = -(x, a, y) = 0$.

In proving the Main Theorem, we try to mimic the proof in [6]. We thus must define an analogue of the function $X(\ )$ considered there.

Let A be an alternative R-algebra. We define

$$X(A) = \{a \in A \mid \operatorname{Ann}(a) \text{ has an } R\text{-infinite } R\text{-subalgebra}\}.$$

In the definition of $X(A)$ we wish to allow for the possibility that $\operatorname{Ann}(a)$ itself might be R-finite while having R-infinite R-subalgebras. Such a situation can occur for non-Noetherian R. An R-algebra B is called a zero-R-algebra if $B^2 = 0$.

LEMMA 2. *Let A be an alternative R-infinite R-algebra. Assume that A has no R-infinite zero-R-subalgebra C. Let $x \in A$ be such that $x^2a = ax^2 = 0$ for all $a \in A$. Then $x \in X(A)$.*

Proof. Let $L = \{xax \mid a \in A\}$. Clearly L is an R-submodule of A. Also, using the identities (2),

$$\begin{aligned}(xa_1x)(xa_2x) &= ((xa_1)x)(xa_2x)\\ &= xa_1(x(xa_2x)) + (xa_1, x, xa_2x)\\ &= (xa_1)(x^2(a_2x)) + (a_1, x, a_2x)x^2 = 0.\end{aligned}$$

So $L^2 = 0$ and thus L is R-finite. The map $t: A \to L: t(a) = xax$ is an R-homomorphism of A and thus, since t is surjective,

$$\ker(t) = \{a \in A \mid xax = 0\}$$

is an R-submodule of A and $\ker(t)$ is R-infinite. Let $M = \{xa \mid a \in \ker(t)\}$. Again, M is an R-submodule of A. Let $a, b \in M$. Then

$$(xa)(xb) = x(a(xb)) + (x, a, xb).$$

Now

$$\begin{aligned}(x, a, xb) &= (x, a, b)x = (x^2, a, b) - x(x, a, b) = -x(x, a, b)\\ &= -x((xa)b) + x(x(ab)) = -x((xa)b).\end{aligned}$$

Hence

$$(xa)(xb) = x(a(xb)) - x((xa)b).$$

Further

$$((xa)(xb))x = (xa, xb, x) = (a, b, x)x^2 = 0.$$

Thus $(xa)(xb) \in M$. Thus M is an R-subalgebra of A. Since $xm = mx = 0$ for all $m \in M$, the lemma is proved if M is R-infinite. If M is R-finite, then $E = \{a \in A \mid xa = 0\}$ is R-infinite. Let $E' = \{ax \mid a \in E\}$. An easy calculation shows that if a, $b \in E$, $(ax)(bx) = 0$. Hence $E'^2 = 0$ and E' is R-finite. The map $f: E \to E'$ defined by $f(a) = ax$ is an R-homomorphism of E onto E'. Thus $\ker(f) = \mathrm{Ann}(x)$ is R-infinite.

If A is an R-algebra and S is a nonempty subset of A, $\mathrm{alg}_R(S)$ denotes the R-subalgebra generated by S.

If I is an R-submodule of an R-module A, then $X(A \bmod I)$ denotes the pre-image of $X(A/I)$ under the natural projection $A \to A/I$.

3. Proof of the Main Theorem

Assume that the Main Theorem is false and let (R, A) be a counterexample.

If $X(A) \neq \{0\}$, choose $x_1 \in X(A) - \{0\}$. Let $H(x_1)$ be an R-infinite R-subalgebra of $\mathrm{Ann}(x_1)$. Let $X_1 = X(H(x_1)) - \mathrm{alg}_R(x_1)$. If X_1 is nonempty, choose $x_2 \in X_1$ and let $H(x_2)$ be an R-infinite R-subalgebra of

$$H(x_1) \cap \mathrm{Ann}(x_2).$$

Having chosen elements $x_1, \ldots, x_n \in A$ and R-subalgebras $H(x_1), \ldots, H(x_n)$ such that $H(x_i) \subseteq H(x_{i-1}) \cap \mathrm{Ann}(x_i)$ $(i > 1)$, we consider

$$X_n = X(H(x_n)) - \mathrm{alg}_R(x_1, \ldots, x_n).$$

If X_n is nonempty let $x_{n+1} \in X_n$ and let $H(x_{n+1})$ be an R-infinite R-subalgebra of $H(x_n) \cap \mathrm{Ann}(x_{n+1})$. Let

$$S_n = \mathrm{alg}_R(x_1, \ldots, x_n).$$

Then S_n is a commutative, associative R-subalgebra of A (since $x_i x_j = 0\ (i \neq j)$), and $x_{n+1} \notin S_n$, so the sequence $x_1, \ldots, x_n, \ldots$ must terminate, otherwise $S = \bigcup_{n=1}^{\infty} S_n$ would be a commutative associative R-infinite R-subalgebra of A.

Write $X_0 = \{0\}$ and $H(x_0) = A$. We have thus established that there exists a nonnegative integer n such that

$$X(H(x_n)) \subseteq \mathrm{alg}_R(x_1, \ldots, x_n).$$

So

$$X(H(x_n)) = \{h \in H(x_n) | hy = yh = 0 \text{ for all } y \in H(x_n)\}.$$

Write $H = H(x_n)$ and $X = X(H)$. Note that X is an ideal of H and that X is a zero-R-subalgebra of H, so all submodules of X are R-subalgebras of A and are R-finite.

(1) *H/X is also a counterexample to the theorem.*

For suppose for the sake of contradiction that H/X has an R-infinite commutative associative R-subalgebra B/X. Let $a \in B - X$ and define $g: B \to B$ by $g(u) = au$. Then g is an R-homomorphism of B. Let $y, z \in B$. The identity

$$3(a, y, z) = (ay, z) - a(y, z) - (a, y)z$$

together with the fact that $(a, y, z), (y, z), (a, y)$ belong to X which annihilates H, forces $3(a, y, z) = (ay, z)$ and thus $0 = 3(a, y, z)a = 3(z, y, az) = (ay, az)$. Also

$$\begin{aligned}(ay)(az) &= (ay)(za)\\ &= a(y(za)) + (a, y, za)\\ &= a(y(za))\\ &= a((yz)a - (y, z, a))\\ &= a((yz)a).\end{aligned}$$

This implies that $g(B)$ is a commutative R-subalgebra of A. Also if $x, y, z \in B$, then $(ax, ay, az) = 0$ and thus $g(B)$ is associative. Hence $g(B)$ is R-finite. Hence $\ker(g)$ is R-infinite. Let $T = \{ba | b \in \ker(g)\}$. If $b_1, b_2 \in \ker(g)$, then $(b_1 a)(b_2 a) = 0$ (by a simple calculation), so $T^2 = \{0\}$ and thus T is R-finite. But now $\operatorname{Ann}(a) \cap B$ is R-infinite and $a \in X$, giving a contradiction.

Next we show

(2) $X(H/X) = 0$.

For let $x \in X(H \bmod X)$. There exists an R-subalgebra V/X of H/X such that $vx, xv \in X$ for all $v \in V$. Let $h: V \to X$ be defined by $h(v) = vx$. Then $\ker(h)$ is R-infinite and $\{xv | v \in \ker(h)\}$ is a zero-R-subalgebra of H and is therefore R-finite. Thus $\operatorname{Ann}(x)$ is R-infinite and $x \in X$.

Let $D = H/X$. Since $X(D) = \{0\}$, Lemma 2 implies

(3) *D has no nonzero nilpotent elements.*

(4) *If $0 \neq a \in D$ and $r \in R$ with $ra = 0$, then $rD = 0$.*

For let $Y = \{rb | b \in D\}$. Now Y is an R-subalgebra of $\mathrm{Ann}(a)$, so Y is R-finite since $X(D) = 0$. For a fixed $b \in D$, $\{c \in A | (rb)c = 0\}$ is R-infinite, so $rb \in X(D) = 0$.

Let $N(R) = \{r \in R | rD = 0\}$. We may regard D as an $R/N(R)$-algebra in the obvious way. It follows immediately from (4) that $R/N(R)$ is an integral domain. Clearly we may assume that $N(R) = 0$.

We now have

(5) D is a *torsion-free R-module and R is an integral domain.*

Next, we show

(6) *D has no nonzero divisors of zero.*

Let $0 \neq x \in D$. Since $\mathrm{alg}_R(x)$ is commutative and associative, it is R-finite. So there exist positive integers k, n and elements $a_1 \neq 0, a_2, \ldots, a_n \in R$ such that

$$(a_1 x + \cdots + a_n x^n)x^k = 0.$$

By (3), $a_1 x + \cdots + a_n x^n = 0$. Let $p(x) = -a_2 x - \cdots - a_n x^{n-1}$. Then $xp(x) = a_1 x$. Let $y \in D$. Then using (1) of Section 2,

$$[p(x)yp(x) - a_1 yp(x)]^2 = [p(x)yp(x) - a_1 p(x)y]^2 = 0,$$

which, by (3), implies that $a_1 yp(x) = a_1 p(x)y$ and thus $yp(x) = p(x)y$ for all $y \in D$.

Suppose now that $xy = 0$. Then, using (1) of Section 2 again, $p(x)p(y) = 0$. Now let $d \in D$. Then

$$p(x)(p(y)d) = -(p(x), p(y), d),$$

so

$$(p(x)(p(y)d))p(x) = -(p(x), p(x)p(y), d) = 0$$

and thus

$$\{p(x)(p(y)d)\}^2 = -(p(x)(p(y)d), p(x), p(y)d) = 0.$$

So, by (3), $p(x)(p(y)d) = 0$.

Let $G = \{p(y)d | d \in D\}$. Since $(p(y), d) = 0$, G is an R-subalgebra of D. Also G is contained in $\mathrm{Ann}(p(x))$, so G is R-finite. Hence $\mathrm{Ann}(p(y))$ is R-infinite and thus $p(y) \in X(D) = \{0\}$. So $y = 0$.

(7) *R is Noetherian.*

For let $0 \neq x \in D$ and let $p(x)$ be defined as above. Suppose J is an ideal of R. Then $Jp(x)$ is an R-subalgebra of D and it is commutative and associative and thus it is R-finite. This forces J to be R-finite.

Let Q be the quotient field of R and let $E = D \otimes_R Q$. From the proof of (6) we have

(8) *E is an alternative division algebra over Q.*

We regard D as an R-subalgebra of E in the obvious way.

(9) *E is finite dimensional over Q.*

Suppose that E has a subfield K which is infinite dimensional over Q. Let $k_1, k_2, \ldots$ be an infinite Q-independent subset of E. There exists $0 \neq r_i \in R$ such that $r_i k_i \in D$. But then $\text{alg}(\{r_i k_i\})$ is an R-infinite commutative associative R-subalgebra of D, giving a contradiction. So all subfields of E are finite dimensional over Q. In particular, the centre (in the nonassociative sense) F of E is finite dimensional. We recall the following result of Kleinfield ([4] if F has characteristic different from 2, [5] if F has characteristic 2):

> Let E be an alternative division algebra with centre F. If E is not associative, then E is an eight-dimensional Cayley–Dickson division algebra over F."

Applying this result, we see that if E is infinite dimensional over Q, then it is associative and hence a maximal subfield of E is infinite dimensional over Q, giving a contradiction. This proves (9).

(10) *D is* not *associative.*

Suppose D is associative. Then E is a finite-dimensional division algebra over Q and is the quotient algebra of D. So D is a prime P.I. algebra whose centre $Z(D)$ is a finitely generated R-module.

We now recall the following result of Formanek [2, Theorem 1 (1)]:

> Let D be an (associative) prime P.I. ring with centre C. Then D can be embedded as a C-submodule of a free C-module of finite rank."

Applying this result, and the fact that R is Noetherian, we see that D is R-finite, giving a contradiction.

Hence by the results of Kleinfeld quoted above:

(11) *E is an (eight-dimensional) Cayley–Dickson division algebra over $Z(E) = F$, say (where $Z(E)$ denotes the centre (in the nonassociative algebra sense) of E).*

We now aim to prove that D is necessarily a finitely generated C-module, where $C = Z(D)$. Suppose first F does not have characteristic 2.

Since E is a Cayley–Dickson division algebra, E has a basis $u_0 = 1, u_1, \ldots, u_7$ such that $u_i^2 \in F$ $(i = 1, 2, \ldots, 7)$ and $u_i u_j + u_j u_i = 0$ $(1 \leqslant i, j \leqslant 7, i \neq j)$. Now there exists $0 \neq z \in R$ such that $z u_i \in D$ $(i = 0, 1, \ldots, 7)$.

Let $d \in D$. We may write

$$d = \sum_{i=0}^{7} \lambda_i u_i$$

for some $\lambda_i \in F$. Then for $0 \neq i \neq j \neq 0$

$$(u_i d + d u_i)u_j + u_j(u_i d + d u_i) = 4\lambda_i u_i^2 u_j,$$

so multiplying by zu_j and noting that then the left-hand side is in D and the right-hand side in F, we get

$$4\lambda_i z^3 u_i^2 u_j^2 \in C.$$

Also

$$u_1 d + d u_1 = 2\lambda_0 u_1 + 2\lambda_1 u_1^2.$$

Hence there exists $0 \neq w \in C$, w independent of d such that $w\lambda_i \in C$ $(i = 0, 1, \ldots, 7)$. But then wD is contained in the C-span of $u_0, u_1, \ldots, u_7$ and thus wD is a finitely generated C-module (note that C must be Noetherian, since otherwise C would not be R-finite). The map $s: D \to wD$ defined by $s(d) = wd$ is injective and C-linear and thus D is a finitely generated C-module.

We thus have

(12) *F has characteristic 2.*

In this case, Kleinfeld [5] has shown that E contains elements a, b, c such that

(i) $(a,b) = (a,c) = (b,c) = (ab,ac) = (ab,bc) = (ac,bc) = 0$,
(ii) a^2, b^2, c^2, (a,b,c) are nonzero elements of F, and
(iii) $\{1, a, b, c, ab, ac, bc, a(bc)\}$ is a basis for E over F.

Let $d \in D$, say

$$d = \lambda_0 + \lambda_1 a + \lambda_2 b + \lambda_3 c + \lambda_4 ab + \lambda_5 ac + \lambda_6 bc + \lambda_7 a(bc).$$

Now, since F has characteristic 2,

$$(ad + da) = \lambda_6(a(bc) + (bc)a) + \lambda_7(a(a(bc)) + (a(bc))a)$$

and using identity 2(c) of Section 2,

$$ad + da = \lambda_6(a,b,c) + \lambda_7(a,b,c)a \cdots \qquad (*)$$

Hence

$$(bc)(ad + da) + (ad + da)bc = \lambda_7(a,b,c)^2.$$

So there exists $0 \neq z_7 \in C$ (z_7 independent of d) such that $z_7\lambda_7 \in C$.

Now, from (*), there exists $0 \neq z_6 \in C$ with $z_6\lambda_6 \in C$ (z_6 independent of d).

By symmetry we find that there exist nonzero elements z_4, z_5 of C (independent of d) with $z_4\lambda_4 \in C$, $z_5\lambda_5 \in C$. Again

$$d(ab) + (ab)d = \lambda_3((ab)c + c(ab)) + \lambda_7((ab)(a(bc)) + (a(bc))(ab))$$

and by a similar argument and symmetry we find that there exist $0 \neq z_i \in C$ ($i = 1, 2, 3$) (z_i independent of d) with $z_i\lambda_i \in C$. Put $v = z_1 z_2 \cdots z_7$. Then $vd = v\lambda_0 + u$ for some $u \in D$. So $v\lambda_0 \in D$. Thus vD is contained in the C-span of 1, a, b, c, ab, ac, bc, $a(bc)$. A contradiction is now obtained as before.

The proof is complete.

COROLLARY 1. *Let A be an infinite alternative ring. Then A has an infinite commutative associative subring C.*

Proof. Regard A as a $\mathbb{Z}$-algebra. If A has a torsion-free element $a \neq 0$, the result is obvious, so we may assume each element of A is a torsion element. The result now follows immediately.

Remark 1. A direct proof of Corollary 1 is given by Bell [1].

COROLLARY 2. *If A is an infinite-dimensional alternative algebra over a field F, then A has an infinite-dimensional commutative associative subalgebra.*

Remark 2. More detailed information about the commutative subalgebras of infinite-dimensional associative algebras over fields can be found in [9].

4. THE SCHMIDT PROBLEM

The well-known Schmidt problem in group theory is:

Determine all infinite groups G all of whose proper subgroups are finite.

In fact if G is such a group and G is abelian, then it is well known that $G = C(p^\infty)$ for some prime p and it is an open problem as to whether any infinite group G with all proper subgroups finite must be abelian. We may formulate the analogous problem for R-algebras, namely: Determine all R-infinite R-algebras A with all proper R-subalgebras R-finite.

In this section we prove the following result:

THEOREM 2. *Let R be a commutative associative ring with identity. Let A be an alternative R-algebra such that A is R-infinite but all proper R-subalgebras of A are R-finite. Let $N = \{r \in R \mid rA = 0\}$ and let $S = R/N$. Then S is an integral domain. Furthermore either $A^2 = 0$ or A is a field. If $A^2 = 0$ then one* *of the following statements holds:*

(1) *There exists an irreducible nonunit $p \in S$ such that A is isomorphic, as an S-module, to*

$$S(p^{-1}) = \left\{ \frac{r}{p^n} \,\middle|\, r \in S, n \geqslant 1 \right\}$$

considered as a subgroup of the quotient field of S.

(2) *There exists a nonzero element $p \in S$ and a sequence x_n of elements of A such that*

(i) $px_1 = 0, px_{n+1} = x_n \ (n \geqslant 1)$,
(ii) $A = \bigcup_{n=1}^{\infty} Sx_n$.

If $A^2 \neq 0$ then one of the following statements holds:

(3) *There exists a nonzero nonunit $p \in S$ such that $S(p^{-1})$ is the quotient field of S. Furthermore A is S-algebra isomorphic to $S(p^{-1})$.*

(4) *S is a field and A is an extension field of S. Regard S as a subfield of A. Then one of the following statements holds:*

- (a) *There exists a sequence $\{d_n\}$ of positive integers such that*
  - (i) *d_n is a proper divisor of d_{n+1} for all $n \geqslant 1$.*
  - (ii) *There exists a sequence $\{x_n\}$ of elements of A such that for all $n \geqslant 1$, $[S(x_n):S] = d_n$.*
  - (iii) $A = \bigcup_{n=1}^{\infty} S(x_n)$.
- (b) *There exists an element $x \in A$ such that $S(x)$ is a finite separable extension of S and A is a purely inseparable extension of $S(x)$. There exists a sequence $\{A_i\}$ of subfields of A containing $S(x)$ such that*
  - (i) $A_0 = S(x)$,
  - (ii) $[A_{i+1}:A_i] = p \ (i \geqslant 0)$ *where p is the characteristic of S,*
  - (iii) $A = \bigcup_{n=1}^{\infty} A_n$.

Proof 1. *$A^2 = 0$ or A is a field.*

If $X(A) = A$, then $A^2 = 0$. Suppose $X(A) \neq A$. The definition of $X(A)$ now forces $X(A) = \operatorname{Ann}(A) = \{a \in A \,|\, aA = Aa = 0\}$. Let $D = A/X(A)$. By the Main Theorem, D is commutative and associative and, because of our hypotheses, D has no nonzero divisors of zero. If $0 \neq a \in D$, then, since D has no nonzero divisors of zero, $aDa = D$ and D is a field. It is now easy to see that A has an identity and thus $\operatorname{Ann}(A) = 0$ and A is a field.

2. *S is an integral domain.*

Suppose $r, s \in S$ and $rs = 0$. Then $r(sA) = 0$, so either sA is R-infinite and equals A forcing $r = 0$ or sA is R-finite, forcing $\{a \in A \,|\, sa = 0\}$ to be R-infinite and thus to be A and thus forcing $s = 0$.

Suppose now that $A^2 = 0$. We will show that (1) or (2) holds.

Suppose first that A is not torsion-free as an S-module. Let $a \in A$, $a \neq 0$, $p \in S$, $p \neq 0$ be such that $pa = 0$. Since $p \neq 0$, $pA \neq 0$. If pA is S-finite, then $\{a \in A \mid pa = 0\}$ is S-infinite and $pA = 0$. Thus pA is R-infinite and thus $pA = A$. Let $x_1 = a$, $x_2 \in A$ with $x_1 = px_2$, $x_3 \in A$ with $x_2 = px_3$, etc.

Let $B = \bigcup_{n=1}^{\infty} Sx_n$. Suppose B is R-finite. Then $Sx_n = Sx_{n+1} = \cdots$ for some n. Hence there exists $r \in R$ such that $x_n = rpx_n$ and thus $(1 - rp)x_n = 0$. Multiplying by p^{n-1} gives $x_1 = 0$, a contradiction. Hence B is R-infinite and thus $B = A$. Suppose now that A is a torsion-free S-module. We first show that S is Noetherian. For suppose this is not so and let T be an S-infinite ideal of S. Let $0 \neq a \in A$. Then Ta is an S-subalgebra of A (since $a^2 = 0$) and Ta is S-infinite, since A is S-torsion-free. Hence $Ta = A$. Hence there exists $t \in T$ with $a = ta$. But then $t = 1$, so $T = S$, a contradiction. Next, since every infinite-dimensional vector space over a field has infinite-dimensional proper subspaces, S is not a field. Since S is Noetherian, S therefore has an irreducible nonunit p. Then $pA = A$ and letting x_1 be any nonzero element of A, we can find a sequence x_n with $x_n = px_{n+1}$ $(n \geqslant 1)$. Let $C = \bigcup_{n=1}^{\infty} Sx_n$. Clearly C is R-infinite and thus $C = A$. This completes the discussion of the case $A^2 = 0$.

Suppose now that A is a field. We embed S in A in the obvious way. Suppose first that S is not a subfield of A. Let $p \in S$ be a nonzero nonunit. Then $\mathrm{alg}_S(p^{-1})$ is not S-finite. Thus conclusion (3) holds.

Suppose now that S is a subfield of A. We note first that each element of A is algebraic over S. Suppose first that A is separable over S. Then conclusion (4a) follows easily. If A is not separable over S, then A has characteristic p for some prime p. Let S_0 be the maximal separable subfield of A. Then $[S_0 : S]$ is finite and A is a purely inseparable extension of S_0. Then conclusion (4b) follows easily. This completes the proof.

COROLLARY 1. *Let A be an R-infinite alternative R-algebra which has no R-infinite proper R-subalgebras. If A has a.c.c. on R-subalgebras, then if $N = \{r \in R \mid rA = 0\}$, $S = R/N$ is an integral domain and there exists a nonzero nonunit $p \in S$ such that $S(p^{-1})$ is a field and A is S-algebra isomorphic to $S(p^{-1})$.*

Proof. Conclusion (3) of Theorem 2 is the only one consistent with the a.c.c. on R-subalgebras.

COROLLARY 2. *Let A be an infinite-dimensional alternative algebra over a field F and suppose that A satisfies the a.c.c. and d.c.c. on commutative associative subalgebras. Then A is finite dimensional.*

Proof. Assume the result is false and let A be a counterexample. Let S be the set of infinite-dimensional subalgebras of A. Since A satisfies the d.c.c. on subalgebras, Zorn's lemma implies that S has a minimal element, C say, under inclusion. Then C is infinite dimensional but all its proper subalgebras are finite dimensional. But C also has the a.c.c. on subalgebras, so Corollary 1 (with $R = F$) gives a contradiction.

We now give some examples of pairs (R, A) satisfying the hypotheses of Theorem 2.

(1) A the field of rational numbers and R the ring of all rationals whose denominators are p-free (p some fixed prime).

(2) A the zero-ring on the quasicyclic group $C(p^\infty)$ and R the ring of integers.

(3) Let p, q be primes and $R = GF(p)$ the field of p elements and $A = \bigcup_{n=1}^{\infty} GF(p^{q^n})$.

(4) Let p be a prime. Let x be transcendental over $GF(p)$ and let $R = GF(p)(x)$ be the rational function field in x over $GF(p)$. Define a sequence y_n of elements of the algebraic closure of R by $y_1^p = x$ and, for $n \geqslant 1$, $y_{n+1}^p = y_n$. Let A be the union of the fields $R(y_n)$ $(n \geqslant 1)$. It is easy to verify that (R, A) satisfies the hypotheses of Theorem 2.

5. Conclusion

It is an interesting problem to determine a minimal (in some sense) class H of R-infinite R-algebras such that every R-infinite alternative R-algebra A must contain as an R-subalgebra at least one element of H. This problem has been solved for infinite rings by Bell [1] and some of his arguments can be extended. When R is a field the results of [8] are relevant to this problem. The Main Theorem of this paper shows that in general we may choose a H consisting entirely of commutative associative R-algebras.

Acknowledgments

We wish to thank Dr. F. J. Gaines for several useful conversations on the material of this paper. We also wish to thank the referee for helpful comments.

References

1. H. E. Bell., Infinite subrings of infinite rings and near-rings; *Pacific J. Math.* **59** (1975), 345–358.
2. E. Formanek, Noetherian P.I. rings, *Comm. Algebra* **1**(1) (1974), 79–86.
3. N. Jacobson, "Structure of Rings," Amer. Math. Soc. Colloq. Publ. 37, Amer. Math. Soc., 1964.
4. E. Kleinfeld, A characterization of the Cayley numbers; *in* "Studies in Modern Algebra," Vol. 2, Math. Assoc. Amer., 1963.

5. E. KLEINFELD, Alternative division algebras of characteristic 2; *Proc. Nat. Acad. Sci.* **37** (1951), 818–820.
6. T. J. LAFFEY, Commutative subrings of infinite rings; *Bull. London Math. Soc.* **4** (1972), 3–5.
7. T. J. LAFFEY, Commutative subalgebras of infinite dimensional algebras; *Bull. London Math. Soc.* **5** (1973), 312–314.
8. T. J. LAFFEY, Infinite rings all of whose proper subrings are finite, *Amer. Math. Monthly* **81** (1974), 270–272.
9. T. J. LAFFEY, On the structure of algebraic algebras; *Pacific J. Math.* **62** (1976), 461–472.
10. T. J. LAFFEY, Commutative subrings of periodic rings, *Math. Scand.* **39** (1976) 161–166.
11. R. D. SCHAFER, "An Introduction to Nonassociative Algebras," Academic Press, New York, 1966.
12. E. WITT, Die Unterringe der freien Lieschen Ringe, *Math. Z.* **64** (1956), 195–216.